VOLUME NINETY FOUR

ADVANCES IN PROTEIN CHEMISTRY AND STRUCTURAL BIOLOGY

VOLUME NINETY FOUR

ADVANCES IN
PROTEIN CHEMISTRY AND STRUCTURAL BIOLOGY

Edited by

ROSSEN DONEV
Singleton Park
Swansea University
Swansea, UK

ELSEVIER

AMSTERDAM • BOSTON • HEIDELBERG • LONDON
NEW YORK • OXFORD • PARIS • SAN DIEGO
SAN FRANCISCO • SINGAPORE • SYDNEY • TOKYO
Academic Press is an imprint of Elsevier

Academic Press is an imprint of Elsevier
The Boulevard, Langford Lane, Kidlington, Oxford, OX5 1GB, UK
32 Jamestown Road, London NW1 7BY, UK
Radarweg 29, PO Box 211, 1000 AE Amsterdam, The Netherlands
225 Wyman Street, Waltham, MA 02451, USA
525 B Street, Suite 1800, San Diego, CA 92101-4495, USA

First edition 2014

ISBN: 978-0-12-800168-4
ISSN: 1876-1623

For information on all Academic Press publications
visit our website at store.elsevier.com

Printed and bound in USA

14 15 16 17 10 9 8 7 6 5 4 3 2 1

CONTENTS

Contributors *ix*

1. Bioinformatics Tools to Identify and Quantify Proteins Using Mass Spectrometry Data **1**

So young Ryu

1. Introduction 2
2. MS Data 2
3. Protein Identification 5
4. Protein Quantification 12
5. Conclusion 15

Acknowledgments 15
References 15

2. Unraveling Oxidation-Induced Modifications in Proteins by Proteomics **19**

Carolina Panis

1. A Very Brief Introduction to Oxidative Stress and Redox Homeostasis 20
2. Proteins as Targets of Oxidative Modification 21
3. Redox Proteomics: The Comprehensive Study of Oxidation-Induced Protein Damage 25
4. The Use of Proteomics for Mapping the Redox Homeostasis Network: Lessons from Cancer Research 29
5. Perspectives and Conclusions 33

References 33

3. Role of Proteomics in Biomarker Discovery: Prognosis and Diagnosis of Neuropsychiatric Disorders **39**

Suman Patel

1. Summary 40
2. Commonly Used Methods for Proteome Characterization 42
3. Proteomics and Biomarker 46
4. Diagnosis of Neuropsychiatric Disorders 50
5. Proteomics of Neuropsychiatric Disorders 55
6. Strength, Weakness, and Future Challenges: Biomarker Discovery 66
7. Concluding Remarks 67

References 68

4. On the Use of Knowledge-Based Potentials for the Evaluation of Models of Protein–Protein, Protein–DNA, and Protein–RNA Interactions **77**

Oriol Fornes, Javier Garcia-Garcia, Jaume Bonet, and Baldo Oliva

1. Introduction 78
2. Knowledge-Based Potentials 80
3. Modeling of Protein Interactions Using Templates 82
4. Modeling Interactions of Proteins Using Docking 85
5. Prediction of Protein-Binding Regions 93
6. Characterization of Transcription Factor-Binding Sites 98
7. Adapting Split-Statistical Potentials for Protein–DNA Interactions 105
8. Conclusions 109

Acknowledgments 110

References 110

5. Algorithms, Applications, and Challenges of Protein Structure Alignment **121**

Jianzhu Ma and Sheng Wang

1. Introduction 122
2. Structural Alphabet 124
3. Protein Pairwise Structure Alignment 130
4. Protein MSA 155
5. Conclusions and Perspectives 167

References 170

6. Application of Evolutionary Based *In Silico* Methods to Predict the Impact of Single Amino Acid Substitutions in Vitelliform Macular Dystrophy **177**

C. George Priya Doss, Chiranjib Chakraborty, N. Monford Paul Abishek, D. Thirumal Kumar, and Vaishnavi Narayanan

1. Introduction 178
2. Materials and Methods 183
3. Results 188
4. Discussion 263
5. Conclusion 263

Acknowledgments 264

References 264

7. **Current State-of-the-Art Molecular Dynamics Methods and Applications** **269**

Dimitrios Vlachakis, Elena Bencurova, Nikitas Papangelopoulos, and Sophia Kossida

1. Introduction 270
2. The Role of Computer Experiments in Modern Science 272
3. Brief History of Computer Molecular Simulations 274
4. Physics in MD 276
5. Algorithms for MD Simulations 278
6. Computational Complexity of MD Simulations and Methods to Increase Efficiency 281
7. High-Performance Parallel Computing 285
8. General-Purpose Computing Using Graphics Processing Units 288
9. Software for MD Simulations 290
10. Force Fields for MD 293
11. Current Limitations of MD 307
12. Conclusions 308

Acknowledgments 309
References 309

8. **Intrinsically Disordered Proteins—Relation to General Model Expressing the Active Role of the Water Environment** **315**

Barbara Kalinowska, Mateusz Banach, Leszek Konieczny, Damian Marchewka, and Irena Roterman

1. Introduction 316
2. Definition of the Fuzzy Oil Drop Model 317
3. Results 324
4. Discussion 339
5. Conclusions 343

Acknowledgments 344
References 344

9. **Conformational Elasticity can Facilitate TALE–DNA Recognition** **347**

Hongxing Lei, Jiya Sun, Enoch P. Baldwin, David J. Segal, and Yong Duan

1. Introduction 348
2. Methods 350

3. Results 352
4. Discussion 360
Acknowledgments 363
References 363

10. Computational Approaches and Resources in Single Amino Acid Substitutions Analysis Toward Clinical Research 365

C. George Priya Doss, Chiranjib Chakraborty, Vaishnavi Narayan, and D. Thirumal Kumar

1. Introduction 366
2. Computational Methods in SAP Analysis 371
3. Database Resources for SAPs 374
4. Molecular Phenotypic Effect Analysis 375
5. Sequence Information Analysis 376
6. Computational Methods for Structure Determination 377
7. Docking 380
8. Types of Docking 391
9. Molecular Dynamics 393
10. Concluding Remarks 396
Acknowledgments 399
References 399

Author Index *425*
Subject Index *461*

CONTRIBUTORS

Enoch P. Baldwin
Department of Molecular and Cellular Biology, University of California, Davis, California, USA

Mateusz Banach
Department of Bioinformatics and Telemedicine, Medical College, and Faculty of Physics, Astronomy and Applied Computer Science - Jagiellonian University, Krakow, Poland

Elena Bencurova
Bioinformatics & Medical Informatics Team, Biomedical Research Foundation, Academy of Athens, Athens, Greece, and Laboratory of Biomedical Microbiology and Immunology, University of Veterinary Medicine and Pharmacy, Kosice, Slovakia

Jaume Bonet
Structural Bioinformatics Lab. (GRIB), Departament de Ciències Experimentals i de la Salut, Universitat Pompeu Fabra, Barcelona, Catalunya, Spain

Chiranjib Chakraborty
Department of Bio-Informatics, School of Computer and Information Sciences, Galgotias University, Greater Noida, Uttar Pradesh, India

Yong Duan
UC Davis Genome Center and Department of Biomedical Engineering, One Shields Avenue, Davis, California, USA

Oriol Fornes
Structural Bioinformatics Lab. (GRIB), Departament de Ciències Experimentals i de la Salut, Universitat Pompeu Fabra, Barcelona, Catalunya, Spain

Javier Garcia-Garcia
Structural Bioinformatics Lab. (GRIB), Departament de Ciències Experimentals i de la Salut, Universitat Pompeu Fabra, Barcelona, Catalunya, Spain

C. George Priya Doss
Medical Biotechnology Division, School of Biosciences and Technology, VIT University, Vellore, Tamil Nadu, India

Barbara Kalinowska
Department of Bioinformatics and Telemedicine, Medical College, and Faculty of Physics, Astronomy and Applied Computer Science - Jagiellonian University, Krakow, Poland

Leszek Konieczny
Chair of Medical Biochemistry, Medical College, Jagiellonian University, Krakow, Poland

Sophia Kossida
Bioinformatics & Medical Informatics Team, Biomedical Research Foundation, Academy of Athens, Athens, Greece

Hongxing Lei
CAS Key Laboratory of Genome Sciences and Information, Beijing Institute of Genomics, Chinese Academy of Sciences, Beijing, China, and UC Davis Genome Center and Department of Biomedical Engineering, One Shields Avenue, Davis, California, USA

Jianzhu Ma
Toyota Technological Institute at Chicago, Chicago, Illinois, USA

Damian Marchewka
Department of Bioinformatics and Telemedicine, Medical College, and Faculty of Physics, Astronomy and Applied Computer Science - Jagiellonian University, Krakow, Poland

N. Monford Paul Abishek
Medical Biotechnology Division, School of Biosciences and Technology, VIT University, Vellore, Tamil Nadu, India

Vaishnavi Narayanan
Biomolecules & Genetics Division, School of Biosciences and Technology, VIT University, Vellore, Tamil Nadu, India

Baldo Oliva
Structural Bioinformatics Lab. (GRIB), Departament de Ciències Experimentals i de la Salut, Universitat Pompeu Fabra, Barcelona, Catalunya, Spain

Carolina Panis
Laboratory of Physiopathology and Free Radicals, Department of General Pathology, State University of Londrina, Londrina, Paraná, Brazil

Nikitas Papangelopoulos
Bioinformatics & Medical Informatics Team, Biomedical Research Foundation, Academy of Athens, Athens, Greece

Suman Patel
CSIR-National Institute of Science, Technology and Developmental Studies, New Delhi, India

Irena Roterman
Department of Bioinformatics and Telemedicine, Medical College, Jagiellonian University, Krakow, Poland

So young Ryu
Stanford University, Stanford, Washington, USA

David J. Segal
Genome Center and Department of Biochemistry and Molecular Medicine, University of California, Davis, California, USA

Jiya Sun
CAS Key Laboratory of Genome Sciences and Information, Beijing Institute of Genomics, Chinese Academy of Sciences, and University of Chinese Academy of Sciences, Beijing, China

D. Thirumal Kumar
Medical Biotechnology Division, School of Biosciences and Technology, VIT University, Vellore, Tamil Nadu, India

Dimitrios Vlachakis
Bioinformatics & Medical Informatics Team, Biomedical Research Foundation, Academy of Athens, Athens, Greece

Sheng Wang
Toyota Technological Institute at Chicago, Chicago, Illinois, USA

CHAPTER ONE

Bioinformatics Tools to Identify and Quantify Proteins Using Mass Spectrometry Data

So Young Ryu[1]
Stanford Genome Technology Center, Biochemistry Department, Stanford University, Stanford, California, USA
[1]Corresponding author: e-mail address: clairesr@stanford.edu

Contents

1. Introduction 2
2. MS Data 2
   2.1 How MS generates data 2
   2.2 Tandem MS 3
   2.3 MS quantification techniques 4
3. Protein Identification 5
   3.1 Bioinformatics tools for peptide identification 5
   3.2 Bioinformatics tools for protein identification 10
4. Protein Quantification 12
   4.1 Bioinformatics tools for ICAT data 12
   4.2 Bioinformatics tools for label-free data 13
5. Conclusion 15
Acknowledgments 15
References 15

Abstracts

Proteomics tries to understand biological function of an organism by studying its protein expressions. Mass spectrometry is used in the field of shotgun proteomics, and it generates mass spectra that are used to identify and quantify proteins in biological samples. In this chapter, we discuss the bioinformatics algorithms to analyze mass spectrometry data. After briefly describing how mass spectrometry generates data, we illustrate the bioinformatics algorithms and software for protein identification such as *de novo* approach and database-searching approach. We also discuss the bioinformatics algorithms and software to quantify proteins and detect the differential proteins using isotope-coded affinity tags and label-free mass spectrometry data.

Advances in Protein Chemistry and Structural Biology, Volume 94
ISSN 1876-1623
http://dx.doi.org/10.1016/B978-0-12-800168-4.00001-9

1. INTRODUCTION

Global protein expression profiles of biological systems in various phenotypic states can reveal the biological roles of proteins (Tyers & Mann, 2003). Mass spectrometry (MS) coupled with liquid chromatography can help us build such a profile by producing an enormous number of spectra. However, building protein expression profiles from these spectra is not straightforward; rather, it involves various sophisticated bioinformatics algorithms. Without bioinformatics tools, interpreting these spectra would be extremely time consuming and less accurate. In this chapter, we introduce currently available bioinformatics tools to analyze MS data.

This chapter contains three sections. In Section 2, we review how MS generates data for protein identification and quantification. In Section 3, we discuss the bioinformatics algorithms and software to identify peptides and proteins. In Section 4, we illustrate the bioinformatics algorithms and software to quantify peptides and proteins.

2. MS DATA

2.1. How MS generates data

The procedure for generating MS data is the following (Fig. 1.1): Proteins in a biological complex mixture are digested by enzymes (i.e., trypsin) into peptides. These peptides are separated based on their hydrophobicity by liquid chromatography and then ionized. Mass-to-charge ratios (m/z) of these ionized peptides are recorded. We call these recorded scans precursor ion spectra or MS1 spectra. Between precursor scans, ionized peptides with relatively high intensities are isolated and fragmented by collision-induced dissociation (CID). We call this isolated ionized peptide a precursor ion, and we call the resulting spectrum tandem mass spectrum or MS/MS spectrum. The instrument produces precursor ion spectra and tandem mass spectra alternatively for a certain period of time, producing thousands of spectra for each experiment run.

Both precursor ion spectra and tandem mass spectra contain several peaks as shown in Fig. 1.1. A peak is characterized by its mass-to-charge ratio (m/z) and intensity value. A precursor ion spectrum contains peaks that represent ionized charge-bearing peptides, while a tandem mass spectrum contains peaks that represent fragment ions of one peptide. (Occasionally, multiple peptides can be fragmented in one tandem mass spectrum. But for

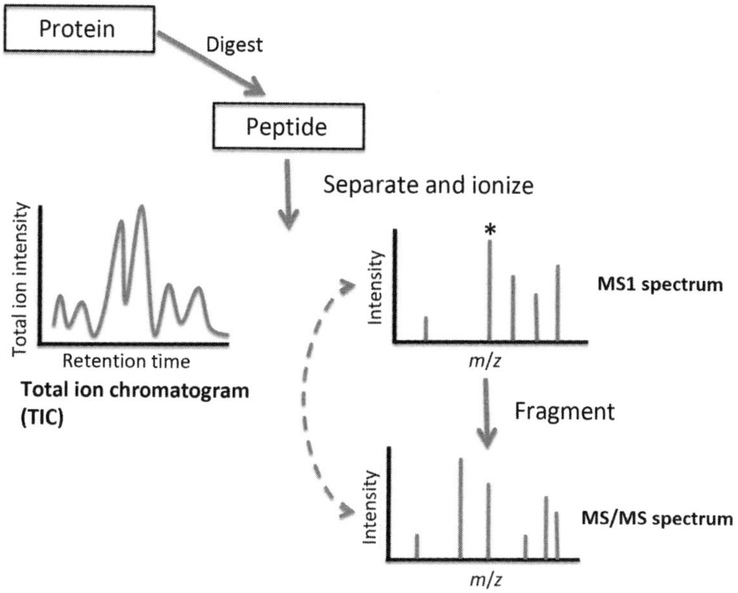

Figure 1.1 General MS-based proteomics workflow. The peak with the asterisk was isolated and fragmented to generate a MS/MS spectrum.

simplicity, let us assume that peaks in one tandem mass spectrum are from one peptide.) Precursor ion spectra are mainly used for peptide quantification, while tandem mass spectra are mainly used for peptide identification.

2.2. Tandem MS

Fragment ions in a tandem mass spectrum are predominantly b- and y-ions. (We ignore other ions such as a- and z-ions for simplicity.) If a tandem mass spectrum is produced by the peptide sequence "PEPTIDE," then b- and y-ions can be expressed as prefixes and suffixes of this peptide sequence, respectively. The sequences of "P," "PE," "PEP," "PEPT," "PEPTI," and "PEPTID" can be generated from b-ions for the given peptide. The sequences of "E," "DE," "IDE," "TIDE," "PTIDE," and "EPTIDE" can be generated from y-ions. Since each amino acid has a unique mass (except amino acids, "I" and "L"), we can easily identify a peptide sequence from a tandem mass spectrum in the following conditions: (1) m/z values of all b- and y-ions are observed, (2) there are no noise peaks, and (3) charge states of peaks can be determined accurately. However, tandem mass spectra do not contain all possible b- and y-ions, but only some of them. Furthermore, they contain noise peaks, and it is not a straightforward task to determine charge states of peaks, especially for low-resolution tandem mass spectra. In the third section, we address how bioinformaticians solve these challenges in identifying peptides from tandem mass spectra.

2.3. MS quantification techniques

There are several MS quantification techniques. For labeling quantification techniques, there are isotope-coded affinity tags (ICATs), isobatic tags for relative and absolute quantitation (Hardt et al., 2005), stable isotope labeling by amino acids in cell culture (Ong et al., 2002), and the ^{18}O labeling universal standard approach (Qian et al., 2009). Also, there are label-free approaches that use either peptide ion current areas or the number of tandem mass spectra identified for a protein of interest. In this section, we briefly review ICAT labeling and label-free approaches.

The ICAT labeling approach utilizes traditional stable isotope dilution theory in the proteomics and measures protein quantitative changes between two different biological samples by MS (Gygi et al., 1999). Prior to MS analysis, two different protein samples are treated with either light- or heavy-ICAT reagents. Then, two samples are combined, digested by trypsin to produce a peptide mixture, and analyzed by LC–MS–MS. The ratios of peptide intensities labeled "light" and "heavy" in precursor ion spectra provide relative quantification of proteins between two biological samples. Disadvantages of ICATs are that it cannot quantify proteins without cysteine residues and that it can perform only pairwise comparisons of two biological samples. To overcome these limitations, the many variants of ICATs were developed (Patterson & Aebersold, 2003).

In contrast to the ICAT quantification technique, the label-free approach using peptide ion current areas does not label biological samples. In this approach, each individual sample is separately analyzed by LC–MS–MS. Thus, the label-free approach makes it possible to compare more than two samples without limiting the number of samples to be compared. Moreover, it reduces experimental cost by not requiring labeling reagents. However, when analyzing label-free samples, it is important to have very robust performance of instruments in measuring peak intensities and elution time. Producing reproducible data is one of the keys in this approach.

As mentioned earlier, the label-free approach using peptide ion current areas uses peptide intensity values measured in MS1 spectra. In contrast, an alternative label-free approach, called spectral counting, uses MS/MS spectra instead of MS1 spectra (Colinge, Chiappe, Lagache, Moniatte, &

Bougueleret, 2005; Liu, Sadygov, & Yates, 2004). This approach sums up all the MS/MS spectra identified for each protein and uses the total number of MS/MS spectra as the relative protein abundance. The MS data for this approach can be prepared in the same way as the label-free approach using peptide ion current areas. The spectral counting approach may not measure protein quantification as accurately as the other quantitation methods, but it does not require highly robust performance as with the label-free approach using peptide ion current areas. Thus, this approach is often used by pilot studies.

In later sections, we introduce the bioinformatics algorithms using MS data produced by the quantitation approaches we discussed in this section.

3. PROTEIN IDENTIFICATION

3.1. Bioinformatics tools for peptide identification

3.1.1 De novo *approach*

There are various approaches to identify peptides from tandem mass spectra. The *de novo* approach is not the most popular way, but it is a natural way to interpret tandem mass spectra considering the mechanism that generates fragment ions such as b- and y-ions. It identifies peptide sequences based on masses of b- and y-ions in MS/MS spectra. If the two smallest observed masses of all the possible b-ions of a peptide are 72 and 171 Da, we know that the first two amino acids of this peptide are "AV" using the theoretical masses of 20 amino acids in Table 1.1. In detail, the mass of the first b-ion matches the mass of Alanine (A) with a monoisotopic residue mass, 71.03 Da in Table 1.1. Since the mass of the second b-ion is 171 Da, the mass of second amino acid is 99 Da (=171 − 72), which is close to the theoretical monoisotopic residue mass 99.07 Da. Thus, the second amino acid of a given peptide is Valine (V). In reality, the manual identification of one tandem mass spectrum using this *de novo* approach might take hours depending on the analyst's experience and the quality of spectra. When the incompleteness of b- and y-ions is too severe and there are many noise peaks, this hard labor might not result in the accurate peptide identification.

Automated *de novo* algorithms are available. PepNovo (Frank & Pevzner, 2005) and PEAKS (Ma et al., 2003) are software using the *de novo* approach. The *de novo* approach may be used when there is no protein database for one's biological samples or when one can produce MS/MS spectra with high quality.

Table 1.1 Monoisotopic residue masses of 20 amino acids

One-letter code	Amino acid	Monoisotopic residue mass (Da)
A	Alanine	71.03711
R	Arginine	156.10111
N	Asparagine	114.04293
D	Aspartic acid	115.02694
C	Cysteine	103.00919
E	Glutamic acid	129.04259
Q	Glutamine	128.05858
G	Glycine	57.02146
H	Histidine	137.05891
I	Isoleucine	113.08406
L	Leucine	113.08406
M	Methionine	128.09496
F	Phenylalanine	147.06841
P	Proline	97.05276
S	Serine	87.03203
T	Threonine	101.04768
W	Tryptophan	186.07931
Y	Tyrosine	163.06333
V	Valine	99.06841

3.1.2 Database-searching approach

The dominant peptide identification paradigm is a database-searching approach. It utilizes protein sequences of an organism of interest. For example, a human protein sequence database is used for human samples. A yeast protein sequence database is used for yeast samples. Therefore, this approach assumes that observed tandem mass spectra are generated from the peptides sequences in the specified protein database.

In detail, the common structure of database-searching algorithms is the following: For a tandem mass spectrum of interest, the charge state and mass of the precursor ion is determined. Next, peptide candidates for this tandem mass spectrum are selected from a user-defined protein sequence database.

Since we can generate all possible peptide sequences from the protein database for a specified enzyme using *in silico* database digestion, we can also compute the theoretical masses of these peptide sequences. Only peptides with theoretical masses that are within a mass tolerance window are considered as candidate peptides for the tandem mass spectrum. Depending on the mass tolerance and the size of the protein sequence database, one can have a few hundred to a few thousand candidate peptides per tandem mass spectrum. For each peptide candidate, a theoretical spectrum is constructed. Then, one measures the similarity between this constructed theoretical spectrum and the observed spectrum. For most database-searching algorithms, the theoretical spectrum contains m/z values of peaks with little variety in their intensities. Thus, their score schemes for measuring the similarity between the observed and the theoretical spectrum mainly depend on the location of peaks (m/z values of peaks). Measuring the similarity score between observed spectra and their corresponding theoretical spectra takes intensive computation time because there are normally tens of thousands of observed MS/MS spectra with thousands of candidate theoretical spectra to be compared. Finally, the theoretical spectrum with the highest similarity to the observed one is assigned to the observed tandem mass spectrum. Since each theoretical spectrum is from one peptide sequence, this is considered as a peptide assignment.

Examples of database-searching software are listed in Table 1.2. Sequest (Eng et al., 1994), Mascot (Perkins et al., 1999), and X!Tandem (Craig & Beavis, 2004) are the most widely used software for peptide identification. In terms of similarity score schemes, Sequest (Eng et al., 1994) uses a cross-correlation to measure the similarity between the observed and theoretical spectra. X!Tandem (Craig & Beavis, 2004) uses a score calculated from the number of matched b- and y-ions between the observed and theoretical spectra and the intensities of those matched ions. Mascot (Perkins et al., 1999) uses a probability-based scoring algorithm, but the details of the algorithm are not publicly known.

At this point, we can see that all observed tandem mass spectra have the corresponding peptide sequences. However, not all peptide assignments are correct assignments. Some tandem mass spectra might contain only noise; thus, their peptide assignments are wrong. Some tandem mass spectra contain fragment ions from real peptides, but because those peptides are not included in the protein database, the resulting peptide assignments are incorrect. Some tandem mass spectra are from real peptides present in the protein database, but incorrect peptide assignments occur. Therefore, only a fraction

Table 1.2 Software to identify and quantify proteins

Software	Description	References
PepNovo	*De novo* sequencing	Frank and Pevzner (2005)
PEAKS	*De novo* sequencing	Ma et al. (2003)
SEQUEST	Database searching	Eng, McCormack, and Yates (1994)
Mascot	Database searching	Perkins, Pappin, Creasy, and Cottrell (1999)
X!Tandem	Database searching	Craig and Beavis (2004)
PeptideProphet	Peptide identification	Choi and Nesvizhskii (2007)
Precolator	Peptide identification	Kall, Canterbury, Weston, Noble, and MacCoss (2007)
SpectraST	Spectral library	Lam et al. (2007)
X! Hunter	Spectral library	Craig, Cortens, Fenyo, and Beavis (2006)
Viper	AMT and protein quantification	Monroe et al. (2007)
TagRecon	*De novo* sequencing and database searching	Dasari et al. (2010)
InsPecT	*De novo* sequencing and database searching	Tanner et al. (2005)
ProteinProphet	Protein identification	Nesvizhskii, Keller, Kolker, and Aebersold (2003)
MaxQuant	Protein quantification	Cox and Mann (2008), Cox et al. (2009)
ASAPRatio	Protein quantification (ICAT)	Li, Zhang, Ranish, and Aebersold (2003)
MSstats	Differential protein detection using peptide ion current area	Clough, Thaminy, Ragg, Aebersold, and Vitek (2012)
QSPEC	Differential protein detection using spectral count	Choi, Fermin, and Nesvizhskii (2008)

of peptide assignments are correct, and it is important to determine the fraction of correctly assigned peptides. In the early proteomics work, one determines the correctly assigned peptides using an arbitrary cutoff value of the database-searching scores. Note that in the peptide assignment step, we have similarity scores measured between an observed and a theoretical spectrum. The scores of all top-ranked peptides and the variants of these scores such as the difference in scores between the top- and second top-ranked peptides reflect the identification confidence for the top-ranked peptides. Thus, taking all peptides with the confidence scores above, user-defined thresholds can provide a rough classification between correct and incorrect peptide identifications. However, this classification does not specify how many false identifications one accepts using user-defined thresholds, and user-defined thresholds might not be optimal. Thus, several scoring schemes to measure uncertainty in peptide identifications are developed (Table 1.2). The software that were developed recently report p-value, E-value, False Discovery Rate (FDR), or q-value for the peptide identification.

PeptideProphet (Choi & Nesvizhskii, 2007) is one of the most widely used approaches to validate peptide assignments of tandem mass spectra. This algorithm uses a mixture model to discriminate the correct peptide assignments from the incorrect ones. The mixture model is based on database search scores obtained from the database search algorithm and information about observed spectra and assigned peptides (i.e., mass differences between theoretical and observed precursor ions, the number of missed cleavage sites). Another popular way to measure uncertainty in peptide assignments is using decoy database-searching results. The decoy database contains random protein sequences. A database that has a shuffled or reversed version of the sequences in the original database is used for the decoy database. This approach uses the decoy database or the combination of the decoy and original databases in the database-searching step. Using a decoy database allows us to estimate false peptide identifications because tandem mass spectra matched to sequences from the decoy database are incorrect peptide identifications. Percolator (Kall et al., 2007) implements the semisupervised approach to measure the uncertainty in peptide identifications. In details, it constructs the null distribution of database-searching scores based on the search results using the decoy database; then, it uses a support vector machine to classify the correct identifications from the incorrect ones.

3.1.3 Other approaches

Besides *de novo* and database-searching paradigms, several peptide identification paradigms are proposed. Spectral library searching is one paradigm that utilizes the previously identified tandem mass spectra and builds the spectral library from them. In contrast to the database-searching paradigm, it searches against the spectral library instead of the protein sequence database.

X! Hunter (Craig et al., 2006) and SpectraST (Lam et al., 2007) are software that uses this approach. One disadvantage of using this approach is that it can identify only peptides that are previously identified and are present in spectral library. But as many tandem mass spectra are deposited in the public domain, this library can grow and become more complete.

Clustering approach is another paradigm that takes an advantage of repeated tandem mass spectra from a same peptide (Frank et al., 2008; Ryu, Goodlett, Noble, & Minin, 2012). Contrary to the database-searching approach, this approach searches clusters of similar MS/MS spectra against the protein sequence database rather than searching individual MS/MS spectra against the database. Since the number of clusters is smaller than the number of spectra, this approach can save the computational time in the peptide identification step. One example of this approach is MS Cluster (Frank et al., 2008), which builds one representative spectrum for each cluster and uses the representative spectra for the database searching.

Accurate mass and time (AMT) tag approach (Monroe et al., 2007) is an approach that can identify peptides present in MS1 spectra whether these charge-bearing peptides are fragmented to generate MS/MS spectra or not. This approach uses a library that contains AMT tags, which are masses and normalized chromatographic elution times of peptide sequences. By matching mass and elution time of features in MS1 spectra to AMT tags in the library, one can identify the peptides. This approach requires the MS1 features to be generated from the high-sensitivity and high-resolution liquid chromatography–MS.

There are also a hybrid approach that combines the *de novo* approach and the database-searching approach (Tanner et al., 2005). This approach first identifies parts of peptide sequences using the *de novo* approach. Then, when applying the database-searching algorithm, it shortens the candidate peptide list by taking only peptides that contain these short sequences found by the *de novo* approach. The software that use this approach are InsPecT (Tanner et al., 2005) and TagRecon (Dasari et al., 2010).

3.2. Bioinformatics tools for protein identification

One major goal in proteomics is to identify as many proteins as possible in biological samples of interest. Thus, after identifying peptides and/or estimating the confidence scores for the peptide identifications, we want to identify proteins present in the samples and to estimate the confidence scores for those identifications. The protein identification is not an easy task due to peptides shared by multiple proteins and the difficulty in combining multiple levels of scores (Nesvizhskii, 2010). One peptide can be shared by multiple proteins because the protein sequence database often contains multiple proteins with similar sequences (i.e., homologous proteins and slice isoforms). This makes it difficult to determine actual proteins present in a sample because we do not usually identify all possible peptides that can be generated from proteins, and because for complex biological samples, the most of proteins have far less than perfect sequence coverage. In the early proteomics studies, only confidently identified peptides that are not shared by multiple proteins are selected and only proteins with at least two of these selected peptides are reported. This approach may provide us the confident protein list but does not provide us any confidence scores of those protein identifications. Moreover, it is very conservative preventing homologous proteins from being identified. So many currently available bioinformatics algorithms group similar proteins and report either proteins (if peptides are not shared by multiple proteins) or protein groups (if peptides are shared by multiple proteins). Normally, they report the smallest number of proteins or protein groups.

Another challenge in protein identification is estimating the confidence scores for the protein identifications by combining peptide-level or spectrum-level scores. Since peptides are substrings of a protein, we may identify multiple unique peptides from one protein. Since peptides can have multiple charge states and modifications, we may identify various forms of one peptide. Since more than one tandem mass spectrum can be generated from a peptide, we may identify one peptide with the same modification and the same charge state multiple times. Thus, several algorithms use the combined peptide evidence approach to estimate the confidence scores for the protein identifications.

ProteinProphet (Nesvizhskii et al., 2003) is a popular software that computes the probability that a protein is present in a sample. First, ProteinProphet (Nesvizhskii et al., 2003) adjusts the peptide identification scores generated by PeptideProphet (Choi & Nesvizhskii, 2007) to avoid the overestimated probability for the protein identification. Next, using these adjusted probabilities, it computes the probability that the protein contains at least one correctly identified peptide. For the peptides shared by multiple proteins, ProteinProphet (Nesvizhskii et al., 2003) adjusts the contribution of each shared peptide to its corresponding protein.

Besides the combined peptide evidence approach, there is a rather simple approach called "best peptide" (Gupta & Pevzner, 2009). This approach works quite well. It uses the best scores of all tandem mass spectra matched to a certain protein for the protein probability score. Another approach is to use a model-based clustering approach (Li, MacCoss, & Stephens, 2010), which measures uncertainty of peptide and protein identifications simultaneously instead of having two-stage models to identify peptides and proteins (i.e., PeptideProphet (Choi & Nesvizhskii, 2007) and ProteinProphet (Nesvizhskii et al., 2003)).

4. PROTEIN QUANTIFICATION

Another important goal in proteomics is to quantify proteins in biological samples. There are two types of relative quantifications. One type is to estimate relative protein abundances in one sample. To determine relative abundances of proteins within one sample, several approaches were proposed. Intensity-based absolute quantification of proteins are based on peptide intensities (Schwanhausser et al., 2011). This approach divides the sum of all identified peptide intensities by the number of theoretically observed peptides and uses them for the protein quantitation (Schwanhausser et al., 2011). There are also normalized spectral abundance factor (Paoletti et al., 2006), normalized spectral index (Griffin et al., 2010), exponentially modified protein abundance index (Ishihama et al., 2005), and minimum acceptable detectability of identified peptide (Tang et al., 2006), which are based on spectral counting results.

Another type of protein quantification is to compare the relative abundances of the same protein across various samples. Biologists are usually more interested in comparing the relative abundances of a certain protein across samples with various genetic or environmental perturbations rather than comparing protein abundances within one sample. In the following subsections, we discuss bioinformatics algorithms to quantify peptides and proteins using ICATs and label-free MS data.

4.1. Bioinformatics tools for ICAT data

Bioinformatics tools for ICAT data uses the light- and heavy isotopes of the MS1 precursor ions to estimate the relative protein abundances. ASAPRatio (Li et al., 2003) is one of software that analyze ICAT data. It processes the data as the following: After the peptide and protein identification, it reconstructs the smoothed single-ion chromatograms of the light- and heavy-isotope peptides by Savitzky–Golay smooth filters. Then, for each identified peptide, the peptide-level ratios between light- and heavy-labeled samples are estimated by (weighted) averaging light versus heavy ratios of the peptides with different charge states. Now, for each protein identification, outliers in peptide-level ratios are removed using Dixon's test and the protein-level ratios between light- and heavy-labeled samples are calculated with their confidence intervals. The *p*-values that measure statistical significance of protein abundances differences between two samples are also reported.

4.2. Bioinformatics tools for label-free data

As mentioned previously, there are label-free data using peptide ion current areas and spectral counting. Due to the advantages of the unlimited number of samples that can be compared, the label free becomes very popular. However, the approach using peptide ion current areas requires reproducible data and sophisticated bioinformatics algorithms. In this section, we first review the necessary bioinformatics steps to analyze the peptide intensity-based label-free data. From now on, we call these data MS1-based label-free data.

There are four steps to analyze the MS1-based label-free data: (1) detecting peptide features in LC–MS, (2) aligning the elution time of peptides across multiple LC–MS data, (3) estimating the protein abundance, and (4) detecting differential proteins between samples.

There are two approaches to detect peptide features and to align the peptide elution times. One is to reconstruct single-ion chromatograms of the identified peptides just like the labeled approach described in previous subsection. This approach (Ryu et al., 2008) is usually extended to detect the peptide features that are not identified in some experiments using their normalized elution times and mass-to-charge ratios. The model to normalize elution times between experiments are trained based on the elution times of identified peptides.

Another way to detect peptide features and to align the peptide elution times is the following (Bellew et al., 2006; Radulovic et al., 2004): First, it distinguishes features from chemical noise prior to the peptide identification. Since peptides elute over time, the algorithms group peaks with very similar mass-to-charge ratios and elution times and assigned them as one feature. Then, the peptide features across all experiments are matched so that the same peptides from different experiments can be compared. Finally, the peptide identifications are assigned to the peptide features based on their elution time and mass-to-charge ratios. In this step, some peptide features might not have the peptide identifications because either they are not fragmented to generate tandem mass spectra or their tandem mass spectra are not identified with the confidence. The benefit of using this approach is that it can quantify the peptides that are never identified in any experiment. The quantified peptides without identification can be further pursued in the future experiment if those peptides are differentially expressed between sample groups.

At this point, we can compare the peptide-level quantifications between sample groups (i.e., normal vs. disease). However, since multiple peptides can be found from one protein, we are more interested in comparing the protein-level quantifications. One naïve way to compute protein quantification is to sum the three most abundant peptide intensities for each protein and use them as the protein quantifications. Then, an ANOVA or a *t*-test can be applied to these protein-level quantification values to detect the differential proteins between different sample groups (Colinge et al., 2005; Liu et al., 2004). The more sophisticated approach is the model-based approach to quantify protein and to test the significance level of the differential proteins. MSstats (Clough et al., 2012) is one of the model-based approaches using a linear mixed-effect model.

One big challenge in protein quantification is missing values of peptide-level quantifications. For example, the peaks matched to a certain peptide sequence are present in only parts of experiments, but are absent in the rest of experiments. One of the reasons for these missing values is the instrument detection level. If peak intensities are below the instrument detection level, then the peaks might be missing. If peak intensities are above the instrument detection level, then peaks might be present. Since these missing data depend on the intensity values of peptides, it is not ignorable. Some research groups proposed statistical models that accounts for the missing values in peptide abundances (Karpievitch et al., 2009).

As mentioned previously, there is another type of label-free data, spectral count. In the earlier studies of spectral count, people used a *t*-test or an ANOVA to detect differential proteins using spectral count (Colinge et al., 2005; Liu et al., 2004). Since the spectral count approach is often used in pilot studies, one might not have enough samples, thus not enough power to detect the differential proteins using *t*-test. QSPEC assumes the Poisson family distribution of spectral count data and calculates the ratio of likelihoods using the information across all proteins; thus, it can be applied to the data with the limited number of technical or biological replicates (Choi et al., 2008).

5. CONCLUSION

The bioinformatics play an important role in proteomics providing the accurate protein identification and quantification. Considering the enormous data produced by the MS and its complexity, the bioinformatics algorithms and software are essential to study the proteomics.

There are some areas (i.e., missing values, posttranslational modification) that needed to be developed further. However, these challenges will attract more bioinformaticians who want to solve the challenging problems using their computational skills and statistical knowledge. Together with technology development, the bioinformatics continually advance in the proteomics field.

ACKNOWLEDGMENTS

This work was supported in part by NIH Grant T32-GM007035 and Shriners Research Grant 85500-BOS.

REFERENCES

Bellew, M., Coram, M., Fitzgibbon, M., Igra, M., Randolph, T., Wang, P., et al. (2006). A suite of algorithms for the comprehensive analysis of complex protein mixtures using high-resolution LC-MS. *Bioinformatics*, *22*(15), 1902–1909. http://dx.doi.org/10.1093/bioinformatics/btl276.

Choi, H., Fermin, D., & Nesvizhskii, A. I. (2008). Significance analysis of spectral count data in label-free shotgun proteomics. *Molecular and Cellular Proteomics*, 7(12), 2373–2385. http://dx.doi.org/10.1074/mcp.M800203-MCP200.

Choi, H., & Nesvizhskii, A. I. (2007). Semisupervised model-based validation of peptide identifications in mass spectrometry-based proteomics. *Journal of Proteome Research*, 7(1), 254–265. http://dx.doi.org/10.1021/pr070542g.

Clough, T., Thaminy, S., Ragg, S., Aebersold, R., & Vitek, O. (2012). Statistical protein quantification and significance analysis in label-free LC-MS experiments with complex designs. *BMC Bioinformatics*, *13*(Suppl. 16), S6.

Colinge, J., Chiappe, D., Lagache, S., Moniatte, M., & Bougueleret, L. (2005). Differential proteomics via probabilistic peptide identification scores. *Analytical Chemistry*, 77(2), 596–606. http://dx.doi.org/10.1021/ac0488513.

Cox, J., & Mann, M. (2008). MaxQuant enables high peptide identification rates, individualized p.p.b.-range mass accuracies and proteome-wide protein quantification. *Nature Biotechnology*, *26*(12), 1367–1372. http://dx.doi.org/10.1038/nbt.1511.

Cox, J., Matic, I., Hilger, M., Nagaraj, N., Selbach, M., Olsen, J. V., et al. (2009). A practical guide to the MaxQuant computational platform for SILAC-based quantitative proteomics. *Nature Protocols*, *4*(5), 698–705. http://dx.doi.org/10.1038/nprot.2009.36.

Craig, R., & Beavis, R. C. (2004). TANDEM: Matching proteins with tandem mass spectra. *Bioinformatics*, *20*(9), 1466–1467. http://dx.doi.org/10.1093/bioinformatics/bth092.

Craig, R., Cortens, J. C., Fenyo, D., & Beavis, R. C. (2006). Using annotated peptide mass spectrum libraries for protein identification. *Journal of Proteome Research*, *5*(8), 1843–1849. http://dx.doi.org/10.1021/pr0602085.

Dasari, S., Chambers, M. C., Slebos, R. J., Zimmerman, L. J., Ham, A.-J. L., & Tabb, D. L. (2010). TagRecon: High-throughput mutation identification through sequence tagging. *Journal of Proteome Research*, *9*(4), 1716–1726. http://dx.doi.org/10.1021/pr900850m.

Eng, J. K., McCormack, A. L., & Yates, J. R. (1994). An approach to correlate tandem mass spectral data of peptides with amino acid sequences in a protein database. *Journal of the American Society for Mass Spectrometry*, *5*(11), 976–989. http://dx.doi.org/10.1016/1044-0305(94)80016-2.

Frank, A. M., Bandeira, N., Shen, Z., Tanner, S., Briggs, S. P., Smith, R. D., et al. (2008). Clustering millions of tandem mass spectra. *Journal of Proteome Research*, 7(1), 113–122. http://dx.doi.org/10.1021/pr070361e.

Frank, A., & Pevzner, P. (2005). PepNovo: De novo peptide sequencing via probabilistic network modeling. *Analytical Chemistry*, 77(4), 964–973.

Griffin, N. M., Yu, J., Long, F., Oh, P., Shore, S., Li, Y., et al. (2010). Label-free, normalized quantification of complex mass spectrometry data for proteomic analysis. *Nature Biotechnology*, *28*(1), 83–89. http://dx.doi.org/10.1038/nbt.1592.

Gupta, N., & Pevzner, P. A. (2009). False discovery rates of protein identifications: A strike against the two-peptide rule. *Journal of Proteome Research*, *8*(9), 4173–4181. http://dx.doi.org/10.1021/pr9004794.

Gygi, S. P., Rist, B., Gerber, S. A., Turecek, F., Gelb, M. H., & Aebersold, R. (1999). Quantitative analysis of complex protein mixtures using isotope-coded affinity tags. *Nature Biotechnology*, *17*(10), 994–999. http://dx.doi.org/10.1038/13690.

Hardt, M., Witkowska, H. E., Webb, S., Thomas, L. R., Dixon, S. E., Hall, S. C., et al. (2005). Assessing the effects of diurnal variation on the composition of human parotid saliva: Quantitative analysis of native peptides using iTRAQ reagents. *Analytical Chemistry*, 77(15), 4947–4954. http://dx.doi.org/10.1021/ac050161r.

Ishihama, Y., Oda, Y., Tabata, T., Sato, T., Nagasu, T., Rappsilber, J., et al. (2005). Exponentially modified protein abundance index (emPAI) for estimation of absolute protein amount in proteomics by the number of sequenced peptides per protein. *Molecular and Cellular Proteomics*, *4*(9), 1265–1272. http://dx.doi.org/10.1074/mcp.M500061-MCP200.

Kall, L., Canterbury, J. D., Weston, J., Noble, W. S., & MacCoss, M. J. (2007). Semi-supervised learning for peptide identification from shotgun proteomics datasets. *Nature Methods*, *4*(11), 923–925. http://dx.doi.org/10.1038/nmeth1113.

Karpievitch, Y., Stanley, J., Taverner, T., Huang, J., Adkins, J. N., Ansong, C., et al. (2009). A statistical framework for protein quantitation in bottom-up MS-based proteomics. *Bioinformatics*, *25*(16), 2028–2034. http://dx.doi.org/10.1093/bioinformatics/btp362.

Lam, H., Deutsch, E. W., Eddes, J. S., Eng, J. K., King, N., Stein, S. E., et al. (2007). Development and validation of a spectral library searching method for peptide identification from MS/MS. *Proteomics*, 7(5), 655–667. http://dx.doi.org/10.1002/pmic.200600625.

Li, Q., MacCoss, M. J., & Stephens, M. (2010). A nested mixture model for protein identification using mass spectrometry. *Annals of Applied Statistics*, *4*(2), 962–987.

Li, X. J., Zhang, H., Ranish, J. A., & Aebersold, R. (2003). Automated statistical analysis of protein abundance ratios from data generated by stable-isotope dilution and tandem mass spectrometry. *Analytical Chemistry*, *75*(23), 6648–6657. http://dx.doi.org/10.1021/ac034633i.

Liu, H., Sadygov, R. G., & Yates, J. R., 3rd. (2004). A model for random sampling and estimation of relative protein abundance in shotgun proteomics. *Analytical Chemistry*, *76*(14), 4193–4201. http://dx.doi.org/10.1021/ac0498563.

Ma, B., Zhang, K., Hendrie, C., Liang, C., Li, M., Doherty-Kirby, A., et al. (2003). PEAKS: Powerful software for peptide de novo sequencing by tandem mass spectrometry. *Rapid Communications in Mass Spectrometry*, *17*(20), 2337–2342. http://dx.doi.org/10.1002/rcm.1196.

Monroe, M. E., Tolić, N., Jaitly, N., Shaw, J. L., Adkins, J. N., & Smith, R. D. (2007). VIPER: An advanced software package to support high-throughput LC-MS peptide identification. *Bioinformatics*, *23*(15), 2021–2023. http://dx.doi.org/10.1093/bioinformatics/btm281.

Nesvizhskii, A. I. (2010). A survey of computational methods and error rate estimation procedures for peptide and protein identification in shotgun proteomics. *Journal of Proteomics*, *73*(11), 2092–2123. http://dx.doi.org/10.1016/j.jprot.2010.08.009.

Nesvizhskii, A. I., Keller, A., Kolker, E., & Aebersold, R. (2003). A statistical model for identifying proteins by tandem mass spectrometry. *Analytical Chemistry*, *75*(17), 4646–4658. http://dx.doi.org/10.1021/ac0341261.

Ong, S. E., Blagoev, B., Kratchmarova, I., Kristensen, D. B., Steen, H., Pandey, A., et al. (2002). Stable isotope labeling by amino acids in cell culture, SILAC, as a simple and accurate approach to expression proteomics. *Molecular and Cellular Proteomics*, *1*(5), 376–386.

Paoletti, A. C., Parmely, T. J., Tomomori-Sato, C., Sato, S., Zhu, D., Conaway, R. C., et al. (2006). Quantitative proteomic analysis of distinct mammalian mediator complexes using normalized spectral abundance factors. *Proceedings of the National Academy of Sciences of the United States of America*, *103*(50), 18928–18933. http://dx.doi.org/10.1073/pnas.0606379103.

Patterson, S. D., & Aebersold, R. H. (2003). Proteomics: The first decade and beyond. *Nature Genetics*, *33*(Suppl.), 311–323. http://dx.doi.org/10.1038/ng1106.

Perkins, D. N., Pappin, D. J. C., Creasy, D. M., & Cottrell, J. S. (1999). Probability-based protein identification by searching sequence databases using mass spectrometry data. *Electrophoresis*, *20*(18), 3551–3567. http://dx.doi.org/10.1002/(SICI)1522-2683(19991201)20:18<3551::AID-ELPS3551>3.0.CO;2-2.

Qian, W. J., Liu, T., Petyuk, V. A., Gritsenko, M. A., Petritis, A. D., & the Host Response to Injury Large Scale Collaborative Research, Program (2009). Large-scale multiplexed quantitative discovery proteomics enabled by the use of an (18)O-labeled "universal" reference sample. *Journal of Proteome Research*, *8*(1), 290–299. http://dx.doi.org/10.1021/pr800467r.

Radulovic, D., Jelveh, S., Ryu, S., Hamilton, T. G., Foss, E., Mao, Y., et al. (2004). Informatics platform for global proteomic profiling and biomarker discovery using liquid chromatography-tandem mass spectrometry. *Molecular and Cellular Proteomics*, *3*(10), 984–997. http://dx.doi.org/10.1074/mcp.M400061-MCP200.

Ryu, S., Gallis, B., Goo, Y. A., Shaffer, S. A., Radulovic, D., & Goodlett, D. R. (2008). Comparison of a label-free quantitative proteomic method based on peptide ion current area to the isotope coded affinity Tag method. *Cancer Informatics*, *6*, 243–255. http://dx.doi.org/10.4137/CIN.S385.

Ryu, S., Goodlett, D. R., Noble, W. S., & Minin, V. N. (2012). A statistical approach to peptide identification from clustered tandem mass spectrometry data. *Proceedings (IEEE International Conference on Bioinformatics and Biomedicine)*, *4*, 648–653. http://dx.doi.org/10.1109/BIBMW.2012.6470214.

Schwanhausser, B., Busse, D., Li, N., Dittmar, G., Schuchhardt, J., Wolf, J., et al. (2011). Global quantification of mammalian gene expression control. *Nature*, *473*(7347), 337–342. http://dx.doi.org/10.1038/nature10098.

Tang, H., Arnold, R. J., Alves, P., Xun, Z., Clemmer, D. E., Novotny, M. V., et al. (2006). A computational approach toward label-free protein quantification using predicted peptide detectability. *Bioinformatics*, *22*(14), e481–e488. http://dx.doi.org/10.1093/bioinformatics/btl237.

Tanner, S., Shu, H., Frank, A., Wang, L.-C., Zandi, E., Mumby, M., et al. (2005). InsPecT: Identification of posttranslationally modified peptides from tandem mass spectra. *Analytical Chemistry*, 77(14), 4626–4639. http://dx.doi.org/10.1021/ac050102d.

Tyers, M., & Mann, M. (2003). From genomics to proteomics. *Nature*, *422*(6928), 193–197. http://dx.doi.org/10.1038/nature01510.

CHAPTER TWO

Unraveling Oxidation-Induced Modifications in Proteins by Proteomics

Carolina Panis[1]

Laboratory of Physiopathology and Free Radicals, Department of General Pathology, State University of Londrina, Londrina, Paraná, Brazil

[1]Corresponding author: e-mail address: carolpanis@sercomtel.com.br; carolpanis@hotmail.com

Contents

1. A Very Brief Introduction to Oxidative Stress and Redox Homeostasis 20
2. Proteins as Targets of Oxidative Modification 21
   2.1 S-nitroso modifications of proteins 22
   2.2 Carbonylation 23
   2.3 Cysteine/disulfide oxidation: Cysteine–thiol chemistry and eletrophilic stress 24
3. Redox Proteomics: The Comprehensive Study of Oxidation-Induced Protein Damage 25
   3.1 Technologies for redox proteomics screening 26
4. The Use of Proteomics for Mapping the Redox Homeostasis Network: Lessons from Cancer Research 29
5. Perspectives and Conclusions 33

References 33

Abstract

Oxidative stress-driven modifications can occur in lipids, proteins, and DNA and form the basis of several chronic pathologies. The metabolites generated during oxidative responses consist of very reactive substances that result in oxidative damage and modulation of redox signaling as the main outcomes. Oxidative modifications occurring in proteins are poorly understood; among the several methods employed to study such modifications, the most promising strategies are based on proteomics approaches. Proteomics has emerged as one of the most powerful and sensitive analytical tools for mapping the oxidative changes present in proteins in a wide range of sample types and disease models. This chapter addresses the main aspects of redox processes, including an overview of oxidative stress and its biological consequences on proteins. Moreover, major proteomic strategies that can be employed as powerful tools for understanding protein oxidative modifications detected in chronic pathologies are discussed, highlighting cancer research as a model.

Advances in Protein Chemistry and Structural Biology, Volume 94
ISSN 1876-1623
http://dx.doi.org/10.1016/B978-0-12-800168-4.00002-0

1. A VERY BRIEF INTRODUCTION TO OXIDATIVE STRESS AND REDOX HOMEOSTASIS

Although all cells need oxygen, the products formed by the incomplete reduction of oxygen in the mitochondrial respiratory chain are extremely reactive and can result in oxidative modifications of lipid, proteins, and DNA, which can either strongly impact redox signaling or promote cell damage. Such products are termed reactive species (RS) or free radicals and are generically defined as chemically reactive molecules bearing an unpaired electron (Halliwell, 1989).

RS are constantly and naturally produced by cells during metabolism, under both physiological and pathological conditions. An imbalance between RS production and their neutralization by the antioxidant system gives rise to a condition defined as oxidative stress. The term "oxidative" mainly refers to species derived from oxygen, but in general, it also includes other RS such as those derived from nitric oxide (NO) and hypochlorous acid. NO itself is a weak RS, but it rapidly reacts with the superoxide anion (a highly reactive oxygen RS, or ROS), yielding a very strong oxidative species called peroxynitrite. Some studies report the impact caused by NO-derived species as "nitrosative stress." The main oxygen-derived species are the superoxide anion, hydroxyl radical, hydrogen peroxide, and oxygen singlet, while the best-known nitrogen-derived RS are peroxynitrite, nitrogen dioxide, and dinitrogen trioxide (Halliwell & Gutteridge, 2007).

To overcome these RS, cells are equipped with sophisticated machinery consisting of enzymatic and nonenzymatic antioxidants, named the antioxidant system. The nonenzymatic system is formed of low molecular weight substances ingested as part of healthy nutrition, such as thiol compounds, ascorbic acid, uric acid, and vitamin E. The enzymatic system comprises powerful enzymes responsible for counteracting RS by converting free radicals in less RS. Some of these enzymes, such as catalase, superoxide dismutase (SOD), and glutathione peroxidase, can act together on common RS, while the nonenzymatic system acts directly on the free radical by capturing its unpaired electrons (Sies, 1993). Both systems are primarily responsible for preventing oxidative damage induced by uncontrolled chain reactions triggered by free radicals.

The constant exposure of cells to oxidative stress can result in either cell adaptation or injury. Adaptive responses occur when continuous RS exposure triggers the mechanism of antioxidant enzyme synthesis at the transcriptional level, regulated by the Kelch-like ECH-associating protein 1 (Keap1)/nuclear factor (erythroid-derived 2)-like 2 (Nrf2)/antioxidant response element (ARE) signaling pathway (Lau, Villeneuve, Sun, Wong, & Zhang, 2008). On the other hand, if there is a failure in this neutralization process, RS can quickly react with cellular structures, resulting in nitrosative/oxidative damage. As consequence, the processes of lipid peroxidation and protein oxidative modifications can be measured as markers of oxidative stress in several disease models (Halliwell & Whiteman, 2004).

2. PROTEINS AS TARGETS OF OXIDATIVE MODIFICATION

After synthesis, proteins can undergo a variety of chemical modifications defined as posttranslational modifications, which are critical for their functionality. Such modifications can be induced by several mechanisms, including oxidative changes in the cell, resulting from sustained oxidative stress conditions (Curtis et al., 2012).

It is well established that certain modifications in protein structure can impair its functioning. As discussed above, oxidative stress resulting from the uncontrolled production of free radicals can result in protein oxidative injury, affecting potentially all systemic functions that are dependent on proteins, such as antibodies, cell receptors, membrane structures, enzymes, and signal transduction proteins. Oxidized proteins can trigger anomalous responses in the organism, as they can be recognized as foreign structures (Halliwell & Gutteridge, 2007).

Several posttranslational modifications are induced by oxidative stress (Gianazza, Crawford, & Miller, 2007). However, proteomics analysis indicates that after oxidative stress exposure, only a small number of proteins are in fact oxidized (Madian & Regnier, 2010a,2010b; Tezel, Yang, & Cai, 2005). Redox modifications of proteins are very complex and depend on the type of RS inducing the oxidation, the location of the protein in cell compartments, and the presence of electrophilic amino acid residues.

There are two main mechanisms of protein oxidative injury: primary damage, induced by the direct attack of free radicals on protein residues, and secondary damage, resulting mainly from metabolites generated by the lipid peroxidation process (low molecular weight aldehydes). In both cases, the protein damage can be reversible (as in nitrosylation) or irreversible, in the case of carbonylation (Baummann & Meri, 2004).

The chemistry of protein-induced oxidative damage is very complex and involves oxygen, nitrogen, and halogen species. It results in the formation of peroxyl radicals, nitro-tyrosine, halogenated tyrosines, and amino acid radicals. Specific amino acid residues are preferentially attacked by RS. The thiol groups of cysteine and methionine are highly oxidizable. It is important to highlight that methionine residues are pivotal to the enzymatic activity of some proteins; therefore, their oxidation can result in impairment of activity. Histidine, proline, tryptophan, and tyrosine are other examples of specific amino acids targeted by free radicals (Ying, Clavreul, Sethuraman, Adachi, & Cohen, 2007).

Oxidized proteins are generally removed from cells by proteolytic mechanisms, since their accumulation can lead to significant cellular dysfunction. These removal mechanisms include cathepsin-mediated hydrolysis, lysosomal degradation, and proteasomal activity. The proteasome is a cytoplasmic and nuclear system that removes oxidized proteins. The subunits of the proteasome recognize target proteins by their linkages to ubiquitin at lysine residues, triggering a chain of ubiquitin enzymes that are responsible for protein degradation (Smith & Schnellmann, 2012).

The loss of protein thiol groups and the generation of carbonyl proteins are the main examples of popularly studied protein oxidative mechanisms. Carbonyl groups result from oxidative reactions derived from hydroxyl, peroxyl, peroxynitrite, and singlet oxygen groups, which oxidize amino acid residues in proteins to form the carbonyl residue (Halliwell & Gutteridge, 2007).

The following topics address the main oxidative modifications that occur in proteins as a result of the formation of oxidative stress-derived species, which can impact protein function and cell homeostasis.

2.1. *S*-nitroso modifications of proteins

Protein nitroso-compounds arise from the interaction between thiol groups and NO-derived species (Broniowska & Hogg, 2012). This mechanism consists of a pivotal nonenzymatic posttranslational modification, especially in proteins involved in signal transduction processes. NO and its derived species can react widely with cellular proteins, resulting in distinct chemical processes such as nitration (nitrite + tyrosine or tryptophan) and nitrosylation (NO-derived species + cysteine or thiol groups). Nitration is irreversible, while nitrosylation can be reversed by denitrosylation (Daiber et al., 2009). The main nitrosating agents proposed are dinitrogen trioxide (N_2O_3, a molecule resulting from the reaction of nitrite with NO) and peroxynitrite (Wink et al., 1998).

Such reactions occur at sites neighboring NO production regions, because this molecule presents a short half-life (Denicola, Souza, & Radi, 1998). Proteins from membranes are incredibly susceptible to such oxidative stress-driven modifications, which probably affect protein assembly on membranes. Some enzymes, like nitrosoglutathione, are capable of promoting the transnitrosylation of proteins, when NO groups are transferred between thiol groups. Studies indicate that these *S*-nitroso modifications are rarely found and suggest that this event may constitute a pivotal signaling mechanism. These *S*-nitroso modifications give rise to a subset of *S*-nitrosoproteins, formed mainly by plasma albumin (*S*-NO-albumin). Albumin can be further nitrosated by *trans*-*S*-nitrosation, serving as a source of thiol residues (Handy & Loscalzo, 2006).

2.2. Carbonylation

The biological impact of RS on lipids is the generation of low molecular weight aldehydes, which are highly reactive with cysteine, histidine, and lysine residues in proteins, resulting in protein carbonylation (Curtis et al., 2012). This reaction can result in protein modifications that are responsible for loss of function, as well as trigger specific signaling pathways involved in redox homeostasis. One of the most studied lipidic metabolites derived from lipid peroxidation is 4-hydroxy-*trans*-2,3-nonenal (4-HNE), which has significant ability to react with proteins by forming adducts during nucleophilic attack on specific amino acid residues (Nikki, 2009).

These modifications of proteins can result in intramolecular and intermolecular protein crosslinks (Sayre, Lin, Yuan, Zhu, & Tang, 2006). Most carbonylation targets result in a decrease or loss of protein function. Protein carbonylation by 4-HNE can result in a wide range of inhibitory biological effects, including inactivation of membrane transporters, insulin resistance, dysregulation of reduced nicotinamide adenine dinucleotide phosphate level, and altered stress signaling (Grimsrud, Xie, Griffin, & Bernlohr, 2008).

On the other hand, a role for 4-HNE-induced carbonylation has been reported in the activation of the Keap1–Nrf2–ARE axis, a pivotal regulator of oxidative stress-mediated responses (Petri, Korner, & Kiaei, 2012). The carbonylation of specific cysteine residues in the Keap1 structure (Cys273 and Cys288) results in Nrf2 release and translocation to the nucleus, and carbonylated Keap1 is degraded by the proteasome system (Levonen et al., 2004).

2.3. Cysteine/disulfide oxidation: Cysteine–thiol chemistry and eletrophilic stress

The occurrence of oxidative stress affects the biological function of cellular components. Redox homeostasis is tightly regulated by several redox-sensitive components, such as thiol groups. Redox homeostasis depends on the balance between thiol oxidation by ROS and its reduction by thiol–disulfide transfer reactions (Stroher & Dietz, 2006).

Dysfunctional redox signaling has been implicated in several chronic pathologies, and thiol systems are important sensors of oxidative danger. Cysteine residues are particularly important during chronic oxidative stress, when the reduced/oxidized cysteine equilibrium affects pivotal cellular networks for controlling cell proliferation and apoptosis (Jones, 2006). Oxidative reactions driven by RS are of importance, since these compounds generally present an unpaired electron that reacts with the sulfuric electrophilic center of the thiol residue. Cysteine is a major target for RS and exhibits high reactivity to peroxynitrite, superoxide, hydroperoxyl, and hypochlorous acid (Held & Gibson, 2012). Thus, modifications in cysteine residues in proteins have been reported as pivotal redox events in biological processes, since they can modify the catalytic and conformational properties of proteins.

The primary type of oxidative modification suffered by cysteine residues is the generation of oxoforms, found in unstressed cells. During chronic oxidative stress, the predominant oxidative forms are the sulfinic and sulfonic forms (Held & Gibson, 2012). Oxidative changes in thiol content have been reported as a pivotal mechanism of redox regulation (Go et al., 2013).

Cysteine residues have been particularly in focus when studying redox homeostasis, due to their nucleophilic properties (Lindahl, Mata-Cabana, & Kieselbach, 2011). The sulfur atom is the nucleophilic center of this molecule, allowing a wide range of reactions that can compromise protein structure. These residues are present in the catalytic center of several important proteins including the ubiquitin–proteasome enzymes, phosphatase and tensin homolog, some caspases, and the peroxiredoxin/thioredoxin families. Modifications in specific cysteine residues can result in both activation and inactivation of enzymes, and such reversible thiol modifications are indicative of the regulatory role of several redox-sensitive proteins (Leichert & Jacob, 2006).

A known example of the relationship between cysteine residue modification and protein activity is caspase-3. This enzyme contains a physiologically nitrosylated cysteine residue in its catalytic site, inhibiting its proteolytic activity. The pro-apoptotic activation of caspase-3 occurs when the NO is removed from the cysteine–thiol residue by thioredoxin-induced transnitrosylation (Benhar, Forrester, Hess, & Stamler, 2008; Mannick et al., 1999). Another example is heat shock protein 33 (Hsp33), a redox-regulated chaperone in which the oxidative formation of intramolecular disulfide bonds converts its reduced/inactive form to the active form (Winter, Linke, Jatzek, & Jakob, 2005).

3. REDOX PROTEOMICS: THE COMPREHENSIVE STUDY OF OXIDATION-INDUCED PROTEIN DAMAGE

In recent years, proteomics has emerged as a powerful analytical strategy for the large-scale study of proteins, mainly in biological systems. The study of proteins in a given disease model is very complex, since the proteome can change according to specific cell types. Advances in technology have allowed the use of proteomics to systematically determine specific posttranslational modifications in proteins, demonstrating that such modifications do not just occur randomly, but have a role in cellular function (Jensen, 2004).

Although the occurrence of oxidative stress is well established in several models of disease, little is known about specific oxidative modifications induced by RS in the proteome. The entire set of oxidative changes occurring in proteins (mainly in cysteine residues) has been defined as the redoxome, and the proteomics approach to studying such changes is termed redox proteomics. The most common oxidative modifications of proteins investigated by proteomics approaches are 4-HNE-induced carbonylation, thiol–cysteine mapping, and *S*-nitrosoproteins. Analytical technologies have included a combination of enrichment and affinity-based methods, multidimensional separation chromatography approaches, and robust mass spectrometry (MS) tools. Although such methods are divided into distinct technologies, the biochemical principle of all modifications covered by redox proteomic tools involves changes in thiol residues.

The main techniques employed to understand such oxidative modifications are described below.

3.1. Technologies for redox proteomics screening

3.1.1 Analysis of protein carbonylation/4-HNE adducts by MS

4-HNE is one of the most studied products that undergoes electrophilic reactions with biological structures due to its enormous reactive capacity. This compound is detoxified by glutathione-*S*-transferases via conjugation with its thiol group (Balogh & Atkins, 2011) but also forms protein adducts with specific amino acid residues.

Although several tools have been developed for tracking oxidative modifications in proteins induced by oxidative stress-derived products, the identification of specific targets of carbonylation is still a challenge. Most of the methods currently available are based on derivatization of carbonyl groups with hydrazine/hydrazide with further identification by western blot, enzyme-linked immunosorbent assay (ELISA), or UV spectrophotometry-based methods.

Two-dimensional (2D) electrophoresis has been used for identifying carbonylated proteins from complex mixtures. Samples are separated on a gel and further identified by immunoblot reactions by using anti-4-HNE antibodies or by labeling the carbonylated residues by nucleophilic hydrazine/hydrazide probes (Grimsrud et al., 2008).

The use of MS approaches is promising for this purpose, especially when based on MS techniques in association with computational biology (Fritz & Petersen, 2011). The use of matrix-assisted laser desorption ionization (MALDI) and electrospray ionization (ESI) has allowed the analysis of specific aldehyde-induced modifications in protein structures and represents a powerful technique for investigating the carbonylation process (Carini, Aldini, & Facino, 2004).

MALDI time-of-flight mass spectrometry (TOF-MS) has been applied to detect modifications induced by the formation of a 4-HNE adduct with a specific peptide. The modified peptide exhibits an increase in the mass-to-charge ratio relative to its native form. Further, this technology provides information regarding the posttranslational modification of the protein, indicating the specific site of adduct formation (Spicket, 2013).

This proteomics approach has been employed *in vitro* to investigate 4-HNE-induced modifications in myoglobin (Liu, Minkler, & Sayre, 2003), cytochrome *c* (Isom et al., 2004), creatine kinase (Eliuk, Renfrow, Shonsey, Barnes, & Kim, 2007), and glyceraldehyde phosphate dehydrogenase (GAPDH) (Ishii, Tatsuda, Kumazawa, Nakayama, & Uchida, 2003). *In vitro* reports include 4-HNE-induced modifications in catalase (D'Souza et al., 2008) and Hsp90 (Carbone, Doorn, Kiebler, & Petersen, 2005).

Advances in serum proteomics have also been reported for mapping such oxidative modifications. Madian and Regnier (2010b) recently published the first proteomics-based approach for identifying oxidized proteins in human plasma. In this study, the authors first isolated carbonylated proteins using avidin-affinity chromatography, with posterior biotinylation of the carbonyls with hydrazide. These fragments were identified by MALDI-MS and ESI. Using this strategy, it was possible to map 24 oxidative modifications in 14 proteins and 15 specific carbonylation sites on 7 proteins, highlighting 20 oxidative modifications in apolipoprotein B-100.

The use of MS technologies has substantially improved the analysis of 4-HNE-induced carbonylation, but some drawbacks remain to be overcome, such as sample oxidation during handling and the identification of each specific site of oxidative modification present in each modified protein. Such information is necessary to advance our understanding of the functional relevance of the carbonylation process. In this context, some proteomic methods for identifying the site and type of carbonyl modification have recently emerged (Grimsrud et al., 2008).

3.1.2 Proteomics targeting the global thiol–cysteine state

Oxidative cysteine modifications are a central mechanism of posttranslational regulation of proteins and constitute the main strategy for evaluating redox proteomics (Leonard & Carrol, 2011). Cysteines present modifications of biological relevance; however, the low occurrence of this residue justifies the existence of specific proteomic strategies for detecting its oxidative modifications (Giron, Dayon, & Sanchez, 2011).

Targeted proteomics strategies have emerged to map and quantify the oxidation of cysteine–thiol residues in electrophile-sensitive proteins (Wall et al., 2013). Technologies for thiol modification mapping include gel electrophoresis or chromatographic methods in association with MS strategies (Stroher & Dietz, 2006). Cysteine-reactive covalent-capture tags are frequently used for the isolation of cysteine-containing peptides by reaction with specific moieties as an enrichment strategy, coupled with MS analysis (Giron, Dayon, Mihala, Sanchez, & Rose, 2009; Giron, Dayon, Sanchez, & Rose, 2009). Enrichment of N-terminal-cysteinyl peptides using the covalent-capture method can facilitate the selective isolation of the free N-terminal cysteine, reducing sample complexity and refining the search for specific less abundant modifications in the thiol content (Giron, Dayon, et al., 2009).

A combined detection approach using fluorescence and MS has also been proposed to measure the redox-sensitive cysteine content in native membrane proteins, using the labeling reagent monobromobimane. This lipophilic compound reacts with free cysteine residues in proteins, leading to the formation of fluorescent derivatives, which can be determined in combination with MALDI detection (Petrotchenko et al., 2006). Recently, the redox fluorescent switch has been used as a thiol-mapping proteomics tool to label cysteine residues that are reversibly oxidized using fluorescent compounds (Izquierdo-Alvarez et al., 2012).

Advances have been reached in the field of thiol proteomics by the introduction of very specific proteomics methods. A successful example is the OxiCAT strategy, a quantitative thiol-trapping technique introduced by Leichert et al. (2008) that employs a powerful MS approach to map thiol content modifications occurring *in vivo*. The OxiCAT technology can be employed to map oxidative thiol modifications induced *in vivo* and *in vitro*. This tool can predict the extent of thiol oxidation using a combined MS global approach with isotope-coded affinity tag (ICAT) technology in association with the thiol-trapping technique in cysteine residues. Therefore, OxiCAT allows determination of the ratio between reduced/oxidized proteins in a sample.

The shotgun OxiCAT technology maps cellular thiol content following treatment of the cells with sublethal concentrations of hydrogen peroxide and hypochlorite, using MS analysis. This technique allows the identification of specific sets of redox-sensitive proteins, as well as of specific cysteine and thiol modifications. Although this approach is new and has not yet been applied to cancer studies, OxiCAT provides a powerful tool to identify, quantify, and monitor oxidative thiol modifications that occur *in vivo*. An extension of the OxiCAT technique has been recently proposed, the NOxICAT method, a highly sensitive and specific tool for the quantitative study of changes induced by oxidative stress on the thiol content of proteins (Lindemann & Leichert, 2012).

The redox proteome presents as a major advantage the fact that this approach can help establish the role of protein oxidation in cell physiology. However, technological advances are necessary, because thiol-redox reactions are very labile and it is difficult to directly determine the type and site of modified residues (Chiappetta et al., 2010).

3.1.3 Nitrosoproteome approach for identifying S-nitrosoproteins

S-nitrosoproteins result from the action of NO-derived RS mainly on thiol residues (Diers, Keszler, & Hogg, 2013). The proteomics study of *S*-nitrosoproteins has been reported using endothelial cells treated with

NO donors as a model (Martinez-Ruiz & Lamas, 2004; Yang & Loscalzo, 2005). Peroxiredoxin, GAPDH, and actin were successfully identified as *S*-nitrosoproteins after 2D electrophoretic separation and peptide sequencing by MS. *In vitro* studies have targeted *S*-nitrosoproteins by investigating *S*-nitroso adducts formed at physiological levels (Handy & Loscalzo, 2006), reinforcing the need for specific MS approaches toward low-abundance proteins. The detection of *S*-nitrosothiols is not simple due to its lability (López-Sànchez et al., 2009).

A recent strategy proposed to map the nitrosoproteome is the use of protein microarrays (Lee et al., 2013). This high-throughput approach employs a high-density protein microarray chip that contains 16,368 unique human proteins. Using this tool, Lee et al. identified 834 potentially *S*-nitrosylated proteins and defined critical sites of NO–cysteine interaction, emphasizing *S*-nitrosylation as a regulatory mechanism of the ubiquitin–proteasome system. Moreover, the most significantly *S*-nitrosylated proteins identified were also investigated by liquid chromatography tandem mass spectrometry (LC–MS/MS) and several *S*-nitrosylated residues located at known functional motifs were identified.

4. THE USE OF PROTEOMICS FOR MAPPING THE REDOX HOMEOSTASIS NETWORK: LESSONS FROM CANCER RESEARCH

Although a significant portion of the functional information regarding cancer is present in the proteome, redox proteomics is still a poorly explored field. Oxidative, nitrosative, and eletrophilic stresses are clearly implicated in several types of human cancer (Halliwell, 2007; Panis, Herrera, Victorino, Aranome, & Cecchini, 2013; Panis, Pizzatti, Herrera, Cecchini, & Abdelhay, 2013; Panis et al., 2012). Therefore, another point of interest beyond investigating oxidative modifications of proteins is the proteomic mapping of the specific components of cell redox homeostasis.

Nowadays, most of the knowledge about the role of oxidative stress in cancer is derived from *in vitro* and experimental studies. Although a growing amount of evidence highlights the impact of redox changes modulated by the cancer process, little has been reported about the use of specific proteomics tools geared toward specific oxidative/nitrosative/eletrophilic changes occurring in the cancer milieu. Further, most of the proteins that form this redox network are underexplored in studies of human cancers;

therefore, an in-depth analysis of specific redox proteins, as well as their oxidative modifications, is still necessary.

Subcellular proteome mapping resulting from the mapping of mitochondrial-specific proteins has brought advances to this field. Mitochondria are the main organelle responsible for cellular homeostasis (Chen, Li, Hou, Xie, & Yang, 2010). During the respiratory electron chain, the mitochondria gives rise to a considerable amount of ROS, making this organelle the main source and target of oxidative stress. Mitochondrial dysfunction can result in oxidative stress, and this fact has been implicated as a pivotal mechanism in several age-related pathologies, including cancer (Da Cruz, Parone, & Martinou, 2005; Verschoor et al., 2013).

Analysis of mitochondrial-derived proteins is still a challenge due to the need for cell protein fractionation and identification of low-abundance proteins in mitochondria (Lopez & Melov, 2002). The use of the ICAT approach coupled with 2D LC–MS/MS analysis and bioinformatics tools can reveal specific mitochondrial proteins. Using this strategy, catalase and AP endonuclease-1 have been reported, as well as their specific locations inside the mitochondrial space (Jiang et al., 2005). The use of proteomics approaches to map mitochondrial components, in combination with bioinformatics tools for identifying and comparing mass spectral datasets, can be useful for the future development of targeted therapeutics in mitochondrial-related diseases (Chen et al., 2010; Kim, Kim, Park, & Joo, 2009).

Several findings regarding changes in redox-related proteins have been reported in cancer studies, as listed in Table 2.1. This indicates that cancer provides a fertile ground for studying the oxidative changes that can modulate redox signaling, as well as other primary protein functions in cell homeostasis.

Through these studies, a wide range of proteomics technologies that can be used to map changes that occur in proteins involved in redox signaling has been reported. As the major samples used for cancer studies are plasma/serum and paraffin-embedded tumor tissue, several analytical strategies have been employed to perform proteomics characterization of cancer samples in human studies, particularly in the areas of tissue immunohistochemistry and western blotting of plasma/serum. In this context, it is possible to search for specific components of the redox network in such samples.

One of the most studied redox enzymes found in cancer is SOD, a component of the enzymatic antioxidant defense system that is frequently overexpressed in cancerous tissue (Fig. 2.1B). SOD expression can also be found in plasma samples (Fig. 2.1C), because this protein has a low molecular

Table 2.1 Examples of redox-related components indentified in cancer studies by proteomic-based approaches

Redox network protein	References
Arg1, Hsp10	Ye et al. (2013)
Peroxiredoxin	Castagna et al. (2004)
SOD	Sarto et al. (2001), Russo et al. (2005), Hamrita et al. (2010), Collet et al. (2011), and Seike et al. (2003)
Glutathione-*S*-transferase Heat shock protein 27 Peroxiredoxins 2, 4, and 6	Hodgkinson et al. (2010)
Protein of mismatch repair PMS2	Panis, Herrera, et al. (2013) and Panis, Pizzatti, et al. (2013)
NADH-ubiquinone oxidoreductase	Yim, Lee, Lee, Kim, and Park (2006)
Heat shock protein 90 Heat shock protein 70 SOD Glutathione-*S*-transferase Peroxiredoxin 1 Thioredoxin	Cecconi et al. (2005)

weight (18 kDa) and can be easily secreted by tumor-derived exosomal vesicles (Gomes, Keller, Altevogt, & Costa, 2007; Rana, Malinowska, & Zoller, 2013). Examples that include changes in components of the redox network are numerous and include a variety of strategies. Molecular analysis of several carcinomas, including renal (Sarto et al., 2001), pancreatic (Cecconi et al., 2005), lung (Huang et al., 2006), thyroid (Russo et al., 2005), breast (Hamrita, Nasr, Chahed, Kabbage, & Chouchane, 2010), glioblastoma (Collet et al., 2011), and intestinal carcinomas (Seike et al., 2003), by proteomics tools implicates the expression of manganese SOD in several aspects of cancer.

Undoubtedly, MS is still the most appropriate tool for high-throughput screening studies in cancer, as shown by the growing list of redox-related components identified by proteomics. Technologies include gel-based methods, labeled and label-free MS approaches, and array-based techniques.

A comparative subproteomic analysis of mitochondria obtained from human hepatocellular carcinoma performed by 2D electrophoresis coupled with MALDI-TOF/TOF technology has been recently reported by Ye and

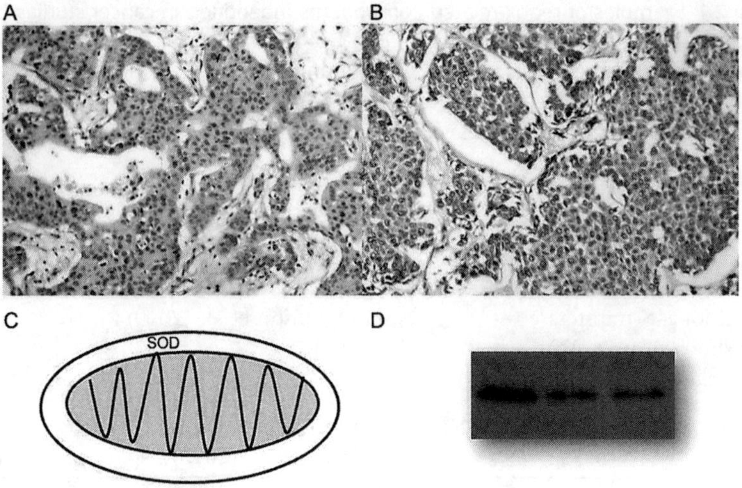

Figure 2.1 Mn-SOD expression in cancerous tissue and plasma samples. In (A), a breast tumor tissue presenting negative immunohistochemical staining for SOD1. A strong positive cytoplasmatic staining is observed in (B). In (C), the representative scheme of SOD localization in mitochondria. SOD1 expression can be also found in plasma samples from tumor patients (D). (For color version of this figure, the reader is referred to the online version of this chapter.)

collaborators. Tumor samples from 20 individuals were analyzed, and revealed 9 downregulated and 6 upregulated proteins, with validation of oxidative stress-related proteins such as Arg1 and Hsp10.

Redox-related proteins, such as glutathione-*S*-transferase, heat shock proteins, and peroxiredoxins, have also been evidenced by MS as pivotal components of chemoresistance (Hodgkinson, Eagle, Drew, Lind, & Cawkwell, 2010). Recent studies have implicated new components of the cancer network in redox-related processes. The use of high-resolution label-free proteomics tools for qualitative and quantitative bidimensional nano-ultrapressure liquid chromatography (nanoUPLC) tandem nano-ESI-MS^E analysis of breast cancer plasma demonstrated, for the first time, the overexpression of the DNA-mismatch repair protein PMS2 in the advanced stages of disease, suggesting deregulation of the oxidative status during metastatic cancer (Panis, Pizzatti, et al., 2013).

A recent work of Karisch and colleagues (2011) described the classical protein-tyrosine kinases (PTPs) phosphatome and redoxome of cancer cells assessed by using a combination of hydrogen peroxide-driven oxidation in combination with monoclonal antibodies. This strategy allowed to distinct the sample proteins in two different pools (oxidized and reduced), which are detected by immunobloting or processed by LC–MS/MS. The results of such study pointed that cancer cells display differential PTPs and its oxidized forms when compared to normal cells. This redoxome-based strategy is novelty in cancer research and opens a new analytical way for the functional discovering the role of oxidative posttranslational modifications in cancer biology.

5. PERSPECTIVES AND CONCLUSIONS

Understanding the role of oxidative stress-induced changes in proteins is crucial, since it appears that small protein modifications can result in significant impact on protein function and are pivotal for cell signaling and survival (Velez et al., 2011). Technologies for mapping the redoxome by studying the cysteine and thiol residue modifications have emerged in the last years in order promising, and should be extensively used to contribute to our knowledge of redox signaling in different disease models. Cancer research can take advantage of such technologies, since tumors are immersed in an oxidative environment and oxidative stress-induced changes are fundamentally involved in this pathology. Further advances are necessary to develop reliable strategies for mapping the specific site and type of oxidative modification, as well as its functional impact on redox signaling.

REFERENCES

Balogh, L. M., & Atkins, W. M. (2011). Interactions of glutathione transferases with 4-hydroxynonenal. *Drug Metabolism Reviews*, *43*, 165–178.

Baummann, M., & Meri, S. (2004). Techniques for studying protein heterogeneity and post-translational modifications. *Expert Reviews in Proteomics*, *1*, 207–217.

Benhar, M., Forrester, M. T., Hess, D. T., & Stamler, J. S. (2008). Regulated protein denitrosylation by cytosolic and mitochondrial thioredoxins. *Science*, *320*, 1050–1054.

Broniowska, K. A., & Hogg, N. (2012). The chemical of S-nitrosothiols. *Antioxidants and Redox Signaling*, *17*, 969–980.

Carbone, D. L., Doorn, J. A., Kiebler, Z., & Petersen, D. R. (2005). Cysteine modification by lipid peroxidation products inhibits protein disulfide isomerase. *Chemical Research in Toxicology*, *18*, 1324–1331.

Carini, M., Aldini, G., & Facino, R. M. (2004). Mass spectrometry for detection of 4-hydroxy-trans-2-nonenal (HNE) adducts with peptides and proteins. *Mass Spectrometry Reviews*, *23*, 281–305.

Castagna, A., Antonioli, P., Astner, H., Hamdan, M., Righetti, S. C., Perego, P., et al. (2004). A proteomic approach to cisplatin resistance in the cervix squamous cell carcinoma cell line A431. *Proteomics*, *4*, 3246–3267.

Cecconi, D., Donadelli, M., Scarpa, A., Milli, A., Palmieri, M., Hamdan, M., et al. (2005). Proteomic analysis of pancreatic ductal carcinoma cells after combined treatment with gemcitabine and trichostatin. *American Journal of Proteome Research*, *4*, 1909–1916.

Chen, X., Li, J., Hou, J., Xie, Z., & Yang, F. (2010). Mammalian mitochondrial proteomcis: Insights into mitochondrial functions and mitochondria-related diseases. *Expert Reviews of Proteomics*, *7*, 333–345.

Chiappetta, G., Ndiaye, S., Igbaria, A., Kumar, C., Vinh, J., & Toledano, M. B. (2010). Proteome screens for Cys residues oxidation: The redoxome. *Methods in Enzymology*, *473*, 199–216.

Collet, B., Guitton, N., Saikali, S., Avril, T., Pineau, C., Hamlat, A., et al. (2011). Differential analysis of glioblastoma multiforme proteome by a 2D-DIGE approach. *Proteome Science*, *9*, 16–30.

Curtis, J. M., Hanh, W. S., Long, E. K., Burrill, J. S., Arriaga, E. A., & Bernlohr, D. A. (2012). Protein carbonylation and metabolic control system. *Trends in Endocrinology and Metabolism*, *23*, 399–406.

Da Cruz, S., Parone, P. A., & Martinou, J. C. (2005). Building the mitochondrial proteome. *Expert Reviews of Proteomics*, *2*, 541–551.

Daiber, A., Schildknecht, S., Muller, J., Kamuf, J., Bachschmid, M. M., & Ulrich, V. (2009). Chemical model systems for cellular nitros (yl) ation reactions. *Free Radical Biology and Medicine*, *47*, 458–467.

Denicola, A., Souza, J. M., & Radi, R. (1998). Diffusion of peroxinitrite across erythrocyte membranes. *Proceedings of the National Academy of Sciences of the United States of America*, *95*, 3566–3571.

Diers, A. R., Keszler, A., & Hogg, N. (2013). Detection of S-nitrosothiols. *Biochimica and Biophysica Acta*, *13*, S0304-4165(13)00339-5.

D'Souza, B. T., Kurien, R., Rodgers, J., Shenoi, S., Kurono, H., Matsumoto, K., et al. (2008). Detection of catalase as a major protein target of the lipid peroxidation product 4-HNE and the lack of its genetic association as a risk factor in SLE. *BMC Medical Genetics*, *9*, 62.

Eliuk, S. M., Renfrow, M. B., Shonsey, E. M., Barnes, S., & Kim, H. (2007). Active site modifications of the brain isoform of creatine kinase by 4-hydroxy-2- nonenal correlate with reduced enzyme activity: Mapping of modified sites by Fourier transform-ion cyclotron resonance mass spectrometry. *Chemical Research in Toxicology*, *20*, 1260–1268.

Fritz, K. S., & Petersen, D. R. (2011). Exploring the biology of lipid peroxidation-derived protein carbonylation. *Chemical Research and Toxicology*, *24*, 1411–1419.

Gianazza, E., Crawford, J., & Miller, I. (2007). Detecting oxidative post-translational modifications in proteins. *Amino Acids*, *33*, 51–56.

Giron, P., Dayon, L., Mihala, N., Sanchez, J. C., & Rose, K. (2009). Cysteine-reactive covalent capture tags for enrichment of cysteine-containing peptides. *Rapid Communications in Mass Spectrometry*, *23*, 3377–3386.

Giron, P., Dayon, L., & Sanchez, J. C. (2011). Cysteine tagging for MS-based proteomics. *Mass Spectrometry Reviews*, *30*, 366–395.

Giron, P., Dayon, L., Sanchez, J. C., & Rose, K. (2009). Enrichment of N-terminal cysteinyl-peptides by covalent capture. *Journal of Proteomics*, *71*, 647–661.

Go, Y. M., Roede, J. R., Walkers, D., Duong, D. M., Seyfried, N. T., Orr, M., et al. (2013). Selective targeting of the cysteine proteome by thioredoxin and glutathione redox systems. *Molecular and Cellular Proteomics*, *12*, 3285–3296. http://dx.doi.org/10.1074/mcp.M113.030437.

Gomes, C., Keller, S., Altevogt, P., & Costa, J. (2007). Evidence for secretion of Cu, Zn superoxide dismutase via exosomes from a cell model of amyotrophic lateral sclerosis. *Neuroscience Letters*, *428*, 43–46.

Grimsrud, P. A., Xie, H., Griffin, T. J., & Bernlohr, D. A. (2008). Oxidative stress and covalent modification of protein with bioactive aldehydes. *The Journal of Biological Chemistry*, *283*, 21837–21841.

Halliwell, B. (1989). Tell me about free radicals, doctor: A review. *Journal of the Royal Society of Medicine*, *82*, 747–752.

Halliwell, B., & Gutteridge, J. M. C. (2007). *Free radicals in biology and medicine* (4th ed.). New York, NY: Oxford University (Chapter 1).

Halliwell, B., & Whiteman, M. (2004). Measuring reactive species and oxidative damage *in vivo* and in cell culture: How should you do it and what do the results mean? *British Journal of Pharmacology*, *142*, 231–255.

Halliwell, B. (2007). Oxidative stress: have we moved forward? *Biochemistry Journal*, *401*, 1–11.

Hamrita, B., Nasr, H. B., Chahed, K., Kabbage, M., & Chouchane, L. (2010). Proteomic analysis of human breast cancer: New technologies and clinical applications for biomarker profiling. *Journal of Proteomics and Bioinformatics*, *3*, 91–98.

Handy, D. E., & Loscalzo, J. (2006). Nitric oxide and post-translational modification of the vascular proteome: S-nitrosation of reactive thiols. *Arteriosclerosis, Thrombosis, and Vascular Biology*, *26*, 1207–1214.

Held, J. M., & Gibson, B. W. (2012). Regulatory control of oxidative damage? Proteomic approaches to interrogate the role of cysteine oxidation status in biological processes. *Molecular and Cellular Proteomics*, *11*, 1–14.

Hodgkinson, V. C., Eagle, G. L., Drew, P. J., Lind, M. J., & Cawkwell, L. (2010). Biomarkers of chemotherapy resistance in breast cancer identified by proteomics: Current status. *Cancer Letters*, *294*, 13–24.

Huang, L. J., Chen, S. X., Luo, W. J., Jiang, H. H., Zhang, P. F., & Yi, H. (2006). Proteomic analysis of secreted proteins of non-small cell lung cancer. *Ai Zheng*, *25*, 1361–1367.

Ishii, T., Tatsuda, E., Kumazawa, S., Nakayama, T., & Uchida, K. (2003). Molecular basis of enzyme inactivation by an endogenous electrophile 4-hydroxy-2-nonenal: Identification of modification sites in glyceraldehyde-3-phosphate dehydrogenase. *Biochemistry*, *42*, 3474–3480.

Isom, A. L., Barnes, S., Wilson, L., Kirk, M., Coward, L., & Darley-Usmar, V. (2004). Modification of Cytochrome c by 4-hydroxy-2-nonenal: Evidence for histidine, lysine, and arginine-aldehyde adducts. *Journal of the American Society for Mass Spectrometry*, *15*, 1136–1147.

Izquierdo-Alvarez, A., Ramos, E., Villanueva, J., Hernansanz-Agustín, P., Fernández-Rodriguez, R., Tello, D., et al. (2012). Differential redox proteomics allows identification of proteins reversibly oxidized at cysteine residues in endothelial cells in response to acute hypoxia. *Journal of Proteomics*, *75*, 5449–5462.

Jensen, O. N. (2004). Modification-specific proteomics: Characterization of post-translational modifications by mass spectrometry. *Current Opinion in Chemical Biology*, *8*, 33–41.

Jiang, X. S., Dai, J., Sheng, Q. H., Zhang, L., Xia, Q. C., Wu, J. R., et al. (2005). A comparative proteomic strategy for subcellular proteome research: ICAT approach coupled with bioinformatics prediction to ascertain rat liver mitochondrial proteins and indication of mitochondrial localization of catalase. *Molecular and Cellular Proteomics*, *4*, 12–34.

Jones, D. P. (2006). Extracellular redox state: Refining the definition of oxidative stress in aging. *Rejuvenation Research*, *9*, 169–181.

Karisch, R., Fernandez, M., Taylor, P., Virtanen, C., St-Germain, J. R., Jin, L. L., et al. (2011). Global proteomic assessment of the classical protein-tyrosine phosphatome and "redoxome". *Cell, 146*, 826–840.

Kim, T., Kim, E., Park, S. J., & Joo, H. (2009). PCHM: A bioinformatics resource for high-throughput human mitochondrial proteome searching and comparison. *Computers in Biology Medicine, 39*, 689–696.

Lau, A., Villeneuve, N. F., Sun, Z., Wong, P. K., & Zhang, D. D. (2008). The roles of Nrf2 in cancer. *Pharmacological Research, 58*, 262–270.

Lee, Y. I., Giovinazzo, D., Kang, H. C., Lee, Y., Jeong, J. S., Doulias, P. T., et al. (2013). Protein microarray characterization of the S-nitrosoproteome. *Molecular and Cellular Proteomics, 13*, 63–72. http://dx.doi.org/10.1074/mcp.M113.032235.

Leichert, L. I., & Jakob, U. (2006). Global methods to monitor the thiol-disulfide state of proteins in vivo. *Antioxidants and Redox Signaling, 8*, 763–772.

Leichert, L. I., Gehrcke, F., Gudiseva, H. V., Blackwell, T., Ilbert, M., Walker, A. K., et al. (2008). Quantifying changes in the thiol redox proteome upon oxidative stress in vivo. *Proceeding of the National Academy of Sciences of the United States of America, 105*, 8197–8202.

Leonard, S. E., & Carrol, K. S. (2011). Chemical "omics" approaches for understanding protein cysteine oxidation in biology. *Current Opinion in Chemical Biology, 15*, 88–102.

Levonen, A. L., Landar, A., Ramachandra, A., Ceaser, E. K., Dickinson, D. A., Zanoni, G., et al. (2004). Cellular mechanisms of redox cell signalling: Role of cysteine modification in controlling antioxidant defences in response to electrophilic lipid oxidation products. *Biochemical Journal, 378*, 373–382.

Lindahl, M., Mata-Cabana, A., & Kieselbach, T. (2011). The dissulfide proteome and other reactive cysteine proteomes: Analysis and functional significance. *Antioxidants and Redox Signaling, 14*, 2581–2642.

Lindemann, C., & Leichert, L. I. (2012). Quantitative redox proteomics: The NOxICAT method. *Methods in Molecular Biology, 893*, 387–403.

Liu, Z., Minkler, P. E., & Sayre, L. M. (2003). Mass spectroscopic characterization of protein modification by 4-hydroxy-2-(E)-nonenal and 4-oxo-2-(E)-nonenal. *Chemical Research in Toxicology, 16*, 901–911.

Lopez, M. F., & Melov, S. (2002). Applied proteomics: Mitochondrial proteins and effect on function. *Circulation Research, 90*, 380–389.

López-Sànchez, L. M., Muntané, J., de la Mata, M., & Rodríguez-Ariza, A. (2009). Unraveling the S-nitrosoproteome: Tools and strategies. *Proteomics, 9*, 808–818.

Madian, A. G., & Regnier, F. E. (2010a). Proteomic identification of carbonylated proteins and their oxidation sites. *Journal of Proteome Research, 9*, 3766–3780.

Madian, A. G., & Regnier, F. E. (2010b). Profiling carbonylated proteins in human plasma. *Journal of Proteome Research, 9*, 1330–1343.

Mannick, J. B., Hausladen, A., Lui, L., Hess, D. T., Zeng, M., Miao, Q. X., et al. (1999). Fas-induced caspase denitrosylation. *Science, 284*, 651–654.

Martinez-Ruiz, A., & Lamas, S. (2004). Detection and proteomic identification of S-nitrosylated proteins in endothelial cells. *Archives of Biochemistry and Biophysics, 423*, 192–199.

Nikki, E. (2009). Lipid peroxidation: Physiological levels and dual biological effects. *Free Radical Biology and Medicine, 47*, 469–484.

Panis, C., Herrera, A. C. S. A., Victorino, V. J., Aranome, A. M., & Cecchini, R. (2013). Screening of circulating TGF-β1 levels and its clinicopathological significance in human breast cancer. *Anticancer Research, 33*, 737–742.

Panis, C., Pizzatti, L., Herrera, A. C. S. A., Cecchini, R., & Abdelhay, E. (2013). Putative circulating markers of the early and advanced stages of breast cancer identified by high-resolution label-free proteomics. *Cancer Letters, 330*, 57–66.

Panis, C., Victorino, V. J., Hererra, A. C. S. A., Freitas, L. F., deRossi, T., Campos, F. C., et al. (2012). Differential oxidative status and imune characterization of the early and advanced stages of human breast cancer. *Breast Cancer Research and Treatment, 133*, 881–888.

Petri, S., Korner, S., & Kiaei, M. (2012). Nrf2/ARE signaling pathway: Key mediator in oxidative stress and potential therapeutic target in ALS. *Neurology Research International, 2012*, 1–8.

Petrotchenko, E. V., Pasek, D., Elms, P., Dokholyan, N. V., Meissner, G., & Borchers, C. H. (2006). Combining fluorescence detection and massspectrometric analysis for comprehensive and quantitative analysis of redox-sensitive cysteines in native membrane proteins. *Analytical Chemistry, 78*, 7959–7966.

Rana, S., Malinowska, K., & Zoller, M. (2013). Exosomal tumor MicroRNA modulates premetastatic organ cells. *Neoplasia, 15*, 281–295.

Russo, D., Bisca, A., Celano, M., Talamo, F., Arturi, F., Scipioni, A., et al. (2005). Proteomic analysis of human thryroid cell lines reveals reduced nuclear localization of Mn-SOD in poorly differentitated thyroid cancer cells. *Journal of Endocrinological Investigation, 28*, 137–144.

Sarto, C., Déon, C., Doro, G., Hochstrasser, D. F., Mocarelli, P., & Sanchez, J. C. (2001). Contribution of proteomics to the molecular analysis of renal cell carcinoma with an emphasis on manganese superoxide dismutase. *Proteomics, 1*, 1288–1294.

Sayre, L. M., Lin, D., Yuan, Q., Zhu, X., & Tang, X. (2006). Protein adducts generated from products of lipid peroxidation: Focus on HNE and one. *Drug Metabolism Reviews, 38*, 651–675.

Seike, M., Kondo, T., Mori, Y., Gemma, A., Kudoh, S., Sakamoto, M., et al. (2003). Proteomic analysis of intestinal epithelial cells expressing stabilized β-catenin. *Cancer Research, 63*, 4641–4647.

Sies, H. (1993). Strategies of antioxidant defense. *European Journal of Biochemistry, 215*, 213–219.

Smith, M. A., & Schnellmann, R. G. (2012). Mitochondrial calpain 10 is degraded by Lon protease after oxidant injury. *Archives of Biochemistry and Biophysics, 517*, 144–152.

Spicket, C. M. (2013). The lipid peroxidation product 4-hydroxy-2-nonenal: Advances in chemistry and analysis. *Redox Biology, 1*, 145–152.

Stroher, E., & Dietz, K. J. (2006). Concepts and approaches towards understanding the cellular redox proteome. *Plant Physiology, 8*, 407–418.

Tezel, G., Yang, X., & Cai, J. (2005). Proteomic identification of oxidatively modified retinal proteins in a chronic pressure-induced rat model of glaucoma. *Invetistigative Ophtalmology and Vision Science, 46*, 3177–3187.

Velez, J. M., Miriyala, S., Nithipongvanitch, R., Noel, T., Plabplueng, C. D., Oberley, T., et al. (2011). p53 regulates oxidative stress-mediated retrograde signaling: A novel mechanism for chemotherapy-induced cardiac injury. *PLoS One, 6*, e18005.

Verschoor, M. L., Ungard, R., Harbottle, A., Jacupciak, J. P., Parr, R. L., & Singh, G. (2013). Mitochondria and cancer: Past, present and future. *Biomedical Research International, 2013*, 612369.

Wall, S. B., Smith, M. R., Ricart, K., Zhou, F., Vayalil, P. K., Oh, J. Y., et al. (2013). Detection of eletrophile-sensitive proteins. *Biochimica Biophysica Acta, 13,* S0304-4165(13) 00378-4.

Wink, D. A., Vodovotz, Y., Laval, J., Laval, F., Dewhirst, M. W., & Mitchell, J. B. (1998). The multifaceted roles of nitric oxide in cancer. *Carcinogenesis, 19*, 711–721.

Winter, J., Linke, K., Jatzek, A., & Jakob, U. (2005). Severe oxidative stress causes inactivation of DnaK and activation of the redox-regulated chaperone Hsp33. *Molecular Cell, 17*, 381–392.

Yang, Y., & Loscalzo, J. (2005). S-nitrosoprotein formation and localization in endothelial cells. *Proceedings of the National Academy of Sciences of the United States of America, 102*, 117–122.

Ye, Y., Huang, A., Huang, C., Liu, J., Wang, B., Lin, K., et al. (2013). Comparative mitochondrial proteomic analysis of hepatocellular carcinoma from patients. *Proteomics Clinical Applications*, 7, 403–415.

Yim, E. K., Lee, S. B., Lee, K. H., Kim, C. J., & Park, J. S. (2006). Analysis of the in vitro synergistic effect of 5-fluorouracil and cisplatin on cervical carcinoma cells. *International Journal of Gynecology and Cancer, 16*, 1321–1329.

Ying, J., Clavreul, N., Sethuraman, M., Adachi, T., & Cohen, R. A. (2007). Thiol oxidation in signaling and responses to stress: Detection and quantification of physiological and pathophysiological thiol modifications. *Free Radicals in Biology and Medicine, 43*, 1099–1108.

CHAPTER THREE

Role of Proteomics in Biomarker Discovery: Prognosis and Diagnosis of Neuropsychiatric Disorders

Suman Patel[1]
CSIR-National Institute of Science, Technology and Developmental Studies, New Delhi, India
[1]Corresponding author: e-mail address: sumanitrc@gmail.com

Contents

1. Summary 40
2. Commonly Used Methods for Proteome Characterization 42
   2.1 Two-dimensional gel electrophoresis and liquid chromatography 42
   2.2 Electrospray ionization, matrix-assisted laser desorption/ionization and surface-enhanced laser desorption/ionization 44
   2.3 *In vitro* chemical labeling 44
   2.4 *In vivo* metabolic labeling 45
3. Proteomics and Biomarker 46
   3.1 Peptide as biomarkers 48
   3.2 Posttranslational modifications as biomarkers 49
4. Diagnosis of Neuropsychiatric Disorders 50
   4.1 Anxiety and depression 50
   4.2 Bipolar disorder 50
   4.3 Schizophrenia 53
   4.4 Alzheimer's disease 54
5. Proteomics of Neuropsychiatric Disorders 55
   5.1 Depression and anxiety 55
   5.2 Schizophrenia 61
   5.3 Alzheimer disease 62
   5.4 Bipolar disorder 65
6. Strength, Weakness, and Future Challenges: Biomarker Discovery 66
7. Concluding Remarks 67
References 68

Abstract

One of the major concerns of modern society is to identify putative biomarkers that serve as a valuable early diagnostic tool to identify a subset of patients with increased risk to develop neuropsychiatric disorders. Today, proteomic approaches have opened

Advances in Protein Chemistry and Structural Biology, Volume 94
ISSN 1876-1623
http://dx.doi.org/10.1016/B978-0-12-800168-4.00003-2

new possibilities in diagnostics of devastating disorders like neuropsychiatric disorders. Proteomics-based technologies for biomarker discovery have been promising because alterations in protein expression and its protein abundance, structure, or function can be used as indicators of pathological abnormalities prior to development of clinical symptoms of neuropsychiatric disorders. This is because using mass spectrometry spectra analysis, it is possible to identify biomarkers of these diseases based on the identification of proteins in body fluids that is easily available, for example, the cerebrospinal fluid, serum, or blood. An ideal biomarker should be present in the blood before the disease is clinically confirmed, have high sensitivity and specificity, and be reproducible. Despite of advances in the proteomic technologies, it has not yielded significant clinical application in neuropsychiatry research. The review discusses overall proteomic approaches for elucidating molecular mechanisms and its applicability for biomarker discovery, diagnosis, and therapeutics of psychiatric disorders such as anxiety, depression, Alzheimer's disease, schizophrenia, and bipolar disorder. In addition, we have also discussed issues and challenges regarding the implementation of proteomic approaches as a routine diagnostic tool in the clinical laboratory in context with neuropsychiatric disorders.

1. SUMMARY

Neuropsychiatry research involving proteomic approaches has made possible to identify which proteins or groups of proteins are responsible for a specific function or neurological disease phenotype. Proteomics constitutes a versatile and constantly evolving disease prognosis and diagnosis tool since identification of differentially expressed protein signatures has the potential to reflect the disease states of the individual. Even though tremendous research has been done on neuropsychiatric disorders, so far there is no reliable prognostic molecular marker available for the disease diagnosis. In addition, medication for psychiatric disorders available today either has side effects or takes a long period to cure the disease and also the response of all patients is not the same to the existing drug treatment (Bystritsky, 2006). Due to the lack of molecular understanding of psychiatric disorders and reliable biomarkers for prognosis and diagnosis of neuropsychiatric disorders, the role of proteomic approaches in resolving these issues cannot be ignored. Stress exposure results in altered behaviors, such as autonomic function, and secretion of multiple hormones, including corticotropin-releasing factor, adrenocorticotropin hormone, and cortisol, through the hypothalamic–pituitary–adrenal axis (Kim & Kim, 2007). Stress is a major factor for the occurrence of psychiatric disorders such as depression, post-traumatic stress disorder, and Alzheimer's disease (AD) (Kim & Kim,

2007). Today, the field of proteomics for neuropsychiatry research is growing rapidly. Neurobiological investigations of normal and diseased brain tissue as well as of cerebrospinal fluid (CSF) are being done using in-depth MS-based approaches including isotope-coded affinity tag (ICAT) labeling, isobaric tag for relative and absolute quantitation (iTRAQ), tandem mass tags, stable isotope labeling with amino acids in cell culture (SILAC), stable isotope labeling in mammals, label-free multiple reaction monitoring, and sequential windowed data-independent acquisition of the total high-resolution and absolute quantification methods (Craft, Chen, & Nairn, 2013). Quantitative proteomic approaches for the comparative analysis of protein changes normal- versus disease-state brain function in specific neuronal cell types may in future provide a better understanding of the disease (Craft et al., 2013). One of the most common factors leading to disability worldwide is major depressive disorder (MDD). This is because understanding of its pathophysiological mechanisms as well as empirical medical tests for accurate diagnosis or for guiding antidepressant treatment strategies is still lacking (Martins-de-Souza, Guest, Harris, et al., 2012). This chapter focuses on understanding the role of proteomic technologies to overcome challenges and future aspects of routine diagnostic tool in the clinical laboratory in the context of psychiatric disorders. Unlike other diseases like cancer that can be predicted based on the biological measurement, it is difficult for diagnosis and treatment of psychiatric disorders (Woods et al., 2012). Therefore, we still need to focus on the development of biomarkers for such complex disorder using advanced and promising approaches like proteomics. Hence, the need for development of biomarkers for neuropsychiatric disorders for improved diagnosis and monitoring treatment response cannot be ignored. It is known that mass spectrometry (MS) techniques have a wide range of applications in proteomics; therefore, it could be used for identification of proteins specific for neuropsychiatric disorders that may often be useful diagnostic and prognostic biomarkers (Xiao, Prieto, Conrads, Veenstra, & Issaq, 2005). Proteomic approaches in neuropsychiatry research may lead to identification of novel drug targets or signaling cascades specific to the disease, hence developing personalized medicine that may not be possible using other modern biological tools. This approach could be very useful for the identification of a panel of ideal or unique protein biomarkers, so this could be one of the most successful methods in terms of high sensitivity and specificity for identification of native or digested peptides in body fluids for a particular neuropsychiatric disease.

2. COMMONLY USED METHODS FOR PROTEOME CHARACTERIZATION

Figure 3.1 describes commonly used proteomic approaches for biomarker discover discovery in neuropsychiatric disorders.

2.1. Two-dimensional gel electrophoresis and liquid chromatography

This method is used for the study of differentially expressed proteins, and separation of proteins is carried out in two steps: first, isoelectric focusing and, second, sodium dodecyl sulfate polyacrylamide gel electrophoresis. Two-dimensional gel electrophoresis (2DGE) method of protein separation is used for quantitative analysis of protein in different samples and has increased sensitivity and reproducibility throughput of proteome analysis (Friedman, 2007; Friedman et al., 2004). The differentially expressed proteins are identified by methods like MS. It is possible not only the

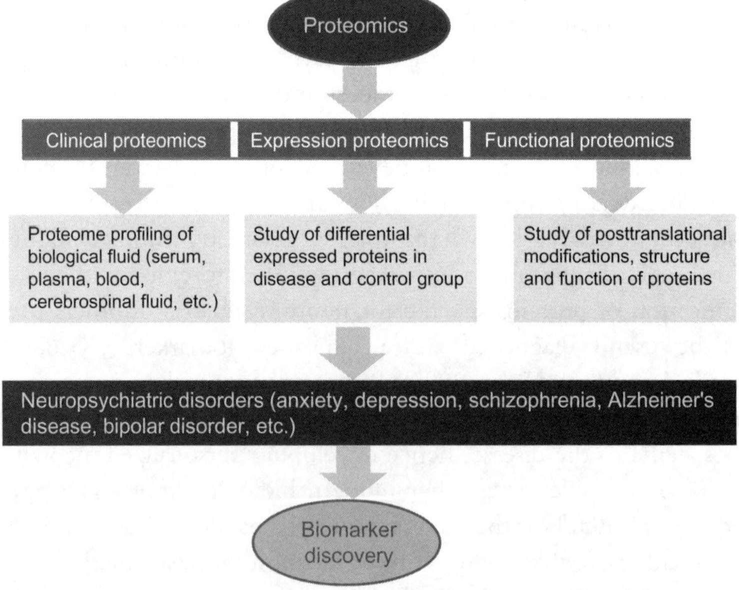

Figure 3.1 Schematic representation of proteomic approaches for biomarker discover discovery in neuropsychiatric disorders. (For color version of this figure, the reader is referred to the online version of this chapter.)

identification but also the quantification of the differentially expressed protein, that is, a given candidate marker via MS method. Previous studies have used expression proteomics to study synaptic plasticity on the basis of differentially expressed synaptic proteins, such as proteins related to spatial learning in the rat hippocampus (Henninger et al., 2007). Since 2DGE is very much suitable for studies involving posttranslational modifications (PTMs) of proteins, therefore, it is utilized for the CSF and plasma biomarkers in neuropsychiatric disorders (Castaño, Roher, Esh, Kokjohn, & Beach, 2006; Davidsson et al., 2002; Finehout, Franck, Choe, Relkin, & Lee, 2007; Hu et al., 2007; Hye et al., 2006; Puchades et al., 2003). However, 2DGE method has a few drawbacks as it is not applicable for proteins/peptides smaller than 10 kDa and is also troubled by comigration issues (e.g., one stained spot may contain more than one protein), and most importantly, its applicability for highly hydrophobic proteins is limited (Thongboonkerd, 2007). Figure 3.2 represents gel picture of separation of extracted proteins from mouse brain by isoelectric point in the first dimension and molecular weight in the second dimension using 2DGE.

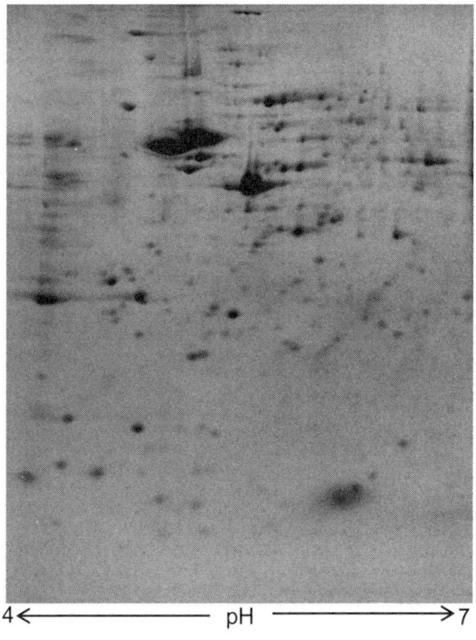

Figure 3.2 Separation of extracted proteins from mouse brain by isoelectric point in the first dimension and molecular weight in the second dimension using two-dimensional gel electrophoresis.

2.2. Electrospray ionization, matrix-assisted laser desorption/ionization and surface-enhanced laser desorption/ionization

MS comprises ion source, mass analyzer, and detection unit. It is categorized into two types: electrospray ionization (ESI) and matrix-assisted laser desorption/ionization (MALDI) instruments based on the difference of ion source specifically in the field of proteomics. Most of the proteomics-based approaches include mass analyzers such as ion trap, triple quadrupole, time-of-flight (TOF), and Fourier transform ion cyclotron, and these differ in their mechanism of ion separation, mass accuracy and resolution, and complementary in protein identification (Chandramouli & Qian, 2009). Today, technologies like ESI, LC-MS, SELDI-TOF-MS (surface-enhanced laser desorption/ionization time-of-flight mass spectrometry), MALDI-MS imaging, laser capture microdissection, or the analysis of PTMs have made possible proteomics of CSF and extracellular fluid that is very important for biomarker discovery in neuropsychiatry research as it contains neurons, astrocytes, oligodendrocytes, and microglia and also because these are the major components of the brain (Maurer, 2010). SELDI is directly connected to protein chip array and this may be in two ways: either a chemically preactivated surface (ionic, hydrophobic, and chelating metal) or a protein-specific surface (antibody and receptor) for selective capture of proteins, and this way, proteins retained on the surface are subsequently ionized and detected by MALDI-TOF-MS. On the other hand, it also has a drawback, that is, it analyzes the absence or presence of a particular signature peak rather than determining the relative abundance of a protein. Another major concern is assessing its reproducibility and also efficiency with which proteins with high molecular weights are analyzed by SELDI-TOF-MS (Thongboonkerd, 2007).

2.3. *In vitro* chemical labeling

2.3.1 Isotope-coded affinity tag

This is a quantitative method of proteome study and is used to label the side chains of cysteinyl residues in two reduced protein samples and uses isotopic light or heavy reagent. Isotopic tags are introduced at specific functional groups on peptides or proteins (Gygi et al., 1999). The comparative study of two proteomes can be done simultaneously using this approach as it allows the identification of sample origin based on the mass signatures and thus making possible the accurate quantification of the protein. ICAT has been widely used for CSF proteome analysis in order to investigate mechanisms associated with neuropsychiatric diseases such as AD (Zhang et al., 2005). Besides, ICAT has limitations too; first, quantification is mostly restricted to cysteine-containing proteins, and second, comparative study of only two conditions simultaneously is possible using ICAT method. However, few of the limitations associated with ICAT may be overcome by other advanced proteomic approaches such as iTRAQ as described below.

2.3.2 Isobaric tag for relative and absolute quantitation

This approach uses isobaric tags for relative and absolute quantification of proteins in different conditions simultaneously (Ross et al., 2004). This includes isobaric tag labeling of digested peptides, glycopeptide enrichment using hydrazide chemistry, protein/peptide identification, and quantification by liquid chromatography-based high-resolution tandem MS, as well as bioinformatic data processing (Shi, Hwang, & Zhang, 2013). In this method, N-terminus of peptides from protein digests from the samples to be compared is chemically tagged. Discovery of protein biomarkers specific to a particular disease has been possible using iTRAQ method as it is developed and optimized for human body fluids such as plasma, serum, and tears (Ernoult, Bourreau, Gamelin, & Guette, 2010; Tonack et al., 2009). The advantage of iTRAQ in comparison with ICAT is that comparative study of samples up to eight conditions simultaneously is possible by this method in addition to the identification and quantification of all peptides. In the past several years, iTRAQ has been utilized for the comparative study of the CSF proteome in neuropsychiatric disease models (Abdi et al., 2006).

2.4. *In vivo* metabolic labeling

2.4.1 Stable isotope labeling with amino acids in cell culture

This method is based on the incorporation of isotopically labeled amino acids into proteins formed by the growing organism. The two most commonly used labeled amino acids are lysine and arginine, and usually, the isotopically labeled amino acids are added to the growth medium (Chen, Smith, & Bradbury, 2000; Ong et al., 2002). The reason for using these amino acids is that the tryptic peptides from proteins synthesized from these amino acids contain an isotopically labeled lysine or arginine, which increases the quantitative coverage of the experiment and, hence, increases the accuracy of the protein quantification. SILAC method has been optimized for primary neurons (Spellman, Deinhardt, Darie, Chao, & Neubert, 2008) and has been applied to compare primary neuron synaptosome proteomes of a mouse model of mental retardation and wild-type controls (Liao, Park, Xu, Vanderklish, & Yates, 2008). According to other investigators, the labeled amino acids may also be generated by the organism through the addition of isotopically labeled *salts* to the growth medium (Oda, Huang, Cross, Cowburn, & Chait, 1999). The comparative study of multiple comparisons within a single experiment is very much possible by using SILAC proteomic method, and it also has the advantage of a predictable mass shift. Since complete incorporation of isotopic amino acids is not the same for all cell lines, therefore, it cannot be used to incorporate certain amino acids in all cell types (Harsha, Molina, & Pandey, 2008).

2.4.2 ^{15}N metabolic labeling

This is a quantitative method of studying proteins that involves the incorporation of stable isotopes ^{15}N during protein biosynthesis, and it is usually a good choice for autotrophic organisms like plants and bacteria. One of the applications of ^{15}N metabolic labeling using a bacterial protein-based diet in animal models of psychopathologies as described earlier is the labeling of the HAB/NAB/LAB mouse model of trait anxiety (Frank et al., 2009). Unlike other methods such as ICAT and SILAC that can be used for comparative study of multiple samples simultaneously, ^{15}N metabolic labeling can be used for only two samples simultaneously within a single experiment. Other complication that is associated with this method is that there are great chances of variations in the number of replaced nitrogen atoms from peptide to peptide, hence, increasing the unpredictability of exact mass shift (Bachi & Bonaldi, 2008). The identification of altered pathways and candidate biomarkers for a particular disease condition is very much possible by using ^{15}N-labeled wild-type animals from the same mouse strain because the models of disease are used as internal standards itself. Schematic representation of proteomic approaches from sample preparation to protein identification by MS analysis for biomarker discovery is shown in Fig. 3.3.

3. PROTEOMICS AND BIOMARKER

The term "proteome" can be defined as the total set of expressed proteins by a cell, tissue, or organism at a given time under a determined condition (Wilkins et al., 1996). The term "biomarker" in neuropsychiatry research has been widely used to diagnose, assess, or predict the course of the disease including the cognitive performance of the individual suffering from the disease (Singh & Rose, 2009). United States Food and Drug

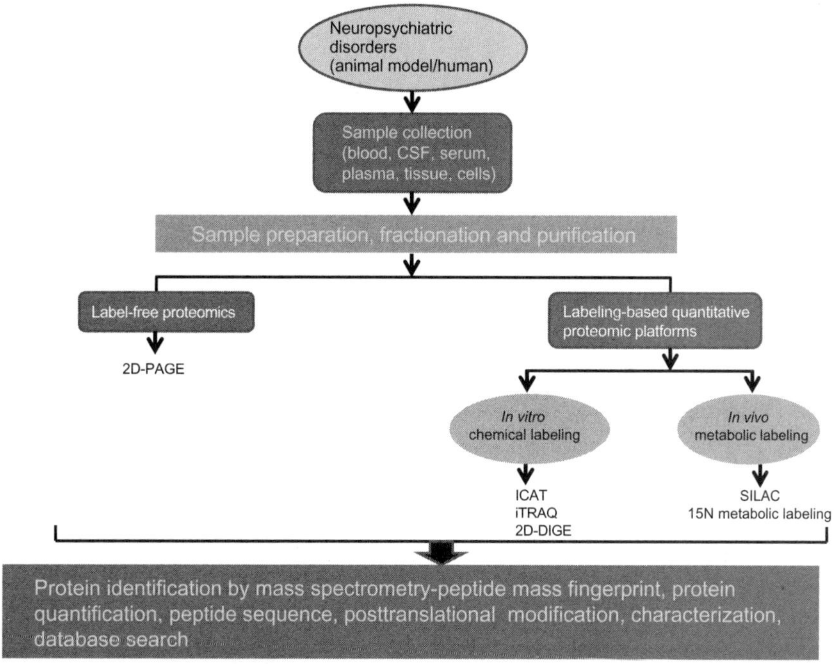

Figure 3.3 Schematic representation of proteomic approaches from sample preparation to protein identification by mass spectrometry analysis for biomarker discovery. (For color version of this figure, the reader is referred to the online version of this chapter.)

Administration (FDA) has defined the term "biomarker" as "a characteristic that is objectively measured and evaluated as an indicator of normal biologic or pathogenic processes or pharmacological responses to a therapeutic intervention" (Chakravarty, 2003).

Genomics-based tools have provided important insights in psychiatry and neuroscience research, but when it comes to clinical use, it has failed to improve markedly the diagnostic and therapeutic options in brain disorders (Hünnerkopf, Grassl, & Thome, 2007). On the contrary, using protein identification based on high-throughput mass spectrometric analysis, it is possible to unravel signal transduction pathways and complex interaction networks on the level of proteins (Hünnerkopf et al., 2007). The importance of proteomic approaches in clinical research cannot be ignored as it facilitates disease diagnosis, ideally at early stages, monitors disease progression, and assesses response to existing and future treatments (Zellner, Veitinger, & Umlauf, 2009). Investigations using proteomic approach have largely improved our understanding of psychiatric disorders and identifying its relevant biomarkers. Proteomics in biomarker discovery includes 2DGE, image analysis, MS, amino acid sequencing, and bioinformatics to comprehensively resolve, quantify, and characterize proteins (Anderson & Anderson, 1998; Jungblut, Holzhütter, Apweiler, & Schlüter, 2008). This approach has enhanced the discovery of potential diagnostic, therapeutic, and disease course biomarkers when it comes to neuropsychiatric disorders such as schizophrenia, bipolar disorder (BD), and depression (Martins-de-Souza, Dias-Neto, et al., 2010; Martins-de-Souza, Harris, Guest, Turck, & Bahn, 2010). Today, novel technologies like proteomics-based approaches have made possible to look into new insights in context with neuropsychiatric disorders, thus overcoming standard targeted methodologies (Martins-de-Souza, Dias-Neto, et al., 2010; Martins-de-Souza, Harris, et al., 2010). Unlike genomics, proteomics can unravel protein expression levels, PTMs, and protein–protein interactions (Hünnerkopf et al., 2007). Quantitative and qualitative identification of protein patterns in postmortem brain tissue, CSF, or serum using proteomic tools has enhanced the knowledge about etiology and path mechanisms of brain diseases such as psychiatric diseases (Hünnerkopf et al., 2007). Interestingly, it is this approach that established the first blood-based test to aid in the diagnosis of schizophrenia based on the identification of set of molecular biomarker assays (Martins-de-Souza, Guest, Harris, et al., 2012; Martins-de-Souza, Guest, Rahmoune, & Martins-de-Souza, Guest, Harris, et al., 2012). Application of MS-based approaches for differential analysis of proteome profile is very prevalent in tissue and body fluids (Frank et al., 2009). Quantitative MS analysis using gold standard methods like stable isotope metabolic labeling of a proteome allows a sensitive and accurate comparative analysis between the respective proteomes (Frank et al., 2009).

3.1. Peptide as biomarkers

The discovery of a set of biomarkers may provide a greater degree of sensitivity and specificity for disease diagnosis rather than single markers (Xiao et al., 2005). One of the crucial points that needs to be considered in studies based on quantitative proteomics is that since quantitative alterations of a modified peptide are different from that of a nonmodified peptide obtained from proteome general profiling, therefore, it is important to focus on unique peptides as compared with intact proteins in clinical proteomics that aims a biomarker search for a specific neuropsychiatric disease (Shi, Caudle, & Zhang, 2009). The potential of proteomics as a tool to elucidate the pathophysiological mechanisms of neuropsychiatric disease mechanisms cannot be ignored (Martins-de-Souza, Dias-Neto, et al., 2010; Martins-de-Souza, Harris, et al., 2010). Proteomic-based genotyping as a biomarker assessment tool for identification of protein alterations is also used to explore metabolic pathways pertinent to the behavioral phenotype (Ditzen et al., 2010). The use of proteomic tools for the development of molecular biomarkers to meet the challenges of translational medicine may definitely provide a great support to critical decision making in drug discovery for antipsychotic treatments (Pich, Vargas, & Domenici, 2012). Such studies may help in the identification of key biomarkers to drive personalized medicine and new treatment development for psychiatric disorders (Pich et al., 2012). Peptidomics, a comprehensive proteomic analysis, has been developed in order to analyze endogenous peptides and small proteins that are often missed by proteomic methods, for example, gel-based approaches or shotgun sequencing approaches of trypsin-digested proteomes (Abdi et al., 2006; Carrette, Burkhard, Hughes, Hochstrasser, & Sanchez, 2005; Jimenez et al., 2007; Yuan & Desiderio, 2005). This method involves a direct *de novo* analysis of endogenous peptides; therefore, it could be considered as an invaluable method for a disease diagnosis.

3.2. Posttranslational modifications as biomarkers

PTMs of proteins are known as key regulators of biological function, activity, localization, and interaction, for example, phosphorylation, ubiquitination, and glycosylation, and involve alteration either in the characteristics of a protein via proteolytic cleavage or covalent addition of a modified group of one or more amino acids. Identification of proteins without PTMs is incomplete since such modifications could be biologically informative and thus provide insights into disease and drug mechanisms of action at the functional level (Martins-de-Souza, Guest, Harris, et al., 2012; Martins-de-Souza, Guest, Rahmoune, et al., 2012). These modifications may serve as important markers for disease activity and prognosis of neuropsychiatric diseases (Ballatore, Lee, & Trojanowski, 2007; Golab, Bauer, Daniel, & Naujokat, 2004; Mechref, Madera, & Novotny, 2008). Protein phosphorylation patterns could be studied based on the immobilized metal ion affinity chromatography (IMAC) for enrichment of phosphoproteins (Martins-de-Souza, Guest, Harris, et al., 2012; Martins-de-Souza, Guest, Rahmoune, et al., 2012). This is then combined with label-free liquid chromatography-mass spectrometry (LC-MS(E)) in order to identify proteins as well as its phosphorylation status

(Martins-de-Souza, Guest, Harris, et al., 2012; Martins-de-Souza, Guest, Rahmoune, et al., 2012). As discussed earlier in the case of peptides as biomarkers, it is challenging to validate the identified PTMs specific for these brain diseases as novel due to the lack of their specific antibody.

4. DIAGNOSIS OF NEUROPSYCHIATRIC DISORDERS

4.1. Anxiety and depression

The characteristic symptoms of depression include low mood, low self-esteem, inappropriate guilt, thoughts of death and suicide, diminished concentration, loss of interest or pleasure in normally enjoyable activities, and disturbance of sleep and appetite (Khawaja, Xu, Liang, & Barrett, 2004; Martins-de-Souza, Dias-Neto, et al., 2010; Martins-de-Souza, Harris, et al., 2010). Table 3.1 shows the list of the characteristic features and treatment for anxiety and depression. Even today, depression in older people remains underdiagnosed and undertreated disorder although it has been past several years in neuropsychiatry research for the clinical and scientific developments toward depression therapeutics (Pitchot, Scantamburlo, & Ansseau, 2012). The occurrence of depressive symptoms is more in older people, and it is often linked with somatic illnesses, with alcoholism, anxiety, and consequently, its effect on their family and the society becomes a serious concern (Pitchot et al., 2012). It has been observed that women are afflicted with depression or anxiety more than men (Schneider & Heuft, 2012). Anxiety symptoms often exist within episodes of depression (Keller & Hanks, 1995). This characteristic feature has become particularly significant for diagnosis and treatment of depression (Keller & Hanks, 1995). They also found anxiety to be a predictor of suicide in depressed patients. Diagnostic and Statistical Manual of Mental Disorders (DSM) or International Classification of Diseases (ICD) criteria have been used for the dimensional diagnosis of anxious depression that is a form of MDD (Ionescu, Niciu, Henter, & Zarate, 2013; Rao & Zisook, 2009; Rush, 2007).

4.2. Bipolar disorder

The hallmark characteristic features of BD previously mentioned include tendency to swing between the two contrasting "poles" of elevated mood and depression, with a return to largely normal functioning in between these episodes (Mitchell, 2013). Others have also described BD as a form of psychiatric illness characterized by manic and depressive episodes (Squassina,

Table 3.1 Representation of psychopharmacological treatments, characteristic features of various neuropsychiatric disorders

Neuropsychiatric disorders	Psychopharmacological treatments	Characteristic features	References
Depression and anxiety	Tianeptine, amitriptyline imipramine fluoxetine, Venlafaxine, benzodiazepines, monoamine oxidase inhibitors (MAOIs), serotonin selective reuptake inhibitors (SSRIs), and tricyclic antidepressants (TCAs)	Low mood, low self-esteem, inappropriate guilt, thoughts of death and suicide, diminished concentration, loss of interest or pleasure in normally enjoyable activities, and disturbance of sleep and appetite	Martins-de-Souza, Dias-Neto, et al. (2010), Martins-de-Souza, Harris, et al. (2010), Khawaja et al. (2004), and Keller and Hanks (1995)
Alzheimer's disease	Acetylcholinesterase inhibitors, *N-methyl-D-aspartate* glutamate receptor antagonists	Aggregates of hyperphosphorylated tau proteins, also called neurofibrillary tangles (NFT), and extracellular deposit of amyloid-beta peptides (Aβ), also called senile plaques	Lehmann et al. (2013)
Schizophrenia	Chlorpromazine, fluphenazine, thioridazine, risperidone, and trifluoperazine	Episodic memory impairments, thought disorder, deterioration of social functioning, delusions, hallucinations, disorganized speech, behavior alterations impaired, sensory processing in patients suffering from schizophrenia get impaired	Li et al. (2012), Matar et al. (2013), Bola et al. (2011), Suzuki et al. (2012), and Said et al. (2013)

Continued

Table 3.1 Representation of psychopharmacological treatments, characteristic features of various neuropsychiatric disorders—cont'd

Neuropsychiatric disorders	Psychopharmacological treatments	Characteristic features	References
Bipolar disorder	Lithium, valproate, carbamazepine, divalproex (Depakote®), clozapine, risperidone and olanzapine, aripiprazole, asenapine, olanzapine, quetiapine, risperidone, chlorpromazine, carbamazepine	Manic and depressive episodes, physical slowing (psychomotor retardation), increased sleep (hypersomnia) and/or increased appetite (hyperphagia), early morning wakening/diurnal mood variation (worse in morning), delusions and hallucinations; excessive (pathological) guilt, lability of mood (interspersed hypomanic symptoms), early onset of first depression (<25 years), multiple prior depressive episodes, positive family history of bipolar disorder	Squassina, Manchia and Del Zompo (2010) and Mitchell (2013)

Manchia, & Del Zompo, 2010). There exists a strong underlying relationship of BD with disturbances in periodic switches in energy, motor activation, and sleep–wake cycles (Robillard, Naismith, & Hickie, 2013). In fact, today treatment of BD also focuses on behavioral or pharmacological therapeutic strategies targeting the sleep–wake and circadian (Robillard et al., 2013). The characteristic features of BD and its treatment are given in Table 3.1. BD has been described (Mitchell, 2013) as follows:

- Bipolar disorder I (BDI)

  Patients with BDI are diagnosed with mania that is characterized by the presence of delusions and/or hallucinations. Mania is prolonged with impaired mood functioning and its diagnosis requires a minimum duration of 7 days.
- Bipolar disorder II (BDII)

  According to DSM-IV, BDII is known to be characterized by recurrent episodes of depression and hypomania (Benazzi, 2007). In this case, mood is distinctly different from normal, but unlike mania, there is no impaired mood functioning. It is diagnosed in a minimum duration of 4 days. Symptoms of hypomania include elevated (euphoric) and/or irritable mood and also at least three of these features: grandiosity, decreased need for sleep, increased talking, racing thoughts, distractibility, overactivity (an increase in goal-directed activity), psychomotor agitation, and excessive involvement in risky activities (Benazzi, 2007).

Based on the recent task force report of International Society for Bipolar Disorders (ISBD), convened to seek consensus recommendations on the use of antidepressants in BD, it is found that the frequency and severity of antidepressant-associated mood elevations were more in BDI than in BDII (Pacchiarotti et al., 2013). Therefore, it has been suggested that prescription of antidepressants for patients suffering from BDI should be restricted only as an adjunct to mood-stabilizing medications (Pacchiarotti et al., 2013).

4.3. Schizophrenia

Schizophrenia has been described as a multifactorial disorder triggered by environmental factors and involves altered protein expression patterns. It is an irreversible neurodevelopment disorder with severe individual, family, and societal burdens and often characterized with impaired performance in various aspects of social cognition, emotion processing, and judgments (Burns, 2004; Bush, Luu, & Posner, 2000; Dunbar, 1998; Jellema, Baker,

Wicker, & Perrett, 2000; Walsh et al., 2008). Others have also described that sensory processing in patients suffering from schizophrenia gets impaired and this affects multisensory integration in them (Postmes et al., 2014). It is a chronic, severe, debilitating mental disorder and has been described as a self-disorder due to perceptual incoherence (Freedman, 2003; Chan, Hsiung, Thompson, Chen, & Hwu, 2007). Discriminate analysis of schizophrenic diagnosis is also done on the basis of prefrontal cortex (PFC) activation (Chuang et al., 2014). Low PFC activation exists in individuals with schizophrenia during verbal fluency tests and also other cognitive tests performed with several neuroimaging methods (Chuang et al., 2014). The characteristic features of schizophrenia and its treatment are given in Table 3.1. Individuals with schizophrenia spectrum disorder possess a history of long-standing subtle issues with social relationships, attention, and behavioral issues, and also they have a true depressive illness in their past (Sikich, 2013).

4.4. Alzheimer's disease

AD is described as an age-related neurological disorder and often characterized by senile plaques (rich in amyloid-β (Aβ) peptide), neuronal fibrillary tangles (rich in phosphorylated tau protein), and synapse loss (Sultana et al., 2013). The characteristic features of AD and its treatment are given in Table 3.1. International work group (IWG) defined AD as a dual clinicobiological entity that could be recognized *in vivo* prior to the onset of the dementia syndrome. This was defined on the basis of two features: (i) a specific core clinical phenotype comprised of an amnestic syndrome of the hippocampal type and (ii) supportive evidence from biomarkers reflecting the location or the nature of Alzheimer-type changes (Dubois et al., 2013). A recent report suggests that imaging biomarkers play a key role in early diagnosis for AD. Its diagnostic accuracy is described to depend on markers such as amyloid imaging, ^{18}F-fluorodeoxyglucose [FDG]-PET, SPECT, and MRI and also how these are measured, that is, "metric": visual, manual, semiautomated, or automated segmentation/computation (Frisoni et al., 2013). The core CSF changes in AD reported earlier studies are decreased amyloid-β(1–42), increased total tau, and increased phospho-tau, probably indicating amyloid plaque accumulation, axonal degeneration, and tangle pathology, respectively (Mattsson et al., 2012). Interestingly, diagnostic performance based on these biomarkers is more useful to identify AD at predementia stage, but this might be affected by age-dependent increase of AD-type brain pathology in cognitively unaffected elderly (Mattsson et al., 2012).

5. PROTEOMICS OF NEUROPSYCHIATRIC DISORDERS

Table 3.2 represents pathways/functions of proteins/possible biomarkers identified in animal model and human-based studies of various neuropsychiatric disorders.

5.1. Depression and anxiety

This has been known as one of the leading causes of years lost due to disability worldwide in women and men (Sibille & French, 2013). Individuals suffering with depression often have a lifelong trajectory of recurring episodes, increasing severity, and progressive treatment resistance (Sibille & French, 2013). The lifetime prevalence of depressive and anxiety disorders is reported to be about 8% and 15%. Whereas unipolar depression is found to have 5% lifetime prevalence, for all depressive disorders, it is 6–8% at a minimum (Keller & Hanks, 1995). MDD has been described as a prevalent debilitating psychiatric mood that contributes to increased rates of disability and suicide (Yang et al., 2013). Anxiety disorders are often known to occur in combination with major depression, alcohol use disorder, or general medical conditions (Sokolowska & Hovatta, 2013). Personal and socioeconomic MDD due to high prevalence of the disorder and the unsatisfying response rate of the available antidepressant treatments has become a major concern (Fabbri, Di Girolamo, & Serretti, 2013). The etiopathology of major chronic mood disorders like anxiety and depression includes repeated exposure to diverse unpredictable stress factors (Chakravarty et al., 2013). A recent report suggests that since drug choice for MDD treatment is based on a trial and error principle with poor clinical efficiency, there is no reliable predictor of treatment efficacy and tolerance in the single patient even today (Fabbri et al., 2013). Meta-analysis of the effects of anxiety and depression on patient adherence indicated that MDD is a leading cause of global disability (DiMatteo, Lepper, & Croghan, 2000). Shotgun proteomic analysis of postmortem dorsolateral PFC brain tissue from 24 MDD patients and 12 matched controls identified few protein signatures (Martins-de-Souza, Guest, Harris, et al., 2012; Martins-de-Souza, Guest, Rahmoune, et al., 2012). This included proteins associated with energy metabolism and synaptic function. The results also identified histidine triad nucleotide-binding protein 1 (HINT1), a protein associated with regulation of mood and behavior (Martins-de-Souza, Guest, Harris, et al., 2012; Martins-de-Souza, Guest, Rahmoune, et al., 2012).

Table 3.2 Representation of pathways/functions of proteins/possible biomarkers identified in animal model and human-based studies of various neuropsychiatric disorders

Neuropsychiatric disorders	Model/ subject	Tissue, blood, cells	Proteins identified (possible biomarkers)	Pathways involved/functions	References
Depression and anxiety	Animal studies	Prefrontal complex (rat)	Glyoxalase 1 and dihydropyrimidinase-related protein 2	Energy and glutathione metabolism	Yang et al. (2013)
		Brain (zebrafish)	PHB2 (prohibitin2), SLC25A5(solute carrier family 25 member 5), VDAC3 (voltage-dependent anion channel 3), and IDH2 (isocitrate dehydrogenase)	Mitochondrial function	Chakravarty et al. (2013)
		Hippocampus (rat)	LASP-1, fascin, and prohibitin	Signal transduction, synaptic vesicles, protein synthesis, cytoskeleton dynamics, and energetic metabolism	Mairesse et al. (2012)
		Brain (mouse)	Serotonin receptor-associated proteins	Synaptic transmission, metabolism, proteolysis, protein biosynthesis and folding, cytoskeletal proteins, brain development and neurogenesis, oxidative stress	Szego et al. (2010)
	Human studies	Prefrontal cortex brain tissue	Histidine triad nucleotide-binding protein 1 (HINT1)	Energy metabolism and synaptic function	Martins-de-Souza, Guest, Harris, et al., 2012

Alzheimer's disease	Animal studies	Hippocampus	α-Enolase, ATP synthase subunit-α, phosphoglycerate mutase 1	Brain energy metabolism	Sultana et al. (2007)
		Frontal cortex	NADH dehydrogenase ubiquinone 75 kDa, tubulin beta 5, glial fibrillary acidic protein, peroxiredoxin II, triose phosphate isomerase 1	Electron transport, respiratory chain, transport	Shiozaki et al. (2004)
		Temporal cortices	Fatty acid-binding protein heart	Transport	Cheon et al. (2003)
	Human studies	Lymphocytes, plasma, CSF	Amyloid precursor protein and tau, amyloid precursor protein and tau, beta-2-microglobulin, brain-derived neurotrophic factor 1 and fragments of cystatin C, thioredoxin-dependent peroxide reductase, myosin light polypeptide 6, and ATP synthase subunit-β	Apoptosis, endocytosis, differentiation, neurogenesis, ATP synthesis, transport	Perneczky et al. (2013), Sultana et al. (2013), and Umlauf et al. (2009)

Continued

Table 3.2 Representation of pathways/functions of proteins/possible biomarkers identified in animal model and human-based studies of various neuropsychiatric disorders—cont'd

Neuropsychiatric disorders	Model/subject	Tissue, blood, cells	Proteins identified (possible biomarkers)	Pathways involved/functions	References
Schizophrenia	Animal studies	Cerebral cortex (rat)	HSP 72 kDa; stathmin; soluble superoxide dismutase; mitochondrial ATP synthase beta subunit; PDH; actin beta; mitochondrial HSP 70 kDa; enolase, mitochondrial HSP 70 kDa; HSP 60, 72 kDa; enolase 1, 2	Brain energy metabolism	Paulson et al. (2003)
	Human studies	Postmortem brain tissue	Aldolase C (ALDOC), gamma enolase (ENO2), aconitase (ACO2), hexokinase (HK1), glyceraldehyde-3-phosphate dehydrogenase (GAPDH), a number of subunits of mitochondrial ATPase	Brain energy metabolism, brain plasticity, and synaptic function	Martins-De-Souza, Dias-Neto, et al. (2010) and Martins-de-Souza, Harris, et al. (2010)
		CSF, red blood cells, and serum	Apolipoprotein A1, apolipoprotein A2 and A4, and transferrin and transthyretin	Cholesterol metabolism, transport	Huang et al. (2008), Levin et al. (2010), Prabakaran et al. (2007), and Nakagawa et al. (2013)

Bipolar disorder	Human studies	Frontal cortex (postmortem study)	Glial fibrillary acidic protein (GFAP); one is dihydropyrimidinase-related protein 2, ubiquinone cytochrome *c* reductase core protein 1, carbonic anhydrase 1, and fructose bisphosphate aldolase C	Glial cell differentiation, astrocyte development, synaptic potentiation	Johnston-Wilson et al. (2000)
		Peripheral blood mononuclear cells (PBMCs) and serum	Cytoskeletal and stress response-associated proteins, apo A-I	Endocytosis	Herberth et al. (2011) and Sussulini et al. (2011)

PFC proteomes of chronic unpredictable mild stress (CUMS) rat model of depression and control rats revealed differential regulation of 29 proteins in the PFC of CUMS rats relative to control rats (Yang et al., 2013). Their study showed significant involvement of biological pathways such as energy and glutathione metabolism in the CUMS rat model of depression (Yang et al., 2013). Proteomic analysis using 2D DIGE and MALDI-TOF in restraint stress-induced rat brain showed differential expression of a number of proteins including downregulated hippocampal cholinergic neurostimulating peptide precursor protein (HCNP-pp) when compared with a nonstress group (Kim & Kim, 2007). Fifteen days of exposure to chronic stressors in zebrafish model appears to induce an anxiety and related mood disorder phenotype (Chakravarty et al., 2013). Brain proteome analyses of chronic stressors-induced zebrafish anxiety and related mood disorder phenotype model revealed differential regulation of 18 proteins that were specific to anxiety and related disorders (Chakravarty et al., 2013). Among these, four proteins were related to mitochondrial function: PHB2, SLC25A5, VDAC3, and IDH2 as reported in rodent and clinical samples. Proteomic analysis in the hippocampus of prenatal restraint stress-induced anxious/depressive phenotype in adult male rats showed changes in the expression profile of a number of proteins, involved in the regulation of signal transduction, synaptic vesicles, protein synthesis, cytoskeleton dynamics, and energetic metabolism (Mairesse et al., 2012). Significant changes in the expression of proteins, such as LASP-1, fascin, and prohibitin, were further confirmed by immunoblot analysis that may play a role in developmental programming triggered by early-life stress (Mairesse et al., 2012). Proteomic and metabolomic profiling of a trait anxiety mouse model showed evidence of anxiety phenotype-specific small molecule biomarkers with particular focus on *myo*-inositol and glutamate as well as the intermediates involved in the tricarboxylic acid cycle (Zhang et al., 2011). Total mouse brain proteome analysis compared to the control suggested alterations in serotonin receptor-associated proteins, carbohydrate metabolism, the cellular redox system, and synaptic docking are all involved in anxiety and 34 of the proteins identified were specific to anxiety, depression, etc. (Szego et al., 2010).

Glyoxalase-I was also identified as a protein biomarker using proteome analysis in mouse model of anxiety- and depression-like behavior (Krömer et al., 2005). Protein patterns from rat hippocampal cytosolic extracts upon administration of antidepressants, that is, venlafaxine or fluoxetine compared to controls in rat hippocampus reflected long-term functional adaptations required for antidepressant activity (Khawaja et al., 2004). These included the modulation of 33 proteins such as differential regulation of proteins associated with neurogenesis (insulin-like growth factor 1 (IGF-1) and glia maturation factor (GMF)-beta), outgrowth/maintenance of neuronal processes (HCNP and PCTAIRE-3), and neural regeneration/axonal guidance collapsin response mediator protein (CRMP-2) systems (Khawaja et al., 2004). Proteome analysis unraveled anxiety- and depression-related pathways in different cellular functions, including synaptic transmission, metabolism, proteolysis, protein biosynthesis and folding, cytoskeletal proteins, brain development and neurogenesis, oxidative stress, and signal transduction (Szego et al., 2010).

5.2. Schizophrenia

A lot of effort has been put in advancements toward drug discovery in mood disorders (Emmett et al., 2013). However, there is still a need for new biomarkers and its validated therapeutic targets for schizophrenia. Integrated biological approaches like proteomics, transcriptomics, metabolomics, and glycomics are being used to deliver new therapeutic targets and biomarkers for personalized medicine (Emmett et al., 2013). MALDI-TOF peptide mass fingerprinting analysis of novel serum markers in first-episode schizophrenics before versus after treatment of risperidone showed a significantly higher content of apolipoprotein A-1 than those before treatment (Li et al., 2012). On the other hand, treatment of the schizophrenics with risperidone resulted in downregulation of haptoglobin protein (Li et al., 2012). LC-MS (E)-based proteomic profiling of the maternal protein restriction rat model for schizophrenia identified translational changes in hormonal signaling pathways and glutamate neurotransmission in frontal cortex (Guest et al., 2012). In addition, alterations in cytoskeletal proteins involved in hormonal secretion and synaptic remodeling were also found in the hypothalamus (Guest et al., 2012). Studies based on differential proteome analysis of five distinct brain regions of postmortem tissue from schizophrenia patients and controls identified proteins involved in energy metabolism, calcium homeostasis, myelinization, and cytoskeleton in schizophrenia brains (Martins-de-Souza, 2011). Recently, protein phosphorylation patterns in serum from schizophrenia patients and healthy controls have identified altered expression of 35 proteins in schizophrenia using LC-MS(E) and IMAC combined approaches (Martins-de-Souza, Guest, Harris, et al., 2012; Martins-de-Souza, Guest, Rahmoune, et al., 2012).

Quantitative proteomic studies have been widely used to delineate pathways of schizophrenia susceptibility (Gokhale et al., 2012). Investigations showed association of 24 proteins with the biogenesis of lysosome-related organelles complex 1 (BLOC-1) could be identified based on large-scale SILAC quantitative proteomics combined with genetic analysis in dysbindin-null mice (*Mus musculus*) and the genome of schizophrenia patients (Gokhale et al., 2012). Proteomics-based reverse genetics from mouse to human has been used to study pathogenesis of schizophrenia. This approach explored the possible relationship between interferon-gamma (IFNG) and schizophrenia. MALDI-TOF/TOF MS analyses of the differentially expressed proteins in the brains of interferon-gamma knockout (Ifng-KO) mice showed five upregulated and five downregulated proteins in Ifng-KO mouse brain (Kim et al., 2012). SELDI-TOF MS-based investigation revealed distinct CSF Aβ peptide signatures in cognitive decline associated with AD and schizophrenia (Albertini et al., 2012). Dysmetabolism of amyloid precursor protein occurs in older schizophrenia patients (Albertini et al., 2012). They also showed that different pathophysiological dynamics in aging brain could be assessed based on the Aβ1–42 in AD and in elderly SCZ patients. Proteome analysis of cerebral cortex from MK-801-treated rats identified the proteins—stathmin, adenosine triphosphate (ATP) synthase, pyruvate dehydrogenase, beta-actin, and alpha-enolase involved in schizophrenia pathology (Paulson et al., 2003). An organelle proteomic method in a neurodevelopmental model of schizophrenia showed the involvement of proteins linked to various neurotransmitter systems—clathrin light chain B, syntaxin-binding protein 1b, and visinin-like protein 1 (Vercauteren et al., 2007). Free liquid chromatography-MS-based proteomic and proton nuclear magnetic resonance-based metabonomic profiling approaches provided the evidence for disease and antipsychotic medication effects in postmortem brain from schizophrenia patients (Chan et al., 2011). Their study revealed novel protein and metabolite changes in low-cumulative-medication subjects associated with synaptogenesis, neuritic dynamics, presynaptic vesicle cycling, amino acid and glutamine metabolism, and energy-buffering systems (Chan et al., 2011).

5.3. Alzheimer disease

Although a number of studies have been done on AD, the search for biomarkers is in progress (Sultana et al., 2013). Even today, proper effective pharmacotherapy for AD does not exist; therefore, this disease has become a burden in our aging society. Multidimensional protein separation techniques involving protein identification by spectrometry analysis have shown the potential to identifying blood-based biomarkers for AD (Henkel et al., 2012). Development of biomarkers is enhanced to a great extent with progress of proteomics-based technologies (Sultana et al., 2013). The pathophysiological lesions of the AD are described to include aggregates of hyperphosphorylated tau proteins, also called neurofibrillary tangles, and extracellular deposit of Aβ peptides, also called senile plaques (Lehmann, Delaby, Touchon, Hirtz, & Gabelle, 2013). Plasma proteomics is widely used approach to search less-invasive biomarkers. Recent investigations have suggested the feasibility of a blood-based biomarker approach in AD diagnostics (Guo, Alexopoulos, Wagenpfeil, Kurz, & Perneczky, 2013). Proteomic analysis has established new panel of markers for AD such as beta-2-microglobulin, brain-derived neurotrophic factor 1, and fragments of cystatin C (Zellner, Veitinger, & Umlauf, 2009). Differential protein expression profile in healthy and disease states using proteomics-based approaches has opened new opportunities so as to establish specific diagnostic and prognostic biomarkers of psychiatric diseases such as AD, drug addiction, schizophrenia, and depression (Hünnerkopf et al., 2007). Proteins involved in the pathophysiology of AD disease or the amyloid/Aβ-peptide processing pathway were identified using this approach. A recent study involving proteomic tool has identified differential expression of thioredoxin-dependent peroxide reductase, myosin light polypeptide 6, and ATP synthase subunit-β that might play an important role in the progression and pathogenesis of AD (Sultana et al., 2013).

Insights into the biochemical pathogenesis as well as biomarker discovery for neurodegenerative diseases like AD have been possible by using approaches like iTRAQ method for quantitative characterization of glycoproteins (Shi, Hwang, & Zhang, 2013). Proteomic based platforms for the development of ideal biomarkers in neuropsychiatric disorders are a very promising approach when it comes to clinical research due to their high sensitivity and specificity since biomarkers should be present in the blood before the onset of these diseases is clinically confirmed (Cocciolo et al., 2012). Recent study involving redox proteomics of AD subjects compared with age-matched controls suggests the involvement of altered expression of extracellular chaperones in exacerbating Aβ toxicity in AD and may be considered a putative marker of disease progression (Cocciolo et al., 2012). One of the challenging issues in context with AD is that its diagnosis is only possible after autopsy and a long prodromal or preclinical phase (Bazenet & Lovestone, 2012). However, there is a great hope that efforts of biomarker development using plasma/serum proteomic approaches may lead to the identification of the much-needed AD biomarkers (Bazenet & Lovestone, 2012). Peptidyl-prolyl *cis/trans* isomerase (Pin1) plays an important role in regulating function of microtubule-associated tau protein (Ando et al., 2013). Proteomic approaches showed that PTM of Pin1 on more than five residues, comprising phosphorylation, N-acetylation, and oxidation, specific in tauopathies could be useful as biomarker for AD (Ando et al., 2013).

The importance of genotype on blood protein profiles in context with age neuropsychiatric disorders is evident from a study that showed association of plasma biomarkers with the apolipoprotein E genotype and AD (Soares et al., 2012). Even redox proteomics is gaining importance for studying the association of neuropsychiatric disorders and protein oxidation since this method can identify and quantify the proteins upon alterations in proteome both in redox signaling and under oxidative stress conditions (Butterfield & Dalle-Donne, 2012). Differential proteome analysis of clinically diagnosed AD patients with nondemented subjects using proteomics approaches the correlation of proteins, namely, apolipoprotein E, apolipoprotein J, complement C4b, hemopexin, and complement factor B interacting partners in Alzheimer's and apoptotic pathways (Manral, Sharma, Hariprasad, Chandralekha, & Srinivasan, 2012). Investigations based on plasma proteomics showed that the differential expression of a set of five plasma proteins were identified in AD dementia group as compared with control with a sensitivity of 89.36% and a specificity of 79.17%. Interestingly, four of five proteins belonged to a common network with amyloid precursor protein and tau, which are known to be the hallmarks of AD (Perneczky et al., 2013).

Increased application of proteomic methods has been used for studying epigenetics of neuropsychiatric disorders. In fact, it has advanced the search for epigenetic biomarkers such as chromatin histone proteins by acetylation as these are deregulated in memory loss and AD. Selected reaction monitoring, which is used to quantitate peptides and proteins in complex biological systems, showed reduced level of histone acetylation in AD temporal lobe compared with age-matched controls (Zhang et al., 2012). Serum proteomic approaches involving 2DE combined with nano-high-performance liquid chromatography electrospray ionization tandem mass spectrometry (nano-HPLC-ESI-MS/MS) followed by peptide fragmentation patterning identified a candidate protein activity-dependent neuroprotector homeobox protein (ADNP) as a diagnostic biomarker for AD (Yang et al., 2012). Their study suggested the involvement of ADNP in slowing the progression of clinical symptoms of AD since it is down-regulated in AD patients (Yang et al., 2012). The differential expression of membrane-enriched proteins, ubiquitin carboxy-terminal hydrolase 1 (UCHL1) and syntaxin-binding protein 1 (Munc-18), was shown based on the analysis of a membrane-enriched proteome from postmortem human brain tissue in AD (Donovan et al., 2012). A novel peptidomic approach to detect markers of AD in CSF identified discriminating peptides in the unbound peptide fraction as VGF nerve growth factor-inducible precursor, complement C4 precursor and discriminating peptides in the protein-bound fraction, VGF nerve growth factor-inducible precursor, and alpha-2-HS-glycoprotein (Wijte et al., 2012).

5.4. Bipolar disorder

Quantitative proteomics is widely used for investigating disease-specific protein signatures of psychiatric disorders as it holds great promise to enhance our current understanding of the molecular mechanisms as well as identification of relevant biomarkers for these diseases (Filiou, Turck, & Martins-de-Souza, 2011). The mechanism of BD is not yet understood; even today, lithium is used as a therapy for this. It is, therefore, necessary to search for its relevant biomarkers in order to facilitate diagnosis of BD or treatment evaluation. The origin of neuropsychiatric disorders, for example, schizophrenia, BD, and MDD, is not yet known and also the molecular markers at the pathological, cellular, or molecular level, for these do not exist (Johnston-Wilson et al., 2000). Proteomics-based identification for biomarkers of BD may facilitate personalized therapy for BD (Squassina, Manchia and Del Zompo (2010). Proteomic approach for characterization of protein abnormalities in the dorsolateral PFC in schizophrenia and BD provided strong association of the septin protein family of proteins in psychiatric disorders (Pennington et al., 2008).

Proteomics study involving response to lithium treatment at the serum protein level proposed apo A-I as a candidate marker for BD (Sussulini et al., 2011). Liquid chromatography-mass spectrometry (LC-MS(E)) and multi-analyte profiling (Human Map(®)) platforms-based proteome profiling of peripheral blood mononuclear cells (PBMCs) and serum from BD patients compared to matched healthy controls identified a molecular fingerprint for predicting and guiding treatment strategies for BD (Herberth et al., 2011). This peripheral fingerprint (e.g., cytoskeletal and stress response-associated proteins in PBMCs and inflammatory response in serum) could distinguish

BD patients from healthy controls despite being in a remission phase as it has detrimental effects on cell function (Herberth et al., 2011). Mood stabilizers are shown to be the effective treatment for acute mania, depression, and maintenance treatment of BD. A postmortem comparative proteomic study in neuropsychiatric disorders demonstrated that the most common proteomic changes occur within the hippocampus in cornu ammonis regions 2 and 3 in schizophrenia and BD, for example, altered expression of septin 11 and proteins involved in clathrin-mediated endocytosis was found in both schizophrenia and BD (Föcking et al., 2011).

SELDI-TOF-MS ProteinChip profiling combined with matrix-assisted laser desorption time-of-flight/post-source decay MS analysis is among the other preferred proteomic technologies used for the discovery of protein biomarkers for BD in the postmortem PFC as it points to changes in protein levels characterizing these diseases (Novikova, He, Cutrufello, & Lidow, 2006). Evidence for disruption of the cytoskeleton and its associated signal transduction proteins in schizophrenia, and to a lesser extent in BD, also came from differential proteomics method using 2D DIGE analysis (Beasley et al., 2006). Proteomics-based approaches involving 2DGE followed by MS analysis identified disease-specific alterations differentially expressed proteins in the frontal cortex—glial fibrillary acidic protein (GFAP), dihydropyrimidinase-related protein 2, ubiquinone cytochrome *c* reductase core protein 1, carbonic anhydrase 1, and fructose bisphosphate aldolase C. Cytoskeletal and mitochondrial dysfunctions play a key role in the neuropathology of the major psychiatric disorders (Beasley et al., 2006). 2DE followed by mass spectrometric sequencing within the anterior cingulate cortex in psychiatric disorders identified 19 distinct proteins associated with the major psychiatric disorders: aconitate hydratase, malate dehydrogenase, fructose bisphosphate aldolase A, ATP synthase, succinyl CoA ketoacid transferase, carbonic anhydrase, alpha- and beta-tubulin, dihydropyrimidinase-related protein-1 and -2, neuronal protein 25, trypsin precursor, glutamate dehydrogenase, glutamine synthetase, sorcin, vacuolar ATPase, creatine kinase, albumin, and guanine nucleotide-binding protein beta subunit (Beasley et al., 2006).

6. STRENGTH, WEAKNESS, AND FUTURE CHALLENGES: BIOMARKER DISCOVERY

One of the challenging issues with biomarker discovery with psychiatric disorders is that it is difficult to prepare homogenous sample with brain tissue due to its complexity (Zellner et al., 2009). Others have also revealed that in spite of the advancements in proteomics-based approaches, even today, the challenges like standardization of sample handling and sample-to-sample variability are not solved (Frank et al., 2009). Other critical steps for biomarker discovery are protein/peptide separation and identification, as well as independent confirmation and validation of the proteomics data (Shi et al., 2009; Zellner et al., 2009). Although proteomic technologies have advanced to a great extent in the past century, however, the validation of differentially expressed proteins is the bottleneck in the application of proteomics for biomarker discovery even today. MS-based approaches have been very useful in clinical neuroscience; however, for reproducibility of the proteomics data across standard operation protocols, experimental setups and quality control schemes are yet to be established (Martins-de-Souza, 2011). One of the biggest challenges in neuropsychiatry research that need to be resolved is to reduce the gap between advances in their pharmacotherapy and biological tests for its diagnosis, which is lacking even today. *In vitro* and *in vivo* differential expression of protein expression patterns due to medication may open up new possibilities so as to design specific pharmaceutical agents with fewer side effects (Hünnerkopf et al., 2007).

The next step after the candidate proteins are identified for these neuropsychiatric disorders in different cohorts of patients/controls needs to be validated by different methods. One of the biomarker testing methods in clinical proteomics is based on multiplex immunoassays like enzyme-linked immunosorbent assay and multi-analyte panel although its major challenge is the availability of specific antibodies specifically for novel proteins identified using proteomic approach (Shi et al., 2009). On the other hand, it is more challenging if the identified protein carries PTMs like phosphorylation and glycosylation or other epigenetic modifications like histone modification. Although many neurodegenerative diseases cannot be cured at the present time, there are often symptomatic treatments available and new drugs are emerging to forestall and/or reverse the onset and/or progress of the diseases (Forman, Trojanowski, & Lee, 2004; Skovronsky, Lee, & Trojanowski, 2006).

7. CONCLUDING REMARKS

We can say that patients suffering from neuropsychiatric disorders can be treated in a better way if we could identify robust biomarkers for the disease. It is possible to identify the patients suffering from specific disease and treat them accordingly, that is, personalized medicine can be subscribed to them for a better monitoring of progression and also response to treatment. One of the major challenges that lie even today for the clinical diagnosis of neuropsychiatric disorders is that since most of these diseases mimic each other or coexist, this creates confusion to the characteristic or symptomatic features of specific brain disorders. Overall, the proteomic research in context with neuropsychiatric disorders reveals the significance of this approach in not only understanding the complexity of the disorder but also the potential for identification of disease-specific biomarkers and development of novel therapeutic strategies. In the future, we still need to explore more on the applicability of implementing proteomic tool as a routine diagnostic tool for neuropsychiatric disorders in the clinical laboratory.

REFERENCES

Abdi, F., Quinn, J. F., Jankovic, J., McIntosh, M., Leverenz, J. B., Peskind, E., et al. (2006). Detection of biomarkers with a multiplex quantitative proteomic platform in cerebrospinal fluid of patients with neurodegenerative disorders. *Journal of Alzheimer's Disease, 9*, 293–348.

Albertini, V., Benussi, L., Paterlini, A., Glionna, M., Prestia, A., Bocchio-Chiavetto, L., et al. (2012). Distinct cerebrospinal fluid amyloid-beta peptide signatures in cognitive decline associated with Alzheimer's disease and schizophrenia. *Electrophoresis, 33*(24), 3738–3744.

Anderson, N. L., & Anderson, N. G. (1998). Proteome and proteomics: New technologies, new concepts, and new words. *Electrophoresis, 19*(11), 1853–1861.

Ando, K., Dourlen, P., Sambo, A. V., Bretteville, A., Bélarbi, K., Vingtdeux, V., et al. (2013). Tau pathology modulates Pin1 post-translational modifications and may be relevant as biomarker. *Neurobiology of Aging, 34*(3), 757–769.

Bachi, A., & Bonaldi, T. (2008). Quantitative proteomics as a new piece of the systems biology puzzle. *Journal of Proteomics, 71*, 357–367.

Ballatore, C., Lee, V. M., & Trojanowski, J. Q. (2007). Tau-mediated neurodegeneration in Alzheimer's disease and related disorders. *Nature Reviews. Neuroscience, 8*, 663–672.

Bazenet, C., & Lovestone, S. (2012). Plasma biomarkers for Alzheimer's disease: Much needed but tough to find. *Neurobiology of Aging, 6*(4), 441–454.

Beasley, C. L., Pennington, K., Behan, A., Wait, R., Dunn, M. J., & Cotter, D. (2006). Proteomic analysis of the anterior cingulate cortex in the major psychiatric disorders: Evidence for disease-associated changes. *Proteomics, 6*(11), 3414–3425.

Benazzi, F. (2007). Bipolar II disorder: Epidemiology, diagnosis and management. *CNS Drugs, 21*(9), 727–740.

Bola, J., Kao, D., & Soydan, H. (2011). Antipsychotic medication for early episode schizophrenia. *Cochrane Database of Systematic Reviews, 6,* CD006374.

Burns, J. K. (2004). An evolutionary theory of schizophrenia: Cortical connectivity, metarepresentation, and the social brain. *The Behavioral and Brain Sciences, 27*(6), 831–885.

Bush, G., Luu, P., & Posner, M. I. (2000). Cognitive and emotional influences in anterior cingulate cortex. *Trends in Cognitive Sciences, 4*(6), 215–222.

Butterfield, D. A., & Dalle-Donne, I. (2012). Redox proteomics. *Antioxidants & Redox Signaling, 17*(11), 1487–1489.

Bystritsky, A. (2006). Treatment-resistant anxiety disorders. *Molecular Psychiatry, 11*, 805–814.
Carrette, O., Burkhard, P. R., Hughes, S., Hochstrasser, D. F., & Sanchez, J. C. (2005). Truncated cystatin C in cerebrospiral fluid: Technical artefact or biological process? *Proteomics, 5*, 3060–3065.
Castaño, E. M., Roher, A. E., Esh, C. L., Kokjohn, T. A., & Beach, T. (2006). Comparative proteomics of cerebrospinal fluid in neuropathologically-confirmed Alzheimer's disease and non-demented elderly subjects. *Neurological Research, 28*(2), 155–163.
Chakravarty, A. (2003). Surrogate markers: Their role in regulatory decision process. Food and Drug Administration. http://www.fda.gov/cder/Offices/Biostatistics/Chakravarty_376/sld016.htm.
Chakravarty, S., Reddy, B. R., Sudhakar, S. R., Saxena, S., Das, T., Meghah, V., et al. (2013). Chronic unpredictable stress (CUS)-induced anxiety and related mood disorders in a zebrafish model: Altered brain proteome profile implicates mitochondrial dysfunction. *PLoS One, 8*(5), e63302.
Chan, S. W., Hsiung, P. C., Thompson, D. R., Chen, S. C., & Hwu, H. G. (2007). Health-related quality of life of Chinese people with schizophrenia in Hong Kong and Taipei: A cross-sectional analysis. *Research in Nursing & Health, 30*, 261–269.
Chan, M. K., Tsang, T. M., Harris, L. W., Guest, P. C., Holmes, E., & Bahn, S. (2011). Evidence for disease and antipsychotic medication effects in post-mortem brain from schizophrenia patients. *Molecular Psychiatry, 16*(12), 1189–1202.
Chandramouli, Kondethimmanahalli, & Qian, Pei-Yuan (2009). Proteomics: Challenges, techniques and possibilities to overcome biological sample complexity. *Human Genomics and Proteomics, 2009*, 239204.
Chen, X., Smith, L. M., & Bradbury, M. E. (2000). Site-specific mass tagging with stable isotopes in proteins for accurate and efficient protein identification. *Analytical Chemistry, 72*, 1134.
Cheon, M. S., Kim, S. H., Fountoulakis, M., & Lubec, G. (2003). Heart type fatty acid binding protein (H-FABP) is decreased in brains of patients with Down syndrome and Alzheimer's disease. *Journal of Neural Transmission. Supplementum*, (67), 225–234.
Chuang, C. C., Nakagome, K., Pu, S., Lan, T. H., Lee, C. Y., & Sun, C. W. (2014). Discriminant analysis of functional optical topography for schizophrenia diagnosis. *Journal of Biomedical Optics, 19*(1), 11006.
Cocciolo, A., Di Domenico, F., Coccia, R., Fiorini, A., Cai, J., Pierce, W. M., et al. (2012). Decreased expression and increased oxidation of plasma haptoglobin in Alzheimer disease: Insights from redox proteomics. *Free Radical Biology & Medicine, 53*(10), 1868–1876.
Craft, G. E., Chen, A., & Nairn, A. C. (2013). Recent advances in quantitative neuroproteomics. *Methods, 61*(3), 186–218.
Davidsson, P., Folkesson, S., Christiansson, M., Lindbjer, M., Dellheden, B., Blennow, K., et al. (2002). Identification of proteins in human cerebrospinal fluid using liquid-phase isoelectric focusing as a prefractionation step followed by two-dimensional gel electrophoresis and matrix-assisted laser desorption/ionisation mass spectrometry. *Rapid Communications in Mass Spectrometry, 16*(22), 2083–2088.
DiMatteo, M. R., Lepper, H. S., & Croghan, T. W. (2000). Depression is a risk factor for noncompliance with medical treatment: Meta-analysis of the effects of anxiety and depression on patient adherence. *Archives of Internal Medicine, 160*, 2101–2107.
Ditzen, C., Varadarajulu, J., Czibere, L., Gonik, M., Targosz, B. S., Hambsch, B., et al. (2010). Proteomic-based genotyping in a mouse model of trait anxiety exposes disease-relevant pathways. *Molecular Psychiatry, 15*(7), 702–711.
Donovan, L. E., Higginbotham, L., Dammer, E. B., Gearing, M., Rees, H. D., Xia, Q., et al. (2012). Analysis of a membrane-enriched proteome from postmortem human brain tissue in Alzheimer's disease. *Proteomics. Clinical Applications, 6*(3–4), 201–211.

Dubois, B., Epelbaum, S., Santos, A., Di Stefano, F., Julian, A., Michon, A., et al. (2013). Alzheimer disease: From biomarkers to diagnosis. *Revue Neurologique (Paris)*, *169*(10), 744–751.

Dunbar, R. I. M. (1998). The social brain hypothesis. *Evolutionary Anthropology*, *6*(5), 178–190.

Emmett, M. R., Kroes, R. A., Moskal, J. R., Conrad, C. A., Priebe, W., Laezza, F., et al. (2013). Integrative biological analysis for neuropsychopharmacology. *Neuropsychopharmacology*, *39*(1), 5–23.

Ernoult, E., Bourreau, A., Gamelin, E., & Guette, C. (2010). A proteomic approach for plasma biomarker discovery with iTRAQ labelling and OFFGEL fractionation. *Journal of Biomedicine & Biotechnology*, *2010*, 927917.

Fabbri, C., Di Girolamo, G., & Serretti, A. (2013). Pharmacogenetics of antidepressant drugs: An update after almost 20 years of research. *American Journal of Medical Genetics. Part B, Neuropsychiatric Genetics*, *162*(6), 487–520.

Filiou, M. D., Turck, C. W., & Martins-de-Souza, D. (2011). Quantitative proteomics for investigating psychiatric disorders. *Proteomics. Clinical Applications*, *5*(1–2), 38–49.

Finehout, E. J., Franck, Z., Choe, L. H., Relkin, N., & Lee, K. H. (2007). Cerebrospinal fluid proteomic biomarkers for Alzheimer's disease. *Annals of Neurology*, *61*(2), 120–129.

Föcking, M., Dicker, P., English, J. A., Schubert, K. O., Dunn, M. J., & Cotter, D. R. (2011). Common proteomic changes in the hippocampus in schizophrenia and bipolar disorder and particular evidence for involvement of cornu ammonis regions 2 and 3. *Archives of General Psychiatry*, *68*(5), 477–488.

Forman, M. S., Trojanowski, J. Q., & Lee, V. M. (2004). Neurodegenerative diseases: A decade of discoveries paves the way for therapeutic breakthroughs. *Nature Medicine*, *10*(10), 1055–1063.

Frank, E., Kessler, M. S., Filiou, M. D., Zhang, Y., Maccarrone, G., Reckow, S., et al. (2009). Stable isotope metabolic labeling with a novel 15N-enriched bacteria diet for improved proteomic analyses of mouse models for psychopathologies. *PLoS One*, *4*(11), e7821.

Freedman, R. (2003). Schizophrenia. *The New England Journal of Medicine*, *349*, 1738–1749.

Friedman, D. B. (2007). Quantitative proteomics for two-dimensional gels using difference gel electrophoresis. *Methods in Molecular Biology*, *367*, 219–239.

Friedman, D. B., Hill, S., Keller, J. W., Merchant, N. B., Levy, S. E., Coffey, R. J., et al. (2004). Proteome analysis of human colon cancer by two-dimensional difference gel electrophoresis and mass spectrometry. *Proteomics*, *4*(3), 793–811.

Frisoni, G. B., Bocchetta, M., Chételat, G., Rabinovici, G. D., de Leon, M. J., Kaye, J., et al. (2013). Imaging markers for Alzheimer disease: Which vs how. *Neurology*, *81*(5), 487–500.

Gokhale, A., Larimore, J., Werner, E., So, L., Moreno-De-Luca, A., Lese-Martin, C., et al. (2012). Quantitative proteomic and genetic analyses of the schizophrenia susceptibility factor dysbindin identify novel roles of the biogenesis of lysosome-related organelles complex 1. *The Journal of Neuroscience*, *32*(11), 3697–3711.

Golab, J., Bauer, T. M., Daniel, V., & Naujokat, C. (2004). Role of the ubiquitin-proteasome pathway in the diagnosis of human diseases. *Clinica Chimica Acta*, *340*, 27–40.

Guest, P. C., Urday, S., Ma, D., Stelzhammer, V., Harris, L. W., Amess, B., et al. (2012). Proteomic analysis of the maternal protein restriction rat model for schizophrenia: Identification of translational changes in hormonal signaling pathways and glutamate neurotransmission. *Proteomics*, *12*(23–24), 3580–3589.

Guo, L. H., Alexopoulos, P., Wagenpfeil, S., Kurz, A., & Perneczky, R. (2013). Plasma proteomics for the identification of Alzheimer disease. *Alzheimer Disease and Associated Disorders*, *27*(4), 337–342.

Gygi, S. P., Rist, B., Gerber, S. A., Turecek, F., Gelb, M. H., & Aebersold, R. (1999). Quantitative analysis of complex protein mixtures using isotope-coded affinity tags. *Nature Biotechnology*, *17*(10), 994–999.

Harsha, H. C., Molina, H., & Pandey, A. (2008). Quantitative proteomics using stable isotope labeling with amino acids in cell culture. *Nature Protocols*, *3*(3), 505–516.

Henkel, A. W., Müller, K., Lewczuk, P., Müller, T., Marcus, K., Kornhuber, J., et al. (2012). Multidimensional plasma protein separation technique for identification of potential Alzheimer's disease plasma biomarkers: A pilot study. *Journal of Neural Transmission*, *119*(7), 779–788.

Henninger, N., Feldmann, R. E., Jr., Fütterer, C. D., Schrempp, C., Maurer, M. H., Waschke, K. F., et al. (2007). Spatial learning induces predominant downregulation of cytosolic proteins in the rat hippocampus. *Genes, Brain, and Behavior*, *6*(2), 128–140.

Herberth, M., Koethe, D., Levin, Y., Schwarz, E., Krzyszton, N. D., Schoeffmann, S., et al. (2011). Peripheral profiling analysis for bipolar disorder reveals markers associated with reduced cell survival. *Proteomics*, *11*(1), 94–105.

Hu, Y., Hosseini, A., Kauwe, J. S., Gross, J., Cairns, N. J., Goate, A. M., et al. (2007). Identification and validation of novel CSF biomarkers for early stages of Alzheimer's disease. *Proteomics. Clinical Applications*, *11*, 1373–1384.

Huang, J. T., Wang, L., Prabakaran, S., Wengenroth, M., Lockstone, H. E., Koethe, D., et al. (2008). Independent protein-profiling studies show a decrease in apolipoprotein A1 levels in schizophrenia CSF, brain and peripheral tissues. *Molecular Psychiatry*, *13*(12), 1118–1128.

Hünnerkopf, R., Grassl, J., & Thome, J. (2007). Proteomics: Biomarker research in psychiatry. *Fortschritte der Neurologie—Psychiatrie*, *75*(10), 579–586.

Hye, A., Lynham, S., Thambisetty, M., Causevic, M., Campbell, J., Byers, H. L., et al. (2006). Proteome-based plasma biomarkers for Alzheimer's disease. *Brain*, *129*, 3042–3050.

Ionescu, D. F., Niciu, M. J., Henter, I. D., & Zarate, C. A. (2013). Defining anxious depression: A review of the literature. *CNS Spectrums*, *18*(5), 252–260.

Jellema, T., Baker, C. I., Wicker, B., & Perrett, D. I. (2000). Neural representation for the perception of the intentionality of actions. *Brain and Cognition*, *44*(2), 280–302.

Jimenez, C. R., El Filali, Z., Knol, J. C., Hoekman, K., Kruyt, F. A., Giaccone, G., et al. (2007). Automated serum peptide profiling using novel magnetic C18 beads off-line coupled to MALDI-TOF-MS. *Proteomics. Clinical Applications*, *1*(6), 598–604.

Johnston-Wilson, N. L., Sims, C. D., Hofmann, J. P., Anderson, L., Shore, A. D., Torrey, E. F., et al. (2000). Disease-specific alterations in frontal cortex brain proteins in schizophrenia, bipolar disorder, and major depressive disorder. The Stanley Neuropathology Consortium. *Molecular Psychiatry*, *5*(2), 142–149.

Jungblut, P. R., Holzhütter, H. G., Apweiler, R., & Schlüter, H. (2008). The speciation of the proteome. *Chemistry Central Journal*, *2*, 16.

Keller, M. B., & Hanks, D. L. (1995). Anxiety symptom relief in depression treatment outcomes. *The Journal of Clinical Psychiatry*, *56*(Suppl. 6), 22–29.

Khawaja, X., Xu, J., Liang, J. J., & Barrett, J. E. (2004). Proteomic analysis of protein changes developing in rat hippocampus after chronic antidepressant treatment: Implications for depressive disorders and future therapies. *Journal of Neuroscience Research*, *75*(4), 451–460.

Kim, H. J., Eom, C. Y., Kwon, J., Joo, J., Lee, S., Nah, S. S., et al. (2012). Roles of interferon-gamma and its target genes in schizophrenia: Proteomics-based reverse genetics from mouse to human. *Proteomics*, *12*(11), 1815–1829.

Kim, H. G., & Kim, K. L. (2007). Decreased hippocampal cholinergic neurostimulating peptide precursor protein associated with stress exposure in rat brain by proteomic analysis. *Journal of Neuroscience Research*, *85*(13), 2898–2908.

Krömer, S. A., Kessler, M. S., Milfay, D., Birg, I. N., Bunck, M., Czibere, L., et al. (2005). Identification of glyoxalase-I as a protein marker in a mouse model of extremes in trait anxiety. *The Journal of Neuroscience*, *25*(17), 4375–4384.

Lehmann, S., Delaby, C., Touchon, J., Hirtz, C., & Gabelle, A. (2013). Biomarkers of Alzheimer's disease: The present and the future. *Revue Neurologique (Paris)*, *169*(10), 719–723.

Levin, Y., Wang, L., Schwarz, E., Koethe, D., Leweke, F. M., & Bahn, S. (2010). Global proteomic profiling reveals altered proteomic signature in schizophrenia serum. *Molecular Psychiatry*, *15*(11), 1088–1100.

Li, X., Song, X. Q., Gao, J. S., Pang, L. J., Li, Y. H., Hao, Y. H., et al. (2012). Proteomic analysis of novel serum markers in first-episode schizophrenics before versus after treatment of risperidone. *Zhonghua Yi Xue Za Zhi*, *92*(45), 3194–3198.

Liao, L. J., Park, S. K., Xu, T., Vanderklish, P., & Yates, J. R. (2008). Quantitative proteomic analysis of primary neurons reveals diverse changes in synaptic protein content in fmr1 knockout mice. *Proceedings of the National Academy of Sciences of the United States of America*, *105*, 15281–15286.

Mairesse, J., Vercoutter-Edouart, A. S., Marrocco, J., Zuena, A. R., Giovine, A., Nicoletti, F., et al. (2012). Proteomic characterization in the hippocampus of prenatally stressed rats. *Journal of Proteomics*, *75*(6), 1764–1770.

Manral, P., Sharma, P., Hariprasad, G., Chandralekha, Tripathi M., & Srinivasan, A. (2012). Can apolipoproteins and complement factors be biomarkers of Alzheimer's disease? *Current Alzheimer Research*, *9*(8), 935–943.

Martins-de-Souza, D. (2011). Proteomics as a tool for understanding schizophrenia. *Clinical Psychopharmacology and Neuroscience*, *9*(3), 95–101.

Martins-de-Souza, D., Dias-Neto, E., Schmitt, A., Falkai, P., Gormanns, P., Maccarrone, G., et al. (2010). Proteome analysis of schizophrenia brain tissue. *The World Journal of Biological Psychiatry*, *2*, 110–120.

Martins-de-Souza, D., Guest, P. C., Harris, L. W., Vanattou-Saifoudine, N., Webster, M. J., Rahmoune, H., et al. (2012). Identification of proteomic signatures associated with depression and psychotic depression in post-mortem brains from major depression patients. *Translational Psychiatry*, *2*, e87.

Martins-de-Souza, D., Guest, P. C., Rahmoune, H., & Bahn, S. (2012). Proteomic approaches to unravel the complexity of schizophrenia. *Expert Review of Proteomics*, *9*(1), 97–108.

Martins-de-Souza, D., Harris, L. W., Guest, P. C., Turck, C. W., & Bahn, S. (2010). The role of proteomics in depression research. *European Archives of Psychiatry and Clinical Neuroscience*, *260*(6), 499–506.

Matar, H. E., Almerie, M. Q., & Sampson, S. (2013). Fluphenazine (oral) versus placebo for schizophrenia. *Schizophrenia Bulletin*, *39*(6), 1187–1188.

Mattsson, N., Rosén, E., Hansson, O., Andreasen, N., Parnetti, L., Jonsson, M., et al. (2012). Age and diagnostic performance of Alzheimer disease CSF biomarkers. *Neurology*, *78*(7), 468–476.

Maurer, M. H. (2010). Proteomics of brain extracellular fluid (ECF) and cerebrospinal fluid (CSF). *Mass Spectrometry Reviews*, *29*(1), 17–28.

Mechref, Y., Madera, M., & Novotny, M. V. (2008). Glycoprotein enrichment through lectin affinity techniques. *Methods in Molecular Biology*, *424*, 373–396.

Mitchell, P. B. (2013). Bipolar disorder. *Australian Family Physician*, *42*(9), 616–619.

Nakagawa, N., Yao, H., Nakahara, T., Inomata, S., Hashimoto, K., & Kuroki, T. (2013). Serum of transthyretin as a treatment-responsive biomarker for schizophrenia. *Brain and Nerve*, *65*(9), 1093–1099.

Novikova, S. I., He, F., Cutrufello, N. J., & Lidow, M. S. (2006). Identification of protein biomarkers for schizophrenia and bipolar disorder in the postmortem prefrontal cortex using SELDI-TOF-MS ProteinChip profiling combined with MALDI-TOF-PSD-MS analysis. *Neurobiology of Disease, 23*(1), 61–76.

Oda, Y., Huang, K., Cross, F. R., Cowburn, D., & Chait, B. T. (1999). Accurate quantitation of protein expression and site-specific phosphorylation. *Proceedings of the National Academy of Sciences of the United States of America, 96*, 6591.

Ong, S. E., Blagoev, B., Kratchmarova, I., Kristensen, D. B., Steen, H., Pandey, A., et al. (2002). Stable isotope labeling by amino acids in cell culture, SILAC, as a simple and accurate approach to expression proteomics. *Molecular and Cellular Proteomics, 1*, 376–386.

Pacchiarotti, I., Bond, D. J., Baldessarini, R. J., Nolen, W. A., Grunze, H., Licht, R. W., et al. (2013). The International Society for Bipolar Disorders (ISBD) task force report on antidepressant use in bipolar disorders. *The American Journal of Psychiatry, 170*(11), 1249–1262.

Paulson, L., Martin, P., Persson, A., Nilsson, C. L., Ljung, E., Westman-Brinkmalm, A., et al. (2003). Comparative genome and proteome analysis of cerebral cortex from MK-801-treated rats. *Journal of Neuroscience Research, 71*, 526–533.

Pennington, K., Beasley, C. L., Dicker, P., Fagan, A., English, J., Pariante, C. M., et al. (2008). Prominent synaptic and metabolic abnormalities revealed by proteomic analysis of the dorsolateral prefrontal cortex in schizophrenia and bipolar disorder. *Molecular Psychiatry, 13*(12), 1102–1117.

Perneczky, R., Guo, L. H., Kagerbauer, S. M., Werle, L., Kurz, A., Martin, J., et al. (2013). Soluble amyloid precursor protein β as blood-based biomarker of Alzheimer's disease. *Translational Psychiatry, 3*, e227.

Pich, E. M., Vargas, G., & Domenici, E. (2012). Biomarkers for antipsychotic therapies. *Handbook of Experimental Pharmacology, 212*, 339–360.

Pitchot, W., Scantamburlo, G., & Ansseau, M. (2012). Clinical and therapeutical approach of depression in old age. *Revue Médicale de Liège, 67*(11), 566–572.

Postmes, L., Sno, H. N., Goedhart, S., van der Stel, J., Heering, H. D., & de Haan, L. (2014). Schizophrenia as a self-disorder due to perceptual incoherence. *Schizophrenia Research, 152*(1), 41–50.

Prabakaran, S., Wengenroth, M., Lockstone, H. E., Lilley, K., Leweke, F. M., & Bahn, S. (2007). 2-D DIGE analysis of liver and red blood cells provides further evidence for oxidative stress in schizophrenia. *Journal of Proteome Research, 6*(1), 141–149.

Puchades, M., Hansson, S. F., Nilsson, C. L., Andreasen, N., Blennow, K., & Davidsson, P. (2003). Proteomic studies of potential cerebrospinal fluid protein markers for Alzheimer's disease. *Brain Research. Molecular Brain Research, 118*(1–2), 140–146.

Rao, S., & Zisook, S. (2009). Anxious depression: Clinical features and treatment. *Current Psychiatry Reports, 11*, 429–436.

Robillard, R., Naismith, S. L., & Hickie, I. B. (2013). Recent advances in sleep-wake cycle and biological rhythms in bipolar disorder. *Current Psychiatry Reports, 15*(10), 402.

Ross, P. L., Huang, Y. N., Marchese, J. N., Williamson, B., Parker, K., Hattan, S., et al. (2004). Multiplexed protein quantitation in Saccharomyces cerevisiae using amine-reactive isobaric tagging reagents. *Molecular & Cellular Proteomics, 3*, 1154–1169.

Rush, A. J. (2007). The varied clinical presentations of major depressive disorder. *The Journal of Clinical Psychiatry, 68*, 4–10.

Said, M. A., Hatim, A., Habil, M. H., Zafidah, W., Haslina, M. Y., Badiah, Y., et al. (2013). Metabolic syndrome and antipsychotic monotherapy treatment among schizophrenia patients in Malaysia. *Preventive Medicine, 57*(Suppl.), S50–S53.

Schneider, G., & Heuft, G. (2012). Anxiety and depression in the elderly. *Zeitschrift für Psychosomatische Medizin und Psychotherapie, 58*(4), 336–356.

Shi, M., Caudle, W. M., & Zhang, J. (2009). Biomarker discovery in neurodegenerative diseases: A proteomic approach. *Neurobiology of Disease, 35*(2), 157–164.

Shi, M., Hwang, H., & Zhang, J. (2013). Quantitative characterization of glycoproteins in neurodegenerative disorders using iTRAQ. *Methods in Molecular Biology*, *951*, 279–296.

Shiozaki, A., Tsuji, T., Kohno, R., Kawamata, J., Uemura, K., Teraoka, H., et al. (2004). Proteome analysis of brain proteins in Alzheimer's disease: subproteomics following sequentially extracted protein preparation. *Journal of Alzheimer's Disease*, *6*(3), 257–268.

Sibille, E., & French, B. (2013). Biological substrates underpinning diagnosis of major depression. *The International Journal of Neuropsychopharmacology*, *16*(8), 1893–1909.

Sikich, Linmarie (2013). Diagnosis and evaluation of hallucinations and other psychotic symptoms in children and adolescents. *Child and Adolescent Psychiatric Clinics of North America*, *22*, 655–673.

Singh, I., & Rose, N. (2009). Biomarkers in psychiatry. *Nature*, *460*(7252), 202–207.

Skovronsky, D. M., Lee, V. M., & Trojanowski, J. Q. (2006). Neurodegenerative diseases: New concepts of pathogenesis and their therapeutic implications. *Annual Review of Pathology*, *1*, 151–170.

Soares, H. D., Potter, W. Z., Pickering, E., Kuhn, M., Immermann, F. W., Shera, D. M., et al. (2012). Plasma biomarkers associated with the apolipoprotein E genotype and Alzheimer disease. *Archives of Neurology*, *69*(10), 1310–1317.

Sokolowska, E., & Hovatta, I. (2013). Anxiety genetics—Findings from cross-species genome-wide approaches. *Biology of Mood and Anxiety Disorders*, *3*(1), 3–9.

Spellman, D. S., Deinhardt, K., Darie, C. C., Chao, M. V., & Neubert, T. A. (2008). Stable isotopic labeling by amino acids in cultured primary neurons. *Molecular & Cellular Proteomics*, 7, 1067–1076.

Squassina, A., Manchia, M., & Del Zompo, M. (2010). Pharmacogenomics of mood stabilizers in the treatment of bipolar disorder. *Human Genomics and Proteomics*, *2010*, 159761.

Sultana, R., Baglioni, M., Cecchetti, R., Cai, J., Klein, J. B., Bastiani, P., et al. (2013). Lymphocyte mitochondria: Toward identification of peripheral biomarkers in the progression of Alzheimer disease. *Free Radical Biology & Medicine*, *65C*, 595–606.

Sultana, R., Boyd-Kimball, D., Cai, J., Pierce, W. M., Klein, J. B., Merchant, M., et al. (2007). Proteomics analysis of the Alzheimer's disease hippocampal proteome. *Journal of Alzheimer's Disease*, *11*(2), 153–164.

Sussulini, A., Dihazi, H., Banzato, C. E., Arruda, M. A., Stühmer, W., Ehrenreich, H., et al. (2011). Apolipoprotein A-I as a candidate serum marker for the response to lithium treatment in bipolar disorder. *Proteomics*, *11*(2), 261–269.

Suzuki, H., Inoue, Y., & Gen, K. (2012). A study of the efficacy and safety of switching from oral risperidone to risperidone long-acting injection in older patients with schizophrenia. *Therapeutic Advances in Psychopharmacology*, *2*(6), 227–234.

Szego, E. M., Janáky, T., Szabó, Z., Csorba, A., Kompagne, H., Müller, G., et al. (2010). A mouse model of anxiety molecularly characterized by altered protein networks in the brain proteome. *European Neuropsychopharmacology*, *20*(2), 96–111.

Thongboonkerd, V. (2007). Proteomics. *Forum of Nutrition*, *60*, 80–90.

Tonack, S., Aspinall-O'Dea, M., Jenkins, R. E., Elliot, V., Murray, S., Lane, C. S., et al. (2009). A technically detailed and pragmatic protocol for quantitative serum proteomics using iTRAQ. *Journal of Proteomics*, *73*, 352–356.

Vercauteren, F. G., Flores, G., Ma, W., Chabot, J. G., Geenen, L., Clerens, S., et al. (2007). An organelle proteomic method to study neurotransmission-related proteins, applied to a neurodevelopmental model of schizophrenia. *Proteomics*, 7(19), 3569–3579.

Walsh, T., McClellan, J. M., McCarthy, S. E., Addington, A. M., Pierce, S. B., Cooper, G. M., et al. (2008). Rare structural variants disrupt multiple genes in neurodevelopmental pathways in schizophrenia. *Science*, *320*(25), 539–543.

Wijte, D., McDonnell, L. A., Balog, C. I., Bossers, K., Deelder, A. M., Swaab, D. F., et al. (2012). A novel peptidomics approach to detect markers of Alzheimer's disease in cerebrospinal fluid. *Methods*, *56*(4), 500–507.

Wilkins, M. R., Pasquali, C., Appel, R. D., Ou, K., Golaz, O., Sanchez, J. C., et al. (1996). From proteins to proteomes: Large scale protein identification by two-dimensional electrophoresis and amino acid analysis. *Biotechnology (New York)*, *14*, 61–65.

Woods, A. G., Sokolowska, I., Taurines, R., Gerlach, M., Dudley, E., Thome, J., et al. (2012). Potential biomarkers in psychiatry: Focus on the cholesterol system. *Journal of Cellular and Molecular Medicine*, *16*(6), 1184–1195.

Xiao, Z., Prieto, D., Conrads, T. P., Veenstra, T. D., & Issaq, H. J. (2005). Proteomic patterns: Their potential for disease diagnosis. *Molecular and Cellular Endocrinology*, *230*(1–2), 95–106.

Yang, M. H., Yang, Y. H., Lu, C. Y., Jong, S. B., Chen, L. J., Lin, Y. F., et al. (2012). Activity-dependent neuroprotector homeobox protein: A candidate protein identified in serum as diagnostic biomarker for Alzheimer's disease. *Journal of Proteomics*, *75*(12), 3617–3629.

Yang, Y., Yang, D., Tang, G., Zhou, C., Cheng, K., Zhou, J., et al. (2013). Proteomics reveals energy and glutathione metabolic dysregulation in the prefrontal cortex of a rat model of depression. *Neuroscience*, *247*, 191–200.

Yuan, X., & Desiderio, D. M. (2005). Human cerebrospinal fluid peptidomics. *Journal of Mass Spectrometry*, *40*, 176–181.

Zellner, M., Veitinger, M., & Umlauf, E. (2009). The role of proteomics in dementia and Alzheimer's disease. *Acta Neuropathologica*, *118*(1), 181–195.

Zhang, Y., Filiou, M. D., Reckow, S., Gormanns, P., Maccarrone, G., Kessler, M. S., et al. (2011). Proteomic and metabolomic profiling of a trait anxiety mouse model implicate affected pathways. *Molecular & Cellular Proteomics*, *10*(12), M111.008110.

Zhang, J., Goodlett, D. R., Quinn, J. F., Peskind, E., Kaye, J. A., Zhou, Y., et al. (2005). Quantitative proteomics of cerebrospinal fluid from patients with Alzheimer disease. *Journal of Alzheimer's Disease*, *7*(2), 125–133.

Zhang, K., Schrag, M., Crofton, A., Trivedi, R., Vinters, H., & Kirsch, W. (2012). Targeted proteomics for quantification of histone acetylation in Alzheimer's disease. *Proteomics*, *12*(8), 1261–1268.

CHAPTER FOUR

On the Use of Knowledge-Based Potentials for the Evaluation of Models of Protein–Protein, Protein–DNA, and Protein–RNA Interactions

Oriol Fornes, Javier Garcia-Garcia, Jaume Bonet, Baldo Oliva[1]

Structural Bioinformatics Lab. (GRIB), Departament de Ciències Experimentals i de la Salut, Universitat Pompeu Fabra, Barcelona, Catalunya, Spain

[1]Corresponding author: e-mail address: baldo.oliva@upf.edu

Contents

1. Introduction 78
2. Knowledge-Based Potentials 80
   2.1 Split-statistical potentials 81
3. Modeling of Protein Interactions Using Templates 82
   3.1 Models of binary complexes 83
   3.2 Models of multimeric complexes 84
4. Modeling Interactions of Proteins Using Docking 85
   4.1 Protein–protein docking 90
   4.2 Protein–nucleic acid docking 92
5. Prediction of Protein-Binding Regions 93
   5.1 Identification of protein interfaces 93
   5.2 Prediction of DNA/RNA-binding proteins 97
6. Characterization of Transcription Factor-Binding Sites 98
   6.1 Application of knowledge-based potentials on DREAM5 targets 100
7. Adapting Split-Statistical Potentials for Protein–DNA Interactions 105
   7.1 Application of split-statistical potentials on DREAM5 targets 106
8. Conclusions 109
Acknowledgments 110
References 110

Abstract

Proteins are the bricks and mortar of cells, playing structural and functional roles. In order to perform their function, they interact with each other as well as with other biomolecules such as DNA or RNA. Therefore, to fathom the function of a protein, we require knowing its partners and the atomic details of its interactions (i.e., the structure

Advances in Protein Chemistry and Structural Biology, Volume 94
ISSN 1876-1623
http://dx.doi.org/10.1016/B978-0-12-800168-4.00004-4

of the complex). However, the amount of protein interactions with an experimentally determined three-dimensional structure is scarce. Therefore, computational techniques such as homology modeling are foremost to fill this gap. Protein interactions can be modeled using as templates the interactions of homologous proteins, if the structure of the complex is known, or using docking methods. In both approaches, the estimation of the quality of models is essential. There are several ways to address this problem. In this review, we focus on the use of knowledge-based potentials for the analysis of protein interactions. We describe the procedure to derive statistical potentials and split them into different energetic terms that can be used for different purposes. We extensively discuss the fields where knowledge-based potentials have been successfully applied to (1) model protein–protein, protein–DNA, and protein–RNA interactions and (2) predict binding sites (in the protein and in the DNA). Moreover, we provide ready-to-use resources for docking and benchmarking protein interactions.

1. INTRODUCTION

During the past decade hundreds of sequenced genomes have come to light, producing a vast amount of protein sequences. Therefore, unraveling the function of these proteins has become one of the major challenges in biology. It is widely accepted that the function of a protein can be predicted from its structure (Watson, Laskowski, & Thornton, 2005). But proteins rarely act alone; instead, they form networks of physical interactions with other biomolecules (i.e., protein–protein, protein–DNA, and protein–RNA interactions). Thus, in order to have a better understanding of the function of a protein, it is also necessary to know with whom it is associated and how, even at atomic level.

The number of proteins with an experimentally determined three-dimensional (3D) structure in the Protein Data Bank (PDB) (Berman et al., 2000) is very low in comparison to the number of known protein sequences, even for well-characterized organisms (Sharan, Ulitsky, & Shamir, 2007), and even lower in the case of protein binary complexes (Kirsanov et al., 2012; Mosca, Céol, Stein, Olivella, & Aloy, 2013). The disproportion between the number of solved 3D structures and protein sequences has encouraged the development of many strategies to model the structure of proteins from their sequence (Dunbrack, 2006; Ginalski, 2006). These strategies have become the basis for the modeling of protein interactions. Protein–protein interactions can be modeled by using as templates complexes of homologous proteins with known structure. This approach relies on the principle that, given a pair of interacting proteins, their homologs will also interact (interologs approach; Garcia-Garcia, Schleker, Klein-Seetharaman, & Oliva, 2012; Matthews et al., 2001), and it is assumed that they will do it in a similar fashion. Occasionally, the models of protein–protein interactions can be constructed by superimposition of the models of the unbound partners over the structure of a template complex. They can also be obtained by docking the structure of one of the two proteins onto the other (Vajda & Kozakov, 2009) via previous modeling of their unbound structures (if necessary). We recently reviewed in detail the modeling of tertiary and quaternary structures of proteins and their role in protein–protein interaction networks (Garcia-Garcia, Bonet, et al., 2012; Planas-Iglesias, Bonet, Feliu, Gursoy, & Oliva, 2012). Similarly, we can also obtain the models of protein–DNA and protein–RNA interactions, which also require to model the nucleotide sequences of interest (Feig, Karanicolas, & Brooks, 2004; Lu & Olson, 2008).

Paired with the modeling of 3D structures, the estimation of their quality has become crucial. In this particular context, several methods have been developed to score models based on energies. One approach to address this problem is based on the derivation of knowledge-based potentials (also referred to as statistical potentials or potentials of mean force) (Sippl, 1990). Knowledge-based potentials have been used to (1) discriminate whether or not a model has the correct fold (Panjkovich, Melo, & Marti-Renom, 2008; Shen & Sali, 2006); (2) detect localized errors in protein structures (Wiederstein & Sippl, 2007); (3) predict the stability of mutant proteins (Zhou & Zhou, 2002); (4) select the closest near-native models from a set of decoys (Aloy & Oliva, 2009; Ferrada & Melo, 2009); (5) model protein–protein interactions (Aloy & Russell, 2003; Lu, Lu, & Skolnick, 2003); (6) analyze the outcome of docking experiments, including protein–protein (analyzed in Moal, Torchala, Bates, & Fernández-Recio, 2013), protein–DNA (Robertson & Varani, 2007; Takeda, Corona, & Guo, 2013; Xu, Yang, Liang, & Zhou, 2009), and protein–RNA (Pérez-Cano, Solernou, Pons, & Fernández-Recio, 2010; Tuszynska & Bujnicki, 2011; Zheng, Robertson, & Varani, 2007); (7) infer the ability of proteins to bind DNA (Gao & Skolnick, 2008, 2009; Zhao, Yang, & Zhou, 2010) and RNA (Zhao, Yang, & Zhou, 2011); (8) recognize the binding regions in proteins (Feliu, Aloy, & Oliva, 2011; Pérez-Cano & Fernández-Recio, 2010); and (9) identify transcription factor-binding sites (Alamanova, Stegmaier, & Kel, 2010; Angarica, Pérez, Vasconcelos, Collado-Vides, & Contreras-Moreira, 2008; Chen, Chien, et al., 2012; Liu, Guo, Li, & Xu, 2008; Xu et al., 2009).

In the following sections, we review the use of knowledge-based potentials for the analysis of protein interactions. In Section 2, we introduce knowledge-based potentials and we split them into different energetic terms. Sections 3–6 are devoted to different fields where knowledge-based potentials have successfully been applied. Specifically, we focus on (1) modeling of protein interactions (including homology modeling and integrative modeling); (2) docking of protein interactions (including protein–protein, protein–DNA, and protein–RNA); (3) prediction of protein-binding regions; (4) characterization of transcription factor-binding sites; and (5) prediction of DNA-binding sites. In Section 7, we adapt the procedure to split-statistical potentials (Aloy & Oliva, 2009) to predict protein–DNA interactions and DNA-binding sites.

2. KNOWLEDGE-BASED POTENTIALS

A knowledge-based potential is an energy function derived from the analysis of known protein structures. There are many methods to obtain such potentials including the quasi-chemical (Miyazawa & Jernigan, 1985) and the potential of mean force (PMF) approximations (Sippl, 1990). We have used the general definition of knowledge-base potential described in Aloy and Oliva (2009) (i.e., Eq. 4.1):

$$\begin{aligned} \mathrm{PMF}(a,b) &= \mathrm{PMF}_{\mathrm{std}}(d_{a,b}) - k_{\mathrm{B}}T\log\left(\frac{P(a,b|d)}{P(a)P(b)}\right) \\ \mathrm{PMF}_{\mathrm{std}}(d_{a,b}) &= k_{\mathrm{B}}T\log\left(\frac{P(d_{a,b})}{\mathrm{weight}_{\mathrm{ref}}}\right) \end{aligned} \tag{4.1}$$

Where "k_{B}" is the Boltzmann constant, "T" is the standard temperature, "$d_{a,b}$" is the pairwise distance between a pair of residues "a,b", being "$P(a)$" and "$P(b)$" their respective probabilities. "$\mathrm{P}(a,b|d_{a,b})$" is the conditional probability of finding residues "a,b" at a maximum distance "$d_{a,b}$" and "$P(d_{a,b})$" is the probability of observing any pair of residues up to that distance. Finally, the "$\mathrm{weight}_{\mathrm{ref}}$" is the reference state function. The probabilities "$P(^*)$" are approximated from the observed frequencies of interactions in a nonredundant set of PDB structures. Moreover, the distance can be calculated as the minimum distance between any pair of heavy atoms or as the pairwise distance between two specific atoms (e.g., between Cβ atoms; Cα for glycine residues). Also, Aloy and Oliva (2009) proved that the reference state function can be neglected for the comparison of decoys (see further in Section 2.1).

The application of Eq. (4.1) over all interacting pairs of residues "a" and "b" in a protein structure results in an estimation of its quality given in terms of energy (i.e., Eq. 4.2):

$$E = \sum_{a,b} \mathrm{PMF}(a, b) \tag{4.2}$$

It has to be noted that, while for protein folding residues "a" and "b" belong to the same protein (or single protein chain), for protein–protein interactions, residues "a" and "b" belong to a pair of interacting proteins "A" and "B" (or different protein chains), respectively.

2.1. Split-statistical potentials

Aloy and Oliva (2009) demonstrated that, using the Bayes theorem, Eq. (4.1) can be decomposed into several energetic terms, one of them including the reference state. We have selected some of these terms as potentials of mean force to score the quality of decoys (i.e., Eq. 4.3):

$$\begin{aligned}
\mathrm{PMF}_{\mathrm{pair}}(a, b) &= -k_{\mathrm{B}} T \log\left(\frac{P(a,b|d_{a,b})}{P(a)P(b)P(d_{a,b})}\right) \\
\mathrm{PMF}_{\mathrm{local}}(a, b) &= k_{\mathrm{B}} T \log\left(\frac{P(a|\theta_a)}{P(a)}\right) + k_{\mathrm{B}} T \log\left(\frac{P(b|\theta_b)}{P(b)}\right) \\
\mathrm{PMF}_{\mathrm{3D}}(a, b) &= k_{\mathrm{B}} T \log(P(d_{a,b})) \\
\mathrm{PMF}_{\mathrm{3DC}}(a, b) &= k_{\mathrm{B}} T \log\left(\frac{P(\theta_a,\theta_b|d_{a,b})}{P(\theta_a, \theta_b)}\right) \\
\mathrm{PMF}_{\mathrm{S3DC}}(a, b) &= -k_{\mathrm{B}} T \log\left(\frac{P(a,b|d_{a,b},\theta_a,\theta_b)P(\theta_a, \theta_b)}{P(a,b|\theta_a,\theta_b)P(\theta_a,\theta_b|d_{a,b})}\right)
\end{aligned} \tag{4.3}$$

Where "θ_a" and "θ_b" are the environments of a pair of residues "a,b", as defined by their hydrophobicity (i.e., polar or nonpolar), degree of exposure (i.e., buried or exposed), and surrounding secondary structure (i.e., α-helix, β-sheet, or coil). As an example, "$P(a,b|d_{ab},\theta_a,\theta_b)$" is the conditional probability of finding residues "a,b", in their respective environments "θ_a" and "θ_b", at a maximum distance "$d_{a,b}$" (see Aloy & Oliva, 2009 for more details).

The statistical potentials "E_{pair}", "E_{local}", "E_{3D}", "E_{3DC}", and "E_{S3DC}" are defined using Eq. (4.2), with the corresponding subscripts between "E_" and "PMF_". We name these potentials "split-statistical potentials". The statistical potential "E_{S3DC}" can be understood as a refinement of the residue-pair statistical potential "E_{pair}". It takes into account not only the residues that interact but also their environments. The statistical potential "E_{3DC}" depends only on the occurrence of interacting environments without considering the specific interacting residues. The score "E_{local}" is distance independent, and it reflects the probability of placing a residue in a specific environment. The energy term "E_{3D}" concerns only the distance at which pairs of residues interact, and it increases together with the number of interacting residue pairs.

The statistical potentials described in Eq. (4.3) differ in order of magnitude and their values cannot be used straightforward for the comparison of conformational decoys. Therefore, they are translated into *Z*-scores. The *Z*-score of an energy (or score within a distribution) is defined as the difference between the energy (i.e., "$E\_$") and the average of energies in the distribution (i.e., "μ"), divided by the standard deviation of the distribution (i.e., "σ"). In general, the background distribution to calculate a *Z*-score uses a random distribution, which in the case of folds or interactions is obtained by shuffling the residues of one or two sequences, respectively. The translation of energies into *Z*-scores neglects the "E_{3D}" term because it is independent of the sequence (i.e., $E_{3D}=\mu$ for any distribution of shuffled sequences). Also, Aloy and Oliva (2009) demonstrated that the distribution of the *Z*-score of the reference state was similar to the random distribution and could be neglected too. Therefore, neither the energy "E_{3D}" nor the reference state were considered when selecting the best model among conformational decoys.

In a recent work, Feliu et al. (2011) modified the split-statistical potentials of Eq. (4.3) for its application to protein–protein interactions. The frequencies of amino acid pairs were extracted from residues belonging to different chains in the interface of protein complexes from the 3DID database (Stein, Céol, & Aloy, 2011). In protein–protein interactions, the *Z*-score of the reference state was still irrelevant. Therefore, the energetic term that included the reference state was assumed to be irrelevant too when ranking decoys of interactions (i.e., docking poses). However, the score "E_{3D}" was associated with the extension of the interacting interface (it is proportional to the number of residues implied in the interface), and it was still valuable in the analysis of protein-docking decoys (see further in Section 4.1).

3. MODELING OF PROTEIN INTERACTIONS USING TEMPLATES

The continuous increase of structural data on protein complexes in the PDB has been exploited for modeling the structure of protein–protein interactions as well as the interactions of proteins with other biomolecules (i.e., protein–DNA or protein–RNA interactions) based on homology. However, when structures of the interaction are not available, docking methods can be used (see further in Section 4). In the past years, new approaches have been developed to assemble large macromolecular complexes by combining different experimental data to apply restraints upon complex assembly.

3.1. Models of binary complexes

The most common way of modeling protein interactions is via comparative modeling (revised in Planas-Iglesias et al., 2012). This approach can only be applied as long as there is a homologous structure of the interaction. Then, applications such as MODELLER (Eswar et al., 2006) are able to directly model the interaction of interest. Nevertheless, homology modeling is limited by those homologs whose structure is too remote to help assigning the correct fold (Rost, 1999). Still, even distantly related proteins may use the same binding regions to interact (Gao & Skolnick, 2010; Tuncbag, Gursoy, Guney, Nussinov, & Keskin, 2008; Zhang, Petrey, Norel, & Honig, 2010), which has been exploited by different authors to model protein–protein interactions. For example, in M-TASSER (Chen & Skolnick, 2008), protein sequences are threaded against a monomer template library. All threading solutions belonging to the same dimer template are then identified. However, if both monomers share less than 30% sequence identity with their templates on the dimer, the threaded dimer is evaluated with statistical potentials (Lu et al., 2003) and, when necessary, discarded. Next, the tertiary structure of each protein is obtained by rearrangement of continuous template fragments (Zhang & Skolnick, 2004). Finally, the quaternary structure is assembled by superimposition of both protein structures over the dimer template.

Recently, three different methods have been proposed for model-building the structure of protein–protein interactions on a genome-wide scale. In PRISM (Tuncbag, Gursoy, Nussinov, & Keskin, 2011), the structures (or models) of two proteins are aligned against a set of known protein–protein interfaces (i.e., template set). If the two complementary sides of a template interface are structurally similar to the proteins (each side to a different protein), then the proteins are predicted to interact and the interaction is modeled using the binding site, as dictated by the template interface. All models produced with this approach are refined to account for flexible changes and finally ranked. In PrePPI (Zhang et al., 2012), the individual structures of the proteins are searched in the PDB or in a database of homology models (i.e., SkyBase (Lee et al., 2010) and ModBase (Pieper et al., 2011)). This step is followed by the identification of close and remote homologs of the two partners. Then, if a PDB structure contains the interaction between the homologs of each partner, it is used as template and the interaction is modeled by superimposition. In order to calculate the reliability of the model, five different empirical structure-based scores are assigned and combined using a Bayesian network, which scores the quality of the structural model of the interaction. Finally, in Interactome3D (Mosca, Céol, & Aloy, 2013), the interaction is modeled in a similar fashion than PrePPI, but it increases the structural coverage of the approach by using templates of interacting domains from 3DID (Mosca, Céol, Stein, et al., 2013). The resulting models are finally evaluated with InterPrets (Aloy & Russell, 2003).

3.2. Models of multimeric complexes

Methods described in Section 3.1 are useful for complexes formed by few molecules. However, the assembly of large macromolecular complexes requires an integrative structural modeling approach. The main idea behind this methodology is to characterize the structural and topological features of the complex in order to reduce the number of plausible solutions. For example, the Integrative Modeling Platform (IMP) (Russel et al., 2012) has been used to describe the yeast nuclear pore complex (Alber et al., 2007) and the structure of chromatin at mega base scale (Baù et al., 2011). The assembly of a complex in IMP is a cyclic procedure involving four different steps (revised in detail in Planas-Iglesias et al., 2012):

(1) Collecting the information regarding the complex. This step includes collecting experimental data from SAXS profiles (Schneidman-Duhovny, Hammel, & Sali, 2011), proteomics data (Alber, Förster, Korkin, Topf, & Sali, 2008), EM images (Lasker, Phillips, et al., 2010), density maps (Lasker, Sali, & Wolfson, 2010), nuclear magnetic resonance (NMR) spectroscopy (Simon, Madl, Mackereth, Nilges, & Sattler, 2010), or even 5C data (Baù et al., 2011). It also implies to include physical–chemical information, such as molecular mechanics force fields (Brooks et al., 1983) and potentials of mean force or statistically derived potentials (Shen & Sali, 2006).

(2) Select a method to represent the data and use the information collected in the previous step, translating it into spatial restraints. IMP uses structures solved with different resolutions. High-resolution structures can be represented by atoms, but low-resolution structures are represented by groups of atoms, such as residues, motifs, or even domains. The translation of information into spatial restraints is used to test the consistency of the model.

(3) Constructing a model that is consistent with the aforementioned spatial restraints. The entire rotational and translational 3D space is searched in order to position and orientate each individual structure inside the complex.

(4) Evaluation of the modeled complex. In theory, if there is only one native state of the complex, we should obtain a single model satisfying all restraints. In contrast, if the data used to encode the restraints is insufficient, more than one possible solution can be obtained or none.

4. MODELING INTERACTIONS OF PROTEINS USING DOCKING

In contrast to the previous methods, which require the structural knowledge of the interaction, docking is used for modeling the structure of an interaction formed by two or more molecules (e.g., two proteins) when the structure of the interaction is not available but the structures of the individual molecules are known (or can be modeled). Docking addresses the problem of finding the best-fit orientation of one molecule with respect to the other. This idea was first introduced 30 years ago by Wodak and Janin (1978). Since then, docking algorithms have largely improved (summarized in Table 4.1). The simplest method of docking two structures is to treat them as rigid bodies, usually using the Fast Fourier Transform technique (e.g., MolFit (Katchalski-Katzir et al., 1992), FTDock (Gabb et al., 1997), PIPER (Kozakov et al., 2006), and ZDOCK (Mintseris et al., 2007)) or geometric matching (e.g., Hex (Ritchie & Kemp, 2000) and FRODOCK (Garzon et al., 2009)). Moreover, several methods have been developed that take into consideration the flexibility of proteins, including Monte Carlo-based methods (e.g., RosettaDock; Gray et al., 2003), the High Ambiguity Driven biomolecular DOCKing (HADDOCK) (Dominguez, Boelens, & Bonvin, 2003), and the use of normal modes describing the changes of conformation suffered upon binding (e.g., SwarmDock; Moal & Bates, 2010). However, it has been shown that for approximately 65% of interactions, proteins suffer little or none conformational changes when they associate, while only for 15% of

Table 4.1 Docking methods

Program	Algorithm	Evaluation	Server	References
Rigid-body docking methods				
ClusPro	FFT	Geometric fit, van der Waals, atomic desolvation energy, electrostatics, and knowledge-based potentials	http://cluspro.bu.edu/login.php	Comeau, Gatchell, Vajda, and Camacho (2004a)
CS	GM	Atomic desolvation energy		Shentu, Al Hasan, Bystroff, and Zaki (2008)
DOT2	FFT	Electrostatics and atomic desolvation energies		Roberts, Thompson, Pique, Perez, and Ten Eyck (2013)
FRODOCK	GM	van der Waals, electrostatics, and atomic desolvation energies	http://frodock.chaconlab.org	Garzon et al. (2009)
FTDock	FFT	Hydrogen bonding, electrostatics, and RPScore (Moont, Gabb, & Sternberg, 1999)		Gabb, Jackson, and Sternberg (1997)
GRAMM-X	FFT	Lennard-Jones potential, evolutionary conservation, knowledge-based potentials, van der Waals, and atomic contact energy	http://vakser.bioinformatics.ku.edu/resources/gramm/grammx	Tovchigrechko and Vakser (2006)
Hex	GM	Geometric fit and electrostatics	http://hexserver.loria.fr	Macindoe, Mavridis, Venkatraman, Devignes, and Ritchie (2010)
LZerD	GM	Geometric fit and atomic desolvation energy		Venkatraman, Yang, Sael, and Kihara (2009)

MolFit	FFT			Katchalski-Katzir et al. (1992)
PatchDock	GM	Geometric fit and atomic desolvation energy	http://bioinfo3d.cs.tau.ac.il/PatchDock	Schneidman-Duhovny, Inbar, Nussinov, and Wolfson (2005)
PIPER	FFT	Geometric fit, electrostatics, and atomic desolvation energy		Kozakov, Brenke, Comeau, and Vajda (2006)
pyDOCK	FFT	Electrostatics, desolvation energies, ODA (Fernandez-Recio, Totrov, Skorodumov, & Abagyan, 2005), and SIPPER (Pons, Talavera, de la Cruz, Orozco, & Fernandez-Recio, 2011)	http://life.bsc.es/servlet/pydock/home	Jiménez-García, Pons, and Fernández-Recio (2013)
shDock	GM	Collision filtering		Gu, Koehl, Hass, and Amenta (2012)
SP-dock	GM	Atomic desolvation energy, electrostatics, hydrophobicity, and Lennard-Jones potential		Axenopoulos, Daras, Papadopoulos, and Houstis (2013)
ZDOCK	FFT	Linear combination of atomistic potentials, and ZRANK2 (Pierce & Weng, 2008)	http://zdock.umassmed.edu	Mintseris et al. (2007)
Flexible docking methods				
3D-Garden	MC	Lennard-Jones potential and electrostatics	http://www.sbg.bio.ic.ac.uk/~3dgarden	Lesk and Sternberg (2008)

Continued

Table 4.1 Docking methods—cont'd

Program	Algorithm	Evaluation	Server	References
ATTRACT	EM	Hydrophobic and hydrophilic contacts		Schneider, Saladin, Fiorucci, Prévost, and Zacharias (2012)
FireDock	EM	van der Waals, electrostatics, atomic desolvation energies, hydrogen and disulfide bonds, π-stacking and aliphatic interactions, rotamer probabilities, etc.	http://bioinfo3d.cs.tau.ac.il/FireDock	Mashiach, Schneidman-Duhovny, Andrusier, Nussinov, and Wolfson (2008)
HADDOCK	MCS	van der Waals and electrostatics	http://haddock.chem.uu.nl	De Vries, van Dijk, and Bonvin (2010)
RosettaDock	MCS	van der Waals, hydrogen bonds, rotamer, knowledge-based potentials, electrostatics, and atomic solvation energies	http://antibody.graylab.jhu.edu	Lyskov and Gray (2008)
SwarmDock	NM	van der Waals and electrostatics	http://bmm.cancerresearchuk.org/~SwarmDock	Torchala, Moal, Chaleil, Fernandez-Recio, and Bates (2013)

EM, energy minimization; FFT, Fast Fourier Transform; GM, geometric matching; MC, marching cubes; MCS, Monte Carlo simulation; NM, normal modes.

interactions, proteins undergo flexible deformations (Stein, Rueda, Panjkovich, Orozco, & Aloy, 2011). As rigid-body docking approaches are in the first step of docking, previous to the introduction of flexibility, we will focus this section on rigid-body docking, which accounts for at least 65% of protein–protein interactions.

A typical docking procedure between two molecules involves several steps (Vajda & Kozakov, 2009). It begins with a rigid-body docking search over the entire rotational and translational 3D space for the orientation and position of one structure (i.e., ligand, usually the smallest structure) with respect to the other (i.e., target or receptor). The resulting conformational predictions (i.e., docking poses or decoys) are then ranked using scoring functions with the objective to assign the higher scores to the poses more similar to the native structure. These poses are named closest to native or near-native structures. The definition of near-native solution relies on the small structural differences of a decoy with respect to the 3D structure of the binary complex (i.e., the native conformation). Several criteria can be used to calculate these structural differences, but the most common measure, as it has been established in the Critical Assessment of Predicted Interactions (CAPRI) (Janin et al., 2003), is to calculate the root mean square deviation (RMSD) by comparing the decoy with the native conformation. However, the selection of residues for the comparison can vary when (1) the whole structure of the receptor is used as reference to superimpose the poses, the RMSD shows the deviation on the location of the ligand (ligand-RMSD) and (2) all residues in the interface of the native structure are selected, the RMSD shows the different disposition of the interface (I-RMSD). In CAPRI, a near-native prediction is achieved if the I-RMSD and the ligand-RMSD are smaller than 2 and 5 Å, respectively. Currently, this implies the prediction of more than 30% of the native residue–residue pairwise contacts and at least 50% of correctly identified contact residues (Janin et al., 2003). The best docking poses are then refined, allowing for conformational changes of the two unbound structures upon binding (Dobbins, Lesk, & Sternberg, 2008; Shen, Paschalidis, Vakili, & Vajda, 2008). Nevertheless, as it has been observed in CAPRI, there are still some difficulties concerning the use of these methods (Janin, 2010; Lensink & Wodak, 2010). On the one hand, programs devoted to rigid-docking do not simulate the conformational changes that can occur during complex formation. On the other hand, each available docking mechanism is highly dependent on its scoring function and none of them can produce a single correct solution among all the predictions (Moal et al., 2013).

4.1. Protein–protein docking

Rigid-docking methods yield a large number of predictions (from 10,000 to more than 50,000), including many false positives. Thus, an important course of action is to identify those docking poses that are closer to the native structure (i.e., near-native) before any refinement takes place. At this point, the number of selected conformations typically spans between 10 and 2000. There are two nonexcluding strategies to perform such selection. The first strategy consists in reranking the docking conformations with a scoring function (e.g., CHARMM (Brooks et al., 1983), AMBER (Cornell et al., 1995), FOLD-X (Guerois, Nielsen, & Serrano, 2002) or ZRANK (Pierce & Weng, 2007, 2008)). The second strategy relies on clustering similar solutions by means of I-RMSD (Comeau, Gatchell, Vajda, & Camacho, 2004b) or ligand-RMSD (Ritchie & Kemp, 2000) in order to reduce redundant solutions and detect energy favorable regions in the surface of the receptor (Moal & Bates, 2010).

4.1.1 Benchmarking

In order to assess the ability of docking approaches to distinguish between near-native and non-near-native structures, several benchmarks have been created (see Table 4.2). These datasets are usually comprised of a non-redundant set of real interactions for which the structure of the interaction and the unbound molecules (in most cases) are available. Benchmark targets are classified in three categories of difficulty based on the best I-RMSD obtained with the unbound conformations of the two proteins: easy, medium, and hard (hard cases usually involve large conformational changes between the bound and the unbound forms of the molecules).

4.1.2 Application of split-statistical potentials to rank docking decoys

In a recent work (Feliu et al., 2011), split-statistical potentials performed better than scoring functions encoding atomistic energy terms when applied to rank protein–protein docking poses from targets of the hard category of difficulty of the protein-docking benchmark version 3.0 (Hwang, Pierce, Mintseris, Janin, & Weng, 2008). Furthermore, the analysis over the whole benchmark revealed that "E_{pair}" and "E_{S3DC}" provided a fair amount of nonoverlapping results. Based on this observation, Feliu et al. (2011) defined a new ranking strategy "MixRank". In this strategy, they first considered the list of decoys ranked by both statistical potentials separately, and they selected the top-scored decoy from each list alternatively. In order to avoid

Table 4.2 Benchmark datasets for docking

References	Interaction type	Description	Benchmark link
Hwang, Vreven, Janin, and Weng (2010)	Protein–protein	176 complexes (121 easy, 30 medium, 25 hard)	http://zlab.umassmed.edu/benchmark/
van Dijk and Bonvin (2008)	Protein–DNA	47 complexes (13 easy, 22 medium, 12 hard)	http://haddock.science.uu.nl/dna/benchmark.html
Kim, Corona, Hong, and Guo (2011)	Protein–DNA	38 complexes for rigid (21 easy, 17 hard) and flexible docking (18 easy, 19 hard)	http://bioinfozen.uncc.edu/tf-dna-benchmarks/
Barik, Nithin, Manasa, and Bahadur (2012)	Protein–RNA	45 complexes	http://www.facweb.iitkgp.ernet.in/~rbahadur/benchmark.html
Pérez-Cano, Jiménez-García, and Fernández-Recio (2012)	Protein–RNA	106 complexes (35 by homology modeling; 64 easy, 24 medium, 18 hard)	http://life.bsc.es/pid/protein-rna-benchmark/
Huang and Zou (2013)	Protein–RNA	72 complexes (49 easy, 12 medium, 7 hard)	http://zoulab.dalton.missouri.edu/RNAbenchmark/

Benchmarks for protein–protein, protein–DNA, and protein–RNA docking.

redundant predictions, they ignored decoys with less than 5 Å ligand-RMSD from any previous selection, which removed redundant solutions and provided a better selection of near-native decoys (Feliu & Oliva, 2010). "MixRank" outperformed, for the medium and hard targets of the benchmark, other ranking methods such as RPScore (Moont et al., 1999), which is another statistical potential, or ZRANK (Mintseris et al., 2007), which is an atomistic-detailed scoring function. The main reason behind this result was due to the use of a rigid-body docking method (i.e., FTDock). Atomistic-detailed scoring functions, such as ZRANK, require an accurate model of the interaction to correctly rank the poses, which implies a flexible docking, while coarse-grained potentials, such as "E_{pair}" and "E_{S3DC}", are less affected by the quality of the model. Recently, Moal et al. (2013) presented an evaluation of 115 different scoring functions for ranking docking poses. Interestingly, "MixRank" and "E_{S3DC}" performed among the best 40 approaches in the analysis of docking decoys generated from the protein–protein docking benchmark version 4.0 (Hwang et al., 2010) with a flexible-docking approach (Moal & Bates, 2010). Still, the best results were obtained by the newest score versions of ZRANK2 (Pierce & Weng, 2008), SIPPER (Pons et al., 2011), and other atomistic potentials.

4.2. Protein–nucleic acid docking

While the field of protein–protein docking is advancing fast, the progress of docking nucleic acids onto proteins lags behind. The flexibility of nucleic acids, and the difficulty to recognize their interaction surface, has limited the number of docking studies involving proteins and nucleic acids (DNA (Knegtel, Antoon, Rullmann, Boelens, & Kaptein, 1994; Poulain, Saladin, Hartmann, & Prévost, 2008; Parisien, Freed, & Sosnick, 2012; van Dijk & Bonvin, 2010; van Dijk, Visscher, Kastritis, & Bonvin, 2013) and RNA (Pérez-Cano et al., 2010)). Similarly, there are only few knowledge-based potentials specifically intended to rank protein–nucleic acid docking solutions.

Regarding the field of protein–DNA docking, Robertson and Varani (2007) and Xu et al. (2009) designed two different all-atom statistical potentials that showed similar results in identifying near-native structures from a set of decoys generated with FTDock. Nevertheless, as shown in the previous section, atomistic-detailed potentials require more accurate conformations to correctly rank docking poses, while residue–residue potentials are coarse-grained and less sensitive to small conformational changes, which allows them to capture the dynamic nature of protein–DNA interactions more accurately (Poulain et al., 2008). In this context, Takeda et al. (2013) derived a residue-pair potential that accommodated the interaction angles between amino acids and nucleotides. Their approach also showed better performance than atomistic potentials in rigid-body docking between protein and DNA.

With respect to protein–RNA docking, Zheng et al. (2007) adapted the statistical potential for scoring protein–DNA interactions (Robertson & Varani, 2007) to protein–RNA interactions. Their potential performed similar to the more complex scoring function for protein–RNA interactions of ROSETTA (Chen, Kortemme, Robertson, Baker, & Varani, 2004). In addition, Tuszynska and Bujnicki (2011) built two statistical potentials dependent on the interaction distance and angles of the contact site of the nucleotide with the amino acids of the protein that penalized for spherical clashes occurring during docking.

5. PREDICTION OF PROTEIN-BINDING REGIONS

One of the major challenges to understand protein interactions is the identification of the specific binding regions (i.e., interfaces). In the previous section, we have seen that docking methods try to find the best possible fitting between two or more molecules by exploring the whole rotational and translational 3D space. Therefore, these methods benefit from the knowledge about the interacting interfaces, which saves computational time and eliminates many potentially wrong solutions. In particular, for protein–DNA and protein–RNA interactions, the problem is two-sided: at the side of the protein and at the side of the nucleic acid. In this section, we will focus on the interface at the side of the protein, either for the interaction with other proteins or for the interaction with nucleic acids. Several approaches have been developed for the prediction of protein-binding regions, but in the case of protein–DNA/RNA binding, the problem has been associated to whether the protein will interact with the nucleic acid or not. In this section, we will split both problems: first on the prediction of binding sites, and second, on the prediction of proteins that bind nucleic acids.

5.1. Identification of protein interfaces

The most straightforward methods to experimentally define the interacting region of a protein are based on the determination of its 3D structure (i.e., X-ray crystallography and NMR spectroscopy). Other experimental approaches such as deletion experiments, alanine-scanning mutations, yeast-two hybrid or protein footprinting can be used to determine which domains are involved in the interaction without the requirement of structure (reviewed in Garcia-Garcia, Bonet, et al., 2012). Alternatively, computational tools provide a significant advantage in terms of time- and cost-effectiveness. We have split these computational tools according to their input requirements into sequence-based and structure-based methods.

5.1.1 Methods based on sequence

It is known that protein interfaces share specific features that distinguish them from the rest of the protein (e.g., there is higher conservation of residues in interface regions due to evolutionary constraints; Valdar & Thornton, 2001). In addition, the physicochemical properties of protein–protein interaction interfaces have shown to bear specific properties due to different amino acid composition propensities (Jones & Thornton, 1997). Moreover, as the conservation of residues is strongly dependent on their structural and functional importance, the degree of conservation has been used not only to predict binding sites but also to infer functional annotation. This is the case of Consurf (Ashkenazy, Erez, Martz, Pupko, & Ben-Tal, 2010), a method that estimates the evolutionary rate of each protein residue derived from multiple sequence alignments using an empirical Bayesian or a maximum likelihood approach. Another method, FINDSITE (Brylinski & Skolnick, 2008), uses a different strategy based on binding-site similarity among superimposed groups of template structures identified by threading, which allows for the analysis of groups with low similarity. The combination of FINDSITE with databases such as DrugBank (Knox et al., 2011) and ChEMBL (Gaulton et al., 2011) has been useful in high-throughput virtual ligand screening (Zhou & Skolnick, 2013). A recent method, PIPE-Sites (Amos-Binks et al., 2011), exploits protein–protein interaction networks to detect reoccurring polypeptide sequences in order to infer specific binding sites. Finally, PSIFR (Pandit et al., 2010) combines different methodologies in a single server, including structure-based prediction tools such as TASSER (Zhang & Skolnick, 2004) and functional inference tools such as FINDSITE, among others.

The observed amino acid conservation in protein–DNA interfaces (Luscombe & Thornton, 2002) has also been exploited by many authors to predict nucleic acid-binding residues of a protein with different machine learning approaches. For example, BindN (Wang & Brown, 2006) predicts DNA- and RNA-binding residues using a support vector machine approach based on biochemical features of nucleic acid-binding amino acids, such as side chain pK_a value, hydrophobicity index, and molecular mass. An evolution of the previous method, BindN+ (Wang, Huang, Yang, & Yang, 2010), incorporates evolutionary information as well. Similarly, DP-Bind (Hwang, Gou, & Kuznetsov, 2007) relies on support vector machine, kernel logistic regression, and penalized logistic regression based on amino acid composition and evolutionary profiles. Another approach, NAPS (Carson, Langlois, & Lu, 2010), combines a decision tree algorithm with bootstrap aggregation and cost-sensitive learning. Finally, metaDBSite (Si, Zhang, Lin, Schroeder, & Huang, 2011) predicts DNA-binding residues by integrating the prediction of six different methods (including BindN and DP-Bind).

5.1.2 Methods based on structure

Methods based on structure use features extracted from known 3D interfaces to predict protein-binding regions. In particular, Fernandez-Recio et al. (2005) used the Optimal Docking Area (ODA) of a protein based on atomic solvation parameters. This method looks for favorable energy changes when the residues involved in the interface become buried upon binding. In addition, a few methods for predicting protein–DNA/RNA-binding regions are based on structure too. For instance, DISPLAR (Tjong & Zhou, 2007) uses neural networks trained on known structures of protein–DNA interactions to predict the residues that contact DNA. The inputs to the neural network include position-specific sequence profiles and solvent accessibilities of each residue and its spatial neighbors. DNABINDPROT (Ozbek, Soner, Erman, & Haliloglu, 2010) exploits Gaussian network models to predict DNA-binding residues, based on the fluctuations of residues in high-frequency modes. In DR_bind (Chen, Wright, & Lim, 2012), the identification of DNA-binding residues is based on electrostatics, sequence conservation, and structural geometry. Regarding the prediction of RNA-binding sites, an evolution of ODA, Optimal Protein-RNA Area (OPRA) (Pérez-Cano & Fernández-Recio, 2010), uses statistical potentials derived from the differential propensities of amino acids at protein–RNA interfaces, weighed by its accessible surface area, to predict RNA-binding regions in proteins. Furthermore, OPRA was used in protein–RNA docking and successfully selected near-native conformations of protein–RNA interactions by simply using the correct prediction of the protein residues involved in the interaction (Pérez-Cano et al., 2010).

5.1.3 Application of split-statistical potentials to predict protein-binding sites

The specific properties exhibited by protein interfaces are present in the split-statistical potentials derived from known interacting domains (Feliu et al., 2011). In fact, the statistical potential "E_{local}" is based on the probability of an amino acid to be in a certain environment, as defined by its hydrophobicity, degree of exposure, and secondary structure (see Section 2.1). In order to show the ability of split-statistical potentials in identifying protein interfaces, we have tested both ODA and the potential "E_{local}" on the unbound structures retrieved from the protein docking benchmark version 3.0 (Hwang et al., 2008). The ODA predictions were obtained using the pyDock software (Cheng, Blundell, & Fernandez-Recio, 2007). In the case of "E_{local}", the prediction of the binding site (i.e., "BS-E_{local}") was obtained by scoring and ranking into a list each residue in the protein surface. The score of a residue in the surface was calculated by averaging the Z-scores of "E_{local}" of the residues within a radius of 15 Å, as defined by the distances between their Cβ atoms (Cα for glycines). Then, binding regions were defined iteratively, starting from the top ranked residue in the list. The first binding site was defined by the surface residues within a radius of 15 Å around the top ranked residue. Residues belonging to a binding site were removed from the list and the iteration was repeated until the next residue in the list had a negative score or there were no more residues left. The score of a binding region was defined as the sum of scores of its residues.

In Fig. 4.1, we show the performance of ODA, "BS-E_{local}", and their combination (i.e., residues predicted by both methods to be in the binding site), in terms of percentage of proteins with a minimum positive predictive value (PPV) of the predicted residues to be involved in the real binding site (see details in the legend). Results were compared with a background distribution of random predictions with similar distribution of binding sites

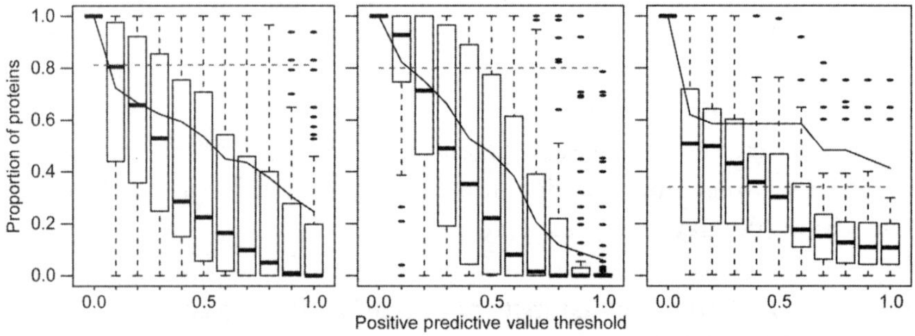

Figure 4.1 Coverage of the prediction of binding sites versus its minimum PPV. The *Y* axes show the ratio of proteins with a PPV equal or greater than a threshold (*X* axes). We have used ODA (Fernandez-Recio et al., 2005) with a minimum pyDockODA score of −10 (A), the prediction based on "BS-E_{local}" with a minimum score of 2 (B), and the binding sites predicted by both (C). The testing dataset contains 85 nonredundant proteins extracted from the docking benchmark 3.0 for which we know the real binding region. PPV is defined as the proportion of correctly predicted residues for each protein over the total number of predicted residues. The binding interface for a protein is defined as the set of residues found to be closer than 12 Å with any other interacting protein reported in the PDB database (Berman et al., 2000), which includes the interacting partner in the benchmark. In order to validate the quality of the prediction, we have calculated the background distribution of obtaining the same PPV thresholds by a random selection of the same number of residues as the actual prediction with ODA (A), "BS-E_{local}" (B), or both (C). The background distribution is shown in boxplots, and it is calculated using sliding windows of the size of each fragment of predicted residues. This definition allows us to compare predictions with similar topology. A horizontal dashed line indicates the applicability of each method (i.e., proportion of proteins with at least one residue predicted in the binding site). (For color version of this figure, the reader is referred to the online version of this chapter.)

along the sequence (i.e., preserved the topology) as the predictions produced by each method. On the one hand, it is noteworthy that the combination of ODA and "BS-E_{local}" yielded predictions that reached PPVs higher than 75% for about 40% of proteins of the benchmark. On the other hand, each method could be applied to more than 80% of proteins of the benchmark, but they achieved PPVs higher than 75% for less than 40% of the proteins. Besides, the individual performances of ODA and "BS-E_{local}" were not strikingly different from random predictions, while the combination of both methods differed considerably from the distribution of topologically similar predictions (i.e., random predictions), thus being more significant.

5.2. Prediction of DNA/RNA-binding proteins

DNA- and RNA-binding proteins can be discriminated from others just from their amino acid sequences using different features, such as amino acid composition (Ahmad, Gromiha, & Sarai, 2004; Yu, Cao, Cai, Shi, & Li, 2006) or evolutionary profiles (Kumar, Gromiha, & Raghava, 2007, 2011; Nimrod, Schushan, Szilágyi, Leslie, & Ben-Tal, 2010). Also, the ability of a protein to bind nucleic acids can be predicted using statistical potentials. For example, DBD-Hunter (Gao & Skolnick, 2008) is a Web server for predicting DNA-binding proteins that combine structural comparisons and evaluation with statistical potentials. Briefly, it scans a given protein structure against a template library composed of 179 protein–DNA complex structures using a structural alignment program (Zhang & Skolnick, 2005). All templates that produce a good structural alignment with the query protein are then evaluated with statistical potentials. Specifically, the statistical potential is applied to score all protein–DNA contacts within a distance of 4.5 Å. The potential also considers whether the contact occurs through the phosphate, sugar, pyrimidine, or imidazole groups. This approach performed better than classical sequence homology-based approaches (i.e., PSI-BLAST; Altschul et al., 1997). An improved version of the previous method, DBD-Threader (Gao & Skolnick, 2009), has the advantage that it only requires the sequence of a protein as input. The sequence is then threaded against the previous template library and, for the best solutions, the interaction score between the threaded sequence and the template DNA is calculated. The exact same procedure of DBD-Hunter, but using an all-atom statistical potential (Xu et al., 2009), has also been applied to predict DNA- (Zhao et al., 2010) as well as RNA-binding proteins (Zhao et al., 2011).

6. CHARACTERIZATION OF TRANSCRIPTION FACTOR-BINDING SITES

In the previous section, we have focused on identifying the binding regions of proteins. However, in protein–DNA/RNA interactions, the nucleic acid also contains specific regions that are recognized by the protein. In particular, transcription factors (TFs) can promote or restrain gene transcription by binding to specific nucleotide sequences (i.e., binding sites) distributed along the genome. Binding sites are often represented with a position weight matrix (PWM) reflecting the observed degeneracy among the recognition sites of TFs. PWMs have been exploited by many methods to search for novel targets of TFs (reviewed in Bulyk, 2003). Therefore, the identification of TF-binding sites is an important step towards the understanding of many biological processes. During the past years, several experimental methods have emerged with the objective to characterize TF-binding sites (reviewed in Xie, Hu, Qian, Blackshaw, & Zhu, 2011). Nevertheless, their application is laborious and expensive and, as a result, they have only been applied to a small fraction of human proteins (Hu et al., 2009). As an alternative, computational tools can be employed to predict TF-binding sites. A well-established procedure consists in searching for over-represented DNA sequences in the promoter regions of genes regulated by a TF with a motif discovery algorithm (analyzed in Das & Dai, 2007), but the success of these approaches depends on the availability of enough sequences for pattern discovery, mainly derived from ChIP-seq, ChIP-exo, and protein-binding microarrays (Grau, Posch, Grosse, & Keilwagen, 2013).

Another successful strategy currently employed is the analysis of TF–DNA complex structures with statistical potentials. Briefly, the TF is put face to face with different DNA sequences and the binding energies of the resulting complexes are analyzed. Those sequences with the best energy are considered to be bound by the TF and are incorporated into a PWM. For example, Angarica et al. (2008) created an algorithm that, given a TF–DNA complex, mutated all nucleotide positions one by one using the 3DNA package (Lu & Olson, 2008), until all possibilities were covered (i.e., A, C, G, and T). The mutated sequences were then scored with a knowledge-based potential and the 50 best oligonucleotides were used to construct a PWM. In another work, Liu et al. (2008) developed a method based on protein–DNA docking coupled with threading of DNA sequences.

They were able to predict 50% of experimentally determined sites for the cAMP regulatory protein (CRP) in the top 1% among all 639,232 possible solutions. They also made a *de novo* prediction by modeling the ferric uptake regulator in complex with DNA, which showed similar results as CRP. Later on, Xu et al. (2009) calculated the PWMs for different TFs by decomposing the binding energies of the FIRE potential into individual contributions of each base. The FIRE potential was first described by Zhou and Zhou (2002), and it was used mostly on homology modeling. Afterwards, FIRE was readjusted so that it could be applied to predict protein–protein and protein–DNA interactions (Zhang, Liu, Zhu, & Zhou, 2005). More recently, Alamanova et al. (2010) used an all-atom statistical potential (Robertson & Varani, 2007) in combination with the MMTSB tool set (Feig et al., 2004) to recover the PWMs of various members from two widely studied families of TFs such as p53 and NF-B. In particular, they were able to create very accurate PWMs for p53 tetramer and p50 dimer as well as for the p50p65 and p50RelB heterodimers. They also obtained very good results with p63 and p73 dimers built by homology modeling using the p53 DNA-binding domain as template. Finally, Chen, Chien, et al. (2012) established a procedure to predict PWM when no protein–DNA complex is available. They superimposed the unbound structure of a TF over the closest homolog TF structure in complex with DNA. Then, the PWM was estimated as in the work of Xu et al. (2009).

Although knowledge-based potentials have been a good alternative to infer TF-binding sites, their application still has some limitations. One of them is the lack of templates due to the small number of TF–DNA complex structures available in the PDB. To avoid any bias, statistical potentials are usually derived from a nonredundant dataset of structures. This redundancy is generally removed on the TF side of the complex. Yet, TFs can recognize different binding sites, and in addition, members of the same family of TFs can bind to distinct DNA sequences (Luscombe & Thornton, 2002). For this reason, the removal of redundancy can generate statistical potentials suffering from low-count and at the same time low diversity of binding patterns. Another problem arises because statistical potentials are applied under the assumption that the contribution of the different DNA base pairs to the binding energy of the complex is independent from each other, which is not true (Benos, Bulyk, & Stormo, 2002). Recently, AlQuraishi and McAdams (2013) addressed the coverage problem by combining TF–DNA structures with experimentally determined PWMs. The inclusion of PWM data adapted the statistical potential to the varying binding preferences of

TFs for different binding sites. Still, they highlighted that the use of PWMs cannot allocate for interposition dependencies among base pairs.

6.1. Application of knowledge-based potentials on DREAM5 targets

In order to evaluate the real capability of statistical potentials in PWM prediction, we have tested two available online methods, 3DTF (Gabdoulline, Eckweiler, Kel, & Stegmaier, 2012) and PiDNA (Lin & Chen, 2013), on 83 mouse TFs from the DREAM5 TF–DNA Motif Recognition Challenge (Weirauch et al., 2013). These two servers only require a TF–DNA complex structure in PDB format as input and return the predicted PWM of the TF as output.

6.1.1 Modeling TF–DNA complexes

Since there were no TF–DNA complex structures available in the PDB for the majority of the DREAM5 targets, we developed a novel modeling protocol that allowed us to obtain a TF–DNA model for a total of 71 DREAM5 targets. An overview of the procedure is shown in Fig. 4.2. In step 1, for each TF target, we searched the best template in a database of TF–DNA complexes using BLAST (Altschul et al., 1997). The database was obtained by selecting from the PDB all TF–DNA complex structures annotated in the TFinDit depository (Turner, Kim, & Guo, 2012) that, according to 3DNA (Lu & Olson, 2008), contained a double-stranded DNA of at least eight base pairs. Then, we identified all dimers in the database by grouping any two protein chains from the same PDB that (1) had at least one common contact with the DNA and (2) had more than five residue–residue contacts between them as to form a binary complex (Mosca, Céol, & Aloy, 2013). In step 2, BLAST hits were filtered according to two criteria: (1) enough percentage of sequence identity and (2) no gaps or insertions in the region of the interface. With respect to the percentage of identity, based on a recent work where we observed that TFs sharing little sequence identity can still bind to the same genes (Gitter et al., 2009), we included distantly related sequences according to Rost's sequence identity curve (Rost, 1999), using parameters adjusted to ensure a 99% precision rate (i.e., $n=5$). In step 3, the template sequence that passed the filter and had the best BLAST *e*-value was realigned with the TF using matcher, from the EMBOSS package (Rice, Longden, & Bleasby, 2000). In step 4, the alignment was used to create a structural model of the TF with MODELLER (Eswar et al., 2006). Models were created applying 3D restraints between Cα atoms. This is the pairwise distance from the Cα atom of each residue to the Cα atoms of any residues within a radius

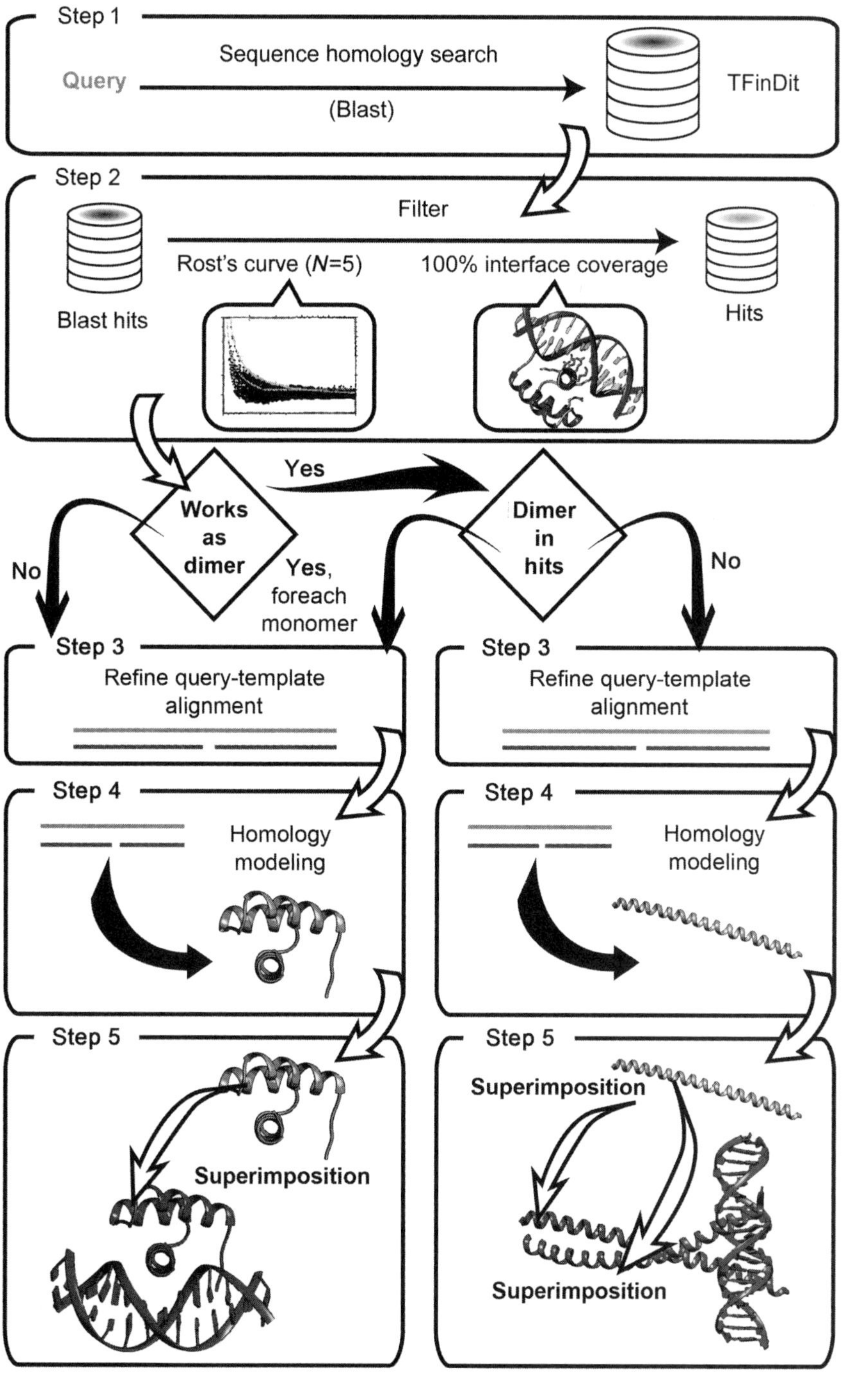

Figure 4.2 Pipeline for modeling transcription factors. Step 1: sequence homology search. Step 2: filter results of step 1 by sequence identity and coverage of the protein–DNA interface. Step 3: optimization of the alignment. Step 4: model-building of the three-dimensional

(Continued)

of 15 Å conserved between the template and the model. In step 5, the final TF–DNA complex was obtained by superimposition of the protein model on the template using PyMOL (Schrödinger, 2010). Additionally, for TFs from the bHLH and bZIP families, since they recognize DNA as homo- or heterodimers, we modeled the dimer: if the two monomers were found among the unfiltered hits in step 2, the dimer was obtained as before (i.e., steps 3–5), but using both template hits; otherwise, if only one monomer could be modeled, it was superimposed as in step 5 on both template chains of the dimer. Table 4.3 shows the 71 DREAM5 targets that could be modeled following this procedure, the percentage of sequence identity, and coverage of the pairwise alignments between the TFs and their templates, and the resulting RMSD of the superimposition.

6.1.2 Analysis of PWM predictions

A first analysis revealed that PiDNA is very sensitive to the format of input files. It occasionally failed even when, for all models we had produced, the DNA molecule had at least eight base pairs in the correct format (according to 3DNA). In contrast, 3DTF could interpret all except one model, but it produced uniform PWMs for 46 TFs (i.e., PWMs with null capacity of discrimination). As a result of this analysis, the applicability of PiDNA (28/71) was slightly better than 3DTF (24/71). Furthermore, out of the 13 different families of TFs taken from the DREAM5 challenge that could be modeled, 3DTF and PiDNA could only make predictions for seven of them. Table 4.3 shows the quality of the predictions by means of comparing the PWMs produced by 3DTF and PiDNA with the real PWMs, using Tomtom (Gupta, Stamatoyannopoulos, Bailey, & Noble, 2007), as distributed in the MEME package (Bailey et al., 2009). Tomtom calculates the similarity between a pair of PWMs by means of a *P*-value. Using a *P*-value threshold of 10^{-3}, PiDNA predicted correctly the PWM for 10 TFs, while 3DTF for 7 (five of which were common to both of them). Besides, we observed that PWM predictions deteriorated together with the alignment between the target and the template. One possible reason is that both 3DTF and PiDNA

Figure 4.2—Cont'd structure of the TF. Step 5: superimposition of the model over the template. If the TF works as a homodimer and only one monomer can be modeled using a heterodimer as template, the model is superimposed on each chain of the template to construct the homodimer. Structural images were created with the UCSF Chimera package (Fraenkel & Pabo, 1998; Glover & Harrison, 1995; Pettersen et al., 2004). (For color version of this figure, the reader is referred to the online version of this chapter.)

Table 4.3 PWM predictions for targets of the DREAM5 challenge

TF	Family	PDB	Chain	%ID	%Cov	RMSD	3DTF	PiDNA	E_{S3DC}
Egr2	C2H2 ZF	1p47	A	94	100	0.10			1.8×10^{-2}
Esr1	NR	1hcq	A	100	98	0.02			2.1×10^{-4}
Esrrb	NR	3dzy	A	36	99	1.08	-	1.5×10^{-3}	3.2×10^{-3}
Esrrg	NR	3dzy	A	36	99	0.34	1.7×10^{-3}	1.4×10^{-4}	-
Foxc2	Forkhead	1vtn	C	72	96	0.01	-	8.4×10^{-3}	-
Foxo1	Forkhead	3co6	C	100	100	1.59		1.4×10^{-4}	1.9×10^{-4}
Foxo3	Forkhead	2uzk	A	100	98	0.04	-	7.7×10^{-6}	6.6×10^{-3}
Foxo4	Forkhead	2uzk	A	83	88	0.04	-	4.5×10^{-4}	4.9×10^{-3}
Foxo6	Forkhead	3co6	C	91	100	1.55		3.9×10^{-3}	3.1×10^{-3}
Gata4	GATA	4hc7	A	86	97	1.11	-	2.9×10^{-3}	-
Hmga2	AT hook	2eze	A	80	80	0.35			8.2×10^{-4}
Klf12	C2H2 ZF	2wbu	A	78	97	0.06	3.4×10^{-4}	1.1×10^{-3}	-
Klf8	C2H2 ZF	2wbu	A	75	97	0.09	7.7×10^{-5}	1.1×10^{-4}	-
Nr2e1	NR	3e00	A	57	23	0.01	1.8×10^{-4}	-	-
Nr2f1	NR	3dzy	A	45	99	2.33	-	3.0×10^{-3}	-
Nr2f6	NR	3e00	A	39	54	4.61	-	3.3×10^{-7}	-

Continued

Table 4.3 PWM predictions for targets of the DREAM5 challenge—cont'd

TF	Family	PDB	Chain	%ID	%Cov	RMSD	3DTF	PiDNA	E_{S3DC}
Pou3f1	Hom	2xsd	C	100	100	0.07		-	1.9×10^{-3}
Sox6	Sox	3f27	D	56	98	0.04	-	3.9×10^{-3}	-
Sp1	C2H2 ZF	2wbu	A	57	97	0.08	-	4.8×10^{-3}	2.6×10^{-4}
Tbx1	T-box	4a04	B	100	98	0.02	9.3×10^{-7}	9.6×10^{-6}	4.5×10^{-3}
Tbx20	T-box	4a04	A	66	99	0.02	7.7×10^{-6}	2.2×10^{-5}	7.7×10^{-4}
Tbx4	T-box	2x6v	A	93	99	0.01	8.2×10^{-7}	1.3×10^{-6}	-
Tbx5	T-box	2x6v	A	100	100	0.01	1.1×10^{-6}	2.2×10^{-6}	-
Tcf3	bHLH	2ql2	C	100	100	0.09	2.2×10^{-3}	-	-
			D	*	*	3.12			
Tcfec	bHLH	4ati	B	88	100	0.18		1.5×10^{-3}	-
			A	85	100	0.10			
Zfp202	C2H2 ZF	2i13	A	51	100	0.38			1.1×10^{-2}

Transcription factors (TF) from the DREAM5 challenge and their families are shown in the first columns. Families "NR" and "Hom" stand for "nuclear receptor" and "homeodomain", respectively. PDB codes and chains of the templates used to model the TFs are shown in the next columns. This is followed by the quality of the model shown by means of the percentage of sequence identity (%ID) and template coverage (%Cov) of the sequence alignment, and the RMSD of the superimposition. For dimers, the information regarding each monomer can be found in separate lines. Asterisks indicate that the homodimer was built by superimposing the model of one chain to both chains of the template heterodimer. The significance of similarity between the predicted and the real PWMs is shown with the *P*-value for 3DTF and PiDNA, and the statistical potential "E_{S3DC}". A hyphen indicates that the *P*-value is not significant and the cell is left empty when the method failed to produce a PWM.
Note: Only TFs with significant predictions are shown.

rely on all-atom statistical potentials and they are sensitive to the wrong orientation of amino acid side chains that could occur upon modeling.

7. ADAPTING SPLIT-STATISTICAL POTENTIALS FOR PROTEIN–DNA INTERACTIONS

As shown in Section 6.1, much improvement is required in the area of TF-binding site prediction based on structure (i.e., via statistical potentials). In this section, we propose a series of changes to the previously described split-statistical potentials for protein folding (Aloy & Oliva, 2009) and protein–protein interactions (Feliu et al., 2011) in order to adapt them to protein–DNA interactions.

The application of split-statistical potentials to protein–DNA interactions requires the definition of an environment for nucleotides. Moreover, in order to address the additivity problem (Benos et al., 2002), we have described statistical potentials for dinucleotides (i.e., two consecutive nucleotides along the DNA sequence). Therefore, the DNA environment of a dinucleotide is defined by its constituting bases (i.e., any combination of two purines and pyrimidines) and three features regarding the interaction between the amino acid and the dinucleotide: (1) the strand (i.e., forward or reverse) that is closer to the amino acid; (2) the DNA groove (i.e., major or minor) where the amino acid is located (or close to); and (3) the closest chemical group of the dinucleotide (i.e., nucleobase or deoxyribose phosphate) to the amino acid (see Fig. 4.4 and Section 7.1.1 for more details).

The definition of environments yields several residue–environment combinations. For amino acids, we consider 20 residues and 6 different environments as before (i.e., helix, coil, or strand, and being buried or exposed). This produces a total of 120 combinations of amino acids and environments. In contrast, we consider 16 dinucleotides (i.e., 4^2 different combinations of two nucleotides) and 8 environments: 2 for the closest strand, 2 for the closest DNA groove, and 2 for the closest chemical group of the dinucleotide. These definitions produce a total of 128 dinucleotide–environment combinations.

Given a particular interaction between an amino acid "*a*" and a dinucleotide "*mn*" (where "*m*" and "*n*" can be any nucleotide), we define the statistical potentials "E_{pair}", "E_{local}", "E_{3D}", "E_{3DC}", and "E_{S3DC}" as in Section 2.1 by replacing "*b*" with "*mn*" in Eqs. (4.2) and (4.3). The contribution of the reference state and the "E_{3D}" potential are ignored, but also the contributions of the "E_{local}" terms. On the one hand, the "E_{local}"

contribution of DNA is not considered because, as long as it is accessible, any nucleotide sequence can be bound by a TF (Urnov, Rebar, Holmes, Zhang, & Gregory, 2010) and, as a result, the environment conditions of the base pairs are not relevant. On the other hand, given a TF, the "E_{local}" term dependent on the protein is always the same when discriminating among different DNA-binding sites and thus, it is irrelevant too. Therefore, we have selected the statistical potential "E_{S3DC}" to evaluate the prediction of DNA-binding sites for the targets of the DREAM5 challenge.

7.1. Application of split-statistical potentials on DREAM5 targets

As a test pilot, we have applied these split-statistical potentials to predict the PWM for the 71 modeled DREAM5 targets in Section 6.1.

7.1.1 Split-statistical potentials for protein–DNA interactions

We derived the potentials from a nonredundant set of templates of the TFinDit repository (Turner et al., 2012) (see Section 6.1.1). Specifically, templates were split into chains and redundancy was removed so that any two chains shared less than 35% of protein–DNA contacts. A contact was defined between an amino acid and a dinucleotide if the Cβ atom of the amino acid (Cα for glycines) was at 15 Å or less from the center of the dinucleotide and its complementary bases (i.e., the geometrical center as defined by the four phosphate atoms of the two nucleotides and its associated partners in the complementary DNA strand; see Fig. 4.3B). In Fig. 4.3, we show how the different details that define the environmental features used on the description of the statistical potential are calculated. We used 3DNA (Lu & Olson, 2008) to define which DNA residues constituted the reference strand (i.e., forward) and which the complementary (i.e., reverse). Moreover, for calculating the potential, we referred to "*mn*" as the pair of nucleotides from the reference strand of the dinucleotide. Also, we used the distances between the Cβ atom of the amino acid and the phosphate atoms of each dinucleotide to decide which of the two DNA strands was the closest (see Fig. 4.3C).

In order to identify the closest DNA groove (see Fig. 4.3A), we adapted a definition of groove widths (El Hassan & Calladine, 1998): First, we selected the closest phosphate from each strand to the Cβ atom of the amino acid; let this be at position "i" for strand S and at position "j" for strand S′. Second, if $i<j$, we calculated the distances between the phosphate atom at position "i" in S and the phosphate atoms at positions "$i+3$" (i.e., D_{i+3}) and "$i+4$" (i.e., D_{i+4}) in S′. Finally, if $D_{i+3}>D_{i+4}$, the amino acid was located in the major groove;

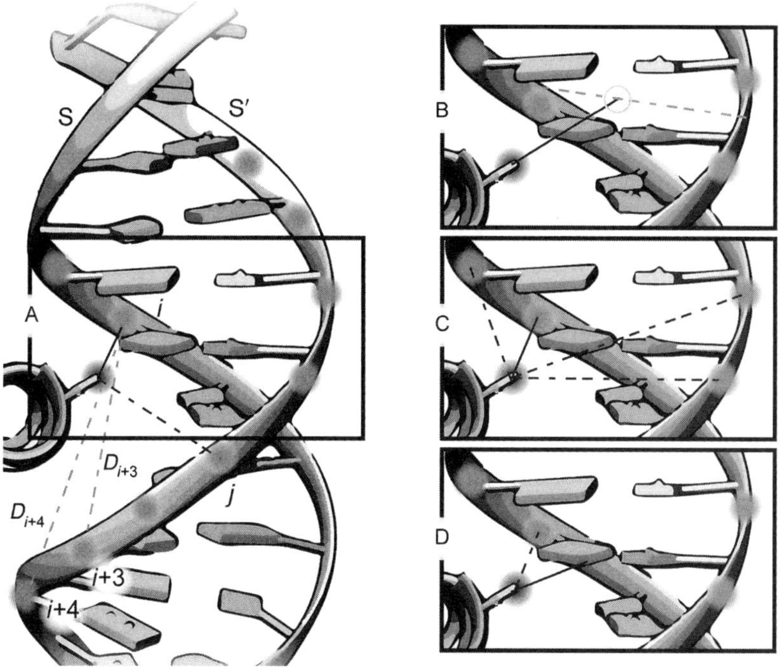

Figure 4.3 Definition of different DNA parameters used for deriving split-statistical potentials. Distances between amino acids and DNA are represented in blue lines; solid when displaying the minimal distance and dashed otherwise. Internal distances in the DNA are shown in orange. Environment features of the DNA for a contact between an amino acid and a dinucleotide at position "*i*" (see details in text for each definition): groove in contact with the amino acid (A); distance between the amino acid and the dinucleotide (B); strand in contact with the amino acid (C); and DNA chemical group in contact with the amino acid (D). Structural images were created with the UCSF Chimera package (Fraenkel & Pabo, 1998; Pettersen et al., 2004). (For interpretation of the references to color in this figure legend, the reader is referred to the online version of this chapter.)

otherwise, it was located in the minor groove. Additionally, if in the second step $i>j$, instead of calculating the distances to positions "$i+3$" and "$i+4$" in S′, we used the distances to "$i-3$" and "$i-4$" in S′ and applied the same criterion to select the DNA groove where the amino acid was located.

The interaction between the amino acid and the DNA could be either with the backbone of the DNA (i.e., any atom of the deoxyribose phosphate) or with the nucleobase (i.e., any atom of the nitrogenous base). This was defined by the minimum distance between the atoms of the amino acid and the atoms of the dinucleotide. If the closest atom of the nucleotide was any atom of the phosphate group or the deoxyribose, the interaction was with the backbone; otherwise, it was with the nucleobase (see Fig. 4.3D).

Finally, to make the potentials independent of the arbitrary designation of forward and reverse strand as defined by 3DNA, we also considered, for each protein–DNA contact, the complementary (i.e., the contact that would have been created if the reference strand was the complementary). For example, the complementary contact of a certain amino acid with two adenosines, through the forward strand, the major groove and the backbone, would be with two thymidines, through the reverse strand, the major groove and the backbone. This increases straight forward the knowledge-base of interactions and improves in a natural way the number of pairs of amino acids and dinucleotides of the structural database.

7.1.2 PWM prediction

The PWM of each TF was calculated by adapting the procedure of Xu et al. (2009) to account for the interaction of amino acids with dinucleotides. We used the scores of the interaction of the protein with a dinucleotide to calculate the probability of a single nucleotide position in the PWM:

$$P(\alpha_i) = \frac{\sum_{m_{i-1}} \exp(-\mathrm{PMF}(a, m_{i-1}\alpha_i)) + \sum_{m_{i+1}} \exp(-\mathrm{PMF}(a, \alpha_i m_{i+1}))}{\sum_{n_i} \left[\sum_{m_{i-1}} \exp(-\mathrm{PMF}(a, m_{i-1} n_i)) + \sum_{m_{i+1}} \exp(-\mathrm{PMF}(a, n_i m_{i+1})) \right]}$$

Where "$P(\alpha_i)$" is the probability of nucleotide "α" at position "i".

The PWMs produced for each modeled TF target were analyzed as in Section 6.1.2 (see Table 4.3). In contrast to 3DTF and PiDNA, we could apply the "E_{S3DC}" score to all modeled DREAM5 targets, which implies that we covered 13 out of 15 families of TFs (six more families than combining both 3DTF and PiDNA). Moreover, we obtained significant results for five different TFs, three of which could not be retrieved with 3DTF nor PiDNA (see Table 4.3). In Fig. 4.4, we compare the logos produced by "E_{S3DC}" with the logos produced by 3DTF and PiDNA for three specific DREAM5 targets (Foxo1, Nr2e1, and Tbx20). As observed in Table 4.3, PiDNA produced significant logos for all the targets, while "E_{S3DC}" and 3DTF predicted significant logos for two TFs: "E_{S3DC}" for Foxo1 and Tbx20, and 3DTF predicted for Nr2e1 and Tbx20. Also, "E_{S3DC}" predicted a logo for Nr2e1 but it was not significant because it failed to predict two out of four nucleotides of the central motif "GTCA" (it could only predict "GT"). The full comparison of "E_{S3DC}" with 3DTF and PiDNA can be

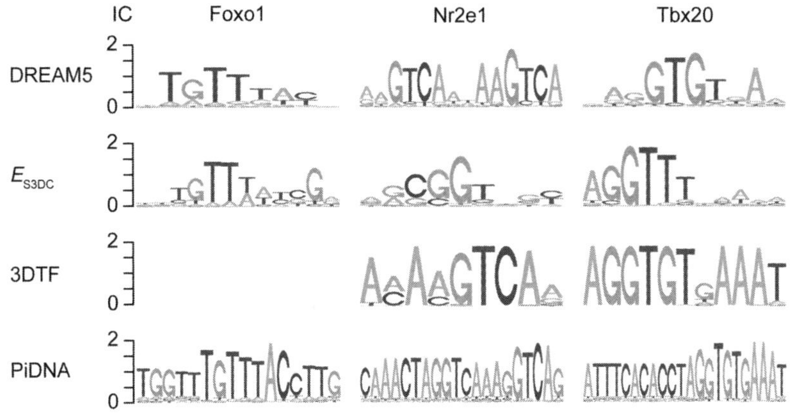

Figure 4.4 Examples of PWM logos. PWM logos for Foxo1, Nr2e1, and Tbx, as described in the DREAM5 challenge compared with the predictions produced by the statistical potential "E_{S3DC}" and the state-of-the-art methods 3DTF and PiDNA. Logos were created with the R software environment (Bembom, 2007; R Core Team, 2013). (For color version of this figure, the reader is referred to the online version of this chapter.)

found in Table 4.3 and it shows a reasonable improvement not only in terms of applicability but also in terms of specificity by means of significant matches of the real PWMs.

8. CONCLUSIONS

We have reviewed the use of knowledge-based potentials as a tool for the analysis of protein–protein, protein–DNA, and protein–RNA interactions. We have explored the general definition of knowledge-based potentials and described the procedure to split them into different energetic terms. We have extensively discussed the application of statistical potentials in (1) the evaluation of protein modeling, including homology and integrative modeling; (2) the evaluation of protein–protein and protein–nucleic acids docking; (3) the prediction of protein-binding regions; and (4) the characterization of TF-binding sites. Finally, we have provided several resources available online for docking, ranking scoring functions of interactions, and benchmark databases.

We have shown that modeling of protein interactions is still limited by the lack of 3D data. Still, this problem can be addressed using docking approaches. We have also shown that docking methods benefit from the prediction of binding sites. In this context, we have proposed two pilots for predicting binding regions, one for proteins and another for DNA, based on split-statistical potentials. On the one hand, we have tested our approach for predicting binding sites in proteins, "BS-E_{local}", and we have compared it with another state-of-art methodology. This test revealed that, while none of the methods yield significant predictions, their combination improve the significance of the binding regions predicted for a protein. We have been able to achieve PPVs higher than 80% for almost one quarter of the proteins tested in our benchmark (for half of them, the methods produced different results, and for half of the proteins for which this was applied, the result was successful). On the other hand, we have proposed a modification of the split-statistical potentials for protein–DNA interactions. We have applied them to predict DNA-binding sites by modeling the structure of the TF and constructing an artificial PWM logo. Our results were comparable to state-of-the-art methods, such as 3DTF and PiDNA; and additionally, we enlarged the application to several targets of the DREAM5 challenge for which 3DTF and PiDNA could not be applied or did not produce significant results.

We have fathomed the main features involved in TF-binging sites, in the modeling of protein–protein and protein–DNA complexes. In conclusion, despite all the advances in the area, there is still a wide-range for improvement in the exploitation of statistical potentials, especially in the field of protein–DNA interactions.

ACKNOWLEDGMENTS

O. F. and B. O. acknowledge the support of FEDER BIO2011-22568 grant from the Spanish Ministry of Science and Innovation (MICINN). J.G.G. acknowledge support by "Departament d'Educació i Universitats de la Generalitat de Catalunya i del Fons Social Europeu" through FI fellowships. J.B. is supported by BIO08-0206 grant from MICINN. We are very grateful to Dr. Jun-tao Guo (UNCC) for providing us a comprehensible list of PDB codes for all transcription factors from the TFinDit depository. We are also thankful to Dr. Fernández Recio (BSC) for providing us the latest version of pyDockOda.

REFERENCES

Ahmad, S., Gromiha, M. M., & Sarai, A. (2004). Analysis and prediction of DNA-binding proteins and their binding residues based on composition, sequence and structural information. *Bioinformatics*, *20*(4), 477–486. http://dx.doi.org/10.1093/bioinformatics/btg432.

Alamanova, D., Stegmaier, P., & Kel, A. (2010). Creating PWMs of transcription factors using 3D structure-based computation of protein-DNA free binding energies. *BMC Bioinformatics*, *11*(1), 225. http://dx.doi.org/10.1186/1471-2105-11-225.

Alber, F., Dokudovskaya, S., Veenhoff, L. M., Zhang, W., Kipper, J., & Devos, D. (2007). Determining the architectures of macromolecular assemblies. *Nature*, *450*(7170), 683–694. http://dx.doi.org/10.1038/nature06404.

Alber, F., Förster, F., Korkin, D., Topf, M., & Sali, A. (2008). Integrating diverse data for structure determination of macromolecular assemblies. *Annual Review of Biochemistry*, *77*(1), 443–477. http://dx.doi.org/10.1146/annurev.biochem.77.060407.135530.

Aloy, P., & Oliva, B. (2009). Splitting statistical potentials into meaningful scoring functions: Testing the prediction of near-native structures from decoy conformations. *BMC Structural Biology*, *9*(1), 71. http://dx.doi.org/10.1186/1472-6807-9-71.

Aloy, P., & Russell, R. B. (2003). InterPreTS: Protein interaction prediction through tertiary structure. *Bioinformatics*, *19*(1), 161–162. http://dx.doi.org/10.1093/bioinformatics/19.1.161.

AlQuraishi, M., & McAdams, H. H. (2013). Three enhancements to the inference of statistical protein-DNA potentials. *Proteins: Structure, Function, and Bioinformatics*, *81*(3), 426–442. http://dx.doi.org/10.1002/prot.24201.

Altschul, S. F., Madden, T. L., Schäffer, A. A., Zhang, J., Zhang, Z., Miller, W., et al. (1997). Gapped BLAST and PSI-BLAST: A new generation of protein database search programs. *Nucleic Acids Research*, *25*(17), 3389–3402. http://dx.doi.org/10.1093/nar/25.17.3389.

Amos-Binks, A., Patulea, C., Pitre, S., Schoenrock, A., Gui, Y., & Green, J. R. (2011). Binding site prediction for protein-protein interactions and novel motif discovery using re-occurring polypeptide sequences. *BMC Bioinformatics*, *12*(1), 225. http://dx.doi.org/10.1186/1471-2105-12-225.

Angarica, V. E., Pérez, A. G., Vasconcelos, A. T., Collado-Vides, J., & Contreras-Moreira, B. (2008). Prediction of TF target sites based on atomistic models of protein-DNA complexes. *BMC Bioinformatics*, *9*(1), 436. http://dx.doi.org/10.1186/1471-2105-9-436.

Ashkenazy, H., Erez, E., Martz, E., Pupko, T., & Ben-Tal, N. (2010). ConSurf 2010: Calculating evolutionary conservation in sequence and structure of proteins and nucleic acids. *Nucleic Acids Research*, *38*(Suppl. 2), W529–W533. http://dx.doi.org/10.1093/nar/gkq399.

Axenopoulos, A., Daras, P., Papadopoulos, G. E., & Houstis, E. N. (2013). SP-dock: Protein-protein docking using shape and physicochemical complementarity. *IEEE/ACM Transactions on Computational Biology and Bioinformatics*, *10*(1), 135–150. http://dx.doi.org/10.1109/TCBB.2012.149.

Bailey, T. L., Boden, M., Buske, F. A., Frith, M., Grant, C. E., & Clementi, L. (2009). MEME Suite: Tools for motif discovery and searching. *Nucleic Acids Research*, *37*(Suppl. 2), W202–W208. http://dx.doi.org/10.1093/nar/gkp335.

Barik, A., Nithin, C., Manasa, P., & Bahadur, R. P. (2012). A protein-RNA docking benchmark (I): Nonredundant cases. *Proteins: Structure, Function, and Bioinformatics*, *80*(7), 1866–1871. http://dx.doi.org/10.1002/prot.24083.

Baù, D., Sanyal, A., Lajoie, B. R., Capriotti, E., Byron, M., & Lawrence, J. B. (2011). The three-dimensional folding of the α-globin gene domain reveals formation of chromatin globules. *Nature Structural & Molecular Biology*, *18*(1), 107–114. http://dx.doi.org/10.1038/nsmb.1936.

Bembom, O. (2007). seqLogo: Sequence logos for DNA sequence alignments.

Benos, P. V., Bulyk, M. L., & Stormo, G. D. (2002). Additivity in protein-DNA interactions: How good an approximation is it? *Nucleic Acids Research*, *30*(20), 4442–4451. http://dx.doi.org/10.1093/nar/gkf578.

Berman, H. M., Westbrook, J., Feng, Z., Gilliland, G., Bhat, T. N., & Weissig, H. (2000). The Protein Data Bank. *Nucleic Acids Research*, *28*(1), 235–242. http://dx.doi.org/10.1093/nar/28.1.235.

Brooks, B. R., Bruccoleri, R. E., Olafson, B. D., States, D. J., Swaminathan, S., & Karplus, M. (1983). CHARMM: A program for macromolecular energy, minimization, and dynamics calculations. *Journal of Computational Chemistry*, *4*(2), 187–217. http://dx.doi.org/10.1002/jcc.540040211.

Brylinski, M., & Skolnick, J. (2008). A threading-based method (FINDSITE) for ligand-binding site prediction and functional annotation. *Proceedings of the National Academy of Sciences of the United States of America*, *105*(1), 129–134. http://dx.doi.org/10.1073/pnas.0707684105.

Bulyk, M. L. (2003). Computational prediction of transcription-factor binding site locations. *Genome Biology*, *5*(1), 201. http://dx.doi.org/10.1186/gb-2003-5-1-201.

Carson, M. B., Langlois, R., & Lu, H. (2010). NAPS: A residue-level nucleic acid-binding prediction server. *Nucleic Acids Research*, *38*(Suppl. 2), W431–W435. http://dx.doi.org/10.1093/nar/gkq361.

Chen, C.-Y., Chien, T.-Y., Lin, C.-K., Lin, C.-W., Weng, Y.-Z., & Chang, D. (2012). Predicting target DNA sequences of DNA-binding proteins based on unbound structures. *PLoS One*, 7(2), e30446. http://dx.doi.org/10.1371/journal.pone.0030446.

Chen, Y., Kortemme, T., Robertson, T., Baker, D., & Varani, G. (2004). A new hydrogen-bonding potential for the design of protein-RNA interactions predicts specific contacts and discriminates decoys. *Nucleic Acids Research*, *32*(17), 5147–5162. http://dx.doi.org/10.1093/nar/gkh785.

Chen, H., & Skolnick, J. (2008). M-TASSER: An algorithm for protein quaternary structure prediction. *Biophysical Journal*, *94*(3), 918–928. http://dx.doi.org/10.1529/biophysj.107.114280.

Chen, Y. C., Wright, J. D., & Lim, C. (2012). DR_bind: A web server for predicting DNA-binding residues from the protein structure based on electrostatics, evolution and geometry. *Nucleic Acids Research*, *40*(W1), W249–W256. http://dx.doi.org/10.1093/nar/gks481.

Cheng, T. M.-K., Blundell, T. L., & Fernandez-Recio, J. (2007). pyDock: Electrostatics and desolvation for effective scoring of rigid-body protein-protein docking. *Proteins: Structure, Function, and Bioinformatics*, *68*(2), 503–515. http://dx.doi.org/10.1002/prot.21419.

Comeau, S. R., Gatchell, D. W., Vajda, S., & Camacho, C. J. (2004a). ClusPro: A fully automated algorithm for protein-protein docking. *Nucleic Acids Research*, *32*(Suppl. 2), W96–W99. http://dx.doi.org/10.1093/nar/gkh354.

Comeau, S. R., Gatchell, D. W., Vajda, S., & Camacho, C. J. (2004b). ClusPro: An automated docking and discrimination method for the prediction of protein complexes. *Bioinformatics*, *20*(1), 45–50. http://dx.doi.org/10.1093/bioinformatics/btg371.

Cornell, W. D., Cieplak, P., Bayly, C. I., Gould, I. R., Merz, K. M., & Ferguson, D. M. (1995). A second generation force field for the simulation of proteins, nucleic acids, and organic molecules. *Journal of the American Chemical Society*, *117*(19), 5179–5197. http://dx.doi.org/10.1021/ja00124a002.

Das, M. K., & Dai, H.-K. (2007). A survey of DNA motif finding algorithms. *BMC Bioinformatics*, *8*(Suppl. 7), S21. http://dx.doi.org/10.1186/1471-2105-8-S7-S21.

De Vries, S. J., van Dijk, M., & Bonvin, A. M. J. J. (2010). The HADDOCK web server for data-driven biomolecular docking. *Nature Protocols*, *5*(5), 883–897. http://dx.doi.org/10.1038/nprot.2010.32.

Dobbins, S. E., Lesk, V. I., & Sternberg, M. J. E. (2008). Insights into protein flexibility: The relationship between normal modes and conformational change upon protein-protein docking. *Proceedings of the National Academy of Sciences of the United States of America*, *105*(30), 10390–10395. http://dx.doi.org/10.1073/pnas.0802496105.

Dominguez, C., Boelens, R., & Bonvin, A. M. J. J. (2003). HADDOCK: A protein-protein docking approach based on biochemical or biophysical information. *Journal of the American Chemical Society*, *125*(7), 1731–1737. http://dx.doi.org/10.1021/ja026939x.

Dunbrack, R. L., Jr. (2006). Sequence comparison and protein structure prediction. *Current Opinion in Structural Biology*, *16*(3), 374–384. http://dx.doi.org/10.1016/j.sbi.2006.05.006.

El Hassan, M., & Calladine, C. (1998). Two distinct modes of protein-induced bending in DNA. *Journal of Molecular Biology*, *282*(2), 331–343. http://dx.doi.org/10.1006/jmbi.1998.1994.

Eswar, N., Webb, B., Marti-Renom, M. A., Madhusudhan, M., Eramian, D., Shen, M.-Y., et al. (2006). Comparative Protein Structure Modeling Using Modeller. *Current Protocols in Bioinformatics, 15*, 5.6.1–5.6.30.

Feig, M., Karanicolas, J., & Brooks, C. L., III (2004). MMTSB Tool Set: Enhanced sampling and multiscale modeling methods for applications in structural biology. *Journal of Molecular Graphics and Modelling, 22*(5), 377–395. http://dx.doi.org/10.1016/j.jmgm.2003.12.005.

Feliu, E., Aloy, P., & Oliva, B. (2011). On the analysis of protein-protein interactions via knowledge-based potentials for the prediction of protein-protein docking. *Protein Science, 20*(3), 529–541. http://dx.doi.org/10.1002/pro.585.

Feliu, E., & Oliva, B. (2010). How different from random are docking predictions when ranked by scoring functions? *Proteins: Structure, Function, and Bioinformatics, 78*(16), 3376–3385. http://dx.doi.org/10.1002/prot.22844.

Fernandez-Recio, J., Totrov, M., Skorodumov, C., & Abagyan, R. (2005). Optimal docking area: A new method for predicting protein-protein interaction sites. *Proteins: Structure, Function, and Bioinformatics, 58*(1), 134–143. http://dx.doi.org/10.1002/prot.20285.

Ferrada, E., & Melo, F. (2009). Effective knowledge-based potentials. *Protein Science, 18*(7), 1469–1485. http://dx.doi.org/10.1002/pro.166.

Fraenkel, E., & Pabo, C. O. (1998). Comparison of X-ray and NMR structures for the Antennapedia homeodomain-DNA complex. *Nature Structural & Molecular Biology, 5*(8), 692–697. http://dx.doi.org/10.1038/1382.

Gabb, H. A., Jackson, R. M., & Sternberg, M. J. E. (1997). Modelling protein docking using shape complementarity, electrostatics and biochemical information. *Journal of Molecular Biology, 272*(1), 106–120. http://dx.doi.org/10.1006/jmbi.1997.1203.

Gabdoulline, R., Eckweiler, D., Kel, A., & Stegmaier, P. (2012). 3DTF: A web server for predicting transcription factor PWMs using 3D structure-based energy calculations. *Nucleic Acids Research, 40*(W1), W180–W185. http://dx.doi.org/10.1093/nar/gks551.

Gao, M., & Skolnick, J. (2008). DBD-Hunter: A knowledge-based method for the prediction of DNA-protein interactions. *Nucleic Acids Research, 36*(12), 3978–3992. http://dx.doi.org/10.1093/nar/gkn332.

Gao, M., & Skolnick, J. (2009). A threading-based method for the prediction of DNA-binding proteins with application to the human genome. *PLoS Computational Biology, 5*(11), e1000567. http://dx.doi.org/10.1371/journal.pcbi.1000567.

Gao, M., & Skolnick, J. (2010). Structural space of protein-protein interfaces is degenerate, close to complete, and highly connected. *Proceedings of the National Academy of Sciences of the United States of America, 107*(52), 22517–22522. http://dx.doi.org/10.1073/pnas.1012820107.

Garcia-Garcia, J., Bonet, J., Guney, E., Fornes, O., Planas, J., & Oliva, B. (2012). Networks of protein-protein interactions: From uncertainty to molecular details. *Molecular Informatics, 31*(5), 342–362. http://dx.doi.org/10.1002/minf.201200005.

Garcia-Garcia, J., Schleker, S., Klein-Seetharaman, J., & Oliva, B. (2012). BIPS: BIANA Interolog Prediction Server. A tool for protein-protein interaction inference. *Nucleic Acids Research, 40*(W1), W147–W151. http://dx.doi.org/10.1093/nar/gks553.

Garzon, J. I., Lopéz-Blanco, J. R., Pons, C., Kovacs, J., Abagyan, R., Fernandez-Recio, J., et al. (2009). FRODOCK: A new approach for fast rotational protein-protein docking. *Bioinformatics, 25*(19), 2544–2551. http://dx.doi.org/10.1093/bioinformatics/btp447.

Gaulton, A., Bellis, L. J., Bento, A. P., Chambers, J., Davies, M., & Hersey, A. (2011). ChEMBL: A large-scale bioactivity database for drug discovery. *Nucleic Acids Research, 40*(D1), D1100–D1107. http://dx.doi.org/10.1093/nar/gkr777.

Ginalski, K. (2006). Comparative modeling for protein structure prediction. *Current Opinion in Structural Biology, 16*(2), 172–177. http://dx.doi.org/10.1016/j.sbi.2006.02.003.

Gitter, A., Siegfried, Z., Klutstein, M., Fornes, O., Oliva, B., Simon, I., et al. (2009). Backup in gene regulatory networks explains differences between binding and knockout results. *Molecular Systems Biology*, *5*(1), 276. http://dx.doi.org/10.1038/msb.2009.33.

Glover, J. N. M., & Harrison, S. C. (1995). Crystal structure of the heterodimeric bZIP transcription factor c-Fos-c-Jun bound to DNA. *Nature*, *373*(6511), 257–261. http://dx.doi.org/10.1038/373257a0.

Grau, J., Posch, S., Grosse, I., & Keilwagen, J. (2013). A general approach for discriminative de novo motif discovery from high-throughput data. *Nucleic Acids Research*, *41*(21), e197. http://dx.doi.org/10.1093/nar/gkt831.

Gray, J. J., Moughon, S., Wang, C., Schueler-Furman, O., Kuhlman, B., Rohl, C. A., et al. (2003). Protein-protein docking with simultaneous optimization of rigid-body displacement and side-chain conformations. *Journal of Molecular Biology*, *331*(1), 281–299. http://dx.doi.org/10.1016/S0022-2836(03)00670-3.

Gu, S., Koehl, P., Hass, J., & Amenta, N. (2012). Surface-histogram: A new shape descriptor for protein-protein docking. *Proteins: Structure, Function, and Bioinformatics*, *80*(1), 221–238. http://dx.doi.org/10.1002/prot.23192.

Guerois, R., Nielsen, J. E., & Serrano, L. (2002). Predicting changes in the stability of proteins and protein complexes: A study of more than 1000 mutations. *Journal of Molecular Biology*, *320*(2), 369–387. http://dx.doi.org/10.1016/S0022-2836(02)00442-4.

Gupta, S., Stamatoyannopoulos, J. A., Bailey, T. L., & Noble, W. S. (2007). Quantifying similarity between motifs. *Genome Biology*, *8*(2), R24. http://dx.doi.org/10.1186/gb-2007-8-2-r24.

Hu, S., Xie, Z., Onishi, A., Yu, X., Jiang, L., & Lin, J. (2009). Profiling the human protein-DNA interactome reveals ERK2 as a transcriptional repressor of interferon signaling. *Cell*, *139*(3), 610–622. http://dx.doi.org/10.1016/j.cell.2009.08.037.

Huang, S.-Y., & Zou, X. (2013). A nonredundant structure dataset for benchmarking protein-RNA computational docking. *Journal of Computational Chemistry*, *34*(4), 311–318. http://dx.doi.org/10.1002/jcc.23149.

Hwang, S., Gou, Z., & Kuznetsov, I. B. (2007). DP-Bind: A web server for sequence-based prediction of DNA-binding residues in DNA-binding proteins. *Bioinformatics*, *23*(5), 634–636. http://dx.doi.org/10.1093/bioinformatics/btl672.

Hwang, H., Pierce, B., Mintseris, J., Janin, J., & Weng, Z. (2008). Protein-protein docking benchmark version 3.0. *Proteins: Structure, Function, and Bioinformatics*, *73*(3), 705–709. http://dx.doi.org/10.1002/prot.22106.

Hwang, H., Vreven, T., Janin, J., & Weng, Z. (2010). Protein-protein docking benchmark version 4.0. *Proteins: Structure, Function, and Bioinformatics*, *78*(15), 3111–3114. http://dx.doi.org/10.1002/prot.22830.

Janin, J. (2010). Protein-protein docking tested in blind predictions: The CAPRI experiment. *Molecular BioSystems*, *6*(12), 2351–2362. http://dx.doi.org/10.1039/C005060C.

Janin, J., Henrick, K., Moult, J., Eyck, L. T., Sternberg, M. J. E., & Vajda, S. (2003). CAPRI: A Critical Assessment of PRedicted Interactions. *Proteins: Structure, Function, and Bioinformatics*, *52*(1), 2–9. http://dx.doi.org/10.1002/prot.10381.

Jiménez-García, B., Pons, C., & Fernández-Recio, J. (2013). pyDockWEB: A web server for rigid-body protein-protein docking using electrostatics and desolvation scoring. *Bioinformatics*, *29*(13), 1698–1699. http://dx.doi.org/10.1093/bioinformatics/btt262.

Jones, S., & Thornton, J. M. (1997). Analysis of protein-protein interaction sites using surface patches. *Journal of Molecular Biology*, *272*(1), 121–132. http://dx.doi.org/10.1006/jmbi.1997.1234.

Katchalski-Katzir, E., Shariv, I., Eisenstein, M., Friesem, A. A., Aflalo, C., & Vakser, I. A. (1992). Molecular surface recognition: Determination of geometric fit between proteins and their ligands by correlation techniques. *Proceedings of the National Academy of Sciences of the United States of America*, *89*(6), 2195–2199.

Kim, R., Corona, R. I., Hong, B., & Guo, J. (2011). Benchmarks for flexible and rigid transcription factor-DNA docking. *BMC Structural Biology*, *11*(1), 45. http://dx.doi.org/10.1186/1472-6807-11-45.

Kirsanov, D. D., Zanegina, O. N., Aksianov, E. A., Spirin, S. A., Karyagina, A. S., & Alexeevski, A. V. (2012). NPIDB: Nucleic acid-protein interaction database. *Nucleic Acids Research*, *41*(D1), D517–D523. http://dx.doi.org/10.1093/nar/gks1199.

Knegtel, R. M. A., Antoon, J., Rullmann, C., Boelens, R., & Kaptein, R. (1994). MONTY: A Monte Carlo approach to protein-DNA recognition. *Journal of Molecular Biology*, *235*(1), 318–324. http://dx.doi.org/10.1016/S0022-2836(05)80035-X.

Knox, C., Law, V., Jewison, T., Liu, P., Ly, S., & Frolkis, A. (2011). DrugBank 3.0: A comprehensive resource for "Omics" research on drugs. *Nucleic Acids Research*, *39*(Suppl. 1), D1035–D1041. http://dx.doi.org/10.1093/nar/gkq1126.

Kozakov, D., Brenke, R., Comeau, S. R., & Vajda, S. (2006). PIPER: An FFT-based protein docking program with pairwise potentials. *Proteins: Structure, Function, and Bioinformatics*, *65*(2), 392–406. http://dx.doi.org/10.1002/prot.21117.

Kumar, M., Gromiha, M. M., & Raghava, G. P. (2007). Identification of DNA-binding proteins using support vector machines and evolutionary profiles. *BMC Bioinformatics*, *8*(1), 463. http://dx.doi.org/10.1186/1471-2105-8-463.

Kumar, M., Gromiha, M. M., & Raghava, G. P. S. (2011). SVM based prediction of RNA-binding proteins using binding residues and evolutionary information. *Journal of Molecular Recognition*, *24*(2), 303–313. http://dx.doi.org/10.1002/jmr.1061.

Lasker, K., Phillips, J. L., Russel, D., Velázquez-Muriel, J., Schneidman-Duhovny, D., & Tjioe, E. (2010). Integrative structure modeling of macromolecular assemblies from proteomics data. *Molecular & Cellular Proteomics*, *9*(8), 1689–1702. http://dx.doi.org/10.1074/mcp.R110.000067.

Lasker, K., Sali, A., & Wolfson, H. J. (2010). Determining macromolecular assembly structures by molecular docking and fitting into an electron density map. *Proteins: Structure, Function, and Bioinformatics*, *78*(15), 3205–3211. http://dx.doi.org/10.1002/prot.22845.

Lee, H., Li, Z., Silkov, A., Fischer, M., Petrey, D., Honig, B., et al. (2010). High-throughput computational structure-based characterization of protein families: START domains and implications for structural genomics. *Journal of Structural and Functional Genomics*, *11*(1), 51–59. http://dx.doi.org/10.1007/s10969-010-9086-7.

Lensink, M. F., & Wodak, S. J. (2010). Docking and scoring protein interactions: CAPRI 2009. *Proteins: Structure, Function, and Bioinformatics*, *78*(15), 3073–3084. http://dx.doi.org/10.1002/prot.22818.

Lesk, V. I., & Sternberg, M. J. E. (2008). 3D-Garden: A system for modelling protein-protein complexes based on conformational refinement of ensembles generated with the marching cubes algorithm. *Bioinformatics*, *24*(9), 1137–1144. http://dx.doi.org/10.1093/bioinformatics/btn093.

Lin, C.-K., & Chen, C.-Y. (2013). PiDNA: Predicting protein-DNA interactions with structural models. *Nucleic Acids Research*, *41*(W1), W523–W530. http://dx.doi.org/10.1093/nar/gkt388.

Liu, Z., Guo, J.-T., Li, T., & Xu, Y. (2008). Structure-based prediction of transcription factor binding sites using a protein-DNA docking approach. *Proteins: Structure, Function, and Bioinformatics*, *72*(4), 1114–1124. http://dx.doi.org/10.1002/prot.22002.

Lu, H., Lu, L., & Skolnick, J. (2003). Development of unified statistical potentials describing protein-protein interactions. *Biophysical Journal*, *84*(3), 1895–1901. http://dx.doi.org/10.1016/S0006-3495(03)74997-2.

Lu, X.-J., & Olson, W. K. (2008). 3DNA: A versatile, integrated software system for the analysis, rebuilding and visualization of three-dimensional nucleic-acid structures. *Nature Protocols*, *3*(7), 1213–1227. http://dx.doi.org/10.1038/nprot.2008.104.

Luscombe, N. M., & Thornton, J. M. (2002). Protein-DNA interactions: Amino acid conservation and the effects of mutations on binding specificity. *Journal of Molecular Biology*, *320*(5), 991–1009. http://dx.doi.org/10.1016/S0022-2836(02)00571-5.

Lyskov, S., & Gray, J. J. (2008). The RosettaDock server for local protein-protein docking. *Nucleic Acids Research*, *36*(Suppl. 2), W233–W238. http://dx.doi.org/10.1093/nar/gkn216.

Macindoe, G., Mavridis, L., Venkatraman, V., Devignes, M.-D., & Ritchie, D. W. (2010). HexServer: An FFT-based protein docking server powered by graphics processors. *Nucleic Acids Research*, *38*(Suppl. 2), W445–W449. http://dx.doi.org/10.1093/nar/gkq311.

Mashiach, E., Schneidman-Duhovny, D., Andrusier, N., Nussinov, R., & Wolfson, H. J. (2008). FireDock: A web server for fast interaction refinement in molecular docking. *Nucleic Acids Research*, *36*(Suppl. 2), W229–W232. http://dx.doi.org/10.1093/nar/gkn186.

Matthews, L. R., Vaglio, P., Reboul, J., Ge, H., Davis, B. P., & Garrels, J. (2001). Identification of potential interaction networks using sequence-based searches for conserved protein-protein interactions or "interologs" *Genome Research*, *11*(12), 2120–2126. http://dx.doi.org/10.1101/gr.205301.

Mintseris, J., Pierce, B., Wiehe, K., Anderson, Robert, Chen, R., & Weng, Z. (2007). Integrating statistical pair potentials into protein complex prediction. *Proteins: Structure, Function, and Bioinformatics*, *69*(3), 511–520. http://dx.doi.org/10.1002/prot.21502.

Miyazawa, S., & Jernigan, R. L. (1985). Estimation of effective interresidue contact energies from protein crystal structures: Quasi-chemical approximation. *Macromolecules*, *18*(3), 534–552. http://dx.doi.org/10.1021/ma00145a039.

Moal, I. H., & Bates, P. A. (2010). SwarmDock and the use of normal modes in protein-protein docking. *International Journal of Molecular Sciences*, *11*(10), 3623–3648. http://dx.doi.org/10.3390/ijms11103623.

Moal, I. H., Torchala, M., Bates, P. A., & Fernández-Recio, J. (2013). The scoring of poses in protein-protein docking: Current capabilities and future directions. *BMC Bioinformatics*, *14*(1), 286. http://dx.doi.org/10.1186/1471-2105-14-286.

Moont, G., Gabb, H. A., & Sternberg, M. J. E. (1999). Use of pair potentials across protein interfaces in screening predicted docked complexes. *Proteins: Structure, Function, and Bioinformatics*, *35*(3), 364–373. http://dx.doi.org/10.1002/(SICI)1097-0134(19990515)35:3<364::AID-PROT11>3.0.CO;2-4.

Mosca, R., Céol, A., & Aloy, P. (2013). Interactome3D: Adding structural details to protein networks. *Nature Methods*, *10*(1), 47–53. http://dx.doi.org/10.1038/nmeth.2289.

Mosca, R., Céol, A., Stein, A., Olivella, R., & Aloy, P. (2013). 3did: A catalog of domain-based interactions of known three-dimensional structure. *Nucleic Acids Research*, *42*(D1), D374–D379. http://dx.doi.org/10.1093/nar/gkt887.

Nimrod, G., Schushan, M., Szilágyi, A., Leslie, C., & Ben-Tal, N. (2010). iDBPs: A web server for the identification of DNA binding proteins. *Bioinformatics*, *26*(5), 692–693. http://dx.doi.org/10.1093/bioinformatics/btq019.

Ozbek, P., Soner, S., Erman, B., & Haliloglu, T. (2010). DNABINDPROT: Fluctuation-based predictor ofDNA-binding residues within a network ofinteracting residues. *Nucleic Acids Research*, *38*(Suppl. 2), W417–W423. http://dx.doi.org/10.1093/nar/gkq396.

Pandit, S. B., Brylinski, M., Zhou, H., Gao, M., Arakaki, A. K., & Skolnick, J. (2010). PSiFR: An integrated resource for prediction of protein structure and function. *Bioinformatics*, *26*(5), 687–688. http://dx.doi.org/10.1093/bioinformatics/btq006.

Panjkovich, A., Melo, F., & Marti-Renom, M. A. (2008). Evolutionary potentials: Structure specific knowledge-based potentials exploiting the evolutionary record of sequence homologs. *Genome Biology*, *9*(4), R68. http://dx.doi.org/10.1186/gb-2008-9-4-r68.

Parisien, M., Freed, K. F., & Sosnick, T. R. (2012). On docking, scoring and assessing protein-DNA complexes in a rigid-body framework. *PLoS One*, 7(2), e32647. http://dx.doi.org/10.1371/journal.pone.0032647.

Pérez-Cano, L., & Fernández-Recio, J. (2010). Optimal protein-RNA area, OPRA: A propensity-based method to identify RNA-binding sites on proteins. *Proteins: Structure, Function, and Bioinformatics*, *78*(1), 25–35. http://dx.doi.org/10.1002/prot.22527.

Pérez-Cano, L., Jiménez-García, B., & Fernández-Recio, J. (2012). A protein-RNA docking benchmark (II): Extended set from experimental and homology modeling data. *Proteins: Structure, Function, and Bioinformatics*, *80*(7), 1872–1882. http://dx.doi.org/10.1002/prot.24075.

Pérez-Cano, L., Solernou, A., Pons, C., & Fernández-Recio, J. (2010). Structural prediction of protein-RNA interaction by computational docking with propensity-based statistical potentials. *Pacific Symposium on Biocomputing*, *15*, 269–280.

Pettersen, E. F., Goddard, T. D., Huang, C. C., Couch, G. S., Greenblatt, D. M., Meng, E. C., et al. (2004). UCSF chimera—A visualization system for exploratory research and analysis. *Journal of Computational Chemistry*, *25*(13), 1605–1612. http://dx.doi.org/10.1002/jcc.20084.

Pieper, U., Webb, B. M., Barkan, D. T., Schneidman-Duhovny, D., Schlessinger, A., & Braberg, H. (2011). ModBase, a database of annotated comparative protein structure models, and associated resources. *Nucleic Acids Research*, *39*(Suppl. 1), D465–D474. http://dx.doi.org/10.1093/nar/gkq1091.

Pierce, B., & Weng, Z. (2007). ZRANK: Reranking protein docking predictions with an optimized energy function. *Proteins: Structure, Function, and Bioinformatics*, *67*(4), 1078–1086. http://dx.doi.org/10.1002/prot.21373.

Pierce, B., & Weng, Z. (2008). A combination of rescoring and refinement significantly improves protein docking performance. *Proteins: Structure, Function, and Bioinformatics*, *72*(1), 270–279. http://dx.doi.org/10.1002/prot.21920.

Planas-Iglesias, J., Bonet, J., Marín-López, M. A., Feliu, E., Gursoy, A., & Oliva, B. (2012). Structural bioinformatics of proteins: Predicting the tertiary and quaternary structure of proteins from sequence. In W. Cai (Ed.), *Protein-protein interactions—Computational and experimental tools.* http://www.intechopen.com/books/protein-protein-interactions-computational-and-experimental-tools/structural-bioinformatics-of-proteins-predicting-the-tertiary-and-quaternary-structure-of-proteins-f.

Pons, C., Talavera, D., de la Cruz, X., Orozco, M., & Fernandez-Recio, J. (2011). Scoring by intermolecular pairwise propensities of exposed residues (SIPPER): A new efficient potential for protein-protein docking. *Journal of Chemical Information and Modeling*, *51*(2), 370–377. http://dx.doi.org/10.1021/ci100353e.

Poulain, P., Saladin, A., Hartmann, B., & Prévost, C. (2008). Insights on protein-DNA recognition by coarse grain modelling. *Journal of Computational Chemistry*, *29*(15), 2582–2592. http://dx.doi.org/10.1002/jcc.21014.

R Core Team, (2013). *R: A language and environment for statistical computing*. Vienna: Austria.

Rice, P., Longden, I., & Bleasby, A. (2000). EMBOSS: The European Molecular Biology Open Software Suite. *Trends in Genetics*, *16*(6), 276–277. http://dx.doi.org/10.1016/S0168-9525(00)02024-2.

Ritchie, D. W., & Kemp, G. J. L. (2000). Protein docking using spherical polar Fourier correlations. *Proteins: Structure, Function, and Bioinformatics*, *39*(2), 178–194. http://dx.doi.org/10.1002/(SICI)1097-0134(20000501)39:2<178::AID-PROT8>3.0.CO;2-6.

Roberts, V. A., Thompson, E. E., Pique, M. E., Perez, M. S., & Ten Eyck, L. F. (2013). DOT2: Macromolecular docking with improved biophysical models. *Journal of Computational Chemistry*, *34*(20), 1743–1758. http://dx.doi.org/10.1002/jcc.23304.

Robertson, T. A., & Varani, G. (2007). An all-atom, distance-dependent scoring function for the prediction of protein–DNA interactions from structure. *Proteins: Structure, Function, and Bioinformatics*, *66*(2), 359–374. http://dx.doi.org/10.1002/prot.21162.

Rost, B. (1999). Twilight zone of protein sequence alignments. *Protein Engineering*, *12*(2), 85–94. http://dx.doi.org/10.1093/protein/12.2.85.

Russel, D., Lasker, K., Webb, B., Velázquez-Muriel, J., Tjioe, E., & Schneidman-Duhovny, D. (2012). Putting the pieces together: Integrative modeling platform software for structure determination of macromolecular assemblies. *PLoS Biology*, *10*(1), e1001244. http://dx.doi.org/10.1371/journal.pbio.1001244.

Schneider, S., Saladin, A., Fiorucci, S., Prévost, C., & Zacharias, M. (2012). ATTRACT and PTOOLS: Open source programs for protein-protein docking. In R. Baron (Ed.), *Computational drug discovery and design* (pp. 221–232). New York: Springer. http://link.springer.com/protocol/10.1007/978-1-61779-465-0_15.

Schneidman-Duhovny, D., Hammel, M., & Sali, A. (2011). Macromolecular docking restrained by a small angle X-ray scattering profile. *Journal of Structural Biology*, *173*(3), 461–471. http://dx.doi.org/10.1016/j.jsb.2010.09.023.

Schneidman-Duhovny, D., Inbar, Y., Nussinov, R., & Wolfson, H. J. (2005). PatchDock and SymmDock: Servers for rigid and symmetric docking. *Nucleic Acids Research*, *33*(Suppl. 2), W363–W367. http://dx.doi.org/10.1093/nar/gki481.

Schrödinger, L. (2010). The PyMOL molecular graphics system (Version 1.3r1).

Sharan, R., Ulitsky, I., & Shamir, R. (2007). Network-based prediction of protein function. *Molecular Systems Biology*, *3*(1). http://dx.doi.org/10.1038/msb4100129.

Shen, Y., Paschalidis, I. C., Vakili, P., & Vajda, S. (2008). Protein docking by the underestimation of free energy funnels in the space of encounter complexes. *PLoS Computational Biology*, *4*(10), e1000191. http://dx.doi.org/10.1371/journal.pcbi.1000191.

Shen, M., & Sali, A. (2006). Statistical potential for assessment and prediction of protein structures. *Protein Science*, *15*(11), 2507–2524. http://dx.doi.org/10.1110/ps.062416606.

Shentu, Z., Al Hasan, M., Bystroff, C., & Zaki, M. J. (2008). Context shapes: Efficient complementary shape matching for protein-protein docking. *Proteins: Structure, Function, and Bioinformatics*, *70*(3), 1056–1073. http://dx.doi.org/10.1002/prot.21600.

Si, J., Zhang, Z., Lin, B., Schroeder, M., & Huang, B. (2011). MetaDBSite: A meta approach to improve protein DNA-binding sites prediction (Report No. Suppl. 1) (p. S7). BioMed Central Ltd. http://www.biomedcentral.com/1752-0509/5/S1/S7/abstract.

Simon, B., Madl, T., Mackereth, C. D., Nilges, M., & Sattler, M. (2010). An efficient protocol for NMR-spectroscopy-based structure determination of protein complexes in solution. *Angewandte Chemie, International Edition*, *49*(11), 1967–1970. http://dx.doi.org/10.1002/anie.200906147.

Sippl, M. J. (1990). Calculation of conformational ensembles from potentials of mean force. An approach to the knowledge-based prediction of local structures in globular proteins. *Journal of Molecular Biology*, *213*(4), 859–883.

Stein, A., Céol, A., & Aloy, P. (2011). 3did: Identification and classification of domain-based interactions of known three-dimensional structure. *Nucleic Acids Research*, *39*(Suppl. 1), D718–D723. http://dx.doi.org/10.1093/nar/gkq962.

Stein, A., Rueda, M., Panjkovich, A., Orozco, M., & Aloy, P. (2011). A systematic study of the energetics involved in structural changes upon association and connectivity in protein interaction networks. *Structure*, *19*(6), 881–889. http://dx.doi.org/10.1016/j.str.2011.03.009.

Takeda, T., Corona, R. I., & Guo, J. (2013). A knowledge-based orientation potential for transcription factor-DNA docking. *Bioinformatics*, *29*(3), 322–330. http://dx.doi.org/10.1093/bioinformatics/bts699.

Tjong, H., & Zhou, H.-X. (2007). DISPLAR: An accurate method for predicting DNA-binding sites on protein surfaces. *Nucleic Acids Research*, *35*(5), 1465–1477. http://dx.doi.org/10.1093/nar/gkm008.

Torchala, M., Moal, I. H., Chaleil, R. A. G., Fernandez-Recio, J., & Bates, P. A. (2013). SwarmDock: A server for flexible protein-protein docking. *Bioinformatics*, *29*(6), 807–809. http://dx.doi.org/10.1093/bioinformatics/btt038.

Tovchigrechko, A., & Vakser, I. A. (2006). GRAMM-X public web server for protein-protein docking. *Nucleic Acids Research*, *34*(Web Server issue), W310–W314. http://dx.doi.org/10.1093/nar/gkl206.

Tuncbag, N., Gursoy, A., Guney, E., Nussinov, R., & Keskin, O. (2008). Architectures and functional coverage of protein-protein interfaces. *Journal of Molecular Biology*, *381*(3), 785–802. http://dx.doi.org/10.1016/j.jmb.2008.04.071.

Tuncbag, N., Gursoy, A., Nussinov, R., & Keskin, O. (2011). Predicting protein-protein interactions on a proteome scale by matching evolutionary and structural similarities at interfaces using PRISM. *Nature Protocols*, *6*(9), 1341–1354. http://dx.doi.org/10.1038/nprot.2011.367.

Turner, D., Kim, R., & Guo, J. (2012). TFinDit: Transcription factor-DNA interaction data depository. *BMC Bioinformatics*, *13*(1), 220. http://dx.doi.org/10.1186/1471-2105-13-220.

Tuszynska, I., & Bujnicki, J. M. (2011). DARS-RNP and QUASI-RNP: New statistical potentials for protein-RNA docking. *BMC Bioinformatics*, *12*(1), 348. http://dx.doi.org/10.1186/1471-2105-12-348.

Urnov, F. D., Rebar, E. J., Holmes, M. C., Zhang, H. S., & Gregory, P. D. (2010). Genome editing with engineered zinc finger nucleases. *Nature Reviews Genetics*, *11*(9), 636–646. http://dx.doi.org/10.1038/nrg2842.

Vajda, S., & Kozakov, D. (2009). Convergence and combination of methods in protein-protein docking. *Current Opinion in Structural Biology*, *19*(2), 164–170. http://dx.doi.org/10.1016/j.sbi.2009.02.008.

Valdar, W. S. J., & Thornton, J. M. (2001). Protein-protein interfaces: Analysis of amino acid conservation in homodimers. *Proteins: Structure, Function, and Bioinformatics*, *42*(1), 108–124. http://dx.doi.org/10.1002/1097-0134(20010101)42:1<108::AID-PROT110->3.0.CO;2-O.

van Dijk, M., & Bonvin, A. M. J. J. (2008). A protein-DNA docking benchmark. *Nucleic Acids Research*, *36*(14), e88. http://dx.doi.org/10.1093/nar/gkn386.

van Dijk, M., & Bonvin, A. M. J. J. (2010). Pushing the limits of what is achievable in protein-DNA docking: Benchmarking HADDOCK's performance. *Nucleic Acids Research*, *38*(17), 5634–5647. http://dx.doi.org/10.1093/nar/gkq222.

van Dijk, M., Visscher, K. M., Kastritis, P. L., & Bonvin, A. M. J. J. (2013). Solvated protein-DNA docking using HADDOCK. *Journal of Biomolecular NMR*, *56*(1), 51–63. http://dx.doi.org/10.1007/s10858-013-9734-x.

Venkatraman, V., Yang, Y. D., Sael, L., & Kihara, D. (2009). Protein-protein docking using region-based 3D Zernike descriptors. *BMC Bioinformatics*, *10*(1), 407. http://dx.doi.org/10.1186/1471-2105-10-407.

Wang, L., & Brown, S. J. (2006). BindN: A web-based tool for efficient prediction of DNA and RNA binding sites in amino acid sequences. *Nucleic Acids Research*, *34*(Suppl. 2), W243–W248. http://dx.doi.org/10.1093/nar/gkl298.

Wang, L., Huang, C., Yang, M. Q., & Yang, J. Y. (2010). BindN+ for accurate prediction of DNA and RNA-binding residues from protein sequence features. *BMC Systems Biology*, *4*(Suppl. 1), S3. http://dx.doi.org/10.1186/1752-0509-4-S1-S3.

Watson, J. D., Laskowski, R. A., & Thornton, J. M. (2005). Predicting protein function from sequence and structural data. *Current Opinion in Structural Biology*, *15*(3), 275–284. http://dx.doi.org/10.1016/j.sbi.2005.04.003.

Weirauch, M. T., Cote, A., Norel, R., Annala, M., Zhao, Y., & Riley, T. R. (2013). Evaluation of methods for modeling transcription factor sequence specificity. *Nature Biotechnology*, *31*(2), 126–134. http://dx.doi.org/10.1038/nbt.2486.

Wiederstein, M., & Sippl, M. J. (2007). ProSA-web: Interactive web service for the recognition of errors in three-dimensional structures of proteins. *Nucleic Acids Research*, *35*(Suppl. 2), W407–W410. http://dx.doi.org/10.1093/nar/gkm290.

Wodak, S. J., & Janin, J. (1978). Computer analysis of protein-protein interaction. *Journal of Molecular Biology*, *124*(2), 323–342. http://dx.doi.org/10.1016/0022-2836(78)90302-9.

Xie, Z., Hu, S., Qian, J., Blackshaw, S., & Zhu, H. (2011). Systematic characterization of protein-DNA interactions. *Cellular and Molecular Life Sciences*, *68*(10), 1657–1668. http://dx.doi.org/10.1007/s00018-010-0617-y.

Xu, B., Yang, Y., Liang, H., & Zhou, Y. (2009). An all-atom knowledge-based energy function for protein-DNA threading, docking decoy discrimination, and prediction of transcription-factor binding profiles. *Proteins: Structure, Function, and Bioinformatics*, *76*(3), 718–730. http://dx.doi.org/10.1002/prot.22384.

Yu, X., Cao, J., Cai, Y., Shi, T., & Li, Y. (2006). Predicting rRNA-, RNA-, and DNA-binding proteins from primary structure with support vector machines. *Journal of Theoretical Biology*, *240*(2), 175–184. http://dx.doi.org/10.1016/j.jtbi.2005.09.018.

Zhang, C., Liu, S., Zhu, Q., & Zhou, Y. (2005). A knowledge-based energy function for protein-ligand, protein-protein, and protein-DNA complexes. *Journal of Medicinal Chemistry*, *48*(7), 2325–2335. http://dx.doi.org/10.1021/jm049314d.

Zhang, Q. C., Petrey, D., Deng, L., Qiang, L., Shi, Y., & Thu, C. A. (2012). Structure-based prediction of protein-protein interactions on a genome-wide scale. *Nature*, *490*(7421), 556–560. http://dx.doi.org/10.1038/nature11503.

Zhang, Q. C., Petrey, D., Norel, R., & Honig, B. H. (2010). Protein interface conservation across structure space. *Proceedings of the National Academy of Sciences of the United States of America*, *107*(24), 10896–10901. http://dx.doi.org/10.1073/pnas.1005894107.

Zhang, Y., & Skolnick, J. (2004). Automated structure prediction of weakly homologous proteins on a genomic scale. *Proceedings of the National Academy of Sciences of the United States of America*, *101*(20), 7594–7599. http://dx.doi.org/10.1073/pnas.0305695101.

Zhang, Y., & Skolnick, J. (2005). TM-align: A protein structure alignment algorithm based on the TM-score. *Nucleic Acids Research*, *33*(7), 2302–2309. http://dx.doi.org/10.1093/nar/gki524.

Zhao, H., Yang, Y., & Zhou, Y. (2010). Structure-based prediction of DNA-binding proteins by structural alignment and a volume-fraction corrected DFIRE-based energy function. *Bioinformatics*, *26*(15), 1857–1863. http://dx.doi.org/10.1093/bioinformatics/btq295.

Zhao, H., Yang, Y., & Zhou, Y. (2011). Structure-based prediction of RNA-binding domains and RNA-binding sites and application to structural genomics targets. *Nucleic Acids Research*, *39*(8), 3017–3025. http://dx.doi.org/10.1093/nar/gkq1266.

Zheng, S., Robertson, T. A., & Varani, G. (2007). A knowledge-based potential function predicts the specificity and relative binding energy of RNA-binding proteins. *FEBS Journal*, *274*(24), 6378–6391. http://dx.doi.org/10.1111/j.1742-4658.2007.06155.x.

Zhou, H., & Skolnick, J. (2013). FINDSITEcomb: A threading/structure-based, proteomic-scale virtual ligand screening approach. *Journal of Chemical Information and Modeling*, *53*(1), 230–240. http://dx.doi.org/10.1021/ci300510n.

Zhou, H., & Zhou, Y. (2002). Distance-scaled, finite ideal-gas reference state improves structure-derived potentials of mean force for structure selection and stability prediction. *Protein Science*, *11*(11), 2714–2726. http://dx.doi.org/10.1110/ps.0217002.

CHAPTER FIVE

Algorithms, Applications, and Challenges of Protein Structure Alignment

Jianzhu Ma, Sheng Wang[1]
Toyota Technological Institute at Chicago, Chicago, Illinois, USA
[1]Corresponding author: e-mail address: wangsheng@ttic.edu

Contents

1. Introduction 122
2. Structural Alphabet 124
   2.1 Introduction of structural alphabet 124
   2.2 Different kinds of structural alphabet 126
   2.3 Conformational letters 128
   2.4 Applications of structural alphabet 128
   2.5 Substitution matrix for structural alphabet 130
3. Protein Pairwise Structure Alignment 130
   3.1 Introduction of protein pairwise structure alignment 130
   3.2 The scoring functions to measure the structure similarity 138
   3.3 The search algorithms for pairwise structure alignment 140
   3.4 DeepAlign: The approach beyond spatial proximity 144
   3.5 Result analysis 147
4. Protein MSA 155
   4.1 Introduction of protein MSA 155
   4.2 The current approaches for MSA 158
   4.3 3DCOMB: The scaffold-first approach for MSA 161
   4.4 Result analysis 165
5. Conclusions and Perspectives 167
References 170

Abstract

As a fundamental problem in computational structure biology, protein structure alignment has attracted the focus of the community for more than 20 years. While the pairwise structure alignment could be applied to measure the similarity between two proteins, which is a first step for homology search and fold space construction, the multiple structure alignment could be used to understand evolutionary conservation and divergence from a family of protein structures. Structure alignment is an NP-hard problem, which is only computationally tractable by using heuristics. Three

Advances in Protein Chemistry and Structural Biology, Volume 94
ISSN 1876-1623
http://dx.doi.org/10.1016/B978-0-12-800168-4.00005-6

levels of heuristics for pairwise structure alignment have been proposed, from the representations of protein structure, the perspectives of viewing protein as a rigid-body or flexible, to the scoring functions as well as the search algorithms for the alignment. For multiple structure alignment, the fourth level of heuristics is applied on how to merge all input structures to a multiple structure alignment. In this review, we first present a small survey of current methods for protein pairwise and multiple alignment, focusing on those that are publicly available as web servers. In more detail, we also discuss the advancements on the development of the new approaches to increase the pairwise alignment accuracy, to efficiently and reliably merge input structures to the multiple structure alignment. Finally, besides broadening the spectrum of the applications of structure alignment for protein template-based prediction, we also list several open problems that need to be solved in the future, such as the large complex alignment and the fast database search.

1. INTRODUCTION

The comparison of protein structures is a very important problem in structural and evolutional biology for more than 20 years. Several excellent reviews could be found in Eidhammer, Jonassen, and Taylor (2000) and Koehl (2006). The applications of pairwise structure alignment include the prediction of the new protein's function (Brylinski & Skolnick, 2008), the homology search against a known database of proteins (Holm & Sander, 1993), the organization and classification of known structures (Holm & Sander, 1994), and the discovery of new structure patterns and their correlation among sequences (Bradley, Kim, & Berger, 2002). The applications of multiple structure alignment (MSA) programs include understanding evolutionary conservation and divergence (Andersen et al., 2001), functional prediction through the identification of structurally conserved active sites in homologous proteins (Irving, Whisstock, & Lesk, 2001), construction ofbenchmark datasets on which to test multiple sequence alignment programs (Edgar & Batzoglou, 2006), and automatic construction of profiles and threading templates for protein structure prediction (Dunbrack, 2006; Panchenko, Marchler-Bauer, & Bryant, 1999).

Structure alignment is an NP-hard problem, which is only computationally tractable by using heuristics (Koehl, 2006). The first level of heuristics is how to represent the protein structure. Four major classes have been proposed. For example, DALI (Holm & Sander, 1993) represents the protein as a C-alpha-based distance map (C-map); VAST (Gibrat, Madej, Spouge, & Bryant, 1997) represents the protein as the secondary structure elements (SSEs), PB-Align (Tyagi, De Brevern, Srinivasan, & Offmann, 2008) transforms the protein structure into 1D string (Profile), and most other methods use C-alpha to represent the protein. The second level of heuristics is the perspective to view the protein as a rigid-body or flexible, such as MATT (Menke, Berger, & Cowen, 2008). The third level is the method for the structure alignment, which has two major components: a scoring function to measure the protein similarity and a search algorithm to optimize the scoring function (Hasegawa & Holm, 2009). For example, as a widely used tool, TMalign (Zhang & Skolnick, 2005) employs the TMscore (Zhang & Skolnick, 2004) as the scoring function and applies dynamic programming (DP) as the search algorithm (Levitt & Gerstein, 1998) to optimize the score.

For MSA, besides all three levels, a fourth level of heuristics is applied to solve the problem that how to merge all input structures to a MSA. The current MSA merging method could be classified into two categories: horizontal-first and vertical-first approaches (Wang & Zheng, 2009). For instance, MUSTANG (Konagurthu, Whisstock, Stuckey, & Lesk, 2006) belongs to the horizontal-first method that it progressively merges pairwise alignments into an MSA along a guide tree; MultiProt (Shatsky, Nussinov, & Wolfson, 2004) belongs to the vertical-first method that it identifies some similar fragment blocks (SFBs) among all proteins and then extends these SFBs to an MSA. In addition to these two approaches, recently Ilinkin, Ye, and Janardan (2010) have developed a consensus-first approach, and Wang, Peng, and Xu (2011) and Wang and Zhengg (2009) have proposed a scaffold-first approach. These new approaches could efficiently and reliably merge input structures to the MSA.

Although many computer programs have been developed, the alignment accuracy of the pairwise alignment programs is still low when judged by manually curated structure alignments, especially on distantly but functionally related proteins (Wang, Ma, Peng, & Xu, 2013). The reason is that almost all pairwise structure alignment methods only capture protein structure similarity using a spatial proximity of the 3D object (Hasegawa & Holm, 2009). However, it is observed that despite conformational changes, protein structure families exhibit a high local flexibility on a smaller scale called phenotypic plasticity (Csaba, Birzele, & Zimmer, 2008), which is grounded on evolutionary events (i.e., mutation, insertion, and deletion). Under the hypothesis that two protein structures are similar, we assume that they share locally similar substructures that are not necessarily restricted to SSEs. This kind of local conformation changes caused by the evolutionary events cannot be accurately quantified by spatial proximity of aligned residues. Instead, evolutionary distance shall be a better measure (Wang et al., 2013).

In order to describe the local structures in proteins beyond the traditional SSEs, a variety of discretized sets, that is, the structural alphabets, have been proposed (see Table 5.1). To quantify the evolutionary distance between these local substructures, a substitution matrix for structural alphabets is constructed by a similar way of BLOSUM (Henikoff & Henikoff, 1992). Therefore, by considering the amino acid (AA) and local structure substitution matrices, a new scoring function for pairwise structure alignment has been developed (Wang et al., 2013). Experimental results show that optimizing this score could generate structure alignments much more consistent with manually curated alignments than other automatic tools especially when proteins under consideration are remote homologs.

In this review, we first briefly introduce the basic concept of structural alphabet and the related substitution matrix. Then, we describe protein pairwise structure alignment. The third part of this review focuses on protein MSA. We make a conclusion as well as propose several applications and address some unsolved problems in the last part.

2. STRUCTURAL ALPHABET

2.1. Introduction of structural alphabet

2.1.1 Why we need a structural alphabet

The protein local structures are classically described as three-state SSE that is composed of two regular states, the alpha-helices and the beta-strands, and one nonregular and variable state, the coil (Pauling, Corey, & Branson, 1951). Although the relevant physical meaning of SSE, the limitations are emerged, such as the uncertainties in the assignment of the boundaries around the helix and sheet regions, as well as about half of all residues are assigned to the coil state even though they have significantly different local conformation (Joseph, Agarwal, et al., 2010; Joseph, Bornot, & de Brevern, 2010). These limitations have stimulated a variety of research teams to focus on abstracting the protein backbone conformation in the localized short fragments, that is, the structural alphabets. Consisting of a discretized set of representatives of the local structures, the structural alphabets are generally defined by clustering local fragments in protein structures through a variety of geometric measures, as well as different segment length (Bystroff & Baker, 1998; Camproux et al., 1999; De Brevern et al., 2000; Kolodny et al., 2002; Pandini et al., 2010; Tung et al., 2007). We have developed a description of

Table 5.1 A selected set of structural alphabets

Name	Len	Size	Features	Available link	References
I-sites	3–19	250	Amino acid profile	http://www.bioinfo.rpi.edu/bystrc/Isites2/	Bystroff and Baker (1998)
HMM-SA	4	27	HMM	http://sa-mot.mti.univ-paris-diderot.fr/main/SA_Mot_Method	Camproux, Tuffery, Chevrolat, Boisvieux, and Hazout (1999)
Protein blocks (PBs)	5	16	Unsupervised classifier (SOM)	http://www.dsimb.inserm.fr/~debrevern/PBs/index.php	De Brevern, Etchebest, and Hazout (2000)
Fragment library	4,5,6,7	4–14, 10–225, 40–300, 50–250	Simulated annealing based on *k*-means	http://csb.stanford.edu/rachel/fragments/	Kolodny, Koehl, Guibas, and Levitt (2002)
3D-BLAST	5	23	Nearest-neighbor clustering	http://3d-blast.life.nctu.edu.tw/sa.php	Tung, Huang, and Yang (2007)
Conformational letter (CLE)	4	17	Gaussian mixture model	http://ttic.uchicago.edu/~majianzhu/PDB_To_CLE_v1.00.tar.gz	Wang and Zheng (2008)
M32K25	4	25	Attractors in conformational space	http://mathbio.nimr.mrc.ac.uk/wiki/Software#Structural_Alphabet_M32K25	Pandini, Fornili, and Kleinjung (2010)
FragBag	9,10,11,12	50–600, 100–600, 100–600, 100–600	Bag of words	http://cs.haifa.ac.il/~ibudowsk/library9_12.html	Budowski-Tal, Nov, and Kolodny (2010)

protein structural alphabet, namely, conformational letter (CLE) (Wang & Zheng, 2008; Zheng, 2008) using pseudo-bond angles of successive four C-alpha atoms. The discretized set of representatives is selected according to the density peaks of probability distribution in the phase space spanned by pseudo-bond angles. Since its development, numerous new research fields using the structural alphabet have been explored, such as fast protein database search (Budowski-Tal et al., 2010; Tung et al., 2007), protein fold recognition (De Brevern et al., 2000), binding-site signature detection (Dudev & Lim, 2007), and most importantly, protein structure alignment (Bornot, Etchebest, & De Brevern, 2009).

2.1.2 Why we need a substitution matrix for structural alphabet

Without a substitution matrix, the use of a structural alphabet is very limited. A substitution matrix called CLESUM (conformational letter substitution matrix) (Zheng & Liu, 2005) is constructed, similar as the way of BLOSUM (Henikoff & Henikoff, 1992), from a representative pairwise aligned structure dataset of the families of structurally similar proteins (FSSPs) (Holm & Sander, 1994). After a protein 3D backbone structure is converted to a 1D sequence of structure alphabet (SA) that is akin to the AA, many tools of sequence analysis can then be applied with certain minor modifications (Wang & Zheng, 2008). As an example of application, fast structure alignment that is based on the CLE is described in the next section. Briefly, structurally similar fragment pairs (SFPs) are found merely by string comparison and a greedy strategy guided by CLESUM similarity scores becomes possible.

2.2. Different kinds of structural alphabet

The principle of a structural alphabet is simple (see Fig. 5.1). A set of average local protein structures is first designed. They approximate every part of the structures. As one residue is associated to one of these prototypes, we can translate the 3D information of the protein structures as a series of prototypes in 1D, as the AA sequence (Zheng, 2008). To see a selected set of structural alphabet, please view Table 5.1. For a more complete review on structural alphabets, see Joseph, Bornot, et al. (2010). Generally, the procedure to deduce finite discrete conformational states from continuous states in a conformational phase space is a clustering analysis. There has been a variety of different ways of clustering using various descriptors for structure. The original descriptor in the structure database Protein Data Bank (PDB) is the Cartesian coordinates of atoms. Many studies to investigate the classification

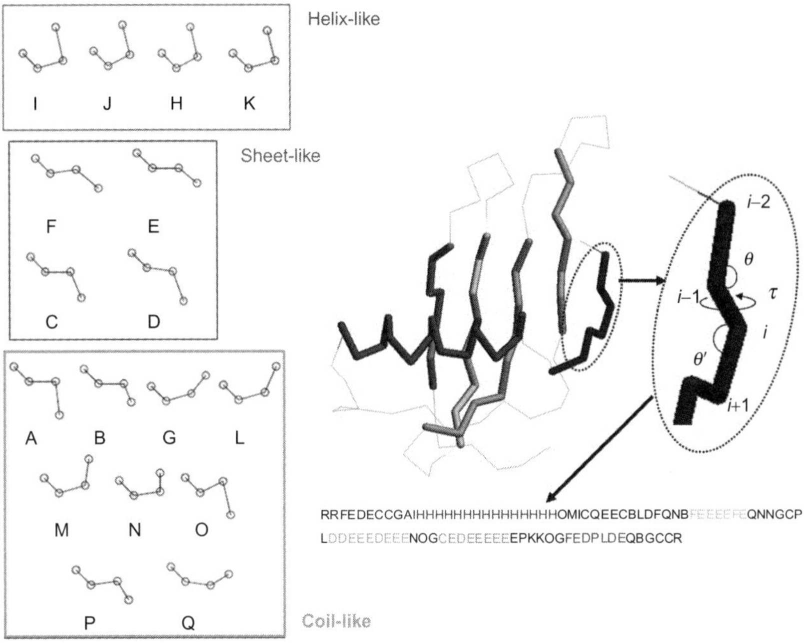

Figure 5.1 Conformational letter (CLE) and one example of assignment. On left shows the 17 CLEs with four contiguous C-alpha in length. They are arranged according to helix-like (in red box), sheet-like (in blue box), and coil-like (in green box). On right shows the example of transforming a 3D protein structure into a 1D CLE string. Each residue i is assigned with a specific CLE by first extracting three angles $x \equiv (\theta,\tau,\theta')$ of four contiguous C-alpha around residue i, and then by converting x to a discretized letter with the maximal probability. (For interpretation of the references to color in this figure legend, the reader is referred to the online version of this chapter.)

of protein fragments use the backbone (ϕ,ψ) dihedral angles, or angles of C-alpha pseudo-bonds or distances derived from the positions of C-alpha atoms (Flocco & Mowbray, 1995; Matsuda, Taniguchi, & Hashimoto, 1997; McCammon, 1977; Robson & Pain, 1974; Rooman, Kocher, & Wodak, 1991). Park and Levitt (1995) represented the polypeptide chain by a sequence of rigid fragments that were chosen from a library of representative fragments and concatenated without any degrees of freedom. De Brevern et al. (2000) proposed a SA called protein blocks (PBs) that are composed of 16 mean protein fragments of 5 residues in length by optimizing 2 goals simultaneously that (i) to obtain a good local structure approximation in (ϕ,ψ) dihedral angles and (ii) to predict local structures from the sequence. Kolodny et al. (2002) constructed libraries that differ in the fragment length (four to seven residues) and number of representative fragments they contain

(25–300), which offer a wide range of optimal fragments suited to different accuracies of fit. Tung et al. (2007) have proposed a SA that consists of 23 representative fragments of length 5 by clustering in the Kappa-alpha plot.

2.3. Conformational letters

Three contiguous C-alpha atoms determine two pseudo-bonds and a bending angle between them. Four contiguous C-alpha atoms, say $i-2$, $i-1$, i, and $i+1$, determine two such bending angles θ, θ' and a torsion angle τ which is the dihedral angle between the two planes of triangles ($i-2$, $i-1, i$) and ($i-1, i, i+1$), as shown in Fig. 5.1. By using a Gaussian mixture model M for the density distribution x of the three angles,

$$P(x|M) = \sum_{k=1}^{C} \pi_k N(\mu_k, \Sigma_k) \tag{5.1}$$

where $x \equiv (\theta, \tau, \theta')$, C is the number of the normal distribution categories in the mixture, π_k is the prior for category c_k, and $N(\mu,\Sigma)$ is the normal distribution; the local structural states have been clustered as 17 discrete CLEs of a protein structural alphabet. The centers μ, inverse covariance matrices Σ^{-1}, and weights π of the clusters for these CLEs in the phase space spanned by the three angles (θ,τ,θ') are listed in Table 5.2. When we convert a continuous point x to its discretized letter k^*, the following equation is applied,

$$k^* = \arg_k \max P(c_k|x) \tag{5.2}$$

where $P(c_k|x) = \pi_k |\Sigma_k|^{-1/2} \exp(1/2(x-\mu_k)\Sigma_k^{-1}(x-\mu_k))$.

2.4. Applications of structural alphabet

The works on PBs have proved their efficiencies in the description and the prediction of long fragments and short loops (de Brevern, Valadié, Hazout, & Etchebest, 2002; Fourrier, Benros, & De Brevern, 2004), to define a reduced AA alphabet dedicated to mutation design (Etchebest, Benros, Bornot, Camproux, & de Brevern, 2007), in the building of a transmembrane protein (De Brevern et al., 2005) and in the binding-site signature analysis (Dudev & Lim, 2007). Tung et al. (2007) have applied their set of SA to a protein structure database search tool, named 3D-BLAST, which has the features of BLAST (Altschul et al., 1997), for analyzing novel structures and can return a ranked list of alignments. Le, Pollastri, and Koehl (2009) have extended the application of structural alphabets to the problem

Table 5.2 The 17 conformational states from the mixture model

			$\boldsymbol{\mu}$			$\boldsymbol{\Sigma}^{-1}$					
State	π	$\lvert\boldsymbol{\Sigma}\rvert^{-1/2}$	θ	τ	θ'	$\theta\theta$	$\tau\theta$	$\tau\tau$	$\theta'\theta$	$\theta'\tau$	$\theta'\theta'$
I	8.2	1881	1.52	0.83	1.52	275.4	−28.3	84.3	106.9	−46.1	214.4
J	7.3	1797	1.58	1.05	1.55	314.3	−10.3	46.0	37.8	−70.0	332.8
H	16.2	10,425	1.55	0.88	1.55	706.6	−93.9	245.5	128.9	−171.8	786.1
K	5.9	254	1.48	0.70	1.43	73.8	−13.7	21.5	15.5	−25.3	75.7
F	4.9	105	1.09	−2.72	0.91	24.1	1.9	10.9	−11.2	−8.8	53.0
E	11.6	109	1.02	−2.98	0.95	34.3	4.2	15.2	−9.3	−22.5	56.8
C	7.5	100	1.01	−1.88	1.14	28.0	4.1	6.2	2.3	−5.1	69.4
D	5.4	78	0.79	−2.30	1.03	56.2	3.8	4.2	−10.8	−2.1	30.1
A	4.3	203	1.02	−2.00	1.55	30.5	9.1	8.7	6.0	5.7	228.6
B	3.9	66	1.06	−2.94	1.34	26.9	4.6	4.9	9.5	−5.0	54.3
G	5.6	133	1.49	2.09	1.05	163.9	0.6	3.8	2.0	−3.7	32.3
L	5.3	40	1.40	0.75	0.84	43.7	2.5	1.4	−7.0	−2.9	34.5
M	3.7	144	1.47	1.64	1.44	72.9	2.1	4.8	1.9	−7.9	72.9
N	3.1	74	1.12	0.14	1.49	25.3	3.2	3.1	9.9	0.9	83.0
O	2.1	247	1.54	−1.89	1.48	170.8	−0.7	3.7	−4.1	3.1	98.7
P	3.2	206	1.24	−2.98	1.49	48.0	8.2	7.3	−4.9	−6.6	155.6
Q	1.7	25	0.86	−0.37	1.01	28.4	1.5	1.2	3.4	0.1	19.5

Angles are in radian. The structural alphabets are sorted according to helix-like, sheet-like, and coil-like.

of protein structure classification that compares with the performance of other two different sequence representation of proteins: the AA sequence and the SSE sequence. For a more detailed review on SA application, please refer Joseph, Agarwal, et al. (2010) and Joseph, Bornot, et al. (2010).

2.5. Substitution matrix for structural alphabet

To use our structural alphabet directly for the structural comparison, a score matrix similar to BLOSUM for AAs is desired. Using the alignments for representative structures in the database FSSP (Holm & Sander, 1994), which contains 2860 sequence families representing 27,181 protein structures, we have constructed a substitution matrix called CLESUM for the CLEs. To the best of our knowledge, CLESUM is the first substitution matrix directly derived from structure alignments for a structural alphabet (Zheng & Liu, 2005). In particular, the structures of the representative set are converted to their CLE sequences. All of the pair alignments of FSSP for the proteins with a sufficient similarity in the representative set are collected for counting aligned pairs of CLEs. The total number of letter pairs is 1,284,750. An entry of the matrix is the log ratio of the observed frequency of the aligned corresponding pair to the expected frequency from a random alignment simply by chance. The substitution matrix (CLESUM) derived in the same way as BLOSUM was shown in Table 5.3. For the other structural alphabet's substitution matrix, please refer to Tung et al. (2007) and Tyagi et al. (2008).

In order to reveal the relationship between sequence and structure, it is interesting to consider both AA and SA in a joint space. However, such space has too many parameters, so the reduction of AA is necessary to bring down the parameter numbers. As a result, we have developed a simple but effective approach called entropic clustering based on selecting the best mutual information between a given reduction of AAs and SAs. The optimized reduction of AA into two groups leads to hydrophobic and hydrophilic. Combined with our CLE of 17 alphabets, we get a joint alphabet called hydropathy conformational letter. A joint substitution matrix with $(17 \times 2) \times (17 \times 2)$ indices is derived (Wang, 2010).

3. PROTEIN PAIRWISE STRUCTURE ALIGNMENT

3.1. Introduction of protein pairwise structure alignment

3.1.1 Why we need the protein pairwise structure alignment

To study a large collection of objects, such as protein structures, we usually start with classifying them according to a given measure of similarity. Protein

Table 5.3 CLESUM: The conformation letter substitution matrix (in units of 0.5 bit)

	I	J	H	K	F	E	C	D	A	B	G	L	M	N	O	P	Q
I	2	2	2	2	−9	−11	−6	−11	−3	−7	−3	−4	0	−1	−3	−5	−6
J	2	4	1	1	−7	−9	−4	−9	−2	−6	−2	−3	2	0	−2	−3	−4
H	2	1	2	1	−11	−13	−8	−12	−5	−10	−5	−6	0	−3	−6	−6	−9
K	2	1	1	5	−7	−8	−3	−8	−2	−5	−1	0	1	3	1	−4	−2
F	−9	−7	−11	−7	5	3	2	2	−2	1	−1	−2	−3	−3	−3	0	0
E	−11	−9	−13	−8	3	4	0	2	−3	2	−2	−2	−5	−4	−5	−1	−1
C	−6	−4	−8	−3	2	0	5	2	1	1	0	0	−1	−1	0	0	3
D	−11	−9	−12	−8	2	2	2	5	−2	1	−3	−2	−5	−4	−4	−1	1
A	−3	−2	−5	−2	−2	−3	1	−2	7	2	−1	−1	0	1	3	4	2
B	−7	−6	−10	−5	1	2	1	1	2	5	1	−1	−1	−1	−1	4	1
G	−3	−2	−5	−1	−1	−2	0	−3	−1	1	7	3	2	−1	0	1	1
L	−4	−3	−6	0	−2	−2	0	−2	−1	−1	3	7	1	1	0	−1	3
M	0	2	0	1	−3	−5	−1	−5	0	−1	2	1	6	1	1	1	−1
N	−1	0	−3	3	−3	−4	−1	−4	1	−1	−1	1	1	9	2	0	3
O	−3	−2	−6	1	−3	−5	0	−4	3	−1	0	0	1	2	10	0	−1
P	−5	−3	−6	−4	0	−1	0	−1	4	4	1	−1	1	0	0	7	1
Q	−6	−4	−9	−2	0	−1	3	1	2	1	1	3	−1	3	−1	1	9

structure similarity is most often detected and quantified by a protein pairwise structure alignment program. The comparison of protein structures, as an extremely important problem in structural and evolutional biology, has a large spectrum of applications. The detection of local or global structural similarity between a new protein and a protein with known function allows the prediction of the new protein's function (Roy, Yang, & Zhang, 2012). Structural comparison methods are useful for organizing and classifying known structures (Orengo et al., 1997), and for discovering structure patterns and their correlation with sequences (Bradley et al., 2002). Recently, protein pairwise structure alignment tools have been applied to generate the training set for a template-based protein structure prediction (i.e., threading) tool that is based on Machine Learning (Ma, Peng, Wang, & Xu, 2012; Ma, Wang, Zhao, & Xu, 2013), as well as been extended to some other applications such as binding-site recognition (Wass, Kelley, & Sternberg, 2010). Since structure alignment is an NP-hard problem, which is only computationally tractable by using heuristics (Koehl, 2006). In the following sections, we discuss three levels of heuristics.

3.1.2 First level of heuristics: What are the representations for protein structures

There are overall four types of representations for protein structure, which are C-alpha, C-map, SSE, and Profile. C-alpha stands for the representation of protein structure by backbone atoms (C-alpha). C-map is contact or distance map-based method. SSE stands for using secondary structure element to represent proteins. Profile is the usage of AA type and/or discrete structural alphabets to represent the protein. For example, DALI (Holm & Sander, 1993) represents the protein as a C-alpha-based distance map (C-map); VAST (Gibrat et al., 1997) represents the protein as the SSEs, PB-Align (Tyagi et al., 2008) transforms the protein structure into 1D PB string (Profile), and most other methods use C-alpha to represent the protein. For a more detailed description, see Table 5.4. Many programs mix several representations, such as Daniluk and Lesyngg (2011) and Kolbeck et al. (2006).

3.1.3 Second level of heuristics: What are the perspectives for protein structures

Although no universally acknowledged definition of what constitutes structural similarity, there are two major perspectives for the spectrum of

Table 5.4 A selected list of pairwise structure alignment tools

Name	Comments	Represent	Perspective	Score[a]	Search algorithm	Order[b]	Available link	References
SSAP	Sequential structure alignment program	SSE	Rigid-body	3D	Double dynamic programming	Y	http://www.biochem.ucl.ac.uk/~orengo/ssap.html	Taylor and Orengo (1989)
DALI	Distance matrix alignment	C-map	Implicit-flexible	2D	Monte Carlo	N	http://ekhidna.biocenter.helsinki.fi/dali_server/start	Holm and Sander (1993)
VAST	Vector alignment search tool	SSE	Rigid-body	3D	Monte Carlo	N	http://www.ncbi.nlm.nih.gov/Structure/VAST/vastsearch.html	Gibrat et al. (1997)
CE	Combinatorial extension on aligned fragment pairs (AFPs)	C-map and C-alpha	Rigid-body	2D and 3D	AFP-chaining	Y	http://source.rcsb.org/jfatcatserver/ceHome.jsp	Shindyalov and Bourne (1998)
STRUCTAL	Reciprocal cRMS minimization by dynamic programming	C-alpha	Rigid-body	3D	Dynamic programming	Y	http://csb.stanford.edu/levitt/Structal/	Levitt and Gerstein (1998)
ProSup	A refined tool for protein structure alignment	C-alpha	Rigid-body	3D	Maximize Ne with constraint RMSD	Y	https://topmatch.services.came.sbg.ac.at/	Lackner, Koppensteiner, Sippl, and Domingues (2000)

Continued

Table 5.4 A selected list of pairwise structure alignment tools—cont'd

Name	Comments	Represent	Perspective	Score	Search algorithm	Order	Available link	References
FlexProt	Flexible alignment of protein structures	C-alpha	Explicit-flexible on fragment	3D	Single-source shortest paths on graph	Y	http://bioinfo3d.cs.tau.ac.il/FlexProt/	Shatsky, Nussinov, and Wolfson (2002)
FATCAT	Flexible alignment by AFP-chaining allow twists	C-alpha	Explicit-flexible on fragment	3D	AFP-chaining	Y	http://fatcat.burnham.org/	Ye and Godzik (2003)
LGA	LGA-score-based protein structure alignment	C-alpha	Rigid-body	3D	Maximize Ne with constraint RMSD	Y	http://proteinmodel.org/AS2TS/LGA/lga.html	Zemla (2003)
SSM	Secondary structure matching	SSE	Rigid-body	3D	Subgraph isomorphism detection	N	http://www.ebi.ac.uk/msd-srv/ssm/	Krissinel and Henrick (2004)
TMalign	TMscore-based protein structure alignment	C-alpha	Rigid-body	3D	Dynamic programming	Y	http://zhanglab.ccmb.med.umich.edu/TM-align/	Zhang and Skolnick (2005)
YAKUSA	Internal coordinates and BLAST type algorithm	Profile	Rigid-body	1D	Blast-like search	Y	http://bioserv.rpbs.jussieu.fr/Yakusa/index.html	Carpentier, Brouillet, and Pothier (2005)

Gangsta	Genetic algorithm for nonsequential and gap structure alignment	SSE and C-map	Rigid-body	3D	Genetic algorithm	N	http://agknapp.chemie.fu-berlin.de/gplus/?page=pair	Kolbeck, May, Schmidt-Goenner, Steinke, and Knapp (2006)
Vorolign	VOROnoi contacts based protein structure alignment	C-map	Implicit-flexible	2D	Double dynamic programming	Y	http://www2.bio.ifi.lmu.de/Vorolign/	Birzele, Gewehr, Csaba, and Zimmer (2007)
3D-Blast	Structural alphabets and BLAST type algorithm	Profile	Implicit-flexible	1D	Blast-like search	Y	http://3d-blast.life.nctu.edu.tw/	Tung et al. (2007)
CLEPAPS	Conformation letter (CLE)-based pairwise alignment of protein structure	Profile and C-alpha	Rigid-body	1D and 3D	Maximize Ne with constraint RMSD	N	http://weblab.cbi.pku.edu.cn/programdoc/html/clepaps/CLePAPS.html	Wang and Zheng (2008)
RAPIDO	Rapid alignment of protein structures with domain move	C-alpha	Explicit-flexible on domain hinge	3D	AFP-chaining	Y	http://webapps.embl-hamburg.de/rapido/	Mosca, Brannetti, and Schneider (2008)
PB-Align	Protein blocks (PB)-based structure alignment	Profile	Implicit-flexible	1D	String dynamic programming	Y	http://www.bo-protscience.fr/pbe/?page_id=12	Tyagi et al. (2008)

Continued

Table 5.4 A selected list of pairwise structure alignment tools—cont'd

Name	Comments	Represent	Perspective	Score	Search algorithm	Order	Available link	References
ALADYN	Dynamics-based alignment by collective movements	Profile	Implicit-flexible	3D	Maximize Ne with constraint RMSD	N	http://aladyn.escience-lab.org/	Potestio, Aleksiev, Pontiggia, Cozzini, and Micheletti (2010)
DEDAL	Descriptor defined alignment	C-map and C-alpha	Implicit-flexible	2D and 3D	Monte Carlo	N	http://bioexploratorium.pl/EP/DEDAL/	Daniluk and Lesyng (2011)
SPalign	A size-independent score-based protein structure alignment	C-alpha	Rigid-body	3D	Dynamic programming	Y	http://sparks.informatics.iupui.edu/yueyang/server/SPalign/	Yang, Zhan, Zhao, and Zhou (2012)
DeepAlign	A score that considers evolutionary information and hydrogen-bonding similarity	Profile and C-alpha	Beyond spatial proximity	1D + 3D	Dynamic programming	Y	http://ttic.uchicago.edu/~jinbo/DeepAlign/DeepAlign_exe_V1.00.tar.gz	Wang et al. (2013)

[a] 1D and 3D indicates that the algorithm applies different scoring functions during different stages. 1D + 3D indicates that the algorithm applies one scoring function that contains 1D and 3D information.

[b] Y indicates that the alignment results are in sequential order. N indicates that the alignment could result in nonsequential order.

structural alignment methods, say rigid and flexible (Hasegawa & Holm, 2009). These perspectives differ in their treatment of structural variations. In particular, the rigid perspective treats the protein structures as rigid three-dimensional objects since there is a strong tradition to visualize structural alignments by least-squares superimposition. On the other hand, the flexible representations are classified into two groups. Implicit-flexible is measured by the difference between distance matrices, examples are DALI and Vorolign (Birzele et al., 2007). Explicit-flexible are the subtle changes in the angle between concatenated fragments, examples are MATT (Menke et al., 2008) and FATCAT (Ye & Godzik, 2003). All flexible representations carry detailed information about internal motions of the protein structures.

3.1.4 Third level of heuristics: How to construct a protein pairwise structure alignment

A protein pairwise structure alignment method consists of two major components: a scoring function to measure the protein similarity and a search algorithm to optimize the scoring function (Hasegawa & Holm, 2009; Wang et al., 2013). It is proposed that different scoring functions can be classified into three types depending on whether the structural representation is three-dimensional, two-dimensional, or one-dimensional. Once a similarity score has been defined, a search algorithm will be applied to find the optimal set of correspondences. An exhaustive search for such correspondence set between two structures is intractable, and various heuristics have been developed. The most commonly used approaches are Monte Carlo search (Daniluk & Lesyng, 2011; Holm & Sander, 1993), fragment assembly including graph extension algorithms (Krissinel & Henrick, 2004; Shatsky et al., 2002; Shindyalov & Bourne, 1998), an initial guess of the rigid-body transformation followed by DP (Levitt & Gerstein, 1998; Yang et al., 2012; Zhang & Skolnick, 2005), maximizing the number of structurally equivalent residues subject to a fixed Euclidean distance cutoff (Lackner et al., 2000; Potestio et al., 2010), and double dynamic programming (Birzele et al., 2007; Taylor & Orengo, 1989). Many programs mix several approaches, such as Ortiz, Strauss, and Olmea (2002), Shindyalov and Bournee (1998), and Wang and Zhengg (2008).

3.1.5 Why we should go beyond the traditional point of view

It should be noted that under the traditional point of view, whatever representations are used, whether rigid or flexible perspective, the scoring functions are based on spatial proximity, which neglects of considering other information such as AA mutation score and local structure substitution potential. It is observed that despite proteins in a family share a similar overall shape, their structures exhibit very high local flexibility due to evolutionary events (i.e., mutation, insertion, and deletion) at the sequence or local substructure level (Csaba et al., 2008). This kind of local conformation change due to evolutionary events cannot be accurately quantified by spatial proximity of aligned residues (e.g., after rigid-body superposition). Instead, evolutionary distance shall be a better measure. Inspired by this observation, we develop a method named DeepAlign that uses hydrogen-bonding similarity, plus the AA and local substructure substitution matrices, which are derived from evolutionarily related protein pairs, to align protein local structures (Wang et al., 2013). Experimental results show that DeepAlign can generate structure alignments much more consistent with manually curated alignments than other automatic tools especially when proteins under consideration are remote homologs. These results imply that in addition to spatial proximity, evolutionary information and hydrogen-bonding similarity are essential to aligning two protein structures.

3.2. The scoring functions to measure the structure similarity

The scoring functions to measure the protein structure similarity could be categorized into three groups that depend on whether the structural representation is three-dimensional, two-dimensional, or one-dimensional or one number characterizing the whole structure. For more details on scoring function, see Hasegawa and Holmm (2009).

3.2.1 Three-dimensional

Upon rigid-body superimposition, the similarity of 3D objects can be measured by the positional deviations of equivalent atoms. Numerous scoring functions have been proposed based on the rigid-body perspective that depends on the balance between the size of the common core (Ne) and residue positional deviations (root mean squared deviation, RMSD) (Hasegawa & Holm, 2009). For protein superposition, the most common metric is the coordinate root mean square deviation (cRMS) defined as,

$$\mathrm{cRMS}(A_i, B_j) = ||A - B|| = \sqrt{\frac{1}{n}\sum_{k=0}^{n-1}\left(a_{i+k} - b_{j+k}\right)^2} \qquad (5.3)$$

where A and B are two-aligned ungapped fragment pairs with length n, begin at a_i, b_j position from two input proteins (or in brief, $\langle i, j\rangle$ for later use). Other metrics that are derived from cRMS consist of STRUCTAL-score (Levitt & Gerstein, 1998), TMscore (Zhang & Skolnick, 2004), GDT-score (Zemla, 2003), just name a few. Instead of a single rigid-body transformation, explicit-flexible perspective-based methods concatenate together a series of fragments, which have tight local superimpositions. Two measurements have been proposed, one seeks to globally minimize the number of bends (Ye & Godzik, 2003), while the other allows flexibilities everywhere between short fragments (Menke et al., 2008).

3.2.2 Two-dimensional

The 2D measurement is based on the similarity of residue–residue interaction patterns between two proteins without superimposition. Such patterns can be described, for example, by representing the protein structures as distance matrices, contact maps, or Voronoi tessellations. Distance matrices can be calculated by intramolecular Cα–Cα distances (Holm & Sander, 1993). The distance root mean squared deviation (dRMS) that compares corresponding distance matrices in the two sets of points is calculated as,

$$\mathrm{dRMS}(A_i, B_j) = \sqrt{\frac{1}{n^2}\sum_{k=0}^{n-1}\sum_{l=0}^{n-1}\left(\left\|a_{i+k} - a_{i+l}\right\| - \left\|b_{j+k} - b_{j+l}\right\|\right)^2} \qquad (5.4)$$

where A and B are two-aligned ungapped fragment pairs with length n, begin at $\langle i, j\rangle$. One popular metric derived from dRMS is DALI-score (Holm & Sander, 1993). Moreover, contact maps can be generated by applying a certain distance threshold on distance matrices. Voronoi tessellations (Birzele et al., 2007), which get rid of the distance threshold to define contacting residues, create a mesh grid inside the protein structure such that every point of space is assigned to the nearest residue. The alignment methods that consider the 2D scoring functions could allow flexibility (i.e., implicit-flexible) as long as the interaction networks are conserved.

3.2.3 One-dimensional

Structural profiles, which classify each residue according to its AA type and/or discrete structural alphabets, could be applied to homology search by fast string comparison (Carpentier et al., 2005; Tung et al., 2007). Although these profiles have limited power to detect structural similarity between proteins that deviate largely, they can provide additional evolutionarily related information that goes beyond the spatial proximity that considers proteins as 3D and/or 2D representation. One example that considers the AA and local substructure substitution matrices is as follows:

$$\mathrm{Local}\left(A_i, B_j\right) = \sum_{k=0}^{n-1} \mathrm{CLESUM}(i+k, j+k) + \omega \cdot \mathrm{BLOSUM}(i+k, j+k) \quad (5.5)$$

where ω is a tunable parameter to judge the contribution of AA evolutionary information.

3.3. The search algorithms for pairwise structure alignment

The current search algorithms could be categorized into five groups: Monte Carlo, AFP-chaining, DP, maximize Ne with constraint RMSD, and double dynamic programming. In this section, we introduce six widely used and/or cited protein structure alignment methods that encompass a variety of representations, perspectives, scoring function types, and search algorithms. They are DALI, CE, TMalign/STRUCTAL, MATT/FATCAT, CLEPAPS/ProSup, and Vorolign. For more available pairwise structure alignment tools, see Table 5.4.

3.3.1 DALI (C-map, implicit-flexible, 2D, Monte Carlo)

The scoring function of DALI (Holm & Sander, 1993), as defined below, is derived from dRMS:

$$\mathrm{DALI}\left(A_i, B_j\right) = \sum_{k=0}^{n-1}\sum_{l=0}^{n-1}\left(0.2 - \frac{\Delta(i,k,j,l)}{d_{i,k,j,l}}\right) e^{-\left(d_{i,k,j,l}/20\right)^2} \quad (5.6)$$

where $\Delta(i,k,j,l) = |(||a_{i+k} - a_{i+l}|| - ||b_{j+k} - b_{j+l}||)|$ and $d_{i,k,j,l} = 0.5 \times |(||a_{i+k} - a_{i+l}|| + ||b_{j+k} - b_{j+l}||)|$. Then DALI uses the Monte Carlo sampling method to search for the best consistent set of SFPs to join into an alignment. The basic step in the Monte Carlo search is addition or deletion of residue equivalence assignments. The native score of DALI is a summation, overall aligned residue pairs in both structures, of a bonus score that is maximal when the inter-residue distances in both structures are equal. DALI uses many initial alignments and runs the searching algorithm in parallel. The final output alignment is the one with the best total score. A later version of DALI introduced an initial fast lookup of common SSEs between the two proteins (Holm & Rosenström, 2010). To determine the statistical significance of output alignments, DALI computes the Z-score of an alignment by using the background distribution of scores from an all-against-all comparison of 225 representative structures with less than 30% sequence identity (Holm & Sander, 1994). As an elastic aligner, DALI could align protein structures in cases where structural flexibilities exist. Due to Monte Carlo methodology, DALI could generate alignments in nonsequential order.

3.3.2 CE (C-map and C-alpha, rigid-body, 2D and 3D, AFP-chaining)

CE (Shindyalov & Bourne, 1998) has two stages during alignment, the initial stage and the refinement stage. For the initial stage, the scoring function of CE is based on dRMS (as shown in Eq. 5.4), and the search algorithm is combinatorial extension to successively join new aligned fragment pairs (AFPs) that is consistent with the previous joined AFPs on the alignment path. Such methodology is called, in brief, AFP-chaining. In particular, the AFPs are pairs of eight-residue fragment, which are considered similar if their corresponding internal distances are similar (of low dRMS). Gaps are allowed between neighboring AFPs, but their length is limited (less than 30 residues) to speedup. CE constructs the alignment by choosing an initial AFP and extending it. Even though using greedy algorithm during search, CE also considers AFPs that are not global optimal with respect to their scoring function to widen and improve the search. For the refinement stage, the scoring function of CE is based on cRMS (as shown in Eq. 5.3), and the search algorithm is to lengthen the alignment without compromising its cRMS. The native score of CE is a Z-score that evaluates the statistical significance of the alignment by considering the probability of finding an alignment of the same number of equivalent residues (Ne), number of gaps (Ngaps), and geometrical distance (measured by dRMS). Due to AFP-chaining methodology, CE could generate alignments in sequential order.

3.3.3 TMalign and STRUCTAL (C-alpha, rigid-body, 3D, DP)

TMalign (Zhang & Skolnick, 2005) is evolved from STRUCTAL (Levitt & Gerstein, 1998). The scoring function for both methods is a reciprocal-like cRMS score (as shown in Eq. 5.7), and the search algorithm is DP to optimize this score. In detail, both methods start from several initial alignments and the output alignment is the one with the best score. To align two protein structures, the methods iteratively get the optimal rigid-body transformation based on the current alignment by Kabsch method (Kabsch, 1976) and then find an optimal alignment based on the current transformation by DP on the reciprocal-like cRMS score matrix. The procedure is repeated till it converges to a local optimum. The other methods that apply DP on protein structure alignment include Yang et al. (2012). Due to DP methodology, TMalign and STRUCTAL could generate alignments in sequential order.

The reciprocal-like cRMS score that applied in TMalign and STRUCTAL is defined as follows:

$$
\begin{aligned}
r\_\mathrm{cRMS}\left(A_i, B_j\right) &= \sum_{k=0}^{n-1} d(i+k, j+k), \quad \text{where} \\
d(i,j) &= \frac{1}{1+\left(a_i - b_j\right)^2 / {d_0}^2}
\end{aligned}
\tag{5.7}
$$

However, the difference between them is that the TMalign score (called TMscore) applies a length-dependent $d_0 = 1.24 \times \sqrt[3]{L_s - 15} - 1.8$, where L_s is the smaller length of the input proteins, while STRUCTAL-score has a fixed value for $d_0 = 5.0\,\text{Å}$. Since the cRMS weights the distances between all residue pairs equally, a small number of local structural deviations could result in a high cRMS, even when the global topologies of the compared structures are similar. TMscore and STRUCTAL-score overcome the problem of cRMS by reweighting the residue pairs with smaller distances relatively stronger than those with larger distances. Therefore, these scores are more sensitive to the global topology than to the local structural variations. Moreover, TMscore is normalized in a way that the score magnitude relative to random structures does not depend on the proteins' size. TMscore around 0.17 means an average pair of randomly related structures (Zhang & Skolnick, 2004), and 0.5 means highly likely that they have similar folds (Xu & Zhang, 2010).

3.3.4 CLEPAPS and ProSup (C-alpha, rigid-body, 3D, maximize Ne with constraint RMSD)

CLEPAPS (Wang, 2009; Wang & Zheng, 2008) has two stages during alignment, the initial stage and the refinement stage. Here, we only focus on the refinement stage of CLEAPAPS, which is evolved from ProSup (Lackner et al., 2000). The goal in this stage is to maximize the number of structurally equivalent residues (i.e., Ne) subject to a fixed Euclidean distance cutoff (i.e., constraint RMSD). The search algorithm basically has four steps: (i) identify a seed SFP in both proteins, (ii) iteratively recruit more SFPs to form initial alignments based on the superposed seed SFP, (iii) iteratively apply a DP to select as many equivalent residues as possible while subject to a fixed distance cutoff, and (iv) evaluate the final alignment based on a certain scoring function. Note that, in the third step, other approaches instead of DP could be applied to maximize the Ne, such as the one used in CLEPAPS. In particular, guided by CLESUM score, it iteratively recruits those SFPs with a smaller length compared to those in the second step, to form the current alignments based on the superposition that is generated from previous alignment. During each iteration, the distance cutoff is lowered down gradually. Such approach, that might be called "Zoom-In," could generate alignments in nonsequential order, while the approach based on DP could result in sequential order alignments. The other methods that apply the similar refinement approach include Ortiz et al. (2002), Shindyalov and Bournee (1998), and Zemla (2003).

3.3.5 MATT and FATCAT (C-alpha, explicit-flexible, 3D, AFP-chaining)

MATT (Menke et al., 2008) is evolved from FATCAT (Ye & Godzik, 2003). Both methods belong to the AFP-chaining method. Like other structural aligners in this class, it first finds AFPs of between five and nine residues in each chain that share very close spatial proximity. Then, they greedily add these AFPs into a final structural alignment and extend these AFPs by adding adjacent residues into the interblock regions. What MATT and FATCAT differs from other AFP-chaining methods is that these two allow flexible (or bend) operations between adjacent AFPs. In particular, MATT allowing these impossible bends, translations, and twists everywhere between AFPs prior to the final extension phase. FATCAT globally minimizes the number of these flexible points. Therefore, MATT and FATCAT are able to detect regions in close spatial contact and incorporate them into an alignment, while other aligners would erroneously disallow such blocks from entering an alignment. Recently, another software ForMATT (Nadimpalli, Daniels, & Cowen, 2012) has been developed as an extension of MATT by taking into consideration primary sequence similarity in aligning two protein structures.

3.3.6 Vorolign (C-map, implicit-flexible, 2D, double dynamic programming)

Vorolign uses Voronoi tessellations to represent each structure. The method aligns protein structures using double dynamic programming (Birzele et al., 2007) and measures the similarity of two residues based on the evolutionary conservation of their corresponding Voronoi contacts in the protein structure. After the neighbor sets of each residue a_i and b_j in the two structures A

and B are defined by Voronoi Tessellation, the first level DP is conducted to calculate the similarity of two neighbor sets. Then the calculated score for $\langle i, j \rangle$ is used to fill the second-level DP that is applied to calculate the similarity between the two proteins. This similarity measurement allows Vorolign to align protein structures in cases where structural flexibilities exist. The other method that applies double dynamic programming on protein structure alignment includes Taylor and Orengoo (1989).

3.4. DeepAlign: The approach beyond spatial proximity

3.4.1 CLEPAPS

DeepAlign (Wang et al., 2013) is evolved from CLEPAPS (Wang & Zheng, 2008) which is the first trial to go beyond spatial proximity. CLEPAPS distinguishes itself from other existing algorithms by the use of a certain type of structural alphabets, say CLE. Guided by the substitution matrix CLESUM that is available to measure the similarity between any two such letter string by simple string comparison, CLEPAPS regards a SFP as an ungapped string pair with a high sum of pairwise CLESUM scores. A highly scored SFP which is spatially consistent, under cRMS measurement, with several other SFPs determines an initial alignment. CLEPAPS then joins consistent SFPs guided by their similarity scores to extend the alignment by several iteration steps (i.e., the "Zoom-In" approach, as described previously).

3.4.2 Scoring function

Instead of using structural alphabets alone, DeepAlign extends the scoring function considering AA mutation score, local substructure substitution potential, hydrogen-bonding similarity, as well as considering a reciprocal-like cRMS to measure the spatial proximity. In particular, the equivalence of two residues a_i and b_j from two input proteins is estimated by the following scoring function:

$$\text{DeepScore}\,(i,j) = (\max(0,\text{BLOSUM}(i,j)) + \text{CLESUM}(i,j)) \times d(i,j) \times v(i,j) \tag{5.8}$$

Meanwhile, BLOSUM and CLESUM measure the evolutionary distance of two proteins at the sequence and local substructure levels, respectively. BLOSUM is the widely used AA substitution matrix BLOSUM62 (Henikoff & Henikoff, 1992), CLESUM is the local structure substitution matrix, $d(i,j)$ measures the spatial proximity of two aligned residues after rigid-body superposition with a reciprocal-like cRMS form (see Eq. 5.7), and $v(i,j)$ measures the hydrogen-bonding similarity,

$$v(i,j) = \frac{1}{3}\left(\sum_{x=\{-1,1\}} \frac{(a_i - a_{i-x})(b_j - b_{j-x})}{|a_i - a_{i-x}||b_i - b_{i-x}|} + \frac{(a_i - a_{i\_cb})(b_j - b_{j\_cb})}{|a_i - a_{i\_cb}||b_j - b_{j\_cb}|}\right) \tag{5.9}$$

where a_{i_cb} denotes the corresponding Cβ atom of a_i, as shown in Fig. 5.2A. This score helps to align hydrogen bonds more accurately. As shown in Fig. 5.2B, the method that optimizes only spatial proximity (e.g., TMscore) leads to a wrong alignment, which can be corrected by incorporating $v(i,j)$ to the scoring function.

In addition to the BLOSUM62 substitution matrix, other matrices (e.g., PAM250) can also be used to measure the evolutionary distance of two proteins at the sequence level (Dayhoff & Schwartz, 1978). The max() function in Eq. (5.8) is used to handle the situation where two proteins to be aligned are distantly related. In this case, we will only rely on CLESUM to measure the evolutionary distance. It is shown that replacing the max() function to a tunable parameter ω like Eq. (5.5) also works well. In the future, we may use sequence profile similarity to measure evolutionary distance, which usually is more sensitive than BLOSUM matrices. Moreover, CLESUM disfavors

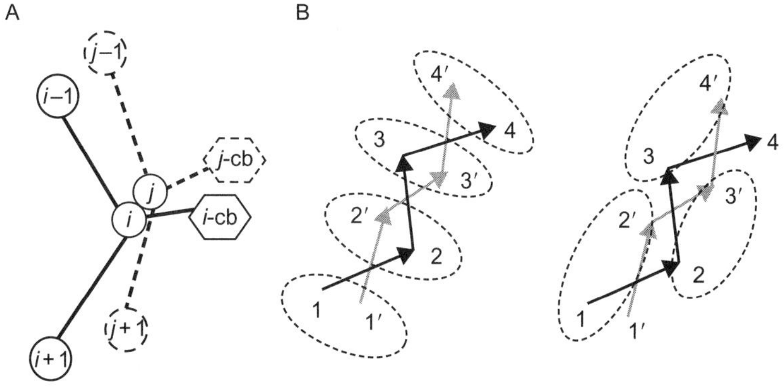

Figure 5.2 The definition and application of the hydrogen-bonding score. (A) This picture shows the three vectors of two proteins used in the hydrogen-bonding score. One protein is represented by solid lines and the other by dashed lines. (B) An illustration of one wrong alignment between two beta-strands. Residues 1, 2, 3, and 4 in dark belong to protein A and residues 1′, 2′, 3′, and 4′ in gray belong to protein B. The aligned residue pairs are in dotted circle. On left is the wrong alignment generated by optimizing only TMscore; while on right is the correct alignment optimizing the product of hydrogen-bonding score and TMscore.

the match of two unrelated helices but favors the alignment of two evolutionarily related loop regions. Loop regions are usually harder to align than alpha-helices and beta-strands if only spatial proximity is used in the scoring function.

3.4.3 Search algorithm

Overview. The DeepAlign algorithm flowchart is illustrated in Fig. 5.3. It consists of the following steps: (i) identifying SFPs using AA and local substructure mutation matrices, (ii) generating an initial alignment from one SFP, (iii) refining alignments by DP, and (iv) gap elimination.

Similar fragment pairs. DeepAlign measures the equivalence of two residues $\langle i, j \rangle$ using AA and local substructure substitution matrices as follows:

$$\text{Similarity}(i,j) = \max(0, \text{BLOSUM}(i,j)) + \text{CLESUM}(i,j) \qquad (5.10)$$

where CLESUM is the local conformation substitution matrix and BLOSUM is the AA substitution matrix as described before. Again, using Eq. (5.5) to calculate the similarity also works well. Using this score, we can identify two evolutionary-related instead of only geometric similar fragments and thus generate better initial alignments. We use two types of SFPs: short SFP with 6–8 residues and long SFP with 9–18 residues. A short SFP, denoted as SFP_s, shall have a similarity score at least 0 while a long SFP, denoted as SFP_L, shall have a similarity score at least 10. It is obvious that each SFP_L must contain at least one SFP_s. We use SFP_L and SFP_s to build coarse-grained and fine-grained initial alignments, respectively. SFP_L is slightly longer than the average length of a helix, while SFP_s has a similar length as a typical beta strand. By combining long and short SFPs, we can speed up our algorithm without losing accuracy. The higher score one SFP has, the more likely it is contained in the best alignment. Therefore,

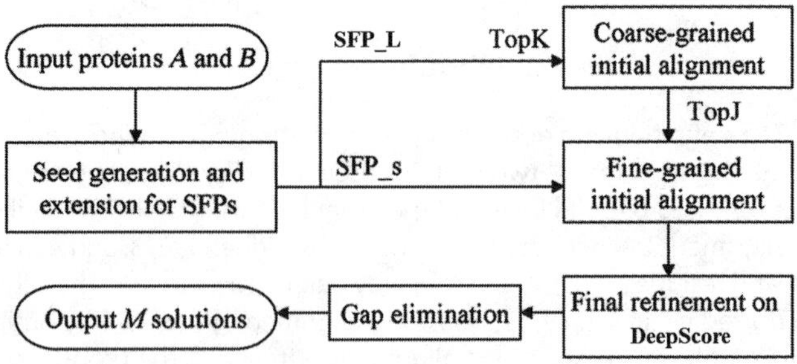

Figure 5.3 The DeepAlign algorithm flowchart.

we sort all SFPs and only keep those top-ranked SFPs. The idea of using two types of SFPs is derived from CLEPAPS (Wang & Zheng, 2008).

Generating initial alignments using SFPs. We select TopK long SFPs (i.e., SFP_L) and from each of them generate one coarse-grained initial alignment. In particular, we first calculate the rotation matrix using the Kabsch method (Kabsch, 1976) to minimize the RMSD of the two fragments in a SFP. Then we use this rotation matrix to transform one protein and generate an initial alignment using DP to maximize the scoring function, subject to the restriction that the distance deviation of two-aligned residues shall be less than $3 \times d_0$ (d_0 is defined in Eq. 5.7, which is length dependent). All these TopK coarse-grained initial alignments are sorted by the alignment score and only top TopJ (<=TopK) are kept for further refinement. Starting from a coarse-grained initial alignment, we recalculate the rotation matrix using the SFP_s contained in the SFP_L and then realign the two proteins using DP to maximize the scoring function, which results in a better initial alignment.

Iterative refinement of alignment. Starting from an initial alignment, an iterative DP refinement procedure is applied to improving the alignment, with the goal to maximizing the scoring function. This procedure is very similar to that in many structure alignment methods such as Lackner et al. (2000), Shindyalov and Bournee (1998), and Zhang and Skolnickk (2005).

Gap elimination. As shown in Holm and Sanderr (1993), an AFP shall not be too short (say less than four residues). However, since our scoring function does not explicitly penalize gap openings, the resultant alignment may have more gap openings than desirable. To deal with this, we use some heuristics to merge one very short AFP (less than four residues) to its neighboring AFPs to reduce the number of gap openings.

3.5. Result analysis

3.5.1 Overview

Common evaluation tests for structure alignment methods measure first, the accuracy of the alignments; second, the ability of the alignment score to discriminate homologous from unrelated proteins in database-wide comparisons (Hasegawa & Holm, 2009). Here, we first evaluate our program DeepAlign using three manually curated alignment databases on a few metrics: LALI (length of alignment), RefAcc (reference-dependent alignment accuracy), RMSD, TMscore, and mutation scores (i.e., BLOSUM and CLESUM). RefAcc is calculated as the percentage of correctly aligned positions as judged by the gold standard (i.e., manual alignments), measuring consistency between automatic alignments and human-curated alignments. We then use a database-wide benchmark to test the performance of DeepAlign in identifying distant homologs and structural analogs. We compare DeepAlign with several popular structure alignment tools such as DALI (Holm & Sander, 1993), TMalign (Zhang & Skolnick, 2005), MATT (Menke et al., 2008), and ForMATT (Nadimpalli et al., 2012), which represent four very different methods. Finally, we also test the performance of multiple solutions of DeepAlign on the three manually curated alignment databases.

3.5.2 The evaluation benchmarks

We use three manually curated benchmarks: (i) A subset of CDD (conserved domain database; Marchler-Bauer et al., 2005) used in Kim and Lee (2007), (ii) MALIDUP (Cheng, Kim, & Grishin, 2008a), and (iii) MALISAM (Cheng, Kim, & Grishin, 2008b) to evaluate the reference-dependent alignment accuracy. In addition, we use (iv) SABmark (Van Walle, Lasters, & Wyns, 2005) to test the performance for identifying distant homologs and structural analogs. The CDD set contains 3591 manually curated pairwise structure alignments. The human-curated alignments for CDD contain only the alignments of core residues. The CDD set has already been used to evaluate a bunch of pairwise structure alignment algorithms, including CE (Shindyalov & Bourne, 1998), FAST (Zhu & Weng, 2005), LOCK2 (Shapiro & Brutlag, 2004), MATRAS (Kawabata, 2003), VAST (Gibrat et al., 1997), and SHEBA (Jung & Lee, 2000). MALIDUP has 241 manually curated pairwise structure alignments for homologous domains originated from internal duplication within the same polypeptide chain. About half of the pairs in MALIDUP are remote homologs. The alignments in these two databases are manually curated, taking into consideration not only geometric similarity but also evolutionary and functional relationship. Therefore, the manually curated alignments make more biological sense, and it is reasonable to use them as reference to judge automatically generated alignments. SABmark (version 1.65) benchmark contains SABmark-sup and SABmark-twi. SABmark-sup is the superfamily set in SABmark, containing 425 protein groups with low to intermediate sequence identity. SABmark-twi is the twilight set in SABmark, containing 209 groups with low sequence identity. Each SABmark-sup (-twi) group contains at most 25 structures sharing a SCOP (Murzin, Brenner, Hubbard, & Chothia, 1995) superfamily (fold). It is believed that if two proteins are in the same SCOP superfamily, it is likely these two proteins are remote homologs. If two proteins share only the same SCOP fold, it is very likely that they are structural analogs instead of remote homologs.

3.5.3 Performance on CDD

DeepAlign obtains the highest reference-dependent alignment accuracy of 93.8% among the five automatic structure alignment methods (Table 5.5). DeepAlign also outperforms the methods evaluated in Kim, Tai, and Lee (2009) in terms of ref-dependent alignment accuracy. That is, DeepAlign is more consistent with human experts than the other programs. In terms of TMscore and RMSD, the TMalign alignments are slightly better than the DeepAlign alignments, but the former are less consistent with manual alignments than the latter. This implies that the geometric similarity score used by TMalign (i.e., TMscore) does not accurately reflect the alignment criteria used by human experts. The DeepAlign alignments also have much better evolutionary scores than the other three programs no MATTer how the mutation scores are calculated. As a control, we also calculate the evolutionary scores of the manual alignments. The manual alignments have much lower mutation score per alignment because only core residues are aligned. However, the manual alignments have the best average mutation scores per aligned position. Note that the manual alignments are not explicitly driven by a specific mutation score. This confirms that human experts indeed take into consideration evolutionary relationship in aligning two protein structures and that TMalign may align many more evolutionarily unrelated residues together than DeepAlign. ForMATT has a similar mutation (i.e., BLOSUM/CLESUM) score per alignment as DeepAlign, but ForMATT has a better average mutation score per aligned position than DeepAlign because ForMATT has a smaller LALI.

3.5.4 Performance on MALIDUP

DeepAlign obtains a reference-dependent alignment accuracy of 92%, greatly exceeding the other three tools (Table 5.5). DeepAlign is 6% better than the second best algorithm DALI. Although the TMalign alignments have a longer alignment length and the MATT alignments have a smaller RMSD, both TMalign and MATT have much lower reference-dependent alignment accuracy. This again implies that the TMalign and MATT scoring functions greatly deviate from what are implicitly used by human experts. In terms of TMscore, DeepAlign is only slightly second to TMalign, but better than the others. However, the DeepAlign alignments have much better evolutionary scores, only second to the manual alignments in terms of the

Table 5.5 Performance of five pairwise structure alignment tools on three benchmarks CDD, MALIDUP, and MALISAM

Method	LALI	RMSD	TMscore	RefAcc	Blosum1	Clesum1	Blosum2	Clesum2
CCD (3591)								
DeepAlign	134.8	2.86	0.667	93.8	0.261	1.782	43.45	243.71
DALI	130.8	2.75	0.663	92.8	0.165	1.684	28.78	225.15
MATT	128.6	2.53	0.655	91.4	0.152	1.728	30.19	229.59
ForMATT	112.3	2.32	0.566	86.4	0.343	1.983	44.11	235.64
TMalign	138.4	2.84	0.686	85.6	0.047	1.531	15.25	211.88
Manual	62.6	1.66	0.345	100.0	0.677	2.499	43.89	157.67
MALIDUP (241)								
DeepAlign	85.5	2.61	0.622	92.0	0.314	1.872	29.31	158.28
DALI	83.5	2.65	0.600	86.4	0.172	1.700	18.63	147.53
MATT	82.3	2.47	0.608	79.8	0.178	1.824	18.84	150.00
ForMATT	70.6	2.19	0.542	86.2	0.344	2.196	28.62	154.66
TMalign	87.0	2.62	0.631	81.0	0.110	1.600	12.50	137.64
Manual	77.9	2.49	0.587	100.0	0.294	1.853	27.67	154.81

MALISAM (130)								
DeepAlign	61.3	2.96	0.521	77.5	−0.601	1.108	−36.48	67.66
DALI	61.0	3.11	0.515	67.7	−0.595	0.925	−35.52	56.28
MATT	56.2	2.74	0.486	51.7	−0.625	1.013	−34.05	56.98
ForMATT	44.9	2.42	0.411	56.3	−0.486	1.489	−21.1	65.69
TMalign	61.1	3.06	0.517	53.7	−0.684	0.739	−40.04	45.65
Manual	56.7	2.92	0.488	100.0	−0.556	1.240	−31.58	70.75

See text for the explanation of LALI, RMSD, TMscore, and RefAcc. "Blosum1 (Clesum1)" is the average mutation score per aligned position, while "Blosum2 (Clesum2)" is the average mutation score per alignment. As a control, the performance of manually curated alignments is also shown in the table.

average mutation score per aligned position. Since the TMalign alignments on average are longer, this again confirms that TMalign may align many more evolutionarily unfavorable residues than DeepAlign. ForMATT performs similarly on this dataset as on CDD.

3.5.5 Performance on MALISAM

DeepAlign obtains the highest ref-dependent alignment accuracy of 77.5% among all the five computer programs (Table 5.5). DeepAlign is 10% better than the second best algorithm DALI. MALISAM is much more challenging than CDD and MALIDUP. In MALISAM, 80 pairs (i.e., 61.5% of the total) contain proteins with different SCOP folds. The DALI and TMalign alignments have similar average alignment lengths as the DeepAlign alignments, but slightly higher RMSD. Furthermore, the DALI, MATT, and TMalign alignments deviate significantly from the manual alignments. In terms of the BLOSUM scores, the difference between the DeepAlign alignments and others is not very significant. This is not unexpected because the proteins in this dataset are only weakly similar at sequence level and BLOSUM is not sensitive enough. However, the DeepAlign alignments have much better CLESUM score than the others, only slightly second to the manual alignments. That is, the DeepAlign alignments are more evolutionarily favorable than others at the local substructure level.

3.5.6 Performance on SABmark

Given a protein structure, we align it to all the proteins in the benchmarks and then rank all the alignments by certain criteria. We examine if the top-ranked protein structures are in the same group as the query protein or not. DeepAlign uses its scoring function, say DeepScore in Eq. (5.8), to rank the proteins. Similarly, TMalign, MATT, and DALI use TMscore, *P*-value, and *Z*-score to rank the alignments, respectively. The ranking results are evaluated by ROC (receiver operator curve) and AUC (area under curve). ForMATT has a very similar result as MATT. As shown in Fig. 5.4A, tested on SABmark-sup, DeepAlign has the best ROC curve, especially at the high specificity area. For example, at the specificity level 0.99, DeepAlign has sensitivity around 0.4, while the other three have sensitivity only around 0.2. We also observe the same trend on the SABmark-twi set. SABmark-twi is more challenging because each group in this set consists of proteins similar at only SCOP fold level. However, DeepAlign outperforms others by an even larger margin. As shown in Fig. 5.4B, DeepAlign has sensitivity 0.6 at specificity 0.96, while the second best algorithm DALI has sensitivity only 0.4 at

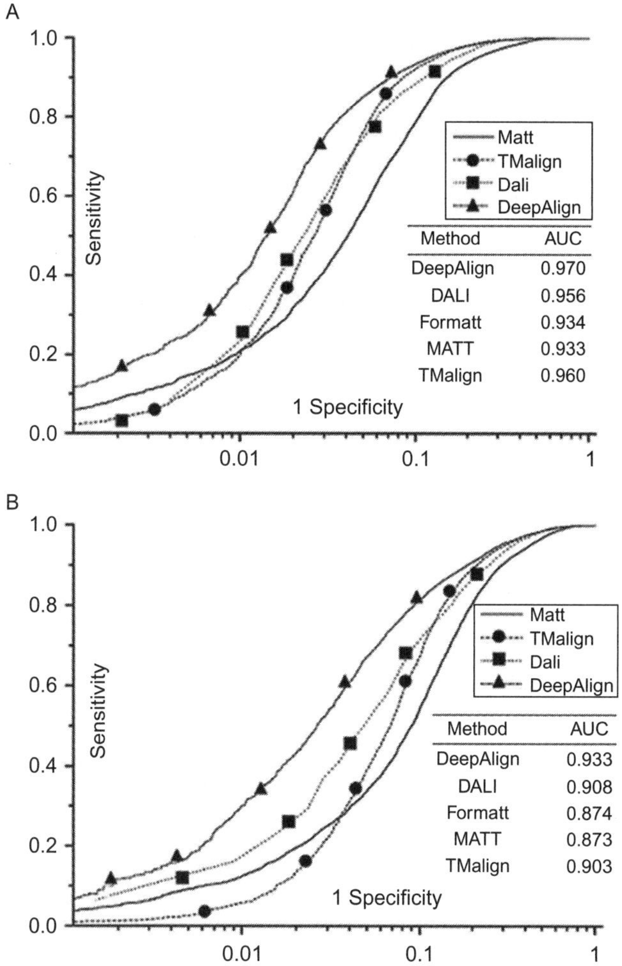

Figure 5.4 The ROC curves and AUC values by DeepAlign, DALI, MATT, and TMalign on (A) SABmark-sup and (B) SABmark-twi. The ROC curves for ForMATT are not in the figure since they are very similar to those for MATT.

the same specificity level. MATT and TMalign have only sensitivity around 0.2 at this specificity level. These results imply that DeepAlign scoring function is better than DALI's *Z*-score, MATT's *P*-value, and TMscore in detecting the superfamily relationship of proteins.

3.5.7 Multiple solutions

Due to symmetry and repeats, some protein pairs may have multiple correct alignments. To deal with this, DeepAlign generates a few alternative alignments for a given protein pair instead of just a single alignment with the highest DeepScore (see Eq. 5.8). DeepAlign fulfills this by searching for the best alignments from many initial alignments and ranking the alignments by DeepScore. For a given protein pair, we generate the topM ($M \leq 10$) alternative alignments using DeepAlign and then pick up the one with the highest ref-dependent alignment accuracy. Table 5.6 lists the average ref-dependent accuracy of DeepAlign on the three benchmarks with respect to M. As shown in this table, we can significantly improve alignment accuracy by generating two alternative alignments. Nevertheless, using ≥ 3 alternative alignments can only slightly improve the accuracy over using two alternatives. When only two alternative alignments are used, DeepAlign obtains alignment accuracy 95.1%, 92.7%, and 87.6% on CDD, MALIDUP, and MALISAM, respectively. As a control, for each protein pair, we also calculate alignment accuracy using the best alignment generated by DeepAlign (only the first-ranked alignment is used), DALI, TMalign, and MATT. It turns out that when only top two alternative alignments are used we can obtain alignment accuracy comparable to the combination of the four tools. Finally, we provide one example to illustrate the application of multiple solutions for the analysis of the symmetry proteins alignment (shown in Fig. 5.5).

Table 5.6 The best ref-dependent alignment accuracy of the topM alternative alignments returned by DeepAlign on three manually curated benchmarks CDD, MALIDUP, and MALISAM

DeepAlign topM	CDD (3591)	MALIDUP (241)	MALISAM (130)
Top1	93.8	92.0	77.5
Top2	95.1	92.7	87.6
Top3	95.2	93.1	88.0
Top4	95.3	93.1	88.9
Top5	95.3	93.1	89.7
Top10	95.3	93.1	90.0
Maximal of four[a]	96.9	94.2	85.6

[a]For each protein pair, the best alignment generated by four pairwise structure alignment tools DALI, TMalign, Matt, and DeepAlign is used to calculate the ref-dependent alignment accuracy.

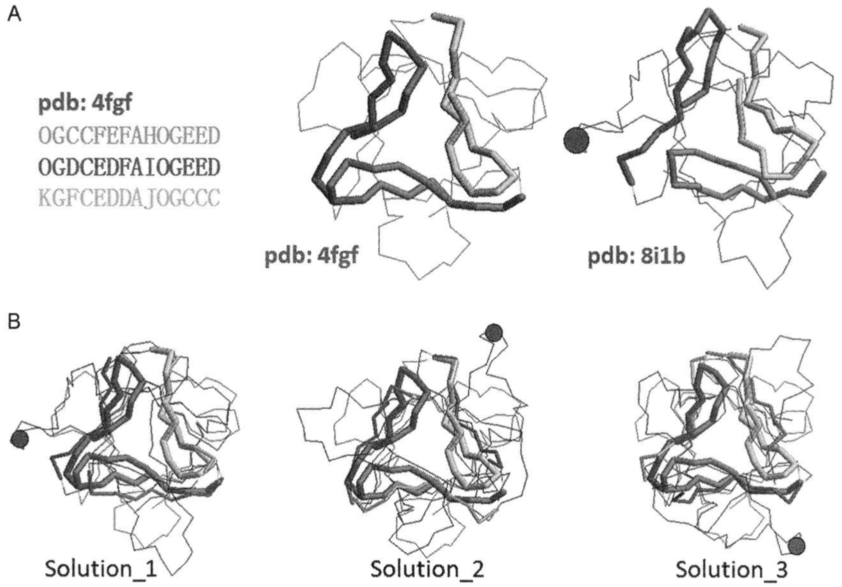

Figure 5.5 Three solutions reported by DeepAlign for the proteins 8i1b (in blue) and 4fgf (in red), that belong to SCOP fold beta-trefoil. (A) This fold exhibits a threefold rotational symmetry, which is confirmed by the fragment shown with bold lines in color orange, purple, and yellow, respectively. The corresponding conformational letter is also shown in left. (B) Superposition of the three solutions with 4fgf fixed. (For interpretation of the references to color in this figure legend, the reader is referred to the online version of this chapter.)

4. PROTEIN MSA

4.1. Introduction of protein MSA

4.1.1 Why we need the protein MSA

MSA carries significantly more information than pairwise alignment and hence has been extensively used for classification, analysis of evolutionary relationship, motif detection, and structure/function prediction (Wang et al., 2011). When proteins are distantly related, sequence methods usually fail to yield accurate alignment. In contrast, structure methods, which exploit geometrical information, may still work well. As more protein structures are experimentally solved, MSA is becoming more useful and important. The applications of MSA programs include understanding evolutionary conservation and divergence (Andersen et al., 2001), functional prediction through the identification of structurally conserved active sites in homologous proteins (Irving et al., 2001), construction of benchmark datasets on which to test multiple sequence alignment programs (Edgar & Batzoglou, 2006), and automatic construction of profiles and threading templates for protein structure prediction (Dunbrack, 2006; Panchenko et al., 1999). However, developing computational methods for accurate MSA, especially of a large set of distantly related protein structures, is still regarded as an open challenge (Wang & Zheng, 2009).

4.1.2 The goal of the protein MSA

The common goal of all MSA methods is to identify a set of residue "columns" from each "row" protein that are structurally similar (Wang & Zheng, 2009). For a given MSA, the aligned blocks, which correspond to contiguous columns, is composed of locally similar fragments. This local similarity within a block may be phrased as "vertical equivalency." The local similarity is necessary to the alignment but is insufficient. For any two structures in the MSA, the transformation to superimpose an AFP in an aligned block should also bring the fragment pairs in other blocks spatially close. This is the "horizontal consistency." In some MSA methods, these aligned blocks are called "cores." The full core (partial core) is defined as the aligned blocks that contain all (some) of the input proteins.

4.1.3 Limitations of the current approaches

A protein MSA method consists of two major components: how the pairwise structure alignment is conducted and how to merge all input structures to form a MSA. The current MSA methods to deal with the merging method could be categorized into horizontal-first and vertical-first approaches. The horizontal-first approaches (Lupyan, Leo-Macias, & Ortiz, 2005; Micheletti & Orland, 2009; Ye & Godzik, 2005) progressively merge pairwise alignments into an MSA, which not only might be slow with the increase of the input protein number but also might be suboptimal since pairwise alignment errors carry over to the final result (Wang et al., 2011; Wang & Zheng, 2009). The vertical-first approaches (Dror, Benyamini, Nussinov, & Wolfson, 2003; Shatsky et al., 2004) identify some SFBs among proteins and then extend the SFB alignments to MSAs. The number of SFBs could grow exponentially with respect to the number of proteins, so these methods may have to examine a large number of SFBs in order to not miss the best MSA, which is usually computationally expensive (Wang et al., 2011). As such, the challenge facing a vertical-first method is to identify only those SFBs, which are very likely contained in the best MSA. Recently,

Ilinkin et al. (2010) have developed a consensus-first approach. Starting from an initial consensus structure, it iteratively pairwise aligns each input structure upon the consensus and updates the consensus based on the current MSA. However, the choice of the initial consensus structure would be biased to suboptimal result. Moreover, the pairwise alignment errors would accumulate to deteriorate the consensus structure.

4.1.4 Solutions for the current limitations

In contrary to the sequence alignment, any insignificant trial alignment for structures can be detected by structure superposition and then excluded. Thus, it is practicable to select from locally SFBs to build up a scaffold, and then upon the scaffold to construct the final multiple alignment. In particular, a scaffold, which is defined as one pivot structure plus several SFBs, contains more information than the initial consensus structure as applied in Ilinkin et al. (2010). By checking the spatial consistency for those SFPs between the pivot and the other structure, we could directly know whether this structure is anchored or not (see Fig. 5.6). For unanchored structures, we could realign them upon the pivot by the guidance of these SFBs, which could reduce the pairwise alignment errors. In conclusion, this approach could be regarded as scaffold-first, and the solutions for the current limitations are (i) using HSFB (highly similar fragment block) to solve the exponential nature of SFB with respect to the number of proteins, (ii) constructing scaffold from ranked HSFBs by spatial consistency to get a better initial start, (iii) aligning each proteins upon this scaffold by checking both horizontal consistency and vertical equivalency to reduce the pairwise alignment errors, and (iv) finally, refining the whole MSA by realigning each input structure to the consensus structure.

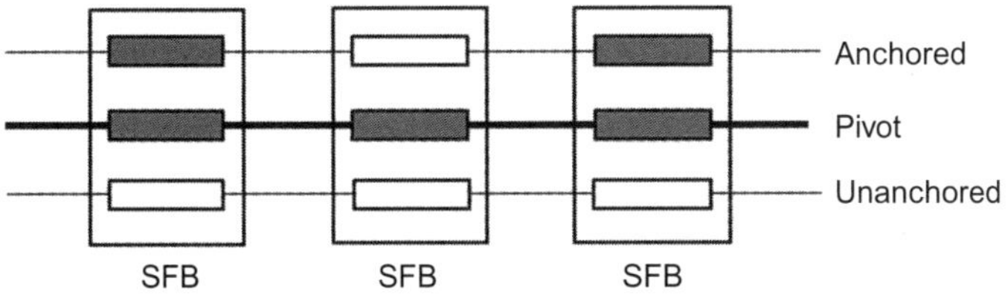

Figure 5.6 This picture shows the basic structure of a scaffold that contains one pivot structure plus several similar fragment blocks (SFBs). Given a superposition, we could check the spatial consistency (dark colored) for those similar fragment pairs (SFPs) between the pivot and the other structure to directly know whether this structure is anchored or not. (For color version of this figure, the reader is referred to the online version of this chapter.)

4.2. The current approaches for MSA

In this section, we introduce four protein MSA methods. They are MUSTANG (Konagurthu et al., 2006), CE-MC (Guda, Lu, Scheeff, Bourne, & Shindyalov, 2004), MASS (Dror et al., 2003), and MAPSCI (Ilinkin et al., 2010). For more tools, see Table 5.7.

4.2.1 MUSTANG (C-map, implicit-flexible, 2D, horizontal-first)

MUSTANG (Konagurthu et al., 2006) uses a progressive pairwise framework that derived from DALI (Holm & Sander, 1993) to build the final MSA. At the core of the method is a robust scoring scheme for pairwise alignments. This makes the use of a simple DP algorithm possible, instead of Monte Carlo, for all pairs of structures in the set, in order to gather accurate pairwise residue–residue equivalences. Before the final multiple alignment is constructed along a guide tree, a special extension phase is undertaken in which the pairwise scores of correspondences are recalculated in the context of all the remaining structures. This procedure significantly reduces the effect caused by making incorrect greedy choices while building the final alignment. The DALI-based scoring scheme lends it the ability to handle conformational flexibilities, such as hinge rotations.

4.2.2 CE-MC (C-alpha, rigid-body, 3D, horizontal-first)

CE-MC (Guda et al., 2004) uses CE (Shindyalov & Bourne, 1998) to perform all pairwise alignments. The algorithm refines a set of pairwise structure alignments using a Monte Carlo optimization technique. In particular, the algorithm iteratively modifies the MSA. The initializations are pairwise alignments of a pivot structure against all other structures, by a random set of moves. The random moves are then accepted with a probability proportional to the gain in the alignment score, which is calculated from the column distances in aligned blocks, using the form of reciprocal-like cRMS score (shown in Eq. 5.7). The iterative process is stopped when the optimal alignment cannot be improved by random moves.

4.2.3 MASS (SSE, rigid-body, 3D, vertical-first)

MASS (Dror et al., 2003) decomposes structures into SSEs. For each element, MASS treats them as vectors in 3D space and applies geometric hashing to calculate sequence order independent alignments. The SSE alignment defines an initial superimposition. Then, MASS extends this SSE alignment to a global optimal alignment by searching C-alpha atom positions close in space. The main advantage of MASS is the running speed since

Table 5.7 A selected list of multiple structure alignment tools

Name	Comments	Represent	Perspective	Score	Merge method	Order[a]	Available link	References
MASS	Multiple alignment by secondary structure	SSE	Rigid-body	3D	Vertical-first	N	http://bioinfo3d.cs.tau.ac.il/MASS/	Dror et al. (2003)
CE-MC	Combinatorial extension Monte Carlo	C-alpha	Rigid-body	3D	Horizontal-first	Y	http://schubert.bio.uniroma1.it/CEMC/	Guda et al. (2004)
MultiProt	Finds the common geometrical cores between the input proteins	C-alpha	Rigid-body	3D	Vertical-first	Y/N	http://bioinfo3d.cs.tau.ac.il/MultiProt/	Shatsky et al. (2004)
POSA	Partial order structure alignment	C-alpha	Explicit-flexible on fragment	3D	Horizontal-first	Y	http://fatcat.burnham.org/POSA/	Ye and Godzik (2005)
Mammoth-Mult	MAMMOTH-based MULTtiple structure alignment	Profile and C-alpha	Rigid-body	1D and 3D	Horizontal-first	Y	http://ub.cbm.uam.es/software/online/mamothmult.php	Lupyan et al. (2005)
MUSTANG	DALI-based multiple structure alignment	C-map	Implicit-flexible	2D	Horizontal-first	Y	http://www.csse.monash.edu.au/~karun/Site/mustang.html	Konagurthu et al. (2006)
MATT	Multiple alignment with translations and twists	C-alpha	Explicit-flexible everywhere	3D	Horizontal-first	Y	http://groups.csail.mit.edu/cb/matt/	Menke et al. (2008)

Continued

Table 5.7 A selected list of multiple structure alignment tools—cont'd

Name	Comments	Represent	Perspective	Score	Merge method	Order	Available link	References
BLOMAPS	Conformation letter block-based multiple alignment	Profile and C-alpha	Rigid-body	1D and 3D	Scaffold-first	N	http://www.itp.ac.cn/~zheng/blomaps.rar	Wang and Zheng (2009)
MISTRAL	Energy-based multiple structural alignment of proteins	Profile and C-alpha	Rigid-body	1D and 3D	Horizontal-first	N	http://eole2.lsce.ipsl.fr/ipht/mistral/protein.php	Micheletti and Orland (2009)
MAPSCI	Multialign of protein structure and consensus Identification	C-alpha	Rigid-body	3D	Consensus-first	Y	http://www.geom-comp.umn.edu/mapsci/index.html	Ilinkin et al. (2010)
3DCOMB	Local and contact pattern with machine learning trained	C-map and C-alpha	Rigid-body	2D and 3D	Scaffold-first	Y	http://ttic.uchicago.edu/~jinbo/3DCOMB/3DCOMB_exe_V1.06.7z	Wang et al. (2011)

[a] Y/N indicates that the algorithm has an option to let the alignment result be in sequential or in nonsequential order.

it reduces the structural complexity to SSEs and applies geometric hashing. The second advantage of MASS is the capability of detecting the partial cores.

4.2.4 MAPSCI (C-alpha, rigid-body, 3D, consensus-first)

MAPSCI (Ilinkin et al., 2010) is a consensus-first approach that contains three major steps. To select the initial consensus structure by four different choices, to compute an initial MSA for all other structures against the consensus structure by a certain pairwise alignment tool, to iteratively get the optimal consensus structure as well as the rigid-body transformations based on the current MSA, and then to update each pairwise alignment between the consensus structure based on the current transformation by DP. MAPSCI considers four choices for initial consensus structure, they are the protein of median length, that minimizes the sum of the pairwise distances to all the other proteins, with the smallest maximum pairwise distance, and that generates the largest initial core.

4.3. 3DCOMB: The scaffold-first approach for MSA

4.3.1 BLOMAPS

3DCOMB (Wang et al., 2011) is evolved from BLOMAPS (Wang & Zheng, 2009), which is the first trial to use the scaffold-first approach for MSA. By means of the CLEs, BLOMAPS turns a structural fragment into a string, and two strings with their CLESUM score being higher than a preset threshold form a SFP. A string from one protein as a seed and its highly similar fragments from other proteins form a SFB. BLOMAPS consists of several steps including finding HSFBs guided by CLESUM, removing block redundancy, constructing scaffold by checking consistency in spatial arrangement among fragments from different blocks, dealing with unanchored structures, and the final step of refinement where the average template for alignment is obtained.

4.3.2 Overview of 3DCOMB

As shown in Fig. 5.7, 3DCOMB first generates a list of pivot structures. By default, this list contains all M input proteins, so TopK is equal to M. For each pivot structure, 3DCOMB uses the CRF model to generate HSFBs, which are ranked by their spatial consistency scores and only the TopJ with the highest scores are extended to initial MSAs (i.e., the scaffold). 3DCOMB identifies those "unanchored" proteins which are not well aligned to the pivot. To improve an initial MSA, 3DCOMB conducts TopF trials to

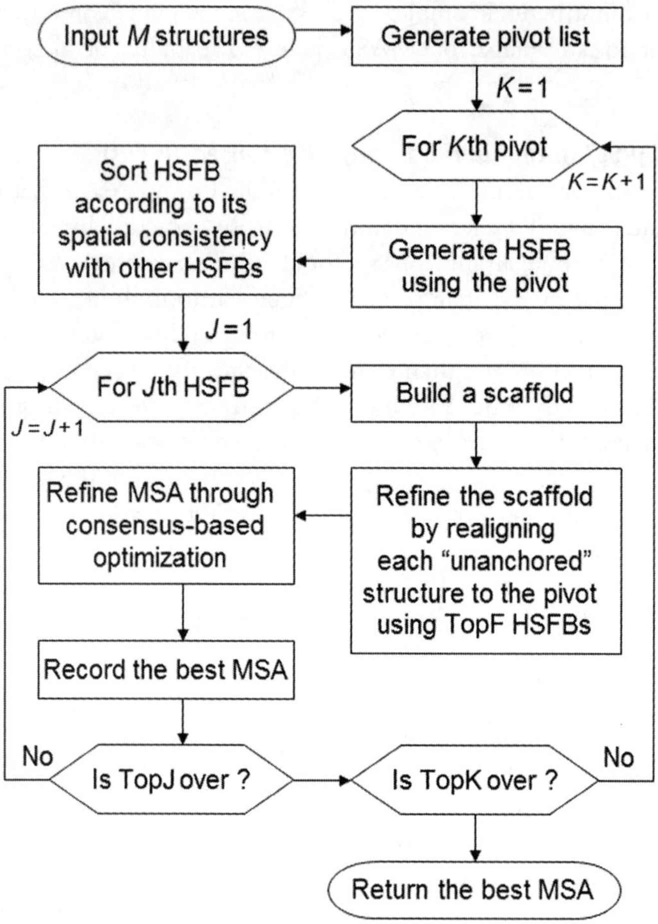

Figure 5.7 The 3DCOMB algorithm flowchart.

realign each of the "unanchored" proteins to the pivot. Finally, 3DCOMB refines the whole MSA based on the consensus structure derived from the MSA.

4.3.3 Detecting HSFBs

Given two protein structures, we can calculate the marginal probability of two fragments being aligned using the forward–backward algorithm (Lafferty, McCallum, & Pereira, 2001). In brief, the marginal probability for each residue pair $\langle i, j \rangle$ is calculated by a CRF model that uses four local features and two global features. The local features measure the ungapped segment pairs with length 9,13,17,21 centering at $\langle i, j \rangle$, while the global features measure the environment fitness sphere with radius 8.0 and 14.0 Å that centers at $\langle i, j \rangle$. This marginal probability is defined as the similarity score of two structure fragments. However, if we directly use CLESUM or (BLOSUM + CLESUM) (see Eq. 5.5 or Eq. 5.10) as the similarity score, we will not lose much accuracy but significantly increases the running speed. Given a short fragment F_1 in protein p_1, let F_i denotes the fragment of the same size in another protein p_i which has the highest similarity score with F_1. All the fragments $F_1, F_2, \ldots, F_M$ form an HSFB with F_1 being the pivot fragment. A protein of size N in total have $N - L + 1$ HSFBs. Note that the "highest similarity" relationship is asymmetric. That is, among all fragments in protein p_i, F_i is the most similar one to F_1 may not imply that among all fragments of p_i, F_1 also is the most similar one to F_i. Therefore, given M proteins with lengths $N_1, N_2, \ldots, N_M$, there are in total $(\sum_{i=1}^{M} N_i) - ML + M$ HSFBs.

4.3.4 Ranking HSFBs by spatial consistency

Two HSFBs may be geometrically inconsistent with each other. That is, we cannot superimpose well the fragments in both HSFBs using a single set of rigid-body transformations. We can calculate the degree to which two HSFBs are geometrically consistent and then rank all the HSFBs according to their consistency with others. The HSFB with the highest consistency score is very likely contained in the best MSA. We use a simple method to estimate the consistency score of one HSFB as follows. For each fragment in the HSFB, we generate a rigid-body transformation (Kabsch, 1976) between this fragment and the pivot fragment. Let $B_1 = \{F_{11}, F_{12}, \ldots, F_{1M}\}$ denotes the HSFB for which we want to calculate its consistency score and F_{11} is the pivot fragment. Let $T_i (i = 2, 3, \ldots, M)$ denotes the rigid-body transformations derived from superimposing F_{1i} to F_{11}. Let $B_2 = \{F_{21}, F_{22}, \ldots, F_{2M}\}$ denotes another HSFB. For any fragment $F_{2i} (i = 2, 3, \ldots, M)$ in B_2, we superimpose F_{2i} to F_{21} using the transformation T_i and then calculate the RMSD between F_{2i} and F_{21}. If the distance is within 3 Å, we increase the consistency score of B_1 by 1, otherwise by 0.

4.3.5 Scoring function for MSA

A good MSA should have a large number of cores (i.e., CORE-LEN) and also a small core RMSD. A core is a fully aligned column, consisting of one residue from each input protein. In addition, pairwise alignments in an MSA should also be of high quality. It is challenging to develop an algorithm that can optimize these criteria simultaneously since sometimes they contradict with one another. For example, a large CORE-LEN usually leads to a large RMSD value. A simple solution is to fix one criterion and then optimize the others, for example, maximizing CORE-LEN while restricting RMSD. This solution is not very flexible in that we have to determine RMSD in advance, not to mention that neither CORE-LEN nor RMSD is the best measure. We use CORE-LEN $\times$ $\overline{\text{TMscore}}$ as the scoring function where $\overline{\text{TMscore}}$ of an MSA is defined as the average TMscore (see Eq. 5.7) of all the pairwise alignments implied in the MSA.

4.3.6 Building an initial MSA from an HSFB for scaffold

Given an HSFB of M structures, we first generate a set of $M-1$ rigid-body transformations by superimposing each fragment in the HSFB to the pivot fragment and minimizing the RMSD of these two fragments. Then we superimpose each structure to the pivot structure using the transformation generated from fragment superimposition, and run DP to generate an alignment of the two structures by maximizing the TMscore. Finally, we assemble the $M-1$ pairwise alignments into an initial MSA using the pivot structure as the anchor.

4.3.7 Adjustment of pairwise alignment

Given the initial MSA, we may refine it by adjusting the pairwise alignment between each input structure and the pivot structure. First, we calculate the TMscore of the pairwise alignment between each input structure and the pivot. In addition, we also calculate the percentage of spatial consistency SFPs between each input structure and the pivot. If TMscore < 0.5 (Xu & Zhang, 2010) or the percentage $<50\%$, this input structure is called "unanchored." We adjust the alignment between each unanchored structure and the pivot using rigid-body transformations derived from other TopF HSFBs. In particular, for each top HSFB, let F_1 and F_2 denote the fragments in the HSFB belonging to the pivot and the unanchored structure, respectively. We realign the unanchored structure to the pivot structure using the rigid-body transformation generated from minimizing the RMSD between F_1 and F_2. The pairwise alignment with the maximum TMscore is kept in the MSA.

4.3.8 Consensus-based MSA refinement

3DCOMB refines an MSA by realigning each input structure to the consensus structure, which is constructed as follows. At each column of this MSA, we calculate the center of all the aligned residues (only Cα is considered).

Second, we merge two neighbor columns into a single one if the following two conditions are satisfied: (A) the total number of aligned residues in these two columns is not more than the total number of input structures and (B) the distance between their two centers is less than 3.0 Å. We use 3.0 Å as the cutoff because in native protein structures more than 99% of Cα–Cα virtual bonds are longer than 3.0 Å. This merge procedure is repeated until no columns can be merged. The consensus structure consists of all the centers. This refinement procedure is repeated until a given number of iterations or the scoring function cannot be improved further.

4.4. Result analysis

4.4.1 Overview

We evaluate our program 3DCOMB using three large-scale benchmarks on three metrics: CORE-LEN (number of fully aligned columns in the MSA), RMSD (root mean squared deviation of the fully aligned columns in the MSA), and $\overline{\text{TMscore}}$ (the average TMscore of all the pairwise alignments in the MSA). We compare 3DCOMB with several MSA tools such as BLOMAPS (Wang & Zheng, 2009), MAPSCI (Ilinkin et al., 2010), MAMMOTH (Lupyan et al., 2005), MUSTANG (Konagurthu et al., 2006), MATT (Menke et al., 2008), and MultiProt (Shatsky et al., 2004), which represent six different methods that include the three major merging methods.

4.4.2 The evaluation benchmarks

We use three benchmarks: HOMSTRAD (Mizuguchi, Deane, Blundell, & Overington, 1998), SABmark-sup, and SABmark-twi (Van Walle et al., 2005). HOMSTRAD contains 398 homologous protein families, each with at least three structures. SABmark-sup is the superfamily set in SABmark (version 1.65), containing 425 families with low to intermediate sequence identity. SABmark-twi represents the twilight set in SABmark, containing 209 families with low sequence identity.

4.4.3 Performance on HOMSTRAD

3DCOMB obtains the largest average CORE-LEN and the third best average core RMSD (slightly larger than MultiProt and MAPSCI). Note that because MultiProt uses a very strict cutoff to determine if an aligned column is a core or not, it always obtains the smallest core RMSD and also very small CORE-LEN on all the benchmarks. As shown in Table 5.8, MAMMOTH and MUSTANG can generate alignments with CORE-LEN comparable to

Table 5.8 Alignment accuracy of seven MSA tools on three benchmarks HOMSTRAD, SABmark-sup, and SABmark-twi

Method	CORE-LEN	RMSD	$\overline{\text{TMscore}}$
HOMSTRAD			
3DCOMB	170.58	2.00	0.800
MAPSCI	162.55	1.87	0.792
MAMMOTH	169.84	3.03	0.786
MATT	169.53	2.00	0.781
BLOMAPS	169.27	2.18	0.779
MUSTANG	169.49	2.66	0.765
MultiProt	140.82	1.33	0.649
SABmark-sup			
3DCOMB	106.66	2.59	0.655
MAPSCI	89.51	2.95	0.627
MAMMOTH	105.50	5.78	0.614
MATT	104.12	2.59	0.613
BLOMAPS	101.82	3.11	0.613
MUSTANG	103.86	4.20	0.583
MultiProt	68.70	1.61	0.404
SABmark-twi			
3DCOMB	71.63	3.02	0.526
MAPSCI	50.11	4.38	0.466
MAMMOTH	64.97	8.31	0.436
MATT	67.08	2.89	0.453
BLOMAPS	67.20	4.22	0.457
MUSTANG	66.89	5.10	0.422
MultiProt	36.38	1.75	0.259

The accuracy is measured by CORE-LEN, RMSD, and $\overline{\text{TMscore}}$. The values in the table are averaged over an individual benchmark.

3DCOMB, MATT, and BLOMAPS, but much larger RMSD. 3DCOMB not only achieves the best average $\overline{\text{TMscore}}$ but also excels others on almost each individual structure group.

4.4.4 Performance on SABmark

As shown in Table 5.8, 3DCOMB obtains the best average CORE-LEN and $\overline{\text{TMscore}}$ for SABmark-sup. By RMSD, 3DCOMB is second to only MultiProt, but MultiProt obtains a much smaller CORE-LEN. By $\overline{\text{TMscore}}$, 3DCOMB is about 4.5% better than the second best method MAPSCI and also outperforms others on almost each individual structure group. For SABmark-twi, it is more challenging because it consists of mostly distantly related proteins. However, 3DCOMB excels others at an even larger margin. By CORE-LEN, 3DCOMB is 6.6% better than the second best method MATT and excels others on a majority of structure groups. By core RMSD, 3DCOMB is second to MultiProt and slightly to MATT. By $\overline{\text{TMscore}}$ 3DCOMB outperforms the second best algorithm MAPSCI by 13% and also excels others on almost each individual structure group.

5. CONCLUSIONS AND PERSPECTIVES

Using secondary structures to describe protein structures has its own limitations. Therefore, protein structural alphabets, which are discretized states of backbone fragment conformations, have been proposed to bridge the gap between the secondary and tertiary structures. Among all the work structural alphabets, our CLEs aptly balance precision with simplicity by a probabilistic perspective (Zheng, 2008). In order to measure the similarity between these discrete letters, a substitution matrix CLESUM is developed by using a similar way as BLOSUM. Based on the newly developed structural alphabets, we invented a protein pairwise structure alignment program (DeepAlign). The scoring function (DeepScore) of our program contains the following key factors: (i) AA mutation score, (ii) local substructure substitution potential, (iii) hydrogen-bonding similarity, and (iv) traditional geometric similarity measure. DeepAlign could generate alignments highly consistent with manually curated alignments. We use manually curated alignments as the reference because they usually make much more biological sense since they are built by human experts taking into consideration evolutionary and functional relationship besides spatial proximity alone. Note that DeepScore (see Eq. 5.8) is the natural combination of four different items and there are no parameters to be fine tuned. Therefore, we do not bias DeepAlign toward a specific performance metric.

It is obvious that MSA will bring more information than pairwise alignment alone. However, the limitation for MSA lies in how to merge the input structures into a MSA. Here, we introduce our method 3DCOMB that applies a novel approach, namely, scaffold-first, to go beyond the current limitations. By using a probabilistic model to combine both local and global structure information, we can accurately identify the most conserved short fragment blocks among proteins to be aligned. These conserved fragment blocks plus a pivot protein form the scaffold, which are very likely contained in the best MSA. Then by pairwise adjusting other proteins with the pivot, 3DCOMB can quickly get the best MSA through a consensus-first approach. 3DCOMB also introduces a novel scoring function to generate an MSA with a large number of cores and high-quality pairwise alignments.

Structure alignment tools have a wide spectrum of applications, ranging from constructing classification database for protein structures (Holm & Sander, 1994; Murzin et al., 1995; Orengo et al., 1997) to protein function annotation (Brylinski & Skolnick, 2008; Roy et al., 2012). Recently, DeepAlign and 3DCOMB have been utilized in numerous different applications, such as (i) constructing the training data for template-based protein structure prediction tools that based on Machine Learning method (Källberg et al., 2012; Ma et al., 2012; Ma, Peng, Wang, & Xu, 2013; Ma, Wang, et al., 2013; Peng, Bo, & Xu, 2009; Shao et al., 2011), for position-specific distance-dependent protein statistical potential (Zhao & Xu, 2012), and for protein residue contact prediction (Wang & Xu, 2013); (ii) searching the ligand database for homology-based binding-site prediction (Källberg et al., 2012); and (iii) building MSA for multiple-template-based threading (Peng & Xu, 2011). These new areas of DeepAlign and 3DCOMB have applied broaden the spectrum of the application list of structure alignment tools.

Despite success, the following open problems for structure alignment need to be solved in the future.

How to deal with the inconsistency problem. Sadowski and Taylorr (2012) have found that even for relatively similar proteins the degree of inconsistency level is high among the structure alignments generated by different tools. One source of inconsistency is found to be around the region near gaps and for proteins with low structural complexity, as well as for helices. Another source of inconsistency is found in those proteins containing periodic or repetitive structures. One possible solution to deal with the periodic or repetitive case is to allow multiple solutions (see Fig. 5.5).

How to process gaps in both pairwise and multiple alignment. As shown in Holm and Sanderr (1993), an AFP shall not be too short (say less than four residues). Also, as mentioned in Sadowski and Taylorr (2012), since indels are responsible for much of the structural change which occurs through evolution, it is necessary to develop a more accurate gap score for structural alignment. Moreover, how to deal with the exponentially growth number of gaps with respect to the number of proteins in MSA remains unsolved.

How to conduct fast and reliable database search. Although a variety of tools have been developed, such as 3D-BALST (Tung et al., 2007) and YAKUSA (Carpentier et al., 2005), a database search approach that could deal with large scale of distant structures is deficiency because of the combinatorial explosion for search space (Hasegawa & Holm, 2009). The ultimate goal is to project an entire protein structure to a fingerprint, that is, one number or histogram. Recently, Harder, Borg, Boomsma, Røgen, and Hamelryck (2012) have applied Gauss Integrals calculated from protein backbones as the protein global descriptor for fast large-scale clustering of protein structures. This methodology might be able to be applied for fast and reliable database search in the future.

How to improve flexible alignment. Although structure comparison methods that allow flexibility and plasticity could generate the most biologically meaningful alignments (Hasegawa & Holm, 2009), it is well known that introducing flexibility into alignments greatly enlarges the searching space, exacerbating problems in generating high-quality alignments for more distantly related proteins (Sadowski & Taylor, 2012). It is also suggested that, instead of allowing flexible "everywhere," we should focus on those functional and/or hinge regions (Sadowski & Taylor, 2012). As a choice, RAPIDO (Mosca et al., 2008), that only lets the region between two domains to be flexible, could serve as a candidate for the purpose. Moreover, Elastic Network Model might be applied to define those highly flexible hinge regions, such as used in Emekli, Schneidman-Duhovny, Wolfson, Nussinov, and Haliloglu (2008) and Yang, Song, and Jernigan (2009).

How to conduct complex–complex alignment. Sippl and Wiedersteinn (2012) pointed out that the comparison of oligomers and molecular complexes differs largely. They also provide several challenging examples that cannot be handled by conventional structure alignment techniques since multiple chains are involved. To deal with this problem, TopMatch (Slater, Castellanos, Sippl, & Melo, 2013) has been developed. The basic idea of TopMatch is that, in a pairwise comparison of two structures, it first concatenates all chains together and then generally yields several basic alignments.

For large proteins and complexes, the number of basic alignments may rise to several hundred. Finally, TopMatch merges these basic alignments, according to a spatial consistency measurement, to form a composite alignment. It should be noted that, in the alignment of multiple chains, the relative order of monomers is arbitrary. This property made TopMatch possible to deal with the complex alignment without considering the chain ordering. However, due to the increase number of large complex protein structures in PDB, the alignment tools with more efficiency and more reliability are still needed in the future.

How to raise the speed and accuracy for large-scale MSA. As mentioned early in Koehl (2006), some of the most accurate MSA methods are still computationally prohibitive to be applied in large scale and continuous experiments. This statement is also true for 3DCOMB. Although has improved accuracy, the running speed of 3DCOMB for large-scale input protein is still far from satisfactory.

REFERENCES

Altschul, S., Madden, T., Schaffer, A., Zhang, J., Zhang, Z., Miller, W., et al. (1997). Gapped BLAST and PSI-BLAST: A new generation of protein database search programs. *Nucleic Acids Research*, *25*(17), 3389.

Andersen, J. N., Mortensen, O. H., Peters, G. H., Drake, P. G., Iversen, L. F., Olsen, O. H., et al. (2001). Structural and evolutionary relationships among protein tyrosine phosphatase domains. *Molecular and Cellular Biology*, *21*(21), 7117–7136.

Birzele, F., Gewehr, J. E., Csaba, G., & Zimmer, R. (2007). Vorolign—Fast structural alignment using Voronoi contacts. *Bioinformatics*, *23*(2), e205–e211. http://dx.doi.org/10.1093/bioinformatics/btl294, 23/2/e205 [pii].

Bornot, A., Etchebest, C., & De Brevern, A. G. (2009). A new prediction strategy for long local protein structures using an original description. *Proteins: Structure, Function, and Bioinformatics*, *76*(3), 570–587.

Bradley, P., Kim, P. S., & Berger, B. (2002). *TRILOGY: Discovery of sequence-structure patterns across diverse proteins*. Paper presented at the Proceedings of the sixth annual international conference on Computational biology.

Brylinski, M., & Skolnick, J. (2008). A threading-based method (FINDSITE) for ligand-binding site prediction and functional annotation. *Proceedings of the National Academy of Sciences*, *105*(1), 129–134.

Budowski-Tal, I., Nov, Y., & Kolodny, R. (2010). FragBag, an accurate representation of protein structure, retrieves structural neighbors from the entire PDB quickly and accurately. *Proceedings of the National Academy of Sciences*, *107*(8), 3481–3486.

Bystroff, C., & Baker, D. (1998). Prediction of local structure in proteins using a library of sequence-structure motifs. *Journal of Molecular Biology*, *281*(3), 565–577.

Camproux, A., Tuffery, P., Chevrolat, J., Boisvieux, J., & Hazout, S. (1999). Hidden Markov model approach for identifying the modular framework of the protein backbone. *Protein Engineering*, *12*(12), 1063–1073.

Carpentier, M., Brouillet, S., & Pothier, J. (2005). YAKUSA: A fast structural database scanning method. *Proteins: Structure, Function, and Bioinformatics*, *61*(1), 137–151.

Cheng, H., Kim, B. H., & Grishin, N. V. (2008a). MALIDUP: A database of manually constructed structure alignments for duplicated domain pairs. *Proteins: Structure, Function, and Bioinformatics*, *70*(4), 1162–1166.

Cheng, H., Kim, B. H., & Grishin, N. V. (2008b). MALISAM: A database of structurally analogous motifs in proteins. *Nucleic Acids Research*, *36*(Suppl. 1), D211.

Csaba, G., Birzele, F., & Zimmer, R. (2008). Protein structure alignment considering phenotypic plasticity. *Bioinformatics*, *24*(16), i98–i104.

Daniluk, P., & Lesyng, B. (2011). A novel method to compare protein structures using local descriptors. *BMC Bioinformatics*, *12*(1), 344.

Dayhoff, M. O., & Schwartz, R. M. (1978). *A model of evolutionary change in proteins*. Paper presented at the In Atlas of protein sequence and structure.

De Brevern, A., Etchebest, C., & Hazout, S. (2000). Bayesian probabilistic approach for predicting backbone structures in terms of protein blocks. *Proteins: Structure, Function, and Bioinformatics*, *41*(3), 271–287.

de Brevern, A. G., Valadié, H., Hazout, S., & Etchebest, C. (2002). Extension of a local backbone description using a structural alphabet: A new approach to the sequence-structure relationship. *Protein Science*, *11*(12), 2871–2886.

De Brevern, A., Wong, H., Tournamille, C., Colin, Y., Le Van Kim, C., & Etchebest, C. (2005). A structural model of a seven-transmembrane helix receptor: The Duffy antigen/receptor for chemokine (DARC). *Biochimica et Biophysica Acta (BBA) General Subjects*, *1724*(3), 288–306.

Dror, O., Benyamini, H., Nussinov, R., & Wolfson, H. (2003). MASS: Multiple structural alignment by secondary structures. *Bioinformatics*, *19*(Suppl. 1), i95–i104.

Dudev, M., & Lim, C. (2007). Discovering structural motifs using a structural alphabet: Application to magnesium-binding sites. *BMC Bioinformatics*, *8*(1), 106.

Dunbrack, R. L., Jr. (2006). Sequence comparison and protein structure prediction. *Current Opinion in Structural Biology*, *16*(3), 374–384.

Edgar, R. C., & Batzoglou, S. (2006). Multiple sequence alignment. *Current Opinion in Structural Biology*, *16*(3), 368–373.

Eidhammer, I., Jonassen, I., & Taylor, W. R. (2000). Structure comparison and structure patterns. *Journal of Computational Biology*, 7(5), 685–716.

Emekli, U., Schneidman-Duhovny, D., Wolfson, H. J., Nussinov, R., & Haliloglu, T. (2008). HingeProt: Automated prediction of hinges in protein structures. *Proteins: Structure, Function, and Bioinformatics*, *70*(4), 1219–1227.

Etchebest, C., Benros, C., Bornot, A., Camproux, A.-C., & de Brevern, A. (2007). A reduced amino acid alphabet for understanding and designing protein adaptation to mutation. *European Biophysics Journal*, *36*(8), 1059–1069.

Flocco, M. M., & Mowbray, S. L. (1995). Cα-based torsion angles: A simple tool to analyze protein conformational changes. *Protein Science*, *4*(10), 2118–2122.

Fourrier, L., Benros, C., & De Brevern, A. G. (2004). Use of a structural alphabet for analysis of short loops connecting repetitive structures. *BMC Bioinformatics*, *5*(1), 58.

Gibrat, J., Madej, T., Spouge, J., & Bryant, S. (1997). The VAST protein structure comparison method. *Biophysical Journal*, *72*, 298.

Guda, C., Lu, S., Scheeff, E. D., Bourne, P. E., & Shindyalov, I. N. (2004). CE-MC: A multiple protein structure alignment server. *Nucleic Acids Research*, *32*(Suppl. 2), W100.

Harder, T., Borg, M., Boomsma, W., Røgen, P., & Hamelryck, T. (2012). Fast large-scale clustering of protein structures using Gauss integrals. *Bioinformatics*, *28*(4), 510–515.

Hasegawa, H., & Holm, L. (2009). Advances and pitfalls of protein structural alignment. *Current Opinion in Structural Biology*, *19*(3), 341–348.

Henikoff, S., & Henikoff, J. G. (1992). Amino acid substitution matrices from protein blocks. *Proceedings of the National Academy of Sciences*, *89*(22), 10915–10919.

Holm, L., & Rosenström, P. (2010). Dali server: Conservation mapping in 3D. *Nucleic Acids Research*, *38*(Suppl. 2), W545–W549.

Holm, L., & Sander, C. (1993). Protein structure comparison by alignment of distance matrices. *Journal of Molecular Biology*, *233*, 123.

Holm, L., & Sander, C. (1994). The FSSP database of structurally aligned protein fold families. *Nucleic Acids Research*, *22*(17), 3600.

Ilinkin, I., Ye, J., & Janardan, R. (2010). Multiple structure alignment and consensus identification for proteins. *BMC Bioinformatics*, *11*, 71. http://dx.doi.org/10.1186/1471-2105-11-71, 1471-2105-11-71 [pii].

Irving, J. A., Whisstock, J. C., & Lesk, A. M. (2001). Protein structural alignments and functional genomics. *Proteins: Structure, Function, and Bioinformatics*, *42*(3), 378–382.

Joseph, A. P., Agarwal, G., Mahajan, S., Gelly, J.-C., Swapna, L. S., Offmann, B., et al. (2010). A short survey on protein blocks. *Biophysical Reviews*, *2*(3), 137–145.

Joseph, A. P., Bornot, A., & de Brevern, A. G. (2010). Local structure alphabets. In H. Rangwala, & G. Karypis (Eds.), *Introduction to protein structure prediction: Methods and algorithms* (pp. 75–106). Hoboken, New Jersey: John Wiley & Sons, Inc.

Jung, J., & Lee, B. (2000). Protein structure alignment using environmental profiles. *Protein Engineering*, *13*(8), 535–543.

Kabsch, W. (1976). A solution for the best rotation to relate two sets of vectors. *Acta Crystallographica. Section A: Crystal Physics, Diffraction, Theoretical and General Crystallography*, *32*(5), 922–923.

Källberg, M., Wang, H., Wang, S., Peng, J., Wang, Z., Lu, H., et al. (2012). Template-based protein structure modeling using the RaptorX web server. *Nature Protocols*, 7(8), 1511–1522.

Kawabata, T. (2003). MATRAS: A program for protein 3D structure comparison. *Nucleic Acids Research*, *31*(13), 3367.

Kim, C., & Lee, B. (2007). Accuracy of structure-based sequence alignment of automatic methods. *BMC Bioinformatics*, *8*(1), 355.

Kim, C., Tai, C. H., & Lee, B. (2009). Iterative refinement of structure-based sequence alignments by seed extension. *BMC Bioinformatics*, *10*(1), 210.

Koehl, P. (2006). Protein structure classification. *Reviews in Computational Chemistry*, *22*, 1.

Kolbeck, B., May, P., Schmidt-Goenner, T., Steinke, T., & Knapp, E.-W. (2006). Connectivity independent protein-structure alignment: A hierarchical approach. *BMC Bioinformatics*, 7(1), 510.

Kolodny, R., Koehl, P., Guibas, L., & Levitt, M. (2002). Small libraries of protein fragments model native protein structures accurately. *Journal of Molecular Biology*, *323*(2), 297–307.

Konagurthu, A. S., Whisstock, J. C., Stuckey, P. J., & Lesk, A. M. (2006). MUSTANG: A multiple structural alignment algorithm. *Proteins: Structure, Function, and Bioinformatics*, *64*(3), 559–574.

Krissinel, E., & Henrick, K. (2004). Secondary-structure matching (SSM), a new tool for fast protein structure alignment in three dimensions. *Acta Crystallographica. Section D: Biological Crystallography*, *60*(12), 2256–2268.

Lackner, P., Koppensteiner, W. A., Sippl, M. J., & Domingues, F. S. (2000). ProSup: A refined tool for protein structure alignment. *Protein Engineering*, *13*(11), 745.

Lafferty, J., McCallum, A., & Pereira, F. (2001). *Conditional random fields: Probabilistic models for segmenting and labeling sequence data* (pp. 282–289). San Francisco, CA, USA: Morgan Kaufmann Publishers Inc.

Le, Q., Pollastri, G., & Koehl, P. (2009). Structural alphabets for protein structure classification: A comparison study. *Journal of Molecular Biology*, *387*(2), 431–450.

Levitt, M., & Gerstein, M. (1998). A unified statistical framework for sequence comparison and structure comparison. *Proceedings of the National Academy of Sciences*, *95*(11), 5913.

Lupyan, D., Leo-Macias, A., & Ortiz, A. R. (2005). A new progressive-iterative algorithm for multiple structure alignment. *Bioinformatics*, *21*(15), 3255.
Ma, J., Peng, J., Wang, S., & Xu, J. (2012). A conditional neural fields model for protein threading. *Bioinformatics*, *28*(12), i59–i66.
Ma, J., Peng, J., Wang, S., & Xu, J. (2013). Estimating the partition function of graphical models using Langevin importance sampling. *JMLR W&CP*, *31*, 433–441.
Ma, J., Wang, S., Zhao, F., & Xu, J. (2013). Protein threading using context-specific alignment potential. *Bioinformatics*, *29*(13), i257–i265.
Marchler-Bauer, A., Anderson, J. B., Cherukuri, P. F., DeWeese-Scott, C., Geer, L. Y., Gwadz, M., et al. (2005). CDD: A conserved domain database for protein classification. *Nucleic Acids Research*, *33*(Suppl. 1), D192.
Matsuda, H., Taniguchi, F., & Hashimoto, A. (1997). *An approach to detection of protein structural motifs using an encoding scheme of backbone conformation*. Paper presented at the In Pacific Symposium on Biocomputing'97.
McCammon, J. A. (1977). Dynamics of folded proteins. *Nature*, *267*, 16.
Menke, M., Berger, B., & Cowen, L. (2008). Matt: Local flexibility aids protein multiple structure alignment. *PLoS Computational Biology*, *4*(1), e10.
Micheletti, C., & Orland, H. (2009). MISTRAL: A tool for energy-based multiple structural alignment of proteins. *Bioinformatics*, *25*(20), 2663–2669.
Mizuguchi, K., Deane, C. M., Blundell, T. L., & Overington, J. P. (1998). HOMSTRAD: A database of protein structure alignments for homologous families. *Protein Sciences*, 7(11), 2469–2471. http://dx.doi.org/10.1002/pro.5560071126.
Mosca, R., Brannetti, B., & Schneider, T. (2008). Alignment of protein structures in the presence of domain motions. *BMC Bioinformatics*, *9*(1), 352.
Murzin, A., Brenner, S., Hubbard, T., & Chothia, C. (1995). SCOP: A structural classification of proteins database for the investigation of sequences and structures. *Journal of Molecular Biology*, *247*(4), 536–540.
Nadimpalli, S., Daniels, N., & Cowen, L. (2012). Formatt: Correcting protein multiple structural alignments by incorporating sequence alignment. *BMC Bioinformatics*, *13*, 259.
Orengo, C. A., Michie, A., Jones, S., Jones, D. T., Swindells, M., & Thornton, J. M. (1997). CATH—A hierarchic classification of protein domain structures. *Structure*, *5*(8), 1093–1109.
Ortiz, A. R., Strauss, C. E., & Olmea, O. (2002). MAMMOTH (matching molecular models obtained from theory): An automated method for model comparison. *Protein Science*, *11*(11), 2606–2621.
Panchenko, A., Marchler-Bauer, A., & Bryant, S. H. (1999). Threading with explicit models for evolutionary conservation of structure and sequence. *Proteins: Structure, Function, and Bioinformatics*, *37*(S3), 133–140.
Pandini, A., Fornili, A., & Kleinjung, J. (2010). Structural alphabets derived from attractors in conformational space. *BMC Bioinformatics*, *11*(1), 97.
Park, B. H., & Levitt, M. (1995). The complexity and accuracy of discrete state models of protein structure. *Journal of Molecular Biology*, *249*(2), 493–507.
Pauling, L., Corey, R. B., & Branson, H. R. (1951). The structure of proteins: Two hydrogen-bonded helical configurations of the polypeptide chain. *Proceedings of the National Academy of Sciences*, *37*(4), 205–211.
Peng, J., Bo, L., & Xu, J. (2009). *Conditional neural fields*. Paper presented at the advances in neural information processing systems.
Peng, J., & Xu, J. (2011). A multiple—Template approach to protein threading. *Proteins: Structure, Function, and Bioinformatics*, *79*(6), 1930–1939.
Potestio, R., Aleksiev, T., Pontiggia, F., Cozzini, S., & Micheletti, C. (2010). ALADYN: A web server for aligning proteins by matching their large-scale motion. *Nucleic Acids Research*, *38*(Suppl. 2), W41–W45.

Robson, B., & Pain, R. H. (1974). Analysis of the code relating sequence to conformation in globular proteins. Development of a stereochemical alphabet on the basis of intra-residue information. *Biochemistry Journal, 141*, 869–882.

Rooman, M. J., Kocher, J.-P. A., & Wodak, S. J. (1991). Prediction of protein backbone conformation based on seven structure assignments: Influence of local interactions. *Journal of Molecular Biology, 221*(3), 961–979.

Roy, A., Yang, J., & Zhang, Y. (2012). COFACTOR: An accurate comparative algorithm for structure-based protein function annotation. *Nucleic Acids Research, 40*(W1), W471–W477.

Sadowski, M. I., & Taylor, W. R. (2012). Evolutionary inaccuracy of pairwise structural alignments. *Bioinformatics, 28*(9), 1209–1215.

Shao, M., Wang, S., Wang, C., Yuan, X., Li, S., Zheng, W., et al. (2011). Incorporating ab initio energy into threading approaches for protein structure prediction. *BMC Bioinformatics, 12*(Suppl. 1), S54.

Shapiro, J., & Brutlag, D. (2004). FoldMiner and LOCK 2: Protein structure comparison and motif discovery on the web. *Nucleic Acids Research, 32*(Suppl. 2), W536.

Shatsky, M., Nussinov, R., & Wolfson, H. J. (2002). Flexible protein alignment and hinge detection. *Proteins: Structure, Function, and Bioinformatics, 48*(2), 242–256.

Shatsky, M., Nussinov, R., & Wolfson, H. J. (2004). A method for simultaneous alignment of multiple protein structures. *Proteins, 56*(1), 143–156. http://dx.doi.org/10.1002/prot.10628.

Shindyalov, I. N., & Bourne, P. E. (1998). Protein structure alignment by incremental combinatorial extension (CE) of the optimal path. *Protein Engineering, 11*(9), 739.

Sippl, M. J., & Wiederstein, M. (2012). Detection of spatial correlations in protein structures and molecular complexes. *Structure, 20*(4), 718–728.

Slater, A. W., Castellanos, J. I., Sippl, M. J., & Melo, F. (2013). Towards the development of standardized methods for comparison, ranking and evaluation of structure alignments. *Bioinformatics, 29*(1), 47–53.

Taylor, W. R., & Orengo, C. A. (1989). Protein structure alignment. *Journal of Molecular Biology, 208*(1), 1–22.

Tung, C.-H., Huang, J.-W., & Yang, J.-M. (2007). Kappa-alpha plot derived structural alphabet and BLOSUM-like substitution matrix for rapid search of protein structure database. *Genome Biology, 8*(3), R31.

Tyagi, M., De Brevern, A., Srinivasan, N., & Offmann, B. (2008). Protein structure mining using a structural alphabet. *Proteins: Structure, Function, and Bioinformatics, 71*(2), 920–937.

Van Walle, I., Lasters, I., & Wyns, L. (2005). SABmark—A benchmark for sequence alignment that covers the entire known fold space. *Bioinformatics, 21*(7), 1267–1268. http://dx.doi.org/10.1093/bioinformatics/bth493, bth493 [pii].

Wang, S. (2009). CLeFAPS: Fast flexible alignment of protein structures based on conformational letters. arXiv, preprint arXiv:0903.0582.

Wang, S. (2010). Hydropathy conformational letter and its substitution matrix HP-CLESUM: An application to protein structural alignment. arXiv, preprint arXiv:1001.2879.

Wang, S., Ma, J., Peng, J., & Xu, J. (2013). Protein structure alignment beyond spatial proximity. *Scientific Reports, 3*, 1448.

Wang, S., Peng, J., & Xu, J. (2011). Alignment of distantly related protein structures: Algorithm, bound and implications to homology modeling. *Bioinformatics, 27*(18), 2537–2545.

Wang, Z., & Xu, J. (2013). Predicting protein contact map using evolutionary and physical constraints by integer programming. *Bioinformatics, 29*(13), i266–i273.

Wang, S., & Zheng, W. M. (2008). CLePAPS: Fast pair alignment of protein structures based on conformational letters. *Journal of Bioinformatics and Computational Biology, 6*(2), 347–366. http://dx.doi.org/10.1142/S0219720008003461 [pii].

Wang, S., & Zheng, W. M. (2009). Fast multiple alignment of protein structures using conformational letter blocks. *Open Bioinformatics Journal, 3*, 69–83.

Wass, M. N., Kelley, L. A., & Sternberg, M. J. (2010). 3DLigandSite: Predicting ligand-binding sites using similar structures. *Nucleic Acids Research, 38*(Suppl. 2), W469–W473.

Xu, J., & Zhang, Y. (2010). How significant is a protein structure similarity with TM-score = 0.5? *Bioinformatics, 26*(7), 889–895.

Yang, L., Song, G., & Jernigan, R. L. (2009). Protein elastic network models and the ranges of cooperativity. *Proceedings of the National Academy of Sciences, 106*(30), 12347–12352.

Yang, Y., Zhan, J., Zhao, H., & Zhou, Y. (2012). A new size-independent score for pairwise protein structure alignment and its application to structure classification and nucleic-acid binding prediction. *Proteins: Structure, Function, and Bioinformatics, 80*(8), 2080–2088.

Ye, Y., & Godzik, A. (2003). Flexible structure alignment by chaining aligned fragment pairs allowing twists. *Bioinformatics, 19*(Suppl. 2), ii246–ii255.

Ye, Y., & Godzik, A. (2005). Multiple flexible structure alignment using partial order graphs. *Bioinformatics, 21*(10), 2362.

Zemla, A. (2003). LGA: A method for finding 3D similarities in protein structures. *Nucleic Acids Research, 31*(13), 3370–3374.

Zhang, Y., & Skolnick, J. (2004). Scoring function for automated assessment of protein structure template quality. *Proteins, 57*, 702–710.

Zhang, Y., & Skolnick, J. (2005). TM-align: A protein structure alignment algorithm based on the TM-score. *Nucleic Acids Research, 33*(7), 2302.

Zhao, F., & Xu, J. (2012). A position-specific distance-dependent statistical potential for protein structure and functional study. *Structure, 20*(6), 1118–1126.

Zheng, W. (2008). Protein conformational alphabets. In L. B. Roswell (Ed.), *Protein conformation: New research* (pp. 1–49). New York: Nova Science Publishers, Inc.

Zheng, W. M., & Liu, X. (2005). A protein structural alphabet and its substitution matrix CLESUM. In S. Istrail, P. Pevzner, & M. Waterman (Eds.), *Transactions on computational systems biology II* (pp. 59–67). Berlin: Springer.

Zhu, J., & Weng, Z. (2005). FAST: A novel protein structure alignment algorithm. *Proteins: Structure, Function, and Bioinformatics, 58*(3), 618–627.

CHAPTER SIX

Application of Evolutionary Based *in Silico* Methods to Predict the Impact of Single Amino Acid Substitutions in Vitelliform Macular Dystrophy

C. George Priya Doss[*,1], **Chiranjib Chakraborty**[†,1], **N. Monford Paul Abishek**[*], **D. Thirumal Kumar**[*], **Vaishnavi Narayanan**[‡]

[*]Medical Biotechnology Division, School of Biosciences and Technology, VIT University, Vellore, Tamil Nadu, India

[†]Department of Bioinformatics, School of Computer and Information Sciences, Galgotias University, Noida, Uttar Pradesh, India

[‡]Biomolecules & Genetics Division, School of Biosciences and Technology, VIT University, Vellore, Tamil Nadu, India

[1]Corresponding authors: e-mail address: georgecp77@yahoo.co.in; georgepriyadoss@vit.ac.in; drchiranjib@yahoo.com

Contents

1. Introduction 178
2. Materials and Methods 183
   2.1 Data information 183
   2.2 *In silico* prediction methods 183
   2.3 Evolutionary-based prediction methods 184
   2.4 Structure-based prediction methods 185
   2.5 Protein stability analysis 186
   2.6 Biophysical validation of SAPs 186
   2.7 Evolutionary conservation profiling 187
   2.8 Functional characterization of SNPs 187
   2.9 Statistical analysis 188
3. Results 188
   3.1 Analysis of deleterious SAPs using evolutionary-based prediction methods 188
   3.2 Analysis of deleterious SAPs using structure-based prediction methods 208
   3.3 Prediction of stability changes by I-MUTANT Suite 208
   3.4 Align-GVGD 209
   3.5 Concordance between the functional consequences of each SAP 209
   3.6 BEST1 and PRPH2 protein sequence analysis 211

Advances in Protein Chemistry and Structural Biology, Volume 94
ISSN 1876-1623
http://dx.doi.org/10.1016/B978-0-12-800168-4.00006-8

3.7 Propensity of each amino acid in native and mutant state of BEST1 and PRPH2 213
3.8 Disulfide bonds 220
3.9 Statistical analysis on the performance of *in silico* prediction methods 222
3.10 Functional SNPs 222
3.11 Ranking of SAPs 233
4. Discussion 233
5. Conclusion 263
Acknowledgments 264
References 264

Abstract

Recent developments in high-throughput discovery and genotyping have generated a tremendous amount of information about the existence of single amino acid polymorphisms (SAPs). Detailed understanding of the SAPs that affect protein structure and function can provide us valuable insight into disease genotype–phenotype correlations. Functional variants of biological importance are likely to be missed in large-scale analysis. Over the past decade, numerous efforts are underway in understanding and characterizing the potential consequences of variants in assessing the risk associated with vitelliform macular dystrophy (VMD). Yet, in spite of this success, we conducted a first SAP analysis via evolutionary-based *in silico* pipeline to unravel functional SAPs from a pool, containing both functional and neutral ones. Furthermore, based on the prediction scores, a ranking system was developed to prioritize the functional SAPs in order to minimize the number of SAPs screened for further genotyping.

1. INTRODUCTION

Vitelliform macular dystrophy (VMD) is an autosomal dominant inherited disorder of eye characterized by gradual loss of vision and even complete blindness. VMD (OMIM 153700) also known as "BEST disease" was first described by Friedrich Best in 1905, which manifests in both childhood and adult forms (Best, 1905). Based on the age of onset and electrooculogram (EOG), VMD is classified as juvenile VMD (VMD2) and adult-onset vitelliform macular dystrophy (AVMD) (Best, 1905; Gass, 1974). VMD2 (OMIM 153700) is the second most common cause of inherited maculopathy that occurs at the age of less than 30 years. Meanwhile, occurrence of AVMD (MIM 608133) is less frequent, and the age of onset is in the mean range of the fifth decade. Decrease in EOG Arden ratio is noticed in VMVD2 and normal or subnormal in AVMD. VMD is associated with mutations in either one or more copy of the *BEST1* and *PRPH2* gene in each cell. *BEST1* gene is mapped to 11q12–q13 and contains 11 exons of which 10 are protein coding. The gene product encodes a 68-kDa protein known as bestrophin 1, which includes 585 aa (Petrukhin et al., 1998). This protein is localized to the basolateral plasma membrane of the retinal pigment epithelium and involved in chloride channel (Cl^-) currents activated by intracellular calcium (Ca^{2+}) (Sun, Tsunenari, Yau, & Nathans, 2002). To date, more than 50% of the mutations of *BEST1* are associated with VMVD2, and the remaining mutations are associated with AVMD, autosomal recessive bestrophinopathy, autosomal dominant familial exudative vitreoretinopathy, and the *BEST1*-associated microcornea, rod-cone dystrophy, cataract, and posterior staphyloma (MRCS) and retinitis pigmentosa (Meunier et al., 2011). *PRPH2* gene also known as *RDS* is a member of the tetraspanin family. The gene is mapped to 6p21.2 and spans three exons (Wrigley, Ahmed, Nevett, & Findlay, 2000). The protein product of the *PRPH2* gene is known as peripherin-2, which contains 346 aa (Wrigley et al., 2000). Likewise, *BEST1* and *PRPH2* mutations are associated with a wide variety of disorders like autosomal dominant retinitis pigmentosa, progressive macular degeneration, macular dystrophy, and retinitis pigmentosa digenic with a mutation in *ROM1* (Wrigley et al., 2000). Peripherin-2 is expressed in rod and cone photoreceptors and plays a significant role in the formation and stability of discs in rods in combination with ROM1 heterodimer (Leroy, Kailasanathan, De Laey, Black, & Manson, 2007). Till date in HGMD Professional (http://www.hgmd.cf.ac.uk/ac/index.php) 2012.4, 146 mutations have been deposited in connection with *BEST1* gene. Majorly, 137 missense mutations in *BEST1* have been identified in connection with best macular dystrophy. So far, 74 missense mutations have been identified in *PRPH2* gene and the majority of them (40) were in association with retinitis pigmentosa, when compared to 20 mutations with AVMD.

Rapid development in various technologies and cost reduction in genotyping have generated a large amount of information regarding the most common type of genetic variation, that is, single nucleotide polymorphisms (SNPs). These advancements have yielded large volume of genomic variant data and revolutionized the field of biology and medicine. Current scenario in medical genetics is to identify and classify the functional or nonfunctional SNPs in complex diseases (cancer, diabetes, obesity, heart disease, etc.), to elucidate the mechanisms through which functional SNPs exert their effects, and ultimately to interrogate this information for association with complex phenotypes. Many large repository databases were made available in the World Wide Web in providing information regarding millions of SNPs such as dbSNP, HGV, Ensembl, and Swiss-Prot. SNPs occur in both coding (within or outside) and noncoding regions. SNPs in noncoding regions may alter the expression level of genes by affecting regulatory elements, and some intronic SNPs activate cryptic splice sites, leading to alternative splicing (Buckland et al., 2004; Cartegni & Krainer, 2002; Pastinen & Hudson, 2004; Savinkova et al., 2009). Considerable efforts have been attempted to explore the functional effects of SNPs in transcriptional regulation by affecting transcription factor binding sites (TFBS) in promoter or intronic enhancer regions (Khan et al., 2006; Mottagui-Tabar et al., 2005) or alternatively splicing regulation by disrupting exonic splicing enhancers or silencers (Buckland, 2006; Pampin & Rodriguez-Rey, 2007). Mostly, SNPs occurring in the noncoding regions are silent mutations. SNPs in the coding region will alter the amino acid sequence of expressed proteins through the creation of missense substitutions or premature termination codon. Nonsynonymous SNPs (nsSNPs) also known as single amino acid polymorphisms (SAPs) result in change of amino acid sequences, which can alter the protein function. SAPs can be categorized into harmful (deleterious) and neutral. Most of the SAPs are not harmful. Therefore, discriminating the harmful ones from neutral SAPs is a substantial challenge in mutational research. This categorization can assist in better understanding of genotype/phenotype relationships and drug response to disease. Recently, many studies have highlighted the involvement of SAPs in affecting the protein function (Kubo et al., 2007; Yoshiura et al., 2006; Sun et al., 2008) or not (Ueki et al., 2010). Usually, SAPs exert their effects by altering the stability of a folded domain, unfolding, and proteolytic degradation by the shift in amino acid residue from hydrophobic core to charged one, affecting catalytically active sites, phosphorylation sites, binding affinities, protein structure, or dynamics and protein folding.

Scientists are facing a primary challenge in identification, functional characterization, and association between SNPs and disease susceptibility from a pool of millions of SNPs containing harmful and neutral ones. Understanding the relationship between SNPs and the disease will allow clinicians to determine whether an individual will respond to a medicine (individualized medicine) or experience severe side effects. The amount of data generated in the public databases tends to increase in near future lending itself into informatics strategy. Proper use of bioinformatic tools requires particular skills and resources usually inaccessible to medical researchers or experimental biologists. The proposed approach will integrate biological information in different ways to push an initiative to speed up mutational research toward personalized medicine.

Differentiating deleterious SAPs (significant phenotypic consequences) from tolerant ones (without phenotypic change) is of immense value in understanding the genetic basis of complex diseases. Generally, this can be achieved by family-based linkage analysis or genome-wide association (GWA) research. Understanding the molecular basis of complex disease by traditional methods is painstaking and time-consuming, and at the structural level often nearly impossible, especially in the case where there are several SAPs causing the disease. Validating the probable outcome of each SAP using *in vitro* studies and animal models is a simple way to reveal the functional effects. Such studies are difficult to mount on a large scale, and their results might not always reflect *in vivo* genotype function in humans. These methods have their own limitations in their prediction analysis in complex diseases, which serves a way to *in silico* methods, which make their predictions based on the characteristics of the SAPs. Alternatively, *in silico* methods have the ability of analyzing the larger data set with greater accuracy and minimizing the number of functional SNPs for further genotyping with their speedy approach. Most of the available online *in silico* methods prioritize functional SNPs by two methods: (i) classifying as deleterious or neutral and (ii) providing the functional significance (FS) such as splicing or transcriptional regulation. These methods incorporate sequence, structure, and amino acid physicochemical characteristics along with multiple sequence alignment (MSA) in making their predictions. To date, sequence- and/or structure-based methods were used to predict the potential impact of amino acid substitutions on protein structure and function (Adzhubei et al., 2010; Ashkenazy, Erez, Martz, Pupko, & Ben-Tal, 2010; Bromberg & Rost, 2007; Calabrese, Capriotti, Fariselli, Martelli, & Casadio, 2009; Capriotti, Calabrese, & Casadio, 2006; Capriotti, Fariselli, Rossi, & Casadio, 2008; Kumar, Henikoff, & Ng, 2009; Mi, Muruganujan, & Thomas, 2013). Disease-causing SAPs usually reside in highly conserved positions. The effect of many SAPs will probably be neutral as natural selection will have removed mutations on essential positions. Assessment of nonneutral SNPs is primarily based on phylogenetic information (i.e., correlation with residue conservation) extended to a certain extent with structural approaches.

Owing to the severity of the disease and its frequency of occurrence, *BEST1* and *PRPH2* gene mutations in VMD were subjected to *in silico* analysis for the first time. We have proposed a simple *in silico* pipeline (Fig. 6.1) for nonbioinformatics researchers to identify deleterious SAPs and functional SNPs in *BEST1* and *PRPH2* systematically. The proposed architecture is as follows:

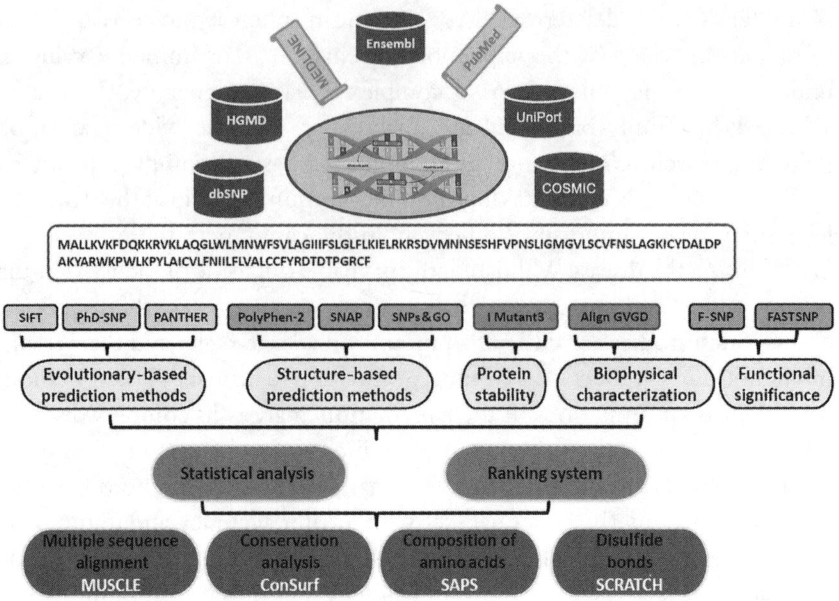

Figure 6.1 *In silico* pipeline integrating various bioinformatics databases and methods to unravel the impact of amino acid substitution in VMD disease spectrum. (For color version of this figure, the reader is referred to the online version of this chapter.)

(i) First, a reliable data set of SNP information in the coding and noncoding regions was constructed by retrieving SNP profile information from dbSNP, Swiss-Prot, HGMD, and Ensembl. This combinatorial SNP profiling will increase the data accuracy.

(ii) Examine the potential association between SAPs and phenotypic variations in the lack of three-dimensional (3D) structures by an evolutionary-based *in silico* approach. In this context, we applied the combination of sequence- and structure-based evolutionary approaches (empirical and SVM): PolyPhen-2 (Adzhubei et al., 2010), sorting intolerant from tolerant (SIFT) (Kumar, Henikoff, & Ng, 2009), PANTHER (Mi et al., 2013), I-MUTANT Suite (Capriotti et al., 2008), PhD-SNP (Capriotti et al., 2006), screening for nonacceptable polymorphisms (SNAP) (Bromberg & Rost, 2007), and SNPs&GO (Calabrese et al., 2009) to prioritize high-risk functional SAPs in a pool containing harmful and neutral ones.

(iii) Evolutionary conservation pattern and biophysical validation of the proposed impact of SAPs on protein structure and function were observed using ConSurf (Ashkenazy et al., 2010) and Align-GVGD (Tavtigian et al., 2006).

(iv) To improve the predictive power of the *in silico* tools, statistical parameters such as true positive (TP), true negative (TN), false negative (FN), false positive (FP), accuracy (ACC), sensitivity, and specificity were employed.

(v) Ranking strategy was developed to prioritize SNPs.

(vi) In the absence of experimental or epidemiologic evidence, the potential functional consequence of an SNP in noncoding SNPs (intronic SNPs and exonic 5′ and 3′ untranslated region (UTR) SNPs) was identified using FASTSNP (Yuan et al., 2006) and F-SNP (Lee & Shatkay, 2008).

In this depth report, we attempted to integrate different *in silico* methods in a pipeline, which can help in better understanding of the VMD disease and also able to prioritize the functional SNPs for further genotyping.

2. MATERIALS AND METHODS

2.1. Data information

We downloaded information regarding the SNPs in coding and noncoding region of *BEST1* and *PRPH2* gene from dbSNP (Sherry et al., 2001), HGV (Fredman et al., 2004), Swiss-Prot (Amos & Rolf, 1996), HGMD (Stenson et al., 2009), and Ensembl (Flicek et al., 2010) databases. We also retrieved information about the functional annotation of each SNP such as exon, intron, 5′ and 3′ UTR of a gene, or upstream or downstream from the dbSNP database (Sherry et al., 2001). In MEDLINE and PubMed (US National Library of Medicine), search terms "VMD" and "VMD-associated mutations" were used to search for relevant publications. References listed in HGMD and Swiss-Prot were cross-checked for mutations with reviewed publications. The protein sequence in FASTA format was retrieved for Human BEST1 and PRPH2 with corresponding IDs, NP_004174.1 (585 aa) and NP_000313.2 (346 aa) for further investigation.

2.2. *In silico* prediction methods

In silico analysis based on sequence homology and physical properties of amino acids can provide the first calculation of prior probability of pathogenicity of the SAPs. The propensity for each observed SAPs in *BEST1* and *PRPH2* genes to be tolerated or not was assessed by PolyPhen-2 (Adzhubei et al., 2010), SIFT (Kumar et al., 2009), PANTHER (Mi et al., 2013), I-MUTANT Suite (Capriotti et al., 2008), PhD-SNP

(Capriotti et al., 2006), SNAP (Bromberg & Rost, 2007), and SNPs&GO (Calabrese et al., 2009). These methods differ in the characteristics of the substitution they take into account, as well as in nature and classification techniques. SIFT, PhD-SNP, and PANTHER are solely based on evolutionary sequence information. Meanwhile, PolyPhen-2, SNAP, and SNPs&GO combine protein structural and/or functional parameters and sequence analysis-derived information. In addition, we used protein stability predictor I-MUTANT Suite. Most of the aforementioned tools are based on machine-learning techniques.

2.3. Evolutionary-based prediction methods

SIFT is a sequence homology-based tool that predicts the functional value of the substituted amino acids based on the alignment of highly similar orthologous and/or paralogous protein sequences. Predictions are primarily based on assumption on whether or not an amino acid is conserved in the protein family, which can be indicative of its importance to the normal function or structure of the expressed protein. We utilized the most recent version of SIFT BLink Beta (http://blocks.fhcrc.org/sift/SIFT.html) for SAP analysis. We submitted query in the form of gene identification (GI) number obtained from NCBI database. The underlying principle of this program is that it generates alignments with a large number of homologous sequences with a medium conservation measure of 3.0 to minimize false-positive and false-negative error. The output of SIFT is a table of probabilities for each amino acid at each position as well as predictions on not tolerated or tolerated amino acids for each position. SIFT assigns scores to each residue, ranging from 0 to 1. Scores close to 0 indicate evolutionary conserved and intolerance to substitution (deleterious), while scores close to one indicate tolerance to substitution (neutral). SIFT scores $\leq$0.05 are predicted by the algorithm to be intolerant or deleterious amino acid substitutions, whereas scores >0.05 are considered tolerant. SIFT scores were classified as intolerant (0.00–0.05), potentially intolerant (0.051–0.10), borderline (0.101–0.20), or tolerant (0.201–1.00) (Ng & Henikoff, 2003). Higher the tolerance index of a particular amino acid substitution, lesser is its possible impact. PANTHER 3.0 (http://www.pantherdb.org/tools/csnpScoreForm.jsp) is a database of gene families, including a phylogenetic tree for each family in which nodes of the tree are annotated with gene attributes. PANTHER estimates the likelihood of a particular SAP causing a functional impact on the protein. PANTHER uses HMM (hidden Markov model)-based statistical modeling techniques and MSAs to perform evolutionary analysis of coding SAPs. PANTHER subPSEC scores vary from 0 (neutral) to about -10 (more likely to be deleterious). Protein sequences having subPSEC value ≤ -3 are said to be deleterious. Predictor of human deleterious SNP (PhD-SNP) uses SVM (support vector machine)-sequence method and SVM profile to classify the mutation into disease-related and neutral polymorphisms. PhD-SNP is available at http://snps.biofold.org/phd-snp/phd-snp.html. For a given protein, its sequence profile is generated in PhD-SNP as follows: (i) For a given mutation, the substitution forms the wild type to the mutant residue, which is encoded in a 20-element vector by defining -1 to the wild-type residue, 1 to newly introduced residue, and 0 to the remaining 18 residues. (ii) A second 20-element vector encoding for the sequence environment is built reporting the occurrence of residues in a window of 19 residues around the mutated residue. It predicts if the given SAP has pathological effect based on the local sequence environment of the mutation. We used the most accurate mode that uses both sequence and evolutionary profiles.

2.4. Structure-based prediction methods

Polymorphism phenotyping (PolyPhen-2) predicts the functional effect of amino acid changes by considering the evolutionary conservation, the physicochemical differences, and the proximity of the substitution to predicted functional domains and/or structural characteristics. PolyPhen-2 is available at http://genetics.bwh.harvard.edu/pph2/. This prediction is based on straightforward empirical rules that are applied to the sequence, phylogenetic, and structural information characterizing the substitution. Another useful feature is that the algorithm calculates a Bayes posterior probability that a given mutation is deleterious. PolyPhen-2 searches for three-dimensional (3D) protein structures, multiple alignments of homologous sequences, and amino acid contact information in several protein structure databases. Then, it calculates position-specific-independent counts (PSIC) scores for each of the two variants and computes the difference of the PSIC scores of the two variants. The greater the PSIC score difference, the greater is the functional impact a particular amino acid substitution is expected to have. A mutation is listed as "probably damaging" if the probabilistic score is above 0.85–1, mutation is classified as "possibly damaging" if the probabilistic score is above 0.15–0.84, and the remaining mutations are classified as benign. SNPs&GO (http://snps-and-go.biocomp.unibo.it/snps-and-go/) is an SVM-based method that predicts disease-associated mutations from protein sequence, evolutionary information, and functions as encoded in the gene ontology (GO) terms. The utilization of functional GO terms is the main aspect of novelty of this tool over other existing bioinformatics tools. From the output page reports, we considered reliability index scoring from 0 (unreliable) to 10 (reliable) to classify mutation as pathogenic/neutral. The probability score higher than 0.5 reveals the disease-related effect of mutation on the parent protein function. SNAP is based on neural network and advanced machine-learning approach to predict the functional change of SNPs on the protein's function. It utilizes sequence, structural features such as secondary structure, solvent accessibility, and residue conservation within sequence families to characterize a position in a protein as gain or loss in protein function. SNAP (http://www.rostlab.org/services/SNAP/) predicts whether the mutation is neutral or nonneutral with required accuracy. SNAP scores range from −100 (strongly predicted as neutral) to 100 (strongly predicted to change function) along with a reliability index and expected accuracy.

2.5. Protein stability analysis

I-MUTANT Suite (http://gpcr.biocomp.unibo.it/~emidio/I-Mutant3.0/I-MutantDDG_Help.html) is also an SVM-based method for the automatic prediction of protein stability changes upon a single point mutation. The output file shows the predicted free energy change (DDG), which is calculated from the unfolding Gibbs free energy change of the mutated protein minus the unfolding free energy value of the native protein (kcal/mol). DDG >0 means that the mutated protein has high stability. We used the sequence-based version of I-MUTANT Suite that classifies the prediction in three classes: neutral mutation ($-0.5 \leq$ kcal/mol), large decrease ($-0.5 <$ kcal/mol), and a large increase ($0.5 >$ kcal/mol).

2.6. Biophysical validation of SAPs

Align-GVGD (http://agvgd.iarc.fr/) provides a class probability based on evolutionary conservation and chemical nature of the amino acid residues to predict whether a mutation is enriched deleterious to enriched neutral. It combines protein MSAs along with side chain composition, polarity, and steric characteristics of the mutant amino acid leading to the modification of the protein structure. Align-GVGD is a combination of Grantham variation or GV and Grantham deviation or GD. GV is an extensive protein

MSA based on quantitative measures of sequence variation at a given position and quantitative measure of the fit between a mutant amino acid and the observed amino acid at its position in the protein MSA (given by GV). If (GV = 0) and (GD > 0), the position of interest is invariant (100% conservation), so any mutation at the position is predicted as deleterious.

2.7. Evolutionary conservation profiling

The conservation pattern of the BEST1 and PRPH2 was calculated by The ConSurf server (http://consurf.tau.ac.il/) to quantify the degree of conservation at each aligned position. This server provides the evolutionary conservation profiles of protein or nucleic acid sequence or structure by first identifying the conserved positions using MSA and then calculating the evolutionary conservation rate using an empirical Bayesian inference. The ConSurf scores range 1–9: 1 is for rapidly evolving (variable) sites, color coded in turquoise; 5 is the average, color coded as white; and 9 is for slowly evolving (evolutionarily conserved) sites, color coded in maroon.

2.8. Functional characterization of SNPs

Functional SNP (F-SNP) database aims to provide a complete collection of functional information about SNPs relating to splicing, transcription, translation, and posttranslation from different bioinformatics tools and databases. F-SNP is available at http://compbio.cs.queensu.ca/F-SNP/. It provides more extensive information about the FS of each SNP quantitatively by measuring the potential deleterious effects on the biomolecular function of their genomic region. The F-SNP-Score system combines assessments from multiple independent computational tools, using a probabilistic framework that takes into account the certainty of each prediction as well as the reliability of various tools. In the new integrative scoring system, F-SNP-Score for neutral SNPs is 0.1764, whereas for disease-related SNPs, the median rises to 0.5–1. Function analysis and selection tool for single nucleotide polymorphisms (FASTSNP) available at http://fastsnp.ibms.sinica.edu.tw was used to predict potential functional effect of SNPs in 5′ UTR, 3′ UTR, and intronic region of *BEST1* and *PRPH2* genes. FASTSNP employs a complete decision tree to assign risk rankings for SNP prioritization. The decision tree will assign risk for SNP prioritization with a ranking of 0, 1, 2, 3, 4, and 5 signifying the level of no, low, medium, high, and very high effects, respectively.

2.9. Statistical analysis

To evaluate the predictive ability of previously mentioned employed methods, TP, TN, FN, FP, sensitivity, specificity, and ACC were calculated. True positive is the percentage of deleterious SAPs correctly predicted to be deleterious, true negative is the percentage of neutral SAPs correctly predicted to be neutral, false negative is the percentage of SAPs incorrectly predicted to be neutral, and false positive is the percentage of neutral SAPs incorrectly predicted to be deleterious. We computed the sensitivity as TP/(TP/FN) (probability of identifying true deleterious mutations) and specificity as TN/(TN/FP) (probability of identifying true neutral mutations).

$$\text{Sensitivity} = \frac{\text{TP}}{\text{TP} + \text{FN}}$$

$$\text{Specificity} = \frac{\text{TP}}{\text{TP} + \text{FN}}$$

3. RESULTS

For this research, we selected SNPs in (i) nonsynonymous coding regions (SAPs), (ii) 5′ and 3′ UTRs (UTR SNPs), and (iii) intronic regions. Out of 1372 SNPs recorded in dbSNP, 275 were nsSNPs (SAPs), 104 SNPs in coding synonymous region, 108 SNPs in UTR, and the remaining 885 SNPs were in the intronic region, respectively. The distributions of dbSNPs in coding and noncoding regions of *BEST1* and *PRPH2* genes were illustrated in Fig. 6.2. Overall, proportion of SNPs in intronic is much higher than the SAPs, csSNPs, and UTR SNPs. Data set for our study on the potential impact of SAPs in both *BEST1* and *PRPH2* genes investigated in this work was retrieved from dbSNP, Swiss-Prot, and HGMD and cross-checked with Ensembl database. Finally, a total of 300 SAPs (213 in *BEST1* and 87 in *PRPH2*) were considered for further investigation by *in silico* methods.

3.1. Analysis of deleterious SAPs using evolutionary-based prediction methods

Protein conservation study was performed using the SIFT algorithm to predict whether an amino acid substitution may have an impact on protein function by aligning related proteins and calculating a score that is used to determine the evolutionary conservation status of the amino acid of

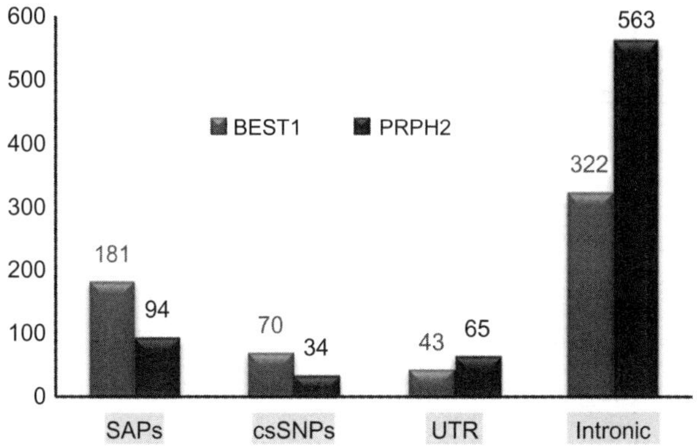

Figure 6.2 Functional annotation of SNPs in BEST1 and PRPH2 genes. (For color version of this figure, the reader is referred to the online version of this chapter.)

interest. Individual protein sequences of BEST1 and PRPH2 were submitted separately to the SIFT program to examine its tolerance index. SIFT scores were obtained for 300 SAPs. Table 6.1 illustrates the distribution of SAPs by SIFT score. SIFT scores were classified as intolerant (0.00–0.05), potentially intolerant (0.051–0.10), borderline (0.101–0.20), or tolerant (0.201–1.00). Higher the tolerance index of a particular amino acid substitution, lesser is its likely impact. Approximately 39.6% of the SAPs exhibit SIFT scores of 0.0. Another 30.3% of the variants have scores between 0.01 and 0.05. Thus, 30% of the SAPs are listed as "intolerant" by SIFT. In addition, we provided interest in variants with more modest impact on protein activity, the SAPs with SIFT scores of 0.051–0.10 have been designated as "potentially intolerant," 0.101–0.2 as "borderline," and 0.201–1.00 as "tolerant."

The protein sequence of BEST1 and PRPH2 was given as an input and further analyzed the 300 SAPs using PANTHER to validate the impact on protein function upon single amino acid substitution. PANTHER estimates the likelihood of a particular SAP by calculating the substitution position-specific evolutionary conservation (subPSEC) score based on an alignment of evolutionarily related proteins, and deleterious effect on protein function is estimated by *P*deleterious, that is, subPSEC score of −3 corresponds to a *P*deleterious of 0.5. Among 300, 78% SAPs were predicted to have a deleterious effect on protein function with subPSEC score ranging from −3 to −10.34534, 10.6% were predicted to nondeleterious with a subPSEC score <−3.0, and 11% do not align to PANTHER library HMM (Supplementary Table 6.1, http://dx.doi.org/10.1016/B978-0-12-800168-4.00006-8).

Table 6.1 Summary of SAPs that were predicted to be deleterious and neutral by eight *in silico* prediction methods in *BEST1* and *PRPH2* genes

Gene ID	rsIDs/variants	Amino acid position	SIFT	PolyPhen-2	PhD-SNP	PANTHER	SNAP	SNPs&GO	I-MUTANT Suite	Align-GVGD
BEST1	rs62637045	M1V	0	0.204	N	NA	NN	N	0.39	Class C0
	VAR_058273	I3T	0	0.549	N	−5.00235	N	N	−0.21	Class C0
	rs28940275	T6P	0.03	0.948	N	−5.12419	NN	D	−0.15	Class C0
	VAR_017366	T6R	0.01	0.942	N	−4.25826	NN	D	−0.04	Class C0
	CM063227	T6A	0.05	0.489	N	−1.72869	NN	N	−0.61	Class C0
	rs199508634	S7N	0.05	1	N	−3.33877	N	N	−0.04	Class C0
	rs28940276	V9M	0	0.784	N	−5.399	NN	D	0.33	Class C0
	HM070108	V9Q	0.05	1	D	NA	NN	D	−2.69	Class C0
	VAR_000831	V9A	0	0.971	N	−4.8871	NN	D	−0.33	Class C0
	VAR_000833	A10T	0	1	N	−5.24261	NN	D	−0.72	Class C0
	VAR_010468	A10V	0.01	0.971	N	−5.53562	NN	D	−0.35	Class C0
	VAR_017367	N11I	0.01	0.998	N	−5.54465	NN	D	0.93	Class C0
	rs62637046	A12P	0.02	0.662	N	−5.96593	NN	D	−0.13	Class C0
	VAR_010469	R13H	0.12	1	N	−4.23205	NN	D	−0.55	Class C0
	CM072887	S16Y	0.03	0.989	N	−3.78351	NN	D	−0.33	Class C0
	VAR_010470	S16F	0.02	0.989	N	−5.51584	NN	D	0.07	Class C0
	VAR_010471	F17C	0	1	N	−5.41555	NN	D	−0.33	Class C0

VAR_000834	L21V	0	1	N	−5.28802	NN	D	−0.77	Class C0
VAR_000835	W24C	0	0.987	D	−6.31508	NN	D	−0.71	Class C0
VAR_000836	R25Q	0	0.993	D	−5.39687	NN	D	−0.3	Class C0
VAR_000837	R25W	0	0.998	D	−8.54054	NN	D	0.17	Class C0
VAR_017368	G26R	0	1	D	−5.12295	NN	D	−0.34	Class C0
VAR_000838	S27R	0	0.26	D	−8.24508	NN	D	−1.4	Class C0
CM071145	Y29C	0	1	D	NA	NN	D	−0.05	Class C0
VAR_017369	Y29H	0.01	0.987	D	−6.88051	NN	D	−0.75	Class C0
VAR_017370	K30R	0.03	1	D	−4.68812	N	D	−0.36	Class C0
HM070111	E35K	0	1	D	NA	NN	D	−1.98	Class C0
rs121918288	L41P	0.01	0.966	D	−5.80563	NN	D	−0.24	Class C0
rs28940278	R47H	0.01	1	D	−4.2523	N	D	−1.07	Class C0
rs62653022	R51K	0.01	0.983	D	NA	NN	D	−1.03	Class C0
rs62637049	L52P	0.9	0.794	D	−5.81799	NN	D	−0.51	Class C0
rs200235532	E57D	0	1	D	−3.01737	N	D	−1.3	Class C0
VAR_000839	Q58L	0.03	0.989	D	−4.2523	NN	D	−1.62	Class C0
rs62637051	F62L	0	0.905	D	−5.77981	N	D	−0.46	Class C15

Continued

Table 6.1 Summary of SAPs that were predicted to be deleterious and neutral by eight *in silico* prediction methods in *BEST1* and *PRPH2* genes—cont'd

Gene ID	rsIDs/variants	Amino acid position	SIFT	PolyPhen-2	PhD-SNP	PANTHER	SNAP	SNPs&GO	I-MUTANT Suite	Align-GVGD
	rs62637053	E63K	0	1	D	−4.74926	N	D	−1.76	Class C55
	rs62637054	K64T	0.04	1	N	−4.74926	NN	D	−1.95	Class C65
	rs62637055	L67P	0.07	1	D	−4.30797	NN	D	−1.7	Class C65
	VAR_000840	L67V	0	0.987	N	−4.21519	N	D	−0.73	Class C25
	rs62641692	C69W	0	1	D	−7.0362	NN	D	−0.14	Class C65
	rs62637056	D70A	0.25	1	D	−4.24579	NN	D	−2.57	Class C65
	rs57132800	S71R	0.26	0.023	D	−2.8083	N	N	−0.13	Class C65
	rs62638168	S71G	0.1	0.994	N	−3.99438	N	D	−1.17	Class C55
	CM010521	I73N	0.42	0.996	D	−3.44136	N	D	0.62	Class C65
	rs62638171	L75H	0.04	1	D	−4.6772	NN	D	−0.41	Class C65
	CM004423	F80L	0.04	1	D	NA	N	D	−0.82	Class C15
	CM991236	L82V	0	1	D	NA	NN	D	−0.58	Class C25
	rs28940274	Y85H	0.04	0.955	D	−6.86621	NN	D	−1.86	Class C65
	rs121918289	V86M	0	0.278	D	−7.2053	NN	D	−0.37	Class C15
	VAR_017374	V89A	0	1	N	−4.25236	NN	D	−0.77	Class C65
	VAR_017375	T91I	0.27	0.987	D	−5.25423	N	D	−2.41	Class C65

VAR_000842	R92S	0	0.854	D	−7.80075	NN	D	−0.56	Class C65
VAR_010474	R92C	0	1	D	−9.32483	NN	D	−0.59	Class C65
VAR_010475	R92H	0	1	D	−7.90737	NN	D	−0.83	Class C25
rs28940273	W93C	0.04	1	D	−8.98379	NN	D	−2.4	Class C65
CM991238	Q96H	0	1	D	NA	NN	D	−0.56	Class C15
CM000837	N99K	0.16	0.939	D	−2.50096	NN	D	−0.02	Class C65
CM001381	L100R	0	0.984	D	−4.08019	NN	D	−0.18	Class C65
VAR_017376	P101T	0	1	D	−5.8653	NN	D	−0.94	Class C35
VAR_017377	W102R	0	1	D	−6.38476	NN	D	−1.28	Class C65
rs62638172	D104A	0.05	0.999	D	−4.51477	NN	D	−2.27	Class C65
VAR_000846	D104E	0.02	0.91	D	−3.65183	NN	D	−1.65	Class C35
VAR_017378	D104H	0	1	D	−6.53791	NN	D	−2.43	Class C65
VAR_025731	R105C	0	1	D	−6.88166	NN	D	−0.99	Class C65
rs62638174	S111A	0	1	D	−3.0545	N	N	−1.81	Class C65
rs62638176	G112S	0	1	D	−3.11827	N	N	−1.71	Class C55
VAR_025732	F113L	0	0.884	D	−2.93399	N	N	0.45	Class C15
rs1805142	E119Q	0.02	0.294	D	−4.4664	N	D	−1.27	Class C25
VAR_017379	N133K	0.02	1	D	0.27157	NN	D	−0.88	Class C65

Continued

Table 6.1 Summary of SAPs that were predicted to be deleterious and neutral by eight *in silico* prediction methods in *BEST1* and *PRPH2* genes—cont'd

Gene ID	rsIDs/variants	Amino acid position	SIFT	PolyPhen-2	PhD-SNP	PANTHER	SNAP	SNPs&GO	I-MUTANT Suite	Align-GVGD
	CM072109	L134V	0.01	1	D	NA	N	D	0.67	Class C25
	VAR_010478	G135S	0.24	1	N	−4.02424	N	N	−0.53	Class C55
	rs267606678	L140V	0.01	0.998	D	−4.41907	NN	D	−0.29	Class C25
	VAR_017380	L140R	0	0.884	D	−5.96333	NN	D	−0.5	Class C65
	rs121918284	R141H	0	0.986	D	−7.07296	NN	D	−2.32	Class C25
	VAR_010479	A146K	0	0.971	N	−5.96333	N	D	−2.42	Class C65
	VAR_043493	P152A	0	0.971	D	−5.81298	NN	D	−3.27	Class C25
	rs62638178	E172Q	0.26	0.999	D	−4.13893	N	N	−0.52	Class C25
	rs187560307	K173Q	0.12	0.023	D	−3.40917	N	N	−0.44	Class C45
	rs62637336	M180V	0.08	1	D	−3.24324	N	D	−1.06	Class C15
	rs62637338	F188L	0	1	N	−3.82255	N	D	0.19	Class C15
	rs62637339	A189T	0.01	0.949	N	−2.87121	N	N	−2.16	Class C55
	CM071144	L191P	0	1	D	NA	NN	D	−0.91	Class C65
	rs62637341	M193V	0.25	0.989	N	−2.93371	N	N	0.36	Class C15
	rs200277476	A195V	0.05	0.966	D	−5.22563	N	D	−0.58	Class C65
	rs200277476	A195P	0.04	1	D	−6.64698	NN	D	−0.99	Class C25

rs62637345	G198R	0.33	1	D	−2.90981	N	D	−0.44	Class C65
rs199529046	I201T	0.03	1	D	NA	NN	D	−1.29	Class C65
rs62637346	R202L	0.14	0.968	D	−4.43884	N	D	−0.36	Class C65
rs267606680	I205T	0.37	1	D	−3.66159	N	N	−1.02	Class C65
rs74653691	L207I	0	1	D	−4.37278	N	D	−0.01	Class C0
rs62637348	Q208E	0	0.999	D	−4.1473	N	D	0.3	Class C25
VAR_000848	S209N	0	1	D	−3.42389	N	D	−1.91	Class C45
rs62637349	N212H	0.01	0.949	D	−5.1447	N	N	−0.66	Class C65
rs138932379	E213K	0	0.992	D	−4.48005	NN	D	−0.96	Class C55
rs62651027	M214L	0.01	1	D	−3.46083	N	N	−0.92	Class C0
VAR_010480	T216I	0	0.987	D	−4.24281	N	D	−0.22	Class C65
CM071147	L217F	0	1	N	NA	N	N	−0.45	Class C15
VAR_000849	R218C	0	0.835	D	−8.63047	NN	D	−0.88	Class C65
VAR_000850	R218Q	0	0.79	D	−5.03005	NN	D	−0.77	Class C35
VAR_000851	R218S	0	0.89	D	−6.69435	NN	D	−1.08	Class C65
VAR_010481	R218H	0.02	1	D	−6.04378	NN	D	−1.13	Class C25
VAR_025735	C221W	0	1	D	−7.3305	NN	D	−0.13	Class C65
CM072110	G222E	0.01	1	D	−3.26537	NN	D	−0.31	Class C65

Continued

Table 6.1 Summary of SAPs that were predicted to be deleterious and neutral by eight *in silico* prediction methods in *BEST1* and *PRPH2* genes—cont'd

Gene ID	rsIDs/variants	Amino acid position	SIFT	PolyPhen-2	PhD-SNP	PANTHER	SNAP	SNPs&GO	I-MUTANT Suite	Align-GVGD
	VAR_025736	G222V	0	1	D	−5.37258	NN	D	−0.56	Class C65
	rs200162075	H223Q	0	1	D	−3.17799	N	N	−1.71	Class C15
	VAR_000852	L224M	0.05	1	D	−5.24341	NN	D	−1.13	Class C0
	VAR_025737	L224P	0.05	1	D	−7.50295	NN	D	−1.59	Class C65
	rs267606677	Y227C	0	0.608	D	−6.08548	NN	D	−0.49	Class C65
	rs28941469	Y227N	0	0.987	N	−7.02001	NN	D	−0.37	Class C65
	CM067982	Y227F	0.15	1	D	NA	NN	D	−1.57	Class C15
	rs267606676	D228N	0.01	0.61	D	−5.86119	NN	D	−2.22	Class C15
	VAR_000855	S231R	0.01	0.998	N	−5.10117	NN	D	0.07	Class C65
	CM000839	S231T	0.2	0.992	D	−2.85939	NN	D	−0.73	Class C55
	CM067981	I232N	0	0.989	D	−4.5185	N	D	−2.1	Class C65
	CM072108	P233Q	0	1	D	NA	NN	D	−2.04	Class C65
	HM070106	L234V	0	1	D	NA	N	D	−0.97	Class C25
	rs267606679	V235A	0.02	0.598	N	−3.4616	NN	D	−2.75	Class C65
	VAR_000856	V235M	0.04	1	N	−4.38806	N	D	−1.04	Class C15
	VAR_010482	V235L	0	1	N	−5.21937	NN	D	−1.38	Class C25

rs121918291	Y236C	0	1	D	−9.02715	NN	D	−0.76	Class C65
CM067983	T237S	0.01	1	D	NA	NN	D	−0.62	Class C55
VAR_000857	T237R	0	0.549	D	−7.00179	NN	D	−0.41	Class C65
rs121918290	V239M	0	1	D	−6.82459	NN	D	−0.09	Class C15
VAR_025738	T241N	0.01	0.884	D	−5.05477	NN	D	−0.85	Class C55
VAR_058277	V242M	0.03	0.971	N	−3.82797	N	N	−1.23	Class C0
rs137853905	A243T	0.04	0.999	D	−5.36649	NN	D	−1.08	Class C0
rs28940570	A243V	0.02	1	D	−4.94649	NN	D	−0.52	Class C0
rs186522420	V244G	0	1	D	−8.51583	NN	D	−1.19	Class C0
rs62637351	F247S	0	0.698	D	−7.5627	NN	D	−1.59	Class C0
rs62637356	G266R	0.29	0.884	D	−4.65276	N	D	−1.31	Class C15
rs62637358	H267N	0.05	1	D	−3.72516	N	D	−2.12	Class C0
rs62639260	E268A	0.27	0.994	N	−3.78157	N	D	−1.45	Class C0
rs62639262	D270N	0	1	D	−5.91033	NN	D	−2.43	Class C0
rs62639263	D270E	0.02	1	D	−5.15311	NN	D	−1.93	Class C0
rs62639264	L271P	0.03	0.672	D	−5.09419	NN	D	−1.77	Class C0
rs62639267	V272G	0	1	D	−3.59777	NN	D	−0.43	Class C0
rs62639269	V273M	0.02	0.971	D	−5.8062	NN	D	−0.5	Class C0

Continued

Table 6.1 Summary of SAPs that were predicted to be deleterious and neutral by eight *in silico* prediction methods in *BEST1* and *PRPH2* genes—cont'd

Gene ID	rsIDs/variants	Amino acid position	SIFT	PolyPhen-2	PhD-SNP	PANTHER	SNAP	SNPs&GO	I-MUTANT Suite	Align-GVGD
	VAR_025740	V275I	0	1	D	−4.94649	N	N	−0.78	Class C0
	rs62639271	F276L	0	1	D	−3.7223	N	D	−1.21	Class C0
	rs62639272	T277A	0	1	N	−5.27678	NN	D	−0.62	Class C0
	CM094327	E292K	0	1	D	NA	N	D	−0.88	Class C55
	VAR_010483	Q293K	0	0.971	D	−4.90343	N	D	−1.84	Class C45
	CM072107	Q293H	0.02	1	D	NA	NN	D	−1.57	Class C15
	VAR_025742	L294V	0.01	1	D	−5.77242	NN	D	−1.2	Class C25
	VAR_025743	I295T	0.01	1	D	−5.23854	NN	D	−0.2	Class C65
	VAR_010484	N296S	0.01	1	D	−6.1431	NN	D	−1.54	Class C45
	VAR_025744	N296H	0	0.971	D	−6.1431	NN	D	−1.88	Class C65
	CM010524	N296K	0	1	D	NA	NN	D	−2.32	Class C65
	CM072886	N296S	0.01	1	D	NA	NN	D	−0.63	Class C45
	CM071143	N296D	0.03	1	D	NA	NN	D	−1.5	Class C15
	rs1805143	P297S	0	1	D	NA	NN	D	−1.85	Class C65
	VAR_000860	P297A	0	1	D	−9.318	NN	D	−2.46	Class C25
	VAR_025745	F298S	0.03	1	D	−7.99005	NN	D	−0.51	Class C65

VAR_000861	G299E	0	1	D	−9.66383	NN	D	−0.95	Class C65
VAR_058313	G299A	0	1	D	−9.0706	NN	D	−1.72	Class C55
CM072111	G299R	0	1	D	NA	NN	D	−2.13	Class C65
VAR_000862	E300K	0	0.987	D	−5.39144	NN	D	−1.45	Class C55
VAR_010486	E300D	0	0.987	D	−4.52199	N	D	−1.25	Class C35
VAR_000863	D301E	0	0.971	D	−5.39144	NN	D	−1.06	Class C35
VAR_000864	D301N	0.01	1	D	−6.28927	NN	D	−1.5	Class C15
VAR_025746	D302G	0.01	1	D	−7.17685	NN	D	−1.78	Class C65
VAR_025747	D302H	0	0.971	D	−7.85511	NN	D	−1.19	Class C65
VAR_025748	D302V	0.03	1	D	−6.84467	NN	D	−0.32	Class C65
VAR_025749	D303E	0	0.884	D	−4.35763	NN	D	−1.04	Class C35
HM070104	D304V	0	1	D	NA	NN	D	−2.01	Class C65
VAR_000865	F305S	0	0.987	D	−8.18427	NN	D	−1.51	Class C65
VAR_025750	E306D	0.13	1	D	−3.92414	NN	D	−2.15	Class C35
VAR_025751	E306G	0	0.884	D	−7.09137	NN	D	−2.96	Class C65
VAR_010487	T307I	0	0.211	D	−4.7011	N	D	0.4	Class C65
VAR_025752	T307A	0.01	0.971	D	−4.7011	N	D	−0.43	Class C55
VAR_025753	N308S	0.01	1	D	−4.95646	NN	D	−1.81	Class C45

Continued

Table 6.1 Summary of SAPs that were predicted to be deleterious and neutral by eight *in silico* prediction methods in *BEST1* and *PRPH2* genes—cont'd

Gene ID	rsIDs/variants	Amino acid position	SIFT	PolyPhen-2	PhD-SNP	PANTHER	SNAP	SNPs&GO	I-MUTANT Suite	Align-GVGD
	VAR_000866	I310T	0	1	D	−6.47996	NN	D	−0.38	Class C65
	VAR_000867	V311G	0	0.971	D	−6.41499	NN	D	−1.07	Class C65
	VAR_000868	D312N	0.01	0.987	D	−6.41499	NN	D	−2.38	Class C15
	rs62640562	R313K	0.04	1	D	−4.29786	N	D	−2.01	Class C25
	rs121918287	V317M	0.02	1	D	−4.39344	N	D	−0.26	Class C15
	rs199960774	L319V	1	0.983	D	−4.51898	N	D	0.1	Class C25
	VAR_043495	M325T	0	1	D	−6.41499	NN	D	−0.24	Class C65
	rs201386186	R331Q	0.01	1	D	−3.14972	N	N	−1.15	Class C35
	rs148326372	P334L	0.01	0.13	D	−3.03722	N	N	−0.23	Class C65
	rs202234687	E342Q	0	1	D	−3.10386	N	N	−0.45	Class C25
	rs147409760	A352T	0	1	D	−4.39855	N	D	−0.86	Class C55
	rs139637557	R355C	0.06	1	D	−3.28235	N	D	−0.66	Class C65
	VAR_043496	A357V	0.01	0.981	N	−3.85764	NN	D	0.33	Class C0
	rs145212203	M360I	0.03	0.999	N	−3.11041	N	N	0.48	Class C0
	rs147228028	T363P	0	1	D	−6.20085	NN	D	−0.95	Class C35
	rs199998058	N378S	0	1	D	−2.62045	N	N	−1.73	Class C45

rs201225558	R392H	0.64	0.149	N	−5.41973	N	D	−0.54	Class C25
rs199890510	S398F	0.01	1	N	−4.59411	N	D	0.63	Class C65
rs149678971	H402Y	0.02	0.442	D	−3.47505	N	N	0.3	Class C35
rs202125490	P404S	0.49	0.013	N	−2.32422	N	N	−0.34	Class C65
rs146689925	T410I	0.21	0.084	D	−4.25635	N	N	0.11	Class C65
rs201210799	G424A	0	1	N	−2.41256	N	N	−1.11	Class C0
rs201299251	Q434E	0.32	0.998	N	−2.41207	N	N	−0.04	Class C25
rs145439032	V436I	0.53	0.013	N	−3.29989	N	N	0.51	Class C0
rs62640563	D441G	0.24	1	D	−2.79348	N	D	−0.83	Class C65
rs148854184	N442T	0.16	0.011	N	−2.68153	N	N	−0.59	Class C0
rs146447431	L472R	0	0.013	N	−2.3919	N	D	−0.03	Class C65
rs140681289	T475N	0.46	0.031	D	−3.11964	N	N	−1.72	Class C55
rs200510369	P480T	0	1	D	−2.29109	N	N	−1.32	Class C35
rs62640564	H490P	0.34	1	D	−2.29236	N	D	−1.36	Class C65
rs111326315	V492I	0.02	1	N	−2.46738	N	N	−0.35	Class C25
rs62640565	D496N	0	1	N	−2.05006	N	D	−1.13	Class C15
rs141071579	S507P	1	0.017	D	−2.88773	N	N	−1.43	Class C65
rs62640566	S507F	0	0.998	N	0.34269	N	N	−1.1	Class C65

Continued

Table 6.1 Summary of SAPs that were predicted to be deleterious and neutral by eight *in silico* prediction methods in *BEST1* and *PRPH2* genes—cont'd

Gene ID	rsIDs/variants	Amino acid position	SIFT	PolyPhen-2	PhD-SNP	PANTHER	SNAP	SNPs&GO	I-MUTANT Suite	Align-GVGD
	rs61747600	D520N	0.51	0.994	N	−2.34187	N	N	−1.75	Class C15
	rs145209035	M524T	0.04	0.868	N	−1.94039	N	N	−0.06	Class C65
	rs200582915	E525A	0.03	1	N	−2.28728	NN	N	−0.59	Class C65
	rs147490956	K535E	0.04	0.442	N	−1.88247	N	D	0.12	Class C55
	rs143671863	E549K	0.12	0.983	D	−1.44199	N	N	−0.45	Class C55
	rs147192139	E557K	0.04	0.007	D	−1.13136	NN	N	−1.73	Class C15
	VAR_010490	T561A	0.2	0.004	D	NA	NN	N	−0.34	Class C0
	rs148060787	L567F	0.14	0.011	N	NA	NN	D	0.53	Class C15
	VAR_009278	E578V	0.05	1	N	NA	NN	N	−0.56	Class C65
PRPH2	rs121918565	M1T	0	1	N	−3.37589	NN	N	−0.8	Class C65
	VAR_006853	R13W	0	1	D	−7.36188	NN	N	−0.09	Class C65
	rs146686238	W25C	0.06	1	D	−4.40303	NN	D	−2.27	Class C65
	rs61755766	S27F	0.7	0.4	D	−4.61574	NN	D	0.08	Class C65
	rs61755767	I32V	0.74	0	N	−3.58534	N	N	−0.05	Class C25
	rs113689552	S36R	0.21	0.74	D	−6.39497	NN	N	−0.61	Class C65
	rs201018137	S36N	0.18	0.59	N	−6.56551	NN	N	−1.07	Class C45

	rs61755770	L45F	0.24	0.99	D	−4.13308	NN	N	−0.2	Class C15
	rs146844134	S49T	0.75	0.64	N	−4.71064	N	N	−0.24	Class C55
	rs61755774	G68R	0	1	N	−9.79341	NN	D	−1.54	Class C65
	rs140227298	A116S	0.17	0.03	D	−4.65997	N	N	−1.4	Class C65
	CM090728	R123W	0.04	0.19	D	−5.71161	NN	N	−0.12	Class C65
	CM090728	S125L	0.3	0.44	D	−2.67121	N	N	−0.41	Class C65
	rs61755779	L126V	0.07	0.97	N	−3.6721	NN	N	−0.79	Class C25
	VAR_006859	L126R	0	1	D	−6.87271	NN	N	−1.47	Class C65
	CM090729	L126P	0.01	1	D	−5.76526	NN	D	0.22	Class C65
	CM083770	G137D	0.07	0.96	N	−4.0645	NN	D	−1.29	Class C65
	rs61755780	Y141H	0.13	1	D	−7.28552	NN	D	−0.98	Class C65
	rs61755781	Y141C	0	1	D	−7.54978	NN	D	−0.3	Class C65
	rs61755783	R142W	0.18	1	D	−6.75386	NN	D	−0.87	Class C65
	rs146703538	M152V	0.41	0.01	N	−3.82948	N	N	−0.87	Class C15
	rs61755785	K153R	0.04	1	D	−4.10762	NN	N	−0.28	Class C25
	rs199572514	T155I	0.45	0.89	D	−3.42813	N	N	−0.03	Class C65
	rs61755787	D157N	0.05	1	D	−4.76518	NN	D	−0.35	Class C15
	rs76989855	I161M	0.54	0.44	D	−3.95552	N	D	0.09	Class C0

Continued

Table 6.1 Summary of SAPs that were predicted to be deleterious and neutral by eight *in silico* prediction methods in *BEST1* and *PRPH2* genes—cont'd

Gene ID	rsIDs/variants	Amino acid position	SIFT	PolyPhen-2	PhD-SNP	PANTHER	SNAP	SNPs&GO	I-MUTANT Suite	Align-GVGD
	CM074464	C165R	0	1	N	NA	NN	D	0	Class C65
	rs61755788	C165Y	0	1	D	NA	NN	D	−0.05	Class C65
	rs61755789	G167D	0	1	D	−9.98725	NN	D	−1.29	Class C65
	VAR_032052	G167S	0	1	D	−8.3126	NN	N	−0.69	Class C55
	rs61755791	G170S	0.97	0.82	D	−3.60889	N	N	−1.49	Class C55
	rs200876455	F171I	0.05	0.92	D	−4.52994	NN	D	−0.03	Class C15
	rs61755792	R172G	0.29	0.8	D	−5.13655	NN	D	−1.44	Class C65
	rs61755793	R172Q	0.56	0.25	D	−3.86894	NN	D	−0.74	Class C35
	CM930639	R172W	0.01	1	D	−5.30179	NN	D	−0.5	Class C65
	rs61755794	D173V	0	1	D	−9.40237	NN	D	−1.58	Class C65
	rs201501819	I177M	0.03	0.87	D	−5.51268	NN	N	−0.11	Class C0
	rs61755795	Q178R	0.25	1	D	−4.00306	NN	D	−1.3	Class C35
	rs61755796	W179R	0	1	D	−6.63422	NN	N	0.13	Class C65
	rs62645926	Y184S	0.03	1	D	−3.42359	NN	N	−0.34	Class C65
	rs121918563	L185P	0.05	1	D	−5.2061	NN	D	−1.55	Class C65
	rs121918567	R195L	0.31	1	D	−3.90835	NN	D	−0.07	Class C65

rs62645931	K197E	0.67	0.76	D	−2.66785	NN	N	−1.5	Class C55
CM063092	S198R	0.26	1	N	−3.48169	NN	D	−1.57	Class C65
rs62645932	V200E	1	0.94	N	−3.60359	NN	N	−0.6	Class C65
rs139185976	G208D	0.34	0.28	D	−4.46855	NN	D	−1.48	Class C65
rs61755797	P210S	0	1	D	−7.00064	NN	N	−0.64	Class C65
rs61755798	P210R	0	1	D	−7.23526	NN	D	−0.35	Class C65
CM011806	P210L	0	1	D	−4.80873	NN	D	−1.36	Class C65
rs61755799	F211L	0.5	1	D	−5.07454	NN	D	−0.47	Class C15
rs61755800	S212G	0	1	N	−7.72847	NN	N	−1.45	Class C55
rs61755801	S212T	0.04	1	N	−5.62785	NN	N	−0.77	Class C55
rs61755802	C213R	0	1	D	−9.71014	NN	N	−0.42	Class C65
rs61755803	C213Y	0	1	D	−7.98527	NN	N	−0.45	Class C65
HM070073	C213F	0	1	D	−9.90769	NN	D	−1.46	Class C65
rs61755804	C214Y	0	1	D	−7.0907	NN	N	−1.52	Class C65
VAR_006880	C214S	0	1	D	−7.49716	NN	N	−1.95	Class C65
rs61755805	P216S	0.61	0.8	N	−4.12563	NN	N	−1.24	Class C65
rs61755806	P216L	0.27	0.06	D	−4.62909	NN	N	−0.79	Class C65
CM063093	P216R	0.56	0.99	D	−3.95344	NN	D	−0.59	Class C65

Continued

Table 6.1 Summary of SAPs that were predicted to be deleterious and neutral by eight *in silico* prediction methods in *BEST1* and *PRPH2* genes—cont'd

Gene ID	rsIDs/variants	Amino acid position	SIFT	PolyPhen-2	PhD-SNP	PANTHER	SNAP	SNPs&GO	I-MUTANT Suite	Align-GVGD
	CM090730	P216A	0.71	0.95	N	−2.98481	NN	N	−0.26	Class C25
	rs149511444	P219T	0.65	0.99	D	−5.10794	NN	D	−0.6	Class C35
	rs61755808	P219R	0.54	0.89	N	−5.22094	NN	N	−0.5	Class C65
	rs61755809	R220W	0.02	0.91	D	−8.25309	NN	D	−0.3	Class C65
	rs61755810	R220Q	0.59	1	D	−5.03055	NN	D	−0.84	Class C35
	CM090731	P221L	0.05	1	N	−4.63071	NN	D	−0.33	Class C65
	rs61755811	Q226E	0.45	0.76	D	NA	N	D	−1.28	Class C25
	rs61755815	N244H	0.09	1	D	−5.37112	NN	D	−0.98	Class C65
	rs61755816	N244K	0.56	1	D	−4.44367	NN	N	−0.77	Class C65
	rs61755817	W246R	0.21	1	D	−4.36827	NN	D	−0.14	Class C65
	CM090732	G249S	0.06	1	N	−3.72797	NN	D	−1.41	Class C65
	CM053390	C250F	0	1	N	−10.3453	NN	D	−0.82	Class C65
	rs62645926	Y258S	0.2	0.98	D	−5.08078	NN	N	−1.38	Class C65
	rs150381599	L261F	0.7	0.84	D	−4.56018	NN	N	0.28	Class C15
	rs62645935	G266D	0.1	1	N	−8.30331	NN	N	−0.31	Class C65
	rs62645936	V268I	0.85	0	D	−3.9444	N	D	−0.55	Class C25

rs140406696	V277G	0.14	0.76	D	−5.64786	NN	N	−3.14	Class C65
rs146659849	I281T	0.14	0	N	4.59277	N	N	−0.65	Class C25
rs62645939	S289L	0.29	0.82	D	−4.64577	NN	N	−1.39	Class C15
rs390659	Q304E	1	1	D	−2.73319	NN	N	−0.43	Class C65
rs61748432	G305D	0.02	1	D	−3.30008	NN	N	−2.09	Class C65
rs425876	K310R	1	0.03	D	−3.52412	NN	N	−1.03	Class C65
rs61748434	P313L	0.28	0.02	N	−3.87946	N	N	−0.05	Class C65
rs202230698	W316G	0.05	0.05	N	−3.71849	N	N	−1.14	Class C65
rs139329966	F319L	0.72	0	N	−2.07626	N	N	0.1	Class C15
rs193921105	E333K	0.1	0.46	D	NA	N	N	−0.07	Class C65
VAR_006895	D338A	0.56	0	D	NA	NN	N	−1.22	Class C0
rs145392651	A339T	0.19	0.85	N	NA	N	N	−0.16	Class C55

rsIDs and variants highlighted in bold are found to be deleterious by all the methods; NA, not available.

Lastly, SAPs with significant scores using SIFT and PANTHER were also analyzed using SVM-based PhD-SNP to predict whether a variant would be potentially deleterious to the function of the protein. PhD-SNP listed 73.6% SAPs as disease and the remaining 26.3% as neutral (Table 6.1).

3.2. Analysis of deleterious SAPs using structure-based prediction methods

We tested the impact of SAPs in protein function using structure-based predictors by applying it to three different methods. The structural levels of alteration were determined by applying the PolyPhen-2 program. Similar data for 300 SAPs submitted to SIFT, PANTHER, and PhD-SNP were also submitted as input to the PolyPhen-2 server. PolyPhen-2 scores were obtained for 300 SAPs (Table 6.1). To give an overview of the distribution of PolyPhen-2 scores, the scores are placed into three groups. PolyPhen-2 scores of 0.85–1, scores required to be "probably damaging" to protein structure and function, account for 79.3% of the SAPs. An additional 12.3% of the variants exhibited PolyPhen-2 scores of 0.2–0.84 indicative of variants that are "possibly damaging" to protein function and the remaining 8.3% SAPs that scored less than 0.02 were designated as "benign." SNPs&GO utilizes sequence and evolutionary information to predict whether a mutation is disease-related or not by exploiting the protein functional annotation. The protein sequences with designated UniProt accession numbers were submitted along with mutational position, wild-type and mutant-type residue as input to the server. Sixty-nine percent of the SAPs were designated as "disease" and the remaining 31% as "neutral." SNAP was used to predict the overall severity of the missense mutations based on neutral network and improved machine-learning methodologies. SNAP classifies the SAPs into deleterious (effect on function) and neutral (no effect) using sequence-based computationally acquired information alone. SNAP computes a series of biochemical properties such as whether there is an inflexible proline into an alpha-helix and mass of wild and mutant residue to construct classification models. In addition, SNAP utilizes Swiss-Prot feature table terms to the final prediction rules. 66.6% of the SAPs were predicted to be nonneutral and 33.3% as neutral (Table 6.1).

3.3. Prediction of stability changes by I-MUTANT Suite

Stability changes are often found for mutated proteins involved in diseases. Predicting the protein stability upon mutation is essential for understanding the structure–function relationship of a protein. Generally, the stability of a protein is represented by the change in the Gibbs free energy upon folding (ΔG), where an increasingly negative number represents greater stability. Single amino acid substitution in a protein sequence can result in a significant change in the protein's stability ($\Delta\Delta G$), where a positive $\Delta\Delta G$ represents a destabilizing mutation and a negative value represents a stabilizing mutation. All the 300 SAPs submitted to six pathogenic prediction tools were also subjected to protein stability analysis by I-MUTANT Suite. Nearly 90.3% of the SAPs were predicted to affect the stability of protein, and the remaining 9.6% were destabilizing mutation (Table 6.1).

3.4. Align-GVGD

Align-GVGD, otherwise a mathematically simple classification of missense substitutions, provides a class probability based on evolutionary conservation and characteristics of the amino acid such as side chain composition, polarity, and steric features to determine the FS of missense variants. This compares the chemical and physical characteristics between exchanging residues with substitution frequencies and with the relevant output the "C-score," which provides seven discrete grades running from C0 to C65 ranging from least likely neutral (class 0) to most likely neutral (class 65) to interfere with the function of the protein. Forty-four percent SAPs occurred at strongly conserved residues ($GV=0$) and had a $GD \geq 65$. These were classified as class 65 of substitutions most likely to interfere with protein function. The remaining SAPs were classified into class 0 (20%), class 15 (10.6%), class 25 (9%), class 35 (5%), class 45 (2.6%), and class 55 (8%) (Table 6.1). Performance of the eight *in silico* prediction methods is depicted in Table 6.2.

3.5. Concordance between the functional consequences of each SAP

The accuracy of deleterious SAPs predicted in this study can be increased by combining various *in silico* methods. We calculated the concordance in three distinct steps: (i) concordance between evolutionary-based methods SIFT, PANTHER, and PhD-SNP; (ii) concordance between structure-based methods PolyPhen-2, SNPs&GO, and SNAP; and (iii) concordance between evolutionary-based, structure-based, I-MUTANT Suite, and Align-GVGD. Lower prediction score in SIFT, PANTHER, and I-Mutant 2.0 classifies an SAP as deleterious, whereas a higher PolyPhen-2 score classifies an SAP as deleterious. Concordances between these combinations were illustrated in Fig. 6.3 and Table 6.3. Out of 213 SAPs in *BEST1*, evolutionary-based techniques in combination, SIFT and PANTHER predicted 64%, PANTHER

Table 6.2 Performance of eight *in silico* methods in their prediction of deleterious and neutral SAPs in *BEST1* and *PRPH2* genes

Predictions	SIFT	PolyPhen-2	PhD-SNP	PANTHER	SNAP	SNPs&GO	I-MUTANT Suite	Align-GVGD
BEST1/VMD2								
Deleterious/decrease/nonneutral	175	200	159	158	129	169	191	75
Tolerated/increase/neutral	38	13	54	28	84	44	22	138
Not predicted	0	0	0	27	0	0	0	0
Total	213	213	213	213	213	213	213	213
PRPH2/RDS								
Deleterious/decrease/nonneutral	35	75	62	76	71	38	80	57
Tolerated/increase/neutral	52	12	25	5	16	49	7	30
Not predicted	0	0	0	6	0	0	0	0
Total	87	87	87	87	87	87	87	87

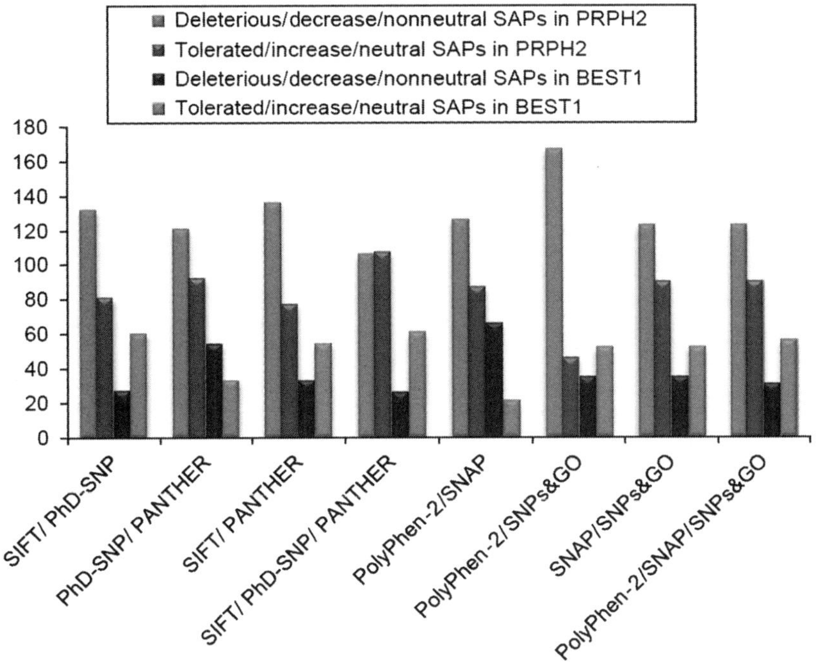

Figure 6.3 Concordance between six *in silico* prediction methods. (For color version of this figure, the reader is referred to the online version of this chapter.)

and PhD-SNP predicted 58%, SIFT and PhD-SNP predicted 62%, and SIFT, PANTHER, and PhD-SNP predicted 50%, respectively. Structure-based methods in combination predicted 56% by PolyPhen-2 and SNAP, 78% by PolyPhen-2 and SNPs&GO, 58% by SNPs&GO and SNAP, and 58% by PolyPhen-2, SNPs&GO, and SNAP, respectively. Among 87 SAPs in *PRPH2*, evolutionary-based methods SIFT and PANTHER predicted 38%, PANTHER and PhD-SNP predicted 62%, SIFT and PhD-SNP predicted 31.3%, and SIFT, PANTHER, and PhD-SNP predicted 30%. Structure-based methods PolyPhen-2 and SNAP predicted 76%, PolyPhen-2 and SNPs&GO predicted 40%, SNPs&GO and SNAP predicted 40%, and PolyPhen-2, SNPs&GO, and SNAP predicted 35%. Overall, the prediction performance of PolyPhen-2 and I-MUTANT Suite is much higher than the remaining six tools.

3.6. BEST1 and PRPH2 protein sequence analysis

To investigate the conserved amino acids of BEST1 and PRPH2 human proteins, we performed MSAs using MUSCLE (multiple sequence

Table 6.3 Concordance between the six *in silico* prediction methods in various combinations

Predictions	SIFT/PhD-SNP	PhD-SNP/PANTHER	SIFT/PANTHER	SIFT/PhD-SNP/PANTHER	PolyPhen-2/SNAP	PolyPhen-2/SNPs&GO	SNAP/SNPs&GO	PolyPhen-2/SNAP/SNPs&GO
BEST1								
Deleterious/decrease/nonneutral	132	121	136	106	126	167	123	123
Tolerated/increase/neutral/NA	81	92	77	107	87	46	90	90
Total	213	213	213	213	213	213	213	213
PRPH2								
Deleterious/decrease/nonneutral	27	54	33	26	66	35	35	31
Tolerated/increase/neutral	60	33	54	61	21	52	52	56
Total	87	87	87	87	87	87	87	87

comparison by log expectation) available at http://www.ebi.ac.uk/Tools/msa/muscle/ (Edgar, 2004). MUSCLE is a Web-based tool to align multiple sequences from various vertebrate species including humans. We searched the protein sequence of BEST1 and PRPH2 against a sequence database to find sequences of homologous proteins. An MSA of the homologous sequences reveals at which positions amino acids are conserved throughout evolutionary time, and these positions can be essential for protein function (Ng & Henikoff, 2003). The information regarding the homologous protein sequences employed in the MUSCLE analysis for BEST1 and PRPH2 protein is provided in Tables 6.4 and 6.5. We aligned the sequences using MUSCLE (Fig. 6.4A and B) and submitted to sequence logo to demonstrate patterns of the sequence alignment. The overall height of the stack of letters indicates the functional conservation at that position and also illustrates the amino acid composition. Generated WebLogo (Crooks, Hon, Chandonia, & Brenner, 2004) of BEST1 and PRPH2 protein shows sequence logos up to 139 and 140 sequences (Fig. 6.5A and B). Importantly, acquired information from the sequence logo of BEST1 and PRPH2 protein indicates the sequences from different species are highly conserved. In this context, we applied ConSurf that utilizes Bayesian study to evaluate the evolutionary significance of the missense variant involved in a protein sequence (Fig. 6.6A and B). Most of the amino acids in BEST1 and PRPH2 were highly conserved by ConSurf and obtained tallest stacks in sequence logos generated from WebLogo. The substitutions of conserved residues are deleterious in nature by any one of the *in silico* prediction methods (Table 6.1).

3.7. Propensity of each amino acid in native and mutant state of BEST1 and PRPH2

For this study, protein sequences were obtained from UniProt, which is a repository of protein properties. BEST1 recorded with longest sequence length of 585 aa whereas PRPH2 with shortest sequence length of 346 aa. We calculated the composition of each amino acid in BEST1 and PRPH2 by statistical analysis of protein sequences (SAPS) (Brendel, Bucher, Nourbakhsh, Blaisdell, & Karlin, 1992). Larger composition of leucine was observed, followed by proline, serine, and glutamine in BEST1 (Fig. 6.7A). Similarly, in PRPH2 sequence, we recognized lager composition of leucine followed by serine, lysine, and glycine (Fig. 6.7A). In addition, we also calculated the propensity of each amino acid in native and mutant state of BEST1 and PRPH2 (Fig. 6.7B). In BEST1, the most common amino acid in the native state was

Table 6.4 List of sequences utilized for MUSCLE and WebLogo analysis in BEST1

Sl. no.	Organism	Length	Protein/gene ID	Other information
1.	*Homo sapiens*	585 aa	GenBank: AAR99654.1	ACCESSION AAR99654 VERSION AAR99654.1 GI:41216873
2.	*Mus musculus*	551 aa	NCBI Reference Sequence: NP_036043.2	ACCESSION NP_036043 XP_129203 VERSION NP_036043.2 GI:160333218
3.	*Gallus gallus*	762 aa	NCBI Reference Sequence: XP_421055.2	ACCESSION XP_421055 VERSION XP_421055.2 GI:118091384
4.	*Canis lupus familiaris*	580 aa	NCBI Reference Sequence: NP_001091014.1	ACCESSION NP_001091014 XP_540912 VERSION NP_001091014.1 GI:148237434
5.	*Bos taurus*	589 aa	NCBI Reference Sequence: NP_001073714.1	ACCESSION NP_001073714 XP_585778 VERSION NP_001073714.1 GI:122692507
6.	*Nomascus leucogenys*	603 aa	NCBI Reference Sequence: XP_003274102.2	ACCESSION XP_003274102 VERSION XP_003274102.2 GI:441604447
7.	*Taeniopygia guttata*	740 aa	NCBI Reference Sequence: XP_002195004.2	ACCESSION XP_002195004 VERSION XP_002195004.2 GI:449503899
8.	*Gorilla gorilla gorilla*	539 aa	NCBI Reference Sequence: XP_004051395.1	ACCESSION XP_004051395 VERSION XP_004051395.1 GI:426368810
9.	*Orcinus orca*	586 aa	NCBI Reference Sequence: XP_004264252.1	ACCESSION XP_004264252 VERSION XP_004264252.1 GI:465976358
10.	*Trichechus manatus latirostris*	653 aa	NCBI Reference Sequence: XP_004384001.1	ACCESSION XP_004384001 VERSION XP_004384001.1 GI:471403445

Table 6.5 List of sequences utilized for MUSCLE and WebLogo analysis in PRPH2

Sl. no.	Organism	Length	Protein/gene ID	Other information
1.	*Homo sapiens*	346 aa	NCBI Reference Sequence: NP_000313.2	ACCESSION NP_000313 VERSION NP_000313.2 GI:118572596
2.	*Mus musculus*	346 aa	NCBI Reference Sequence: NP_032964.1	ACCESSION NP_032964 VERSION NP_032964.1 GI:7110699
3.	*Gallus gallus*	354 aa	NCBI Reference Sequence: NP_990369.1	ACCESSION NP_990369 VERSION NP_990369.1 GI:45384290
4.	*Canis lupus familiaris*	346 aa	NCBI Reference Sequence: NP_001003289.1	ACCESSION NP_001003289 VERSION NP_001003289.1 GI:50979108
5.	*Bos taurus*	346 aa	NCBI Reference Sequence: NP_001159959.1	ACCESSION NP_001159959 XP_589639 VERSION NP_001159959.1 GI:262050660
6.	*Nomascus leucogenys*	346 aa	NCBI Reference Sequence: XP_004090911.1	ACCESSION XP_004090911 VERSION XP_004090911.1 GI:441648819
7.	*Taeniopygia guttata*	354 aa	NCBI Reference Sequence: XP_002195201.1	ACCESSION XP_002195201 VERSION XP_002195201.1 GI:224047194
8.	*Gorilla gorilla gorilla*	346 aa	NCBI Reference Sequence: XP_004044072.1	ACCESSION XP_004044072 VERSION XP_004044072.1 GI:426353173
9.	*Orcinus orca*	346 aa	NCBI Reference Sequence: XP_004267646.1	ACCESSION XP_004267646 VERSION XP_004267646.1 GI:465996575
10.	*Trichechus manatus latirostris*	346 aa	ACCESSION XP_004379516 VERSION XP_004379516.1 GI:471388991	ACCESSION XP_004379516 VERSION XP_004379516.1 GI:471388991

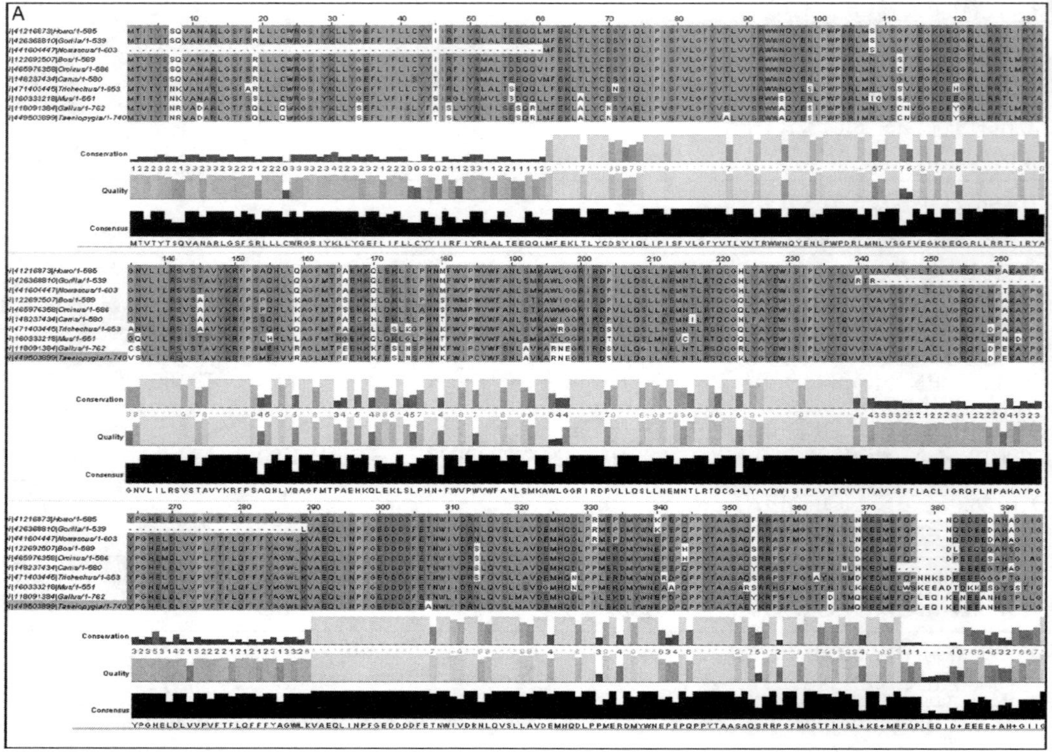
A
Conservation
Quality
Consensus

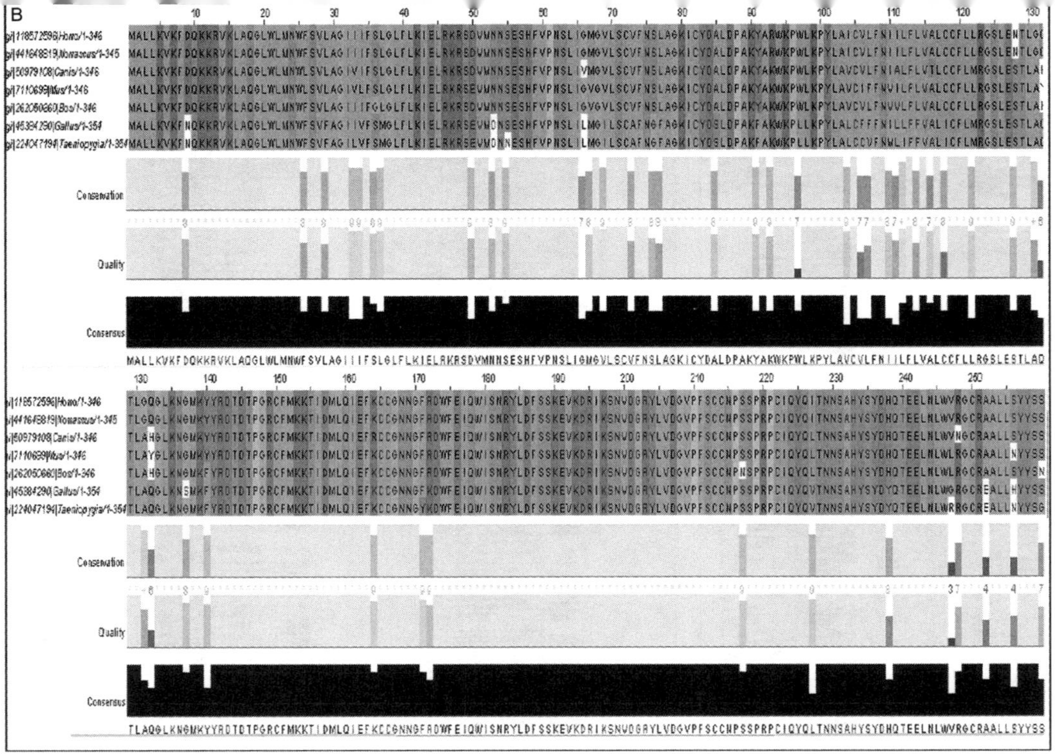

Figure 6.4 (A) Multiple sequence alignment of amino acids of selected 10 sequences in BEST1. The alignment of the sequence was obtained using MUSCLE. Accession number area as follows: *Homo sapiens* (AAR99654), *Mus musculus* (NP_036043), *Gallus gallus* (NP_001091014), *Canis lupus familiaris* (NP_001073714), *Bos taurus* (XP_003274102), *Nomascus leucogenys* (XP_421055), *Taeniopygia guttata* (XP_002195004), *Gorilla gorilla gorilla* (XP_004051395), *Orcinus orca* (XP_004264252), and *Trichechus manatus latirostris* (XP_004384001). (B) Multiple sequence alignment of amino acids of selected 10 sequences in PRPH2. The alignment of the sequence was obtained using MUSCLE. Accession number area as follows: *Homo sapiens* (NP_000313), *Mus musculus* (NP_032964), *Gallus gallus* (NP_990369), *Canis lupus familiaris* (NP_001003289), *Bos taurus* (NP_001159959), *Nomascus leucogenys* (XP_004090911), *Taeniopygia guttata* (XP_002195201), *Gorilla gorilla gorilla* (XP_004044072), *Orcinus orca* (XP_004267646), and *Trichechus manatus latirostris* (XP_004379516). (For color version of this figure, the reader is referred to the online version of this chapter.)

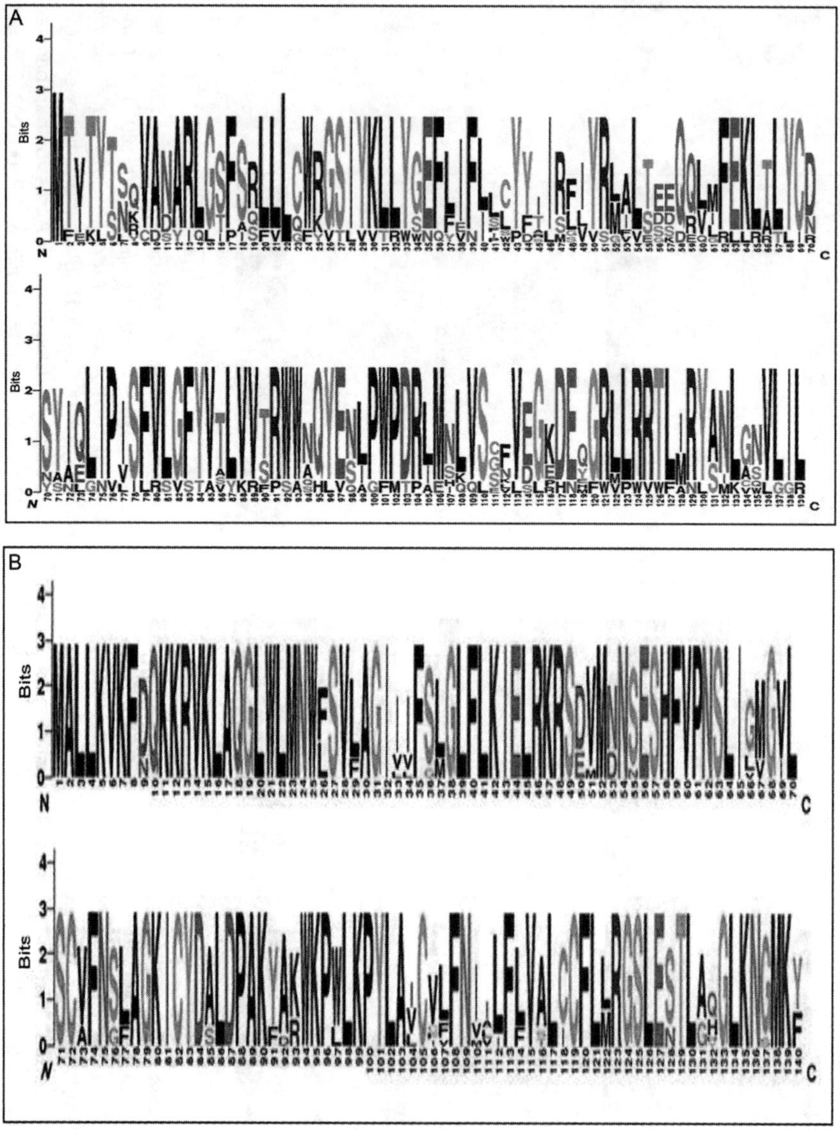

Figure 6.5 (A) Generated WebLogo of selected BEST1. It shows sequence logos up to 139 sequences and sequences from different species are highly conserved. (B) Generated WebLogo of selected PRPH2. It shows sequence logos up to 140 sequences and sequences from different species are highly conserved. (For color version of this figure, the reader is referred to the online version of this chapter.)

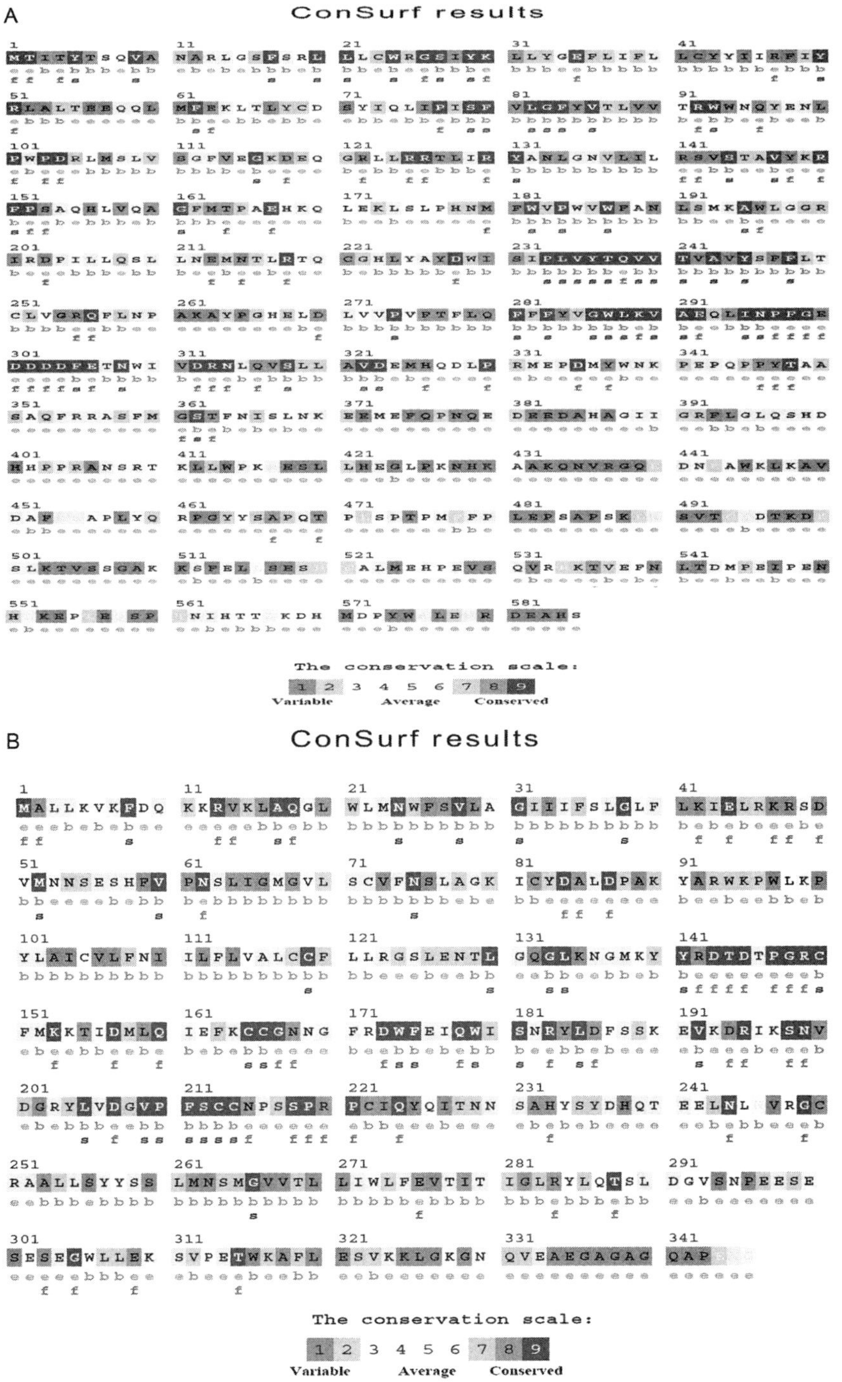

Figure 6.6 (A) Conservation analysis using ConSurf in BEST1 protein sequence from 1 to 581 amino acids. The amino acids are colored by conservation grades, and conservation levels are assigned to each amino acid. Most of the amino acids are highly conserved

(Continued)

leucine, while valine was the most common amino acid of substitution in PRPH2 (Table 6.1). Leu → Val, Val → Met, Asp → Asn, Glu → Lys, and Leu → Pro were the most frequent substitution due to SAPs. Eight substitutions from Leu → Val and seven substitutions from Val → Met were predicted to be pathogenic by PolyPhen-2 and SNPs&GO, and SIFT, PolyPhen-2, and PANTHER. Six substitutions from Asp → Arg, Glu → Lys, and Leu → Pro were predicted to be pathogenic by PolyPhen-2 and I-MUTANT Suite, PhD-SNP and I-MUTANT Suite, and PolyPhen-2, PhD-SNP, SNAP, SNPs&GO, and I-MUTANT Suite, respectively. For PRPH2, the most common amino acid of wild type was Pro, while Arg was the most common amino acid of substitution. R → W, G → D, P → L, R → W, C → Y, G → S and P → R were the most frequent substitution due to SAPs. From Table 6.1, we can infer that occurrence of five substitutions from Arg → Trp and Gly → Asp was predicted to be pathogenic by PhD-SNP, PANTHER, SNAP, I-MUTANT Suite, and Align-GVGD and PANTHER, SNAP, I-MUTANT Suite, and Align-GVGD, respectively. Four substitutions from Pro → Leu were predicted to be pathogenic by PANTHER, I-MUTANT Suite, and Align-GVGD, and three substitutions from Cys → Tyr, Gly → Ser, and Pro → Arg were predicted to be pathogenic by SIFT, PolyPhen-2, PhD-SNP, SNAP, I-MUTANT Suite, and Align-GVGD; PolyPhen-2, PANTHER, I-MUTANT Suite, and Align-GVGD; and PolyPhen-2, PANTHER, SNAP, I-MUTANT Suite, and Align-GVGD, respectively.

3.8. Disulfide bonds

Disulfide bond formation in the BEST1 and PRPH2 proteins was calculated using SCRATCH protein predictor (Cheng, Randall, Sweredoski, & Baldi, 2005). In BEST1, five Cys residues in their corresponding positions 23, 42, 69, 221, and 251 were involved in disulfide bond formation between 221 and 251 and 23 and 42. Similarly, in PRPH2, Cys in the following positions 72, 82, 150, 165, 213, 214, 222, and 250 was predicted to form the disulfide bond between 72 and 82, 213 and 250, 214 and 222, and 150 and 165. Cys residues in BEST1 wild type and SAPs, namely, C69W, R218C, F17C, W24C, Y29C R92C, R105C, C221W, Y227C, Y236C, R355C, and

Figure 6.6—Cont'd and are in maroon (dark red). (B) Conservation analysis using ConSurf in PRPH2 protein sequence from 1 to 346 amino acids. The amino acids are colored by conservation grades, and conservation levels are assigned to each amino acid. Most of the amino acids are highly conserved and are in maroon (dark red). (For interpretation of the references to color in this figure legend, the reader is referred to the online version of this chapter.)

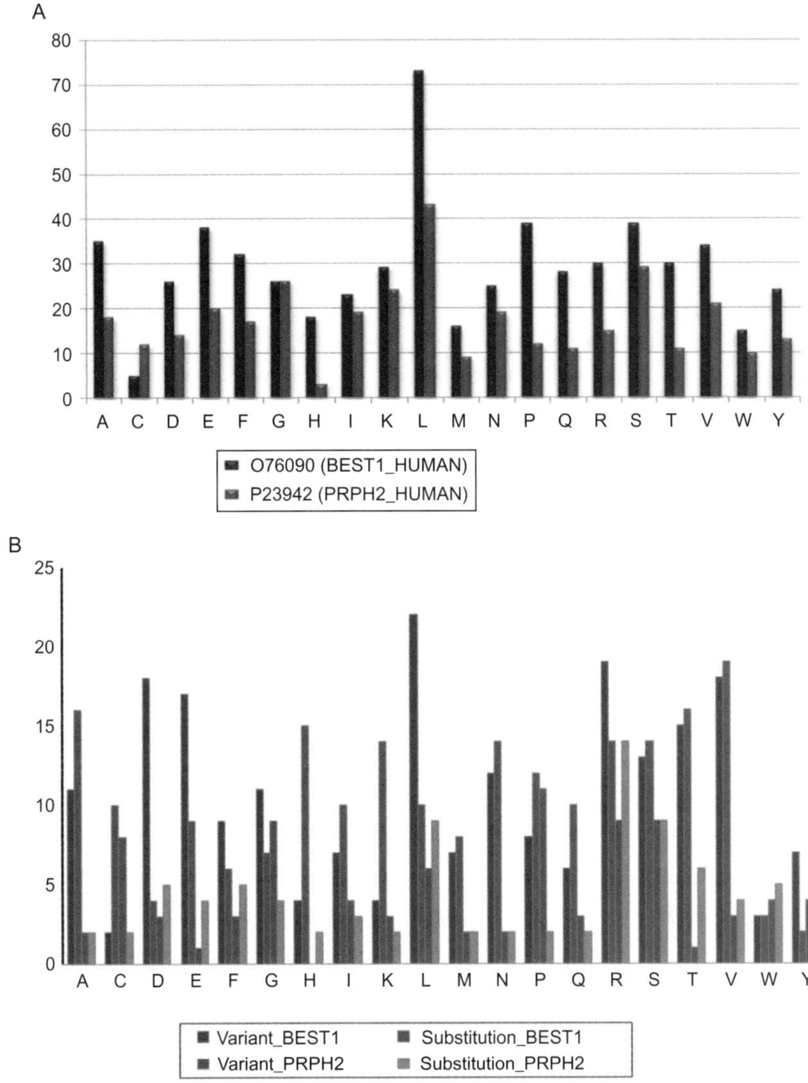

Figure 6.7 (A) Composition of amino acids in BEST1 and PRPH2 by statistical analysis of protein sequences. (B) Composition of amino acids in variant and substitution position BEST1 and PRPH2. (For color version of this figure, the reader is referred to the online version of this chapter.)

W93C, were predicted to be deleterious by PolyPhen-2, PANTHER, SNPs&GO, and I-Mutatnt3. Similarly, in PRPH2, wild-type Cys residues, namely, C165Y, C165R, C213R, C213F, C213Y, C214Y, C214S, and C250F, and SAPS W25C and Y141C, were predicted to be deleterious by PolyPhen-2, SNAP, I-Mutatnt3, and Align-GVGD.

3.9. Statistical analysis on the performance of *in silico* prediction methods

To evaluate the performance of deleterious SAPs by *in silico* predicting tools, we used five statistical measures: accuracy, specificity, sensitivity, positive predictive value, and negative predictive value. The test data set of 108 experimentally determined pathogenic SAPs (*BEST1*-86 and *PRPH2*-22 SAPs) was obtained from Swiss-Prot database, HGMD, and literatures. Based on the *in silico* prediction scores, the test data set was evaluated to obtain TP (true positive), TN (true negative), FP (false positive), and FN (false negative) values in order to calculate the static measures (Table 6.6). Among the eight *in silico* prediction methods, Polyphen-2 and PANTHER performed best in terms of sensitivity (0.95 and 0.96), Align-GVGD performed best in terms of specificity (0.57) and accuracy (0.54), and PANTHER performed best in terms of NVP (0.88). In contrast to these performances, PolyPhen-2 performed worst in terms of accuracy (0.37) and specificity (0.11), and Align-GVGD performed worst in terms of sensitivity (0.47).

3.10. Functional SNPs

The functional SNPs in the regulatory regions were analyzed and scored according to location of the SNPs (splice site, ESE, TFBS, coding region) and known results in diseases using FASTSNP and F-SNP. FASTSNP classifies and prioritizes phenotypic risk and deleterious effects of SNPs in coding and noncoding regions based upon their influence over determining 3D structure, pre-mRNA\splicing, deviation in transcriptional levels of the sequence, alterations in the premature translation termination, deviations in the sites at the promoter region for transcription factor binding, etc. Functional importance of the SNPs in the 5′ and 3′ of BEST1 and PRPH2 genes was not predicted by FASTSNP. By FASTSNP in *BEST1* gene, 22 SNPs in intronic region were predicted with the functional impact of intronic enhancer with risk ranking of 1–2, 1 SNP in intronic region was predicted with the functional importance of splicing site with risk ranking of 3–4, 1 SAP was predicted with functional importance of splicing site with risk ranking of 3–4, 2 SAPs were predicted with FS of splicing regulation with risk ranking of 3–4, 26 SAPs were predicted with FS of splicing regulation with risk ranking of 2–3, and 10 csSNPs were predicted with FS of splicing regulation with risk ranking of 2–3, respectively (Table 6.7). Similarly, in *PRPH2* gene, almost 8 SNPs (nearGene-5) were predicted with FS of promoter/regulatory region with risk ranking of 1–3, 82 SNPs (intronic)

Table 6.6 Statistical measures of eight *in silico* prediction tools

Parameters	SIFT	Polyphen-2	PhD-SNP	PANTHER	SNAP	SNPs&GO	I-MUTANT Suite	Align-GVGD
TP	78	89	76	86	77	77	62	44
TN	75	22	60	30	82	76	73	117
FP	131	184	146	146	124	130	123	89
FN	16	5	18	4	17	17	32	50
Total	300	300	300	266	300	300	290	300
N/A	0	0	0	34	0	0	10	0
TP–FN	94	94	94	90	94	94	94	94
TN–FP	206	206	206	176	206	206	196	206
Cases	94/206	94/206	94/206	90/176	94/206	94/206	94/196	94/206
Sensitivity	0.83	0.95	0.81	0.96	0.82	0.82	0.66	0.47
Specificity	0.36	0.11	0.29	0.17	0.4	0.37	0.37	0.57
Positive predictive value	0.37	0.33	0.34	0.37	0.38	0.37	0.34	0.33
Negative predictive value	0.82	0.81	0.77	0.88	0.83	0.82	0.7	0.7
Accuracy	0.51	0.37	0.45	0.44	0.53	0.51	0.47	0.54

Table 6.7 List of SNPs found to be functional significance by FASTSNP

Gene name	rsIDs	Functional annotation	Functional significance	Risk range
PRPH2	rs34799119	nearGene-5	Promoter/regulatory region	1–3
	rs11754813	nearGene-5	Promoter/regulatory region	1–3
	rs35039266	nearGene-5	Promoter/regulatory region	1–3
	rs35039266	nearGene-5	Promoter/regulatory region	1–3
	rs56168732	nearGene-5	Promoter/regulatory region	1–3
	rs56188317	nearGene-5	Promoter/regulatory region	1–3
	rs73426482	nearGene-5	Promoter/regulatory region	1–3
	rs4714621	nearGene-5	Promoter/regulatory region	1–3
	rs433286	Intron	Intronic enhancer	1–2
	rs10606150	Intron	Intronic enhancer	1–2
	rs380246	Intron	Intronic enhancer	1–2
	rs378835	Intron	Intronic enhancer	1–2
	rs73426429	Intron	Intronic enhancer	1–2
	rs668736	Intron	Intronic enhancer	1–2
	rs71679374	Intron	Intronic enhancer	1–2
	rs438995	Intron	Intronic enhancer	1–2
	rs2268687	Intron	Intronic enhancer	1–2
	rs2268686	Intron	Intronic enhancer	1–2
	rs70990127	Intron	Intronic enhancer	1–2
	rs9471898	Intron	Intronic enhancer	1–2
	rs9471900	Intron	Intronic enhancer	1–2
	rs9471901	Intron	Intronic enhancer	1–2
	rs9471902	Intron	Intronic enhancer	1–2
	rs11753241	Intron	Intronic enhancer	1–2
	rs9394918	Intron	Intronic enhancer	1–2
	rs9471904	Intron	Intronic enhancer	1–2
	rs9471905	Intron	Intronic enhancer	1–2
	rs61435457	Intron	Intronic enhancer	1–2

Table 6.7 List of SNPs found to be functional significance by FASTSNP—cont'd

Gene name	rsIDs	Functional annotation	Functional significance	Risk range
	rs9394919	Intron	Intronic enhancer	1–2
	rs58317577	Intron	Intronic enhancer	1–2
	rs9369377	Intron	Intronic enhancer	1–2
	rs12198221	Intron	Intronic enhancer	1–2
	rs9369378	Intron	Intronic enhancer	1–2
	rs67162942	Intron	Intronic enhancer	1–2
	rs72862655	Intron	Intronic enhancer	1–2
	rs66510729	Intron	Intronic enhancer	1–2
	rs72088922	Intron	Intronic enhancer	1–2
	rs67865363	Intron	Intronic enhancer	1–2
	rs59389246	Intron	Intronic enhancer	1–2
	rs72290946	Intron	Intronic enhancer	1–2
	rs3818086	Intron	Intronic enhancer	1–2
	rs34722725	Intron	Intronic enhancer	1–2
	rs649472	Intron	Intronic enhancer	1–2
	rs34700033	Intron	Intronic enhancer	1–2
	rs9369379	Intron	Intronic enhancer	1–2
	rs59665557	Intron	Intronic enhancer	1–2
	rs675825	Intron	Intronic enhancer	1–2
	rs12207176	Intron	Intronic enhancer	1–2
	rs12207123	Intron	Intronic enhancer	1–2
	rs58035225	Intron	Intronic enhancer	1–2
	rs57674056	Intron	Intronic enhancer	1–2
	rs9357402	Intron	Intronic enhancer	1–2
	rs9349239	Intron	Intronic enhancer	1–2
	rs9369380	Intron	Intronic enhancer	1–2
	rs9471907	Intron	Intronic enhancer	1–2
	rs9367162	Intron	Intronic enhancer	1–2

Continued

Table 6.7 List of SNPs found to be functional significance by FASTSNP—cont'd

Gene name	rsIDs	Functional annotation	Functional significance	Risk range
	rs9357403	Intron	Intronic enhancer	1–2
	rs9349240	Intron	Intronic enhancer	1–2
	rs72390300	Intron	Intronic enhancer	1–2
	rs72142676	Intron	Intronic enhancer	1–2
	rs66913157	Intron	Intronic enhancer	1–2
	rs57333654	Intron	Intronic enhancer	1–2
	rs72088434	Intron	Intronic enhancer	1–2
	rs58632063	Intron	Intronic enhancer	1–2
	rs9471908	Intron	Intronic enhancer	1–2
	rs2268684	Intron	Intronic enhancer	1–2
	rs34139829	Intron	Intronic enhancer	1–2
	rs6915637	Intron	Intronic enhancer	1–2
	rs12111487	Intron	Intronic enhancer	1–2
	rs2268683	Intron	Intronic enhancer	1–2
	rs2268682	Intron	Intronic enhancer	1–2
	rs73426469	Intron	Intronic enhancer	1–2
	rs28669148	Intron	Intronic enhancer	1–2
	rs12660132	Intron	Intronic enhancer	1–2
	rs9471909	Intron	Intronic enhancer	1–2
	rs6458301	Intron	Intronic enhancer	1–2
	rs3793007	Intron	Intronic enhancer	1–2
	rs12207113	Intron	Intronic enhancer	1–2
	rs3846893	Intron	Intronic enhancer	1–2
	rs12214120	Intron	Intronic enhancer	1–2
	rs35800607	Intron	Intronic enhancer	1–2
	rs7757333	Intron	Intronic enhancer	1–2
	rs71560776	Intron	Intronic enhancer	1–2
	rs9394921	Intron	Intronic enhancer	1–2

Table 6.7 List of SNPs found to be functional significance by FASTSNP—cont'd

Gene name	rsIDs	Functional annotation	Functional significance	Risk range
	rs7768690	Intron	Intronic enhancer	1–2
	rs62416278	Intron	Intronic enhancer	1–2
	rs1985910	Intron	Intronic enhancer	1–2
	rs1985909	Intron	Intronic enhancer	1–2
	rs3805927	Intron	Intronic enhancer	1–2
	rs6928781	Intron	Intronic enhancer	1–2
	rs61748432	nsSNPs	Missense (conservative); splicing regulation	2–3
	rs390659	nsSNPs	Missense (conservative); splicing regulation	2–3
	rs61755818	nsSNPs	Missense (conservative); splicing regulation	2–3
	rs61755817	nsSNPs	Missense (conservative); splicing regulation	2–3
	rs61755816	nsSNPs	Missense (conservative); splicing regulation	2–3
	rs61755815	nsSNPs	Missense (conservative); splicing regulation	2–3
	rs61755813	nsSNPs	Missense (conservative); splicing regulation	2–3
	rs61755811	nsSNPs	Missense (conservative); splicing regulation	2–3
	rs61755810	nsSNPs	Missense (conservative); splicing regulation	2–3
	rs61755807	nsSNPs	Missense (conservative); splicing regulation	2–3
	rs61755809	nsSNPs	Missense (conservative); splicing regulation	2–3
	rs61755808	nsSNPs	Missense (conservative); splicing regulation	2–3
	rs61755806	nsSNPs	Missense (conservative); splicing regulation	2–3

Continued

Table 6.7 List of SNPs found to be functional significance by FASTSNP—cont'd

Gene name	rsIDs	Functional annotation	Functional significance	Risk range
	rs61755805	nsSNPs	Missense (conservative); splicing regulation	2–3
	rs61755804	nsSNPs	Missense (conservative); splicing regulation	2–3
	rs61755801	nsSNPs	Missense (conservative); splicing regulation	2–3
	rs61755800	nsSNPs	Missense (conservative); splicing regulation	2–3
	rs61755799	nsSNPs	Missense (conservative); splicing regulation	2–3
	rs61755798	nsSNPs	Missense (conservative); splicing regulation	2–3
	rs61755797	nsSNPs	Missense (conservative); splicing regulation	2–3
	rs61755796	nsSNPs	Missense (conservative); splicing regulation	2–3
	rs61755795	nsSNPs	Missense (conservative); splicing regulation	2–3
	rs61755794	nsSNPs	Missense (conservative); splicing regulation	2–3
	rs61755793	nsSNPs	Missense (conservative); splicing regulation	2–3
	rs61755791	nsSNPs	Missense (conservative); splicing regulation	2–3
	rs61755790	nsSNPs	Missense (conservative); splicing regulation	2–3
	rs61755789	nsSNPs	Missense (conservative); splicing regulation	2–3
	rs61755788	nsSNPs	Missense (conservative); splicing regulation	2–3
	rs61755787	nsSNPs	Missense (conservative); splicing regulation	2–3
	rs61755785	nsSNPs	Missense (conservative); splicing regulation	2–3

Table 6.7 List of SNPs found to be functional significance by FASTSNP—cont'd

Gene name	rsIDs	Functional annotation	Functional significance	Risk range
	rs61755783	nsSNPs	Missense (conservative); splicing regulation	2–3
	rs61755781	nsSNPs	Missense (conservative); splicing regulation	2–3
	rs61755780	nsSNPs	Missense (conservative); splicing regulation	2–3
	rs61755779	nsSNPs	Missense (conservative); splicing regulation	2–3
	rs7764439	csSNPs	Sense/synonymous; splicing regulation	2–3
	rs61755776	csSNPs	Sense/synonymous; splicing regulation	2–3
	rs61755775	csSNPs	Sense/synonymous; splicing regulation	2–3
	rs61755774	nsSNPs	Missense (conservative); splicing regulation	2–3
	rs61755767	nsSNPs	Missense (conservative); splicing regulation	2–3
	rs61755766	nsSNPs	Missense (conservative); splicing regulation	2–3
	rs61754402	nsSNPs	Missense (conservative); splicing regulation	2–3
BEST1	rs62639356	Intron	Splicing site	3–4
	rs73493205	Intron	Intronic enhancer	1–2
	rs72503400	Intron	Intronic enhancer	1–2
	rs11825719	Intron	Intronic enhancer	1–2
	rs57890952	Intron	Intronic enhancer	1–2
	rs168991	Intron	Intronic enhancer	1–2
	rs2727272	Intron	Intronic enhancer	1–2
	rs195166	Intron	Intronic enhancer	1–2
	rs2736594	Intron	Intronic enhancer	1–2
	rs62639352	Intron	Intronic enhancer	1–2

Continued

Table 6.7 List of SNPs found to be functional significance by FASTSNP—cont'd

Gene name	rsIDs	Functional annotation	Functional significance	Risk range
	rs62639354	Intron	Intronic enhancer	1–2
	rs195164	Intron	Intronic enhancer	1–2
	rs909268	Intron	Intronic enhancer	1–2
	rs760306	Intron	Intronic enhancer	1–2
	rs195161	Intron	Intronic enhancer	1–2
	rs741886	Intron	Intronic enhancer	1–2
	rs2668898	Intron	Intronic enhancer	1–2
	rs2524295	Intron	Intronic enhancer	1–2
	rs195160	Intron	Intronic enhancer	1–2
	rs195158	Intron	Intronic enhancer	1–2
	rs62640560	Intron	Intronic enhancer	1–2
	rs73493223	Intron	Intronic enhancer	1–2
	rs467446	Intron	Intronic enhancer	1–2
	rs62637349	nsSNPs	Splicing site	3–4
	rs28941469	nsSNPs	Missense (nonconservative); splicing regulation	3–4
	rs28940570	nsSNPs	Missense (nonconservative); splicing regulation	3–4
	rs62637051	nsSNPs	Missense (conservative); splicing regulation	2–3
	rs62637053	nsSNPs	Missense (conservative); splicing regulation	2–3
	rs1801393	nsSNPs	Sense/synonymous; splicing regulation	2–3
	rs62641692	nsSNPs	Missense (conservative); splicing regulation	2–3
	rs62637056	nsSNPs	Missense (conservative); splicing regulation	2–3
	rs57132800	csSNPs	Sense/synonymous; splicing regulation	2–3

Table 6.7 List of SNPs found to be functional significance by FASTSNP—cont'd

Gene name	rsIDs	Functional annotation	Functional significance	Risk range
	rs62638170	csSNPs	Sense/synonymous; splicing regulation	2–3
	rs1109748	csSNPs	Sense/synonymous; splicing regulation	2–3
	rs1805141	csSNPs	Sense/synonymous; splicing regulation	2–3
	rs62639279	csSNPs	Sense/synonymous; splicing regulation	2–3
	rs1800995	nsSNPs	Missense (conservative); splicing regulation	2–3
	rs62637335	csSNPs	Sense/synonymous; splicing regulation	2–3
	rs62639355	nsSNPs	Missense (conservative); splicing regulation	2–3
	rs62637336	nsSNPs	Missense (conservative); splicing regulation	2–3
	rs62639339	nsSNPs	Missense (conservative); splicing regulation	2–3
	rs62637345	nsSNPs	Missense (conservative); splicing regulation	2–3
	rs62637347	nsSNPs	Missense (conservative); splicing regulation	2–3
	rs62637348	nsSNPs	Missense (conservative); splicing regulation	
	rs62637357	csSNPs	Sense/synonymous; splicing regulation	2–3
	rs62639261	csSNPs	Sense/synonymous; splicing regulation	2–3
	rs62639267	nsSNPs	Missense (conservative); splicing regulation	2–3
	rs62639271	nsSNPs	Missense (conservative); splicing regulation	2–3
	rs62639272	nsSNPs	Missense (conservative); splicing regulation	2–3

Continued

Table 6.7 List of SNPs found to be functional significance by FASTSNP—cont'd

Gene name	rsIDs	Functional annotation	Functional significance	Risk range
	rs62639273	nsSNPs	Missense (conservative); splicing regulation	2–3
	rs62639274	csSNPs	Sense/synonymous; splicing regulation	2–3
	rs62639275	nsSNPs	Missense (conservative); splicing regulation	2–3
	rs62639277	nsSNPs	Missense (conservative); splicing regulation	2–3
	rs28941468	nsSNPs	Missense (conservative); splicing regulation	2–3
	rs1805144	nsSNPs	Missense (conservative); splicing regulation	2–3
	rs62640562	nsSNPs	Missense (conservative); splicing regulation	2–3
	rs1801390	nsSNPs	Sense/synonymous; splicing regulation	2–3
	rs17854138	nsSNPs	Missense (conservative); splicing regulation	2–3
	rs62640563	nsSNPs	Missense (conservative); splicing regulation	2–3
	rs149698	nsSNPs	Sense/synonymous; splicing regulation	2–3
	rs62640566	nsSNPs	Missense (conservative); splicing regulation	2–3
	rs17854139	csSNPs	Sense/synonymous; splicing regulation	2–3
	rs62640567	csSNPs	Sense/synonymous; splicing regulation	2–3
	rs1800010	nsSNPs	Missense (conservative); splicing regulation	2–3
	rs17185413	nsSNPs	Missense (conservative); splicing regulation	2–3
	rs4963437	nsSNPs	Missense (conservative); splicing regulation	2–3

SNPs highlighted in bold were found to be functional by F-SNP and underlined are predicted to be functional by any one of the prediction tool.

were predicted with FS of intronic enhancer with risk ranking of 1–2, 38 SAPs were predicted with FS of splicing regulation with risk ranking of 2–3, and 3 csSNPs were predicted with FS of splicing regulation with risk ranking of 2–3, respectively (Table 6.7). To locate and predict the SNPs in *BEST1* and *PRPH2* genes within TFBS, tools like TFSearch and ConSite and for the exonic splicing enhancer's tools like ESEfinder, ESRSearch, and PESX were utilized by F-SNP and predict with FS score ranging from 0.05 to 1 (Table 6.8). In *BEST1*, 46 SNPs were located within putative TFBS with the functional category transcriptional regulation and 13 SNPs with the functional category splicing regulation. Conversely, in *PRPH2*, 30 SNPs were located within putative TFBS with the functional category transcriptional regulation and 4 SNPs with the functional category splicing regulation, respectively.

3.11. Ranking of SAPs

We adopted a ranking system to classify the SAPs associated in VMD based on the scores obtained from SIFT, PolyPhen-2, SNAP, PhD-SNP, I-Mutant, SNPs&GO, and Align-GVGD (Table 6.9). We did not consider the PANTHER subPSEC score for ranking order (33 SAPs did not align to PANTHER library HMM). Combining the scores of all the previously mentioned seven techniques, we have assigned the ranking from 1 to 4 and designated them overall pathogenic (six to seven tools predicted pathogenic), probably pathogenic (four to five of seven tools predicted pathogenic), possibly pathogenic (two to three of seven tools predicted pathogenic), and probably benign (zero to one tool predicted pathogenic).

4. DISCUSSION

Completion of human genome project, International HapMap Project, and GWA studies (GWASs) and ongoing 1000 genomes has led to the detailed exploration of DNA sequence and revolutionized the field of human genetics (The 1000 Genome Consortium, 2010). Due to this, massive amount of human genome sequence and genotype data has been documented. Genomic variants are usually listed, as small insertions and deletions, structural variants, and SNPs. Among them, SNPs found to be the most general form of genomic variation occurring one SNP for every 290 bp (Ng et al., 2008). Availability of 15 million SNPs in public databases such as dbSNP along with functional annotation may provide a great hold in understanding the relationship between the genotype and phenotypic

Table 6.8 List of SNPs found to be functional significant by F-SNP

Gene	rsIDs	Functional category	Prediction tool	Prediction result	FS score
BEST1	rs974121	Transcriptional_regulation	TFSearch	Changed	0.176
	rs3758976	Transcriptional_regulation	TFSearch, Consite	Changed	0.208
	rs972355	Transcriptional_regulation	TFSearch, Consite	Changed	0.208
	rs972354	Transcriptional_regulation	Consite	Changed	0.05
	rs2736597	Transcriptional_regulation	TFSearch	Changed	0.176
	rs11230840	Transcriptional_regulation	Consite	Changed	0.05
	rs2736596	Transcriptional_regulation	TFSearch	Changed	0.176
	rs11825719	Transcriptional_regulation	TFSearch, Consite	Changed	0.208
	rs1800007	Splicing_regulation	ESEfinder, ESRSearch, PESX	Changed	0.568
	rs168991	Transcriptional_regulation	TFSearch, Consite	Changed	0.208
	rs195169	Transcriptional_regulation	TFSearch, Consite	Changed	0.208
	rs2727272	Transcriptional_regulation	TFSearch	Changed	0.176
	rs2524294	Transcriptional_regulation	TFSearch, Consite	Changed	0.208
	rs195168	Transcriptional_regulation	TFSearch, Consite	Changed	0.208
	rs195166	Transcriptional_regulation	TFSearch	Changed	0.176
	rs2736594	Transcriptional_regulation	TFSearch	Changed	0.176

	rs1109748	Splicing_regulation	ESRSearch, ESEfinder	Changed	0.354
	rs195165	Transcriptional_regulation	TFSearch, Consite, GoldenPath	Changed	0.268
	rs28940274	Splicing_regulation	PolyPhen, SNPs3D, ESEfinder, ESRSearch, PESX	Changed	0.626
	rs28940273	Splicing_regulation	PolyPhen, SNPs3D, ESEfinder, ESRSearch, PESX	Changed	0.585
	rs195164	Transcriptional_regulation	TFSearch, Consite	Changed	0.208
	rs909268	Transcriptional_regulation	TFSearch	Changed	0.176
	rs168990	Transcriptional_regulation	Consite	Changed	0.05
	rs760306	Transcriptional_regulation	TFSearch, Consite	Changed	0.208
	rs195162	Transcriptional_regulation	TFSearch, Consite	Changed	0.263
	rs195161	Transcriptional_regulation	TFSearch, Consite	Changed	0.208
	rs741886	Transcriptional_regulation	TFSearch, Consite	Changed	0.208
	rs28941469	Splicing_regulation	ESEfinder, ESRSearch, PESX	Changed	0.569
	rs2955684	Transcriptional_regulation	Consite	Changed	0.05
	rs2955683	Transcriptional_regulation	Consite	Changed	0.05
	rs2668898	Transcriptional_regulation	TFSearch, Consite	Changed	0.208
	rs28940570	Splicing_regulation	ESEfinder, ESRSearch	Changed	0.466
	rs2524295	Transcriptional_regulation	TFSearch, Consite	Changed	0.208

Continued

Table 6.8 List of SNPs found to be functional significant by F-SNP—cont'd

Gene	rsIDs	Functional category	Prediction tool	Prediction result	FS score
	rs195160	Transcriptional_regulation	TFSearch, Consite	Changed	0.208
	rs3815046	Transcriptional_regulation	TFSearch	Changed	0.176
	rs195159	Transcriptional_regulation	Consite	Changed	0.05
	rs195158	Transcriptional_regulation	TFSearch, Consite	Changed	0.208
	rs1801390	Splicing_regulation	ESEfinder	Changed	0.121
	rs17854138	Splicing_regulation	ESEfinder, ESRSearch	Changed	0.25
	rs195157	Transcriptional_regulation	Consite	Changed	0.05
	rs17156602	Transcriptional_regulation	TFSearch, Consite	Changed	0.208
	rs195156	Transcriptional_regulation	Consite	Changed	0.05
	rs467446	Transcriptional_regulation	TFSearch	Changed	0.176
	rs149698	Splicing_regulation	ESEfinder, ESRSearch, PESX	Changed	0.594
	rs17854139	Splicing_regulation	ESEfinder, ESRSearch, PESX, RESCUE_ESE	Changed	0.5
	rs17185413	Transcriptional_regulation	TFSearch, Consite	Changed	0.25
		Splicing_regulation	ESEfinder, ESRSearch	Changed	0.25
	rs4963437	Transcriptional_regulation	TFSearch	Changed	0.5
		Splicing_regulation	ESEfinder, ESRSearch, PESX, RESCUE_ESE	Changed	0.5

	rs2668897	Transcriptional_regulation	TFSearch	Changed	0.158
	rs195155	Transcriptional_regulation	TFSearch, Consite	Changed	0.208
	rs17156609	Transcriptional_regulation	TFSearch, Consite	Changed	0.208
	rs3180123	Transcriptional_regulation	TFSearch	Changed	0.217
	rs11554846	Transcriptional_regulation	TFSearch, Consite	Changed	0.765
	rs3180122	Transcriptional_regulation	TFSearch	Changed	0.352
	rs1052603	Transcriptional_regulation	TFSearch	Changed	0.176
	rs1064740	Transcriptional_regulation	TFSearch	Changed	0.5
	rs1155484	Transcriptional_regulation	TFSearch, Consite	Changed	1
		Splicing_regulation	ESEfinder	Changed	1
		Tr-46, sp-13			
PRPH2	rs405043	Transcriptional_regulation	TFSearch	Changed	0.5
	rs405059	Transcriptional_regulation	TFSearch	Changed	0.5
	rs1758213	Transcriptional_regulation	TFSearch	Changed	0.5
	rs1042562	Transcriptional_regulation	TFSearch	Changed	0.5
	rs835	Transcriptional_regulation	TFSearch	Changed	0.516
	rs361524	Transcriptional_regulation	TFSearch	Changed	0.643

Continued

Table 6.8 List of SNPs found to be functional significant by F-SNP—cont'd

Gene	rsIDs	Functional category	Prediction tool	Prediction result	FS score
	rs434102	Splicing_regulation	ESEfinder, ESRSearch	Changed	0.749
	rs425876	Splicing_regulation	ESRSearch	Changed	0.103
	rs390659	Splicing_regulation	ESEfinder, ESRSearch, PESX	Changed	0.749
	rs389999	Transcriptional_regulation	TFSearch	Changed	0.5
	rs433286	Transcriptional_regulation	TFSearch	Changed	0.5
	rs415906	Transcriptional_regulation	TFSearch	Changed	0.5
	rs665368	Transcriptional_regulation	TFSearch	Changed	0.5
	rs665442	Transcriptional_regulation	TFSearch	Changed	0.5
	rs371323	Transcriptional_regulation	TFSearch	Changed	0.5
	rs380246	Transcriptional_regulation	TFSearch	Changed	0.5
	rs378835	Transcriptional_regulation	TFSearch	Changed	0.5
	rs668736	Transcriptional_regulation	TFSearch	Changed	0.5
	rs377118	Transcriptional_regulation	TFSearch	Changed	0.5
	rs438995	Transcriptional_regulation	TFSearch	Changed	0.5
	rs393034	Transcriptional_regulation	TFSearch	Changed	0.5
	rs9394921	Transcriptional_regulation	TFSearch	Changed	0.5
	rs7768690	Transcriptional_regulation	TFSearch	Changed	0.5
	rs7752619	Transcriptional_regulation	TFSearch	Changed	0.5

	rs1985910	Transcriptional_regulation	TFSearch	Changed	0.5
	rs1985909	Transcriptional_regulation	TFSearch	Changed	0.5
	rs3805927	Transcriptional_regulation	TFSearch	Changed	0.5
	rs6928781	Transcriptional_regulation	TFSearch	Changed	0.5
	rs7764439	Splicing_regulation	ESEfinder, ESRSearch, PESX, RESCUE_ESE	Changed	0.5
	rs4714621	Transcriptional_regulation	TFSearch	Changed	0.5
	rs4714622	Transcriptional_regulation	TFSearch	Changed	0.5
	rs10948044	Transcriptional_regulation	TFSearch	Changed	0.5
	rs12180958	Transcriptional_regulation	TFSearch	Changed	0.5
	rs11754813	Transcriptional_regulation	TFSearch	Changed	0.5

SNPs highlighted in bold were found to be functionally significant by FASTSNP.

Table 6.9 Ranking system based on the prediction scores of seven methods

rsIDs/variants	Amino acid position	SIFT	PolyPhen-2	I-MUTANT Suite	PhD-SNP	SNAP	SNPs&GO	Align-GVGD	Rank
VAR_025751	E306G	0	0.88	−3.0	D	NN	D	Class C65	1
VAR_017378	D104H	0	1.00	−2.4	D	NN	D	Class C65	1
rs28940273	W93C	0.04	1.00	−2.4	D	NN	D	Class C65	1
rs62638172	D104A	0.05	1.00	−2.3	D	NN	D	Class C65	1
VAR_025744	N296H	0	0.97	−1.9	D	NN	D	Class C65	1
rs28940274	Y85H	0.04	0.96	−1.9	D	NN	D	Class C65	1
VAR_025746	D302G	0.01	1.00	−1.8	D	NN	D	Class C65	1
VAR_025737	L224P	0.05	1.00	−1.6	D	NN	D	Class C65	1
VAR_000865	F305S	0	0.99	−1.5	D	NN	D	Class C65	1
VAR_017377	W102R	0	1.00	−1.3	D	NN	D	Class C65	1
VAR_025747	D302H	0	0.97	−1.2	D	NN	D	Class C65	1
VAR_000851	R218S	0	0.89	−1.1	D	NN	D	Class C65	1
VAR_000867	V311G	0	0.97	−1.1	D	NN	D	Class C65	1
VAR_025731	R105C	0	1.00	−1.0	D	NN	D	Class C65	1
VAR_000861	G299E	0	1.00	−1.0	D	NN	D	Class C65	1
VAR_017379	N133K	0.02	1.00	−0.9	D	NN	D	Class C65	1

VAR_010474	R92C	0	1.00	−0.6	D	NN	D	Class C65	1
VAR_000842	R92S	0	0.85	−0.6	D	NN	D	Class C65	1
VAR_025745	F298S	0.03	1.00	−0.5	D	NN	D	Class C65	1
VAR_000849	R218C	0	0.84	−0.9	D	NN	D	Class C65	1
rs61755794	D173V	0	1.00	−1.6	D	NN	D	Class C65	1
rs121918563	L185P	0.05	1.00	−1.6	D	NN	D	Class C65	1
HM070073	C213F	0	1.00	−1.5	D	NN	D	Class C65	1
CM011806	P210L	0	1.00	−1.4	D	NN	D	Class C65	1
rs61755789	G167D	0	1.00	−1.3	D	NN	D	Class C65	1
VAR_025753	N308S	0.01	1.00	−1.8	D	NN	D	Class C45	1
VAR_058313	G299A	0	1.00	−1.7	D	NN	D	Class C55	1
VAR_010484	N296S	0.01	1.00	−1.5	D	NN	D	Class C45	1
VAR_000862	E300K	0	0.99	−1.5	D	NN	D	Class C55	1
rs138932379	E213K	0	0.99	−1.0	D	NN	D	Class C55	1
VAR_025738	T241N	0.01	0.88	−0.9	D	NN	D	Class C55	1
VAR_025736	G222V	0	1.00	−0.6	D	NN	D	Class C65	1
VAR_043493	P152A	0	0.97	−3.3	D	NN	D	Class C25	1

Continued

Table 6.9 Ranking system based on the prediction scores of seven methods—cont'd

rsIDs/variants	Amino acid position	SIFT	PolyPhen-2	I-MUTANT Suite	PhD-SNP	SNAP	SNPs&GO	Align-GVGD	Rank
VAR_000860	P297A	0	1.00	−2.5	D	NN	D	Class C25	1
VAR_000868	D312N	0.01	0.99	−2.4	D	NN	D	Class C15	1
VAR_000846	D104E	0.02	0.91	−1.7	D	NN	D	Class C35	1
CM072107	Q293H	0.02	1.00	−1.6	D	NN	D	Class C15	1
VAR_000864	D301N	0.01	1.00	−1.5	D	NN	D	Class C15	1
VAR_025742	L294V	0.01	1.00	−1.2	D	NN	D	Class C25	1
VAR_010481	R218H	0.02	1.00	−1.1	D	NN	D	Class C25	1
VAR_000863	D301E	0	0.97	−1.1	D	NN	D	Class C35	1
VAR_025749	D303E	0	0.88	−1.0	D	NN	D	Class C35	1
rs200277476	A195P	0.04	1.00	−1.0	D	NN	D	Class C25	1
rs147228028	T363P	0	1.00	−1.0	D	NN	D	Class C35	1
VAR_017376	P101T	0	1.00	−0.9	D	NN	D	Class C35	1
VAR_010475	R92H	0	1.00	−0.8	D	NN	D	Class C25	1
rs62639262	D270N	0	1.00	−2.4	D	NN	D	Class C0	1
rs62639263	D270E	0.02	1.00	−1.9	D	NN	D	Class C0	1
VAR_000839	Q58L	0.03	0.99	−1.6	D	NN	D	Class C0	1
rs186522420	V244G	0	1.00	−1.2	D	NN	D	Class C0	1

VAR_000852	L224M	0.05	1.00	−1.1	D	NN	D	Class C0	1
VAR_017369	Y29H	0.01	0.99	−0.8	D	NN	D	Class C0	1
VAR_000835	W24C	0	0.99	−0.7	D	NN	D	Class C0	1
rs28940570	A243V	0.02	1.00	−0.5	D	NN	D	Class C0	1
rs62639269	V273M	0.02	0.97	−0.5	D	NN	D	Class C0	1
rs267606676	D228N	0.01	0.61	−2.2	D	NN	D	Class C15	1
VAR_000850	R218Q	0	0.79	−0.8	D	NN	D	Class C35	1
rs62639264	L271P	0.03	0.67	−1.8	D	NN	D	Class C0	1
rs62637351	F247S	0	0.70	−1.6	D	NN	D	Class C0	1
VAR_000838	S27R	0	0.26	−1.4	D	NN	D	Class C0	1
rs61748432	G305D	0.02	1.00	−2.1	D	NN	N	Class C65	1
VAR_006880	C214S	0	1.00	−2.0	D	NN	N	Class C65	1
rs61755804	C214Y	0	1.00	−1.5	D	NN	N	Class C65	1
VAR_006859	L126R	0	1.00	−1.5	D	NN	N	Class C65	1
rs61755797	P210S	0	1.00	−0.6	D	NN	N	Class C65	1
VAR_032052	G167S	0	1.00	−0.7	D	NN	N	Class C55	2
rs200277476	A195V	0.05	0.97	−0.6	D	N	D	Class C65	1

Continued

Table 6.9 Ranking system based on the prediction scores of seven methods—cont'd

rsIDs/variants	Amino acid position	SIFT	PolyPhen-2	I-MUTANT Suite	PhD-SNP	SNAP	SNPs&GO	Align-GVGD	Rank
VAR_000848	S209N	0	1.00	−1.9	D	N	D	Class C45	2
VAR_010483	Q293K	0	0.97	−1.8	D	N	D	Class C45	2
rs62637053	E63K	0	1.00	−1.8	D	N	D	Class C55	2
CM094327	E292K	0	1.00	−0.9	D	N	D	Class C55	2
rs147409760	A352T	0	1.00	−0.9	D	N	D	Class C55	2
rs62640562	R313K	0.04	1.00	−2.0	D	N	D	Class C25	2
VAR_010486	E300D	0	0.99	−1.3	D	N	D	Class C35	2
rs62637358	H267N	0.05	1.00	−2.1	D	N	D	Class C0	2
rs200235532	E57D	0	1.00	−1.3	D	N	D	Class C0	2
rs62639271	F276L	0	1.00	−1.2	D	N	D	Class C0	2
rs28940278	R47H	0.01	1.00	−1.1	D	N	D	Class C0	2
rs1805142	E119Q	0.02	0.29	−1.3	D	N	D	Class C25	2
rs62638174	S111A	0	1.00	−1.8	D	N	N	Class C65	2
rs62637349	N212H	0.01	0.95	−0.7	D	N	N	Class C65	2
rs62638176	G112S	0	1.00	−1.7	D	N	N	Class C55	2
rs200162075	H223Q	0	1.00	−1.7	D	N	N	Class C15	2
rs201386186	R331Q	0.01	1.00	−1.2	D	N	N	Class C35	2

rs62651027	M214L	0.01	1.00	−0.9	D	N	N	Class C0	2
VAR_025740	V275I	0	1.00	−0.8	D	N	N	Class C0	2
CM072111	G299R	0	1.00	−2.1	D	NN	D	Class C65	1
CM072108	P233Q	0	1.00	−2.0	D	NN	D	Class C65	1
rs1805143	P297S	0	1.00	−1.9	D	NN	D	Class C65	1
rs199529046	I201T	0.03	1.00	−1.3	D	NN	D	Class C65	1
CM071144	L191P	0	1.00	−0.9	D	NN	D	Class C65	1
CM930639	R172W	0.01	1.00	−0.5	D	NN	D	Class C65	1
CM991236	L82V	0	1.00	−0.6	D	NN	D	Class C25	1
CM991238	Q96H	0	1.00	−0.6	D	NN	D	Class C15	1
HM070108	V9Q	0.05	1.00	−2.7	D	NN	D	Class C0	1
rs62653022	R51K	0.01	0.98	−1.0	D	NN	D	Class C0	1
HM070106	L234V	0	1.00	−1.0	D	N	D	Class C25	2
rs199998058	N378S	0	1.00	−1.7	D	N	N	Class C45	2
rs200510369	P480T	0	1.00	−1.3	D	N	N	Class C35	2
rs62637054	K64T	0.04	1.00	−2.0	N	NN	D	Class C65	1
VAR_017374	V89A	0	1.00	−0.8	N	NN	D	Class C65	1

Continued

Table 6.9 Ranking system based on the prediction scores of seven methods—cont'd

rsIDs/variants	Amino acid position	SIFT	PolyPhen-2	I-MUTANT Suite	PhD-SNP	SNAP	SNPs&GO	Align-GVGD	Rank
rs267606679	V235A	0.02	0.60	−2.8	N	NN	D	Class C65	1
rs61755774	G68R	0	1.00	−1.5	N	NN	D	Class C65	1
CM053390	C250F	0	1.00	−0.8	N	NN	D	Class C65	1
VAR_010482	V235L	0	1.00	−1.4	N	NN	D	Class C25	2
VAR_000834	L21V	0	1.00	−0.8	N	NN	D	Class C0	2
VAR_000833	A10T	0	1.00	−0.7	N	NN	D	Class C0	2
rs62639272	T277A	0	1.00	−0.6	N	NN	D	Class C0	2
VAR_017367	N11I	0.01	1.00	0.9	N	NN	D	Class C0	2
rs121918565	M1T	0	1.00	−0.8	N	NN	N	Class C65	2
rs61755800	S212G	0	1.00	−1.5	N	NN	N	Class C55	2
rs61755801	S212T	0.04	1.00	−0.8	N	NN	N	Class C55	2
VAR_010479	A146K	0	0.97	−2.4	N	N	D	Class C65	2
rs199890510	S398F	0.01	1.00	0.6	N	N	D	Class C65	2
VAR_000856	V235M	0.04	1.00	−1.0	N	N	D	Class C15	2
VAR_000840	L67V	0	0.99	−0.7	N	N	D	Class C25	2
VAR_058277	V242M	0.03	0.97	−1.2	N	N	N	Class C0	3
rs200582915	E525A	0.03	1.00	−0.6	N	NN	N	Class C65	2

VAR_009278	E578V	0.05	1.00	−0.6	N	NN	N	Class C65	2
rs62640565	D496N	0	1.00	−1.1	N	N	D	Class C15	2
rs62640566	S507F	0	1.00	−1.1	N	N	N	Class C65	2
rs62637339	A189T	0.01	0.95	−2.2	N	N	N	Class C55	3
rs201210799	G424A	0	1.00	−1.1	N	N	N	Class C0	3
rs121918291	Y236C	0	1.00	−0.8	D	NN	D	Class C65	1
VAR_017380	L140R	0	0.88	−0.5	D	NN	D	Class C65	1
rs62638171	L75H	0.04	1.00	−0.4	D	NN	D	Class C65	1
VAR_000866	I310T	0	1.00	−0.4	D	NN	D	Class C65	1
VAR_025748	D302V	0.03	1.00	−0.3	D	NN	D	Class C65	1
CM072110	G222E	0.01	1.00	−0.3	D	NN	D	Class C65	1
VAR_025743	I295T	0.01	1.00	−0.2	D	NN	D	Class C65	1
CM001381	L100R	0	0.98	−0.2	D	NN	D	Class C65	1
rs62641692	C69W	0	1.00	−0.1	D	NN	D	Class C65	1
VAR_025735	C221W	0	1.00	−0.1	D	NN	D	Class C65	1
rs267606677	Y227C	0	0.61	−0.5	D	NN	D	Class C65	1
VAR_000857	T237R	0	0.55	−0.4	D	NN	D	Class C65	1

Continued

Table 6.9 Ranking system based on the prediction scores of seven methods—cont'd

rsIDs/variants	Amino acid position	SIFT	PolyPhen-2	I-MUTANT Suite	PhD-SNP	SNAP	SNPs&GO	Align-GVGD	Rank
rs61755798	P210R	0	1.00	−0.4	D	NN	D	Class C65	1
rs61755781	Y141C	0	1.00	−0.3	D	NN	D	Class C65	1
rs61755809	R220W	0.02	0.91	−0.3	D	NN	D	Class C65	1
CM090729	L126P	0.01	1.00	0.2	D	NN	D	Class C65	1
VAR_043495	M325T	0	1.00	−0.2	D	NN	D	Class C65	2
rs121918284	R141H	0	0.99	−2.3	D	NN	D	Class C25	2
rs267606678	L140V	0.01	1.00	−0.3	D	NN	D	Class C25	2
rs121918290	V239M	0	1.00	−0.1	D	NN	D	Class C15	2
rs137853905	A243T	0.04	1.00	−1.1	D	NN	D	Class C0	2
rs62639267	V272G	0	1.00	−0.4	D	NN	D	Class C0	2
VAR_017368	G26R	0	1.00	−0.3	D	NN	D	Class C0	2
VAR_000836	R25Q	0	0.99	−0.3	D	NN	D	Class C0	2
rs121918288	L41P	0.01	0.97	−0.2	D	NN	D	Class C0	2
VAR_000837	R25W	0	1.00	0.2	D	NN	D	Class C0	2
rs121918289	V86M	0	0.28	−0.4	D	NN	D	Class C15	2
rs61755787	D157N	0.05	1.00	−0.4	D	NN	D	Class C15	2
rs200876455	F171I	0.05	0.92	0.0	D	NN	D	Class C15	2

rs61755803	C213Y	0	1.00	−0.5	D	NN	N	Class C65	2
rs61755802	C213R	0	1.00	−0.4	D	NN	N	Class C65	2
rs62645926	Y184S	0.03	1.00	−0.3	D	NN	N	Class C65	2
VAR_006853	R13W	0	1.00	−0.1	D	NN	N	Class C65	2
rs61755796	W179R	0	1.00	0.1	D	NN	N	Class C65	2
CM090728	R123W	0.04	0.19	−0.1	D	NN	N	Class C65	2
rs61755785	K153R	0.04	1.00	−0.3	D	NN	N	Class C25	2
rs201501819	I177M	0.03	0.87	−0.1	D	NN	N	Class C0	2
CM067981	I232N	0	0.99	−2.1	D	N	D	Class C65	2
VAR_010480	T216I	0	0.99	−0.2	D	N	D	Class C65	2
VAR_010487	T307I	0	0.21	0.4	D	N	D	Class C65	2
VAR_025752	T307A	0.01	0.97	−0.4	D	N	D	Class C55	2
rs62637051	F62L	0	0.91	−0.5	D	N	D	Class C15	2
rs121918287	V317M	0.02	1.00	−0.3	D	N	D	Class C15	2
rs62637348	Q208E	0	1.00	0.3	D	N	D	Class C25	2
VAR_017370	K30R	0.03	1.00	−0.4	D	N	D	Class C0	2
rs74653691	L207I	0	1.00	0.0	D	N	D	Class C0	2

Continued

Table 6.9 Ranking system based on the prediction scores of seven methods—cont'd

rsIDs/variants	Amino acid position	SIFT	PolyPhen-2	I-MUTANT Suite	PhD-SNP	SNAP	SNPs&GO	Align-GVGD	Rank
rs202234687	E342Q	0	1.00	−0.5	D	N	N	Class C25	3
rs149678971	H402Y	0.02	0.44	0.3	D	N	N	Class C35	3
CM010524	N296K	0	1.00	−2.3	D	NN	D	Class C65	1
HM070104	D304V	0	1.00	−2.0	D	NN	D	Class C65	1
CM067983	T237S	0.01	1.00	−0.6	D	NN	D	Class C55	1
rs61755788	C165Y	0	1.00	−0.1	D	NN	D	Class C65	1
CM072886	N296S	0.01	1.00	−0.6	D	NN	D	Class C45	2
CM071143	N296D	0.03	1.00	−1.5	D	NN	D	Class C15	2
HM070111	E35K	0	1.00	−2.0	D	NN	D	Class C0	2
CM071145	Y29C	0	1.00	−0.1	D	NN	D	Class C0	2
CM004423	F80L	0.04	1.00	−0.8	D	N	D	Class C15	2
CM072109	L134V	0.01	1.00	0.7	D	N	D	Class C25	2
VAR_025732	F113L	0	0.88	0.5	D	N	N	Class C15	2
rs28941469	Y227N	0	0.99	−0.4	N	NN	D	Class C65	2
VAR_000855	S231R	0.01	1.00	0.1	N	NN	D	Class C65	2
CM090731	P221L	0.05	1.00	−0.3	N	NN	D	Class C65	2
VAR_010468	A10V	0.01	0.97	−0.4	N	NN	D	Class C0	2

VAR_010471	F17C	0	1.00	−0.3	N	NN	D	Class C0	2
CM072887	S16Y	0.03	0.99	−0.3	N	NN	D	Class C0	2
VAR_000831	V9A	0	0.97	−0.3	N	NN	D	Class C0	2
rs28940275	T6P	0.03	0.95	−0.2	N	NN	D	Class C0	2
VAR_010470	S16F	0.02	0.99	0.1	N	NN	D	Class C0	2
VAR_043496	A357V	0.01	0.98	0.3	N	NN	D	Class C0	2
rs62637046	A12P	0.02	0.66	−0.1	N	NN	D	Class C0	2
rs28940276	V9M	0	0.78	0.3	N	NN	D	Class C0	2
rs62637338	F188L	0	1.00	0.2	N	N	D	Class C15	3
rs199508634	S7N	0.05	1.00	0.0	N	N	N	Class C0	3
rs145212203	M360I	0.03	1.00	0.5	N	N	N	Class C0	3
VAR_058273	I3T	0	0.55	−0.2	N	N	N	Class C0	3
CM074464	C165R	0	1.00	0.0	N	NN	D	Class C65	2
rs62637045	M1V	0	0.20	0.4	N	NN	N	Class C0	3
rs147490956	K535E	0.04	0.44	0.1	N	N	D	Class C55	3
rs145209035	M524T	0.04	0.87	−0.1	N	N	N	Class C65	3
CM071147	L217F	0	1.00	−0.5	N	N	N	Class C15	3

Continued

Table 6.9 Ranking system based on the prediction scores of seven methods—cont'd

rsIDs/variants	Amino acid position	SIFT	PolyPhen-2	I-MUTANT Suite	PhD-SNP	SNAP	SNPs&GO	Align-GVGD	Rank
rs111326315	V492I	0.02	1.00	−0.4	N	N	N	Class C25	3
rs147192139	E557K	0.04	0.01	−1.7	D	NN	N	Class C15	2
rs202230698	W316G	0.05	0.05	−1.1	N	N	N	Class C65	3
CM063227	T6A	0.05	0.49	−0.6	N	NN	N	Class C0	2
rs148326372	P334L	0.01	0.13	−0.2	D	N	N	Class C65	3
rs146447431	L472R	0	0.01	0.0	N	N	D	Class C65	3
rs62637055	L67P	0.07	1.00	−1.7	D	NN	D	Class C65	1
rs62637056	D70A	0.25	1.00	−2.6	D	NN	D	Class C65	1
rs146686238	W25C	0.06	1.00	−2.3	D	NN	D	Class C65	1
rs61755815	N244H	0.09	1.00	−1.0	D	NN	D	Class C65	1
rs61755780	Y141H	0.13	1.00	−1.0	D	NN	D	Class C65	1
rs61755783	R142W	0.18	1.00	−0.9	D	NN	D	Class C65	1
CM063093	P216R	0.56	0.99	−0.6	D	NN	D	Class C65	1
rs121918567	R195L	0.31	1.00	−0.1	D	NN	D	Class C65	1
rs61755792	R172G	0.29	0.80	−1.4	D	NN	D	Class C65	1
VAR_025750	E306D	0.13	1.00	−2.2	D	NN	D	Class C35	2
rs62637049	L52P	0.9	0.79	−0.5	D	NN	D	Class C0	2

rs61755795	Q178R	0.25	1.00	−1.3	D	NN	D	Class C35	2
rs61755810	R220Q	0.59	1.00	−0.8	D	NN	D	Class C35	2
rs149511444	P219T	0.65	0.99	−0.6	D	NN	D	Class C35	2
rs62645926	Y258S	0.2	0.98	−1.4	D	NN	N	Class C65	2
rs140406696	V277G	0.14	0.76	−3.1	D	NN	N	Class C65	2
rs61755816	N244K	0.56	1.00	−0.8	D	NN	N	Class C65	2
rs113689552	S36R	0.21	0.74	−0.6	D	NN	N	Class C65	2
rs61755806	P216L	0.27	0.06	−0.8	D	NN	N	Class C65	2
rs62645939	S289L	0.29	0.82	−1.4	D	NN	N	Class C15	2
rs139637557	R355C	0.06	1.00	−0.7	D	N	D	Class C65	2
VAR_017375	T91I	0.27	0.99	−2.4	D	N	D	Class C65	2
rs62637336	M180V	0.08	1.00	−1.1	D	N	D	Class C15	2
rs62637356	G266R	0.29	0.88	−1.3	D	N	D	Class C15	2
rs267606680	I205T	0.37	1.00	−1.0	D	N	N	Class C65	2
rs62638178	E172Q	0.26	1.00	−0.5	D	N	N	Class C25	3
rs61755791	G170S	0.97	0.82	−1.5	D	N	N	Class C55	3
CM067982	Y227F	0.15	1.00	−1.6	D	NN	D	Class C15	2

Continued

Table 6.9 Ranking system based on the prediction scores of seven methods—cont'd

rsIDs/variants	Amino acid position	SIFT	PolyPhen-2	I-MUTANT Suite	PhD-SNP	SNAP	SNPs&GO	Align-GVGD	Rank
CM063092	S198R	0.26	1.00	−1.6	N	NN	D	Class C65	2
rs62645931	K197E	0.67	0.76	−1.5	D	NN	N	Class C55	2
rs62640564	H490P	0.34	1.00	−1.4	D	N	D	Class C65	2
rs62640563	D441G	0.24	1.00	−0.8	D	N	D	Class C65	2
rs61755811	Q226E	0.45	0.76	−1.3	D	N	D	Class C25	2
CM090732	G249S	0.06	1.00	−1.4	N	NN	D	Class C65	2
CM083770	G137D	0.07	0.96	−1.3	N	NN	D	Class C65	2
VAR_010469	R13H	0.12	1.00	−0.6	N	NN	D	Class C0	2
rs62645932	V200E	1	0.94	−0.6	N	NN	N	Class C65	2
rs61755808	P219R	0.54	0.89	−0.5	N	NN	N	Class C65	2
rs61755805	P216S	0.61	0.80	−1.2	N	NN	N	Class C65	2
rs61755779	L126V	0.07	0.97	−0.8	N	NN	N	Class C25	3
rs201018137	S36N	0.18	0.59	−1.1	N	NN	N	Class C45	3
rs62638168	S71G	0.1	0.99	−1.2	N	N	D	Class C55	2
rs62639260	E268A	0.27	0.99	−1.5	N	N	D	Class C0	3
rs201225558	R392H	0.64	0.15	−0.5	N	N	D	Class C25	3
VAR_010478	G135S	0.24	1.00	−0.5	N	N	N	Class C55	3

rs61747600	D520N	0.51	0.99	−1.8	N	N	N	Class C15	3
rs61755817	W246R	0.21	1.00	−0.1	D	NN	D	Class C65	2
rs61755766	S27F	0.7	0.40	0.1	D	NN	D	Class C65	2
rs61755799	F211L	0.5	1.00	−0.5	D	NN	D	Class C15	2
rs61755770	L45F	0.24	0.99	−0.2	D	NN	N	Class C15	3
rs150381599	L261F	0.7	0.84	0.3	D	NN	N	Class C15	3
rs62637346	R202L	0.14	0.97	−0.4	D	N	D	Class C65	2
rs199960774	L319V	1	0.98	0.1	D	N	D	Class C25	3
rs76989855	I161M	0.54	0.44	0.1	D	N	D	Class C0	3
rs199572514	T155I	0.45	0.89	0.0	D	N	N	Class C65	3
CM000837	N99K	0.16	0.94	0.0	D	NN	D	Class C65	2
CM000839	S231T	0.2	0.99	−0.7	D	NN	D	Class C55	2
rs390659	Q304E	1	1.00	−0.4	D	NN	N	Class C65	2
rs62637345	G198R	0.33	1.00	−0.4	D	N	D	Class C65	2
CM010521	I73N	0.42	1.00	0.6	D	N	D	Class C65	2
rs193921105	E333K	0.1	0.46	−0.1	D	N	N	Class C65	3
rs143671863	E549K	0.12	0.98	−0.5	D	N	N	Class C55	3

Continued

Table 6.9 Ranking system based on the prediction scores of seven methods—cont'd

rsIDs/variants	Amino acid position	SIFT	PolyPhen-2	I-MUTANT Suite	PhD-SNP	SNAP	SNPs&GO	Align-GVGD	Rank
VAR_017366	T6R	0.01	0.94	0.0	N	NN	D	Class C0	3
rs62645935	G266D	0.1	1.00	−0.3	N	NN	N	Class C65	3
rs146844134	S49T	0.75	0.64	−0.2	N	N	N	Class C55	4
rs62637341	M193V	0.25	0.99	0.4	N	N	N	Class C15	3
CM090730	P216A	0.71	0.95	−0.3	N	NN	N	Class C25	3
rs201299251	Q434E	0.32	1.00	0.0	N	N	N	Class C25	4
rs145392651	A339T	0.19	0.85	−0.2	N	N	N	Class C55	4
rs139185976	G208D	0.34	0.28	−1.5	D	NN	D	Class C65	2
rs61755793	R172Q	0.56	0.25	−0.7	D	NN	D	Class C35	2
rs425876	K310R	1	0.03	−1.0	D	NN	N	Class C65	2
rs62645936	V268I	0.85	0.00	−0.6	D	N	D	Class C25	3
rs140227298	A116S	0.17	0.03	−1.4	D	N	N	Class C65	3
rs140681289	T475N	0.46	0.03	−1.7	D	N	N	Class C55	3
VAR_006895	D338A	0.56	0.00	−1.2	D	NN	N	Class C0	3
rs141071579	S507P	1	0.02	−1.4	D	N	N	Class C65	3
rs145439032	V436I	0.53	0.01	0.5	N	N	N	Class C0	4
rs146659849	I281T	0.14	0.00	−0.7	N	N	N	Class C25	4

rs146703538	M152V	0.41	0.01	−0.9	N	N	N	Class C15	4
rs148060787	L567F	0.14	0.01	0.5	N	NN	D	Class C15	3
rs148854184	N442T	0.16	0.01	−0.6	N	N	N	Class C0	4
rs146689925	T410I	0.21	0.08	0.1	D	N	N	Class C65	3
rs187560307	K173Q	0.12	0.02	−0.4	D	N	N	Class C45	4
VAR_010490	T561A	0.2	0.00	−0.3	D	NN	N	Class C0	3
rs57132800	S71R	0.26	0.02	−0.1	D	N	N	Class C65	3
CM090728	S125L	0.30	0.44	−0.4	D	N	N	Class C65	3
rs61748434	P313L	0.28	0.02	−0.1	N	N	N	Class C65	4
rs61755767	I32V	0.74	0.00	−0.1	N	N	N	Class C25	4
rs202125490	P404S	0.49	0.01	−0.3	N	N	N	Class C65	4
rs139329966	F319L	0.72	0.00	0.1	N	N	N	Class C15	4

Rank 1: Overall pathogenic (six to seven tools predicted pathogenic).
Rank 2: Probably pathogenic (four to five of seven tools predicted pathogenic).
Rank 3: Possibly pathogenic (two to three of seven tools predicted pathogenic).
Rank 4: Probably benign (zero to one tool predicted pathogenic).

variations. To facilitate this, bioinformatics acts as a central interface in modern genetics by implementation of computational algorithms and software tools to determine the coding variants that actually cause or confer susceptibility to disease. Discriminating the pathogenic variants from the neutral ones by traditional genetic methods in large-scale analysis remains as a daunting task.

SNPs can be used as markers to identify disease-causing genes, disease susceptibility, and interindividual variability in drug response (Shastry, 2003). In addition, SNPs can be used to understand the molecular mechanisms of sequence evolution. Among the SNPs, SAPs lead to substitution of single amino acid and potentially affect protein structure and function that are subjected to natural selection. On the other hand, SNPs that do not change encoded amino acids are called synonymous or csSNPs and are not subjected to natural selection (Kimura, 1983). Recent evidence from the 1000 Genomes project suggests that each individual human genome typically carries approximately 10,000–11,000 SAPs and 10,000–12,000 csSNPs (synonymous) (Kimura, 1983; Ng et al., 2008; Shastry, 2003; The 1000 Genome Consortium, 2010; Wu & Jiang, 2013). There are reports suggesting that more than 50% of the mutations known to be involved in human-inherited diseases are attributed to SAPs (Stenson et al., 2009). Differentiating deleterious SAPs with significant phenotypic consequences from tolerant ones without phenotypic change is of immense value in understanding the molecular basis of complex diseases. Generally, this can be achieved by family-based linkage analysis or GWA research. Understanding the molecular basis of complex diseases by traditional methods is painstaking and lingering, and at the structural level often almost impossible, especially in case where there are many SAPs causing the disease. Validating the outcomes of each SAP using *in vitro* studies and animal models is the simple way to reveal the functional consequences. Such studies are difficult to mount on a large scale, and their results might not always reflect *in vivo* genotype function in humans. These methods have their own limitations in their prediction analysis in complex diseases and serve a way to computational techniques, which make their predictions based on the characteristics of the SAPs. Alternatively, computational techniques have the potential to identify and prioritize the SAPs susceptible to diseases in large-scale study in an accurate way. Several computational methods are available to predict the impact of mutation on protein function by using dirichlet mixtures, logistic regression (Kjong-Van & Ting, 2013), neural networks (SNAP), Bayesian models (Schwarz, Rödelsperger,

Schuelke, & Seelow, 2010) (MutationTaster), HMMs, rule-based methods (Ramensky, Bork, & Sunyaev, 2002), and naive Bayes classifier (Adzhubei et al., 2010; Bromberg & Rost, 2007; Calabrese et al., 2009; Capriotti et al., 2006) for classification. They take evolutionary sequence conservation, structure information, and combination of sequence and structure information into consideration for making their predictions. The structure of a protein can be distorted in various ways due to the dissimilarity in physicochemical properties of the amino acid variant such as acidic, basic, or hydrophobic and locality of the variant in the protein sequence.

All the previously mentioned methods described here follow a similar procedure in which each SAP is first labeled with the characteristics related to damage it may cause on protein structure and function. The resulting feature vector is then employed to determine whether a single-residue substitution has either no effect or any effect on protein function. Analyzing SAPs based on the amino acid properties is generally considered to be an essential characteristic in defining the protein folding, stability, and protein function (Teng, Madej, Panchenko, & Alexov, 2009). Proteins with mutations do not always have crystallized 3D structures that are deposited in protein data bank. As a result, it is necessary to locate the mutation and design the 3D structures. This is a simple way of monitoring what level of adverse effects a mutation can have on a protein function and structure. Information about the 3D structure of a gene product is of great help in understanding the protein function and its role in causing disease. Therefore, a 3D structural detail of a protein is an important feature in predicting the deleteriousness of SAPs and also provides information about the environment of the mutation. Many groups have tried to assess the deleteriousness of SAPs along with protein 3D structure information by *in silico* analysis (Karchin, Diekhan, et al., 2005; Karchin, Kelly, & Sali, 2005). For most of the proteins, 3D structural information is not yet available. Therefore, it is an inevitable trend to predict the deleterious variations in proteins by sequence-based and position-specific evolutionary information (Balasubramanian, Xia, Freinkman, & Gerstein, 2005; Saunders & Baker, 2002; Sunyaev et al., 2001). Similarly, in this study, unavailability of the 3D structures and 3D model in BEST1 and PRPH2 protein insisted us to develop a coherent *in silico* method in order to prioritize the overall amino acid spectrum of VMD disease mutations using evolutionary information. It is assumed that SAPs in the protein sequences that are observed among living organisms have survived natural selection. Disease-causing or deleterious mutations are most likely to correspond to evolutionarily conserved positions in protein sequence due to their functional importance (Tavtigian, Greenblatt, Lesueur, & Byrnes, 2008; Thusberg & Vihinen, 2009). Therefore, our hypothesis was that amino acids conserved across species are more likely to be functionally significant. This illustrates that application of the molecular evolutionary approach might hold a strong upper hand in prioritizing SNPs to be genotyped in future molecular epidemiological studies. Several studies have confirmed this assumption in human disease-causing genes like *p53*, *p16*, *CFTR*, *G6PD*, *L1CAM*, and *PAH* (Greenblatt et al., 2003; Miller & Kumar, 2001; Mooney & Klein, 2002; Rishishwar et al., 2012; Sunyaev, Ramensky, & Bork, 2000). Therefore, we feel that this study provides useful information in selecting SNPs that are expected to have potential functional impact in creating VMD disease susceptibility.

The main goal of the work reported here is to identify and prioritize the overall amino acid spectrum of VMD disease mutations either pathogenic or neutral using eight different evolutionary (sequence and structural)-based prediction tools SIFT, PhD-SNP, PANTHER, PolyPhen-2, SNAP, SNPs&GO, I-MUTANT Suite, and Align-GVGD. Sequence-based methods have the added advantage over structure-based approaches in their predictions as they can be applied to any proteins with known relatives. Similarly, structure-based methods incorporate physicochemical properties of amino acids along with known 3D structures to make their predictions. Both the methods have their own limitations in making prediction scores. Sequence-based predictions based on homology and evolutionary conservation are unable to reveal the underlying mechanisms of how SNPs result in changed protein phenotypes. Additionally, sequence-based methods like PANTHER fail to generate prediction scores due to the poor sequence alignments or a variant in a particular position is not located in the majority of species. Major limitation of the structure-based methods lies in a context with the unavailability of the 3D structures for proteins, which makes it with limited applicability. Integrating the predictions of sequence and structure resources has added advantage to assess the reliability of the prediction results by cross-referencing the results from both methods. Most of the existing prediction tools extract information from MSAs within the homologous sequences to provide more information about the extent of conservation based on the input generated internally (SIFT, PANTHER, SNPs&GO) or submitted by the user (PolyPhen-2 and Align-GVGD). Tools like PolyPhen-2 and PhD-SNP have been trained using only human proteins in their prediction study. Meanwhile, SIFT (sequence-based) and SNAP (structure-based) were trained using variant proteins from diverse organisms.

Out of 213 SAPs in PRPH2, 82%, 94%, 75%, 74%, 61%, 79%, 90%, and 34% were specific to SIFT, PolyPhen-2, PhD-SNP, PANTHER, SNAP, SNP&GO, I-MUTANT Suite, and Align-GVGD, respectively, and 15% were predicted to be functionally significant by all eight methods. Similarly 87 SAPs in BEST1 exhibited 40%, 86%, 71%, 87%, 82%, 44%, 92%, and 65% by SIFT, PolyPhen-2, PhD-SNP, PANTHER, SNAP, SNP&GO, I-MUTANT Suite, and Align-GVGD, and in combination 13% were functionally significant by all eight methods. Among the eight *in silico* tools, PolyPhen-2 and PANTHER performed best in terms of sensitivity (0.95 and 0.96), and I-MUTANT Suite, performed worst in terms of sensitivity (0.47). In terms of specificity, Align-GVGD performed best (0.57) and PolyPhen-2 performed worst (0.11). PolyPhen-2 prediction performed worst in terms of accuracy (0.37) when compared to other seven tools with accuracy ranging from 0.44 to 0.054. Most of the commonly used *in silico* prediction tools available in the World Wide Web are benchmarked by the curators with their own data sets and shown to perform well. There are few studies that compared the prediction scores from a set of tools with the same data set. Chan et al. (2007)) compared the performance of four tools: SIFT, PolyPhen, Align-GVGD, and the BLOSUM62 matrix with accuracy ranging from 73% (BLOSUM62) to 82% (SIFT). When all the predictions were in agreement (62.7% of variants), predictive value improved to 88.1%. Notably, SIFT had high sensitivity value of 82% with lower specificity of 77%, whereas BLOSUM62 had lower sensitivity value of 75% with higher specificity of 85%. Chun and Fay (2009) in their study suggested a difference in the prediction performance of the *in silico* prediction techniques may be required to the difference in the sequence or MSA to identify evolutionary conserved mutations. Wei, Wang, Wang, Kruger, and Dunbrack (2010) utilized six tools SIFT, PolyPhen, PMut, SNPs3D, PhD-SNP, and nsSNPAnalyzer in their comparative study and found SIFT and PolyPhen to be best predictors. Schwarz et al. (2010) in their study validated the performance of MutationTaster against PolyPhen, PolyPhen-2, SNAP, PANTHER, and PMut and found MutationTaster to be best predictor with the overall accuracy of 86%. Similarly, Thusberg, Olatubosun, and Vihinen (2011) compared the performance of nine tools MutPred, nsSNPAnalyzer, PANTHER, PhD-SNP, PolyPhen, PolyPhen-2, SIFT, SNAP, and SNPs&GO and found SNPs&GO and MutPred to be best predictors with the overall accuracy of 82% and 81%. These mounting studies insist us that no single method could be rated as the best predictors in deleterious SAPs.

Combining of techniques with sequence and structure information using different algorithms and methodologies may provide a wider coverage and accurate prediction in the study of SAPs to be deleterious or neutral.

In addition, we adopted a ranking strategy to prioritize the SAPs based on the prediction scores of these *in silico* methods. This ranking system can be applied to any number of genes or proteins or diseases in which there is no 3D structure information and biochemical characterization available. Such a ranking scoring system will assist in prioritizing functional SAPs in large scale before further experimental investigation. HMM-based PANTHER could not make their subPSEC prediction for 27 and 6 SAPs in *BEST1* and *PRPH2* genes. Based on this, PANTHER predictions were not considered for ranking analysis. Combining the scores of all the previously mentioned seven methods (Supplementary Table 6.5, http://dx.doi.org/10.1016/B978-0-12-800168-4.00006-8), we have assigned the ranking from 1 to 4 and designated 110 SAPs (37%) as rank 1 or overall pathogenic (six to seven tools predicted pathogenic), 132 SAPs (44%) as rank 2 or probably pathogenic (four to five of seven tools predicted pathogenic), 46 SAPs (15%) as rank 3 or possibly pathogenic (two to three of seven tools predicted pathogenic), and 12 SAPs (4%) as rank 4 or probably benign (zero to one tool predicted pathogenic).

Protein primary sequence provides the most direct and readily available information regarding the clues for functional mutation sites that can be extracted from the amino acid sequence in cases where no structural information is available. A comparative analysis of amino acid conservation from multiple species by protein sequence alignments gives an indication of which amino acid residues are truly conserved and which of them represent localized the evolution. Conversely, most of the substituted amino acids in BEST1 and PRPH2 were deleterious in nature by any one of the *in silico* prediction methods, obtained tallest stacks in sequence logos generated from WebLogo and also located in highly conserved region by ConSurf. Population genetic studies describe that a significant fraction of functional SAPs was present in a conserved region. Residues that evolve under strong selective pressure are found to be significantly associated with human diseases (Arbiza et al., 2006). There were quite a lot of research indicating the function of substituted amino acids in causing diseases (Vitkup, Sander, & Church, 2003) and Trp and Cys in determining protein stability (Arbiza et al., 2006). Few studies have illustrated the importance of Cys residues in a protein sequence since most of the proteins folding are dependent on disulfide bonds (Song et al., 2009).

SNP-associated residue changes to or from Cys will likely destabilize the protein structure. Taking all this into consideration, we extended our analysis by highlighting the propensity of each amino acid in native and substituted states along with the function of Cys residues in the formation of disulfide bonds in BEST1 and PRPH2 and how it affects its function. Lastly, we defined the role of each SNP present in coding and noncoding region of BEST1 and PRPH2 using FASTSNP and F-SNP. The vast amount of data generated from this cost-effective study can provide a clue to the biologist to understand the relationship between SNPs and VMD disease. The main aim of this proposed work is to make aware of the benefits to biologists in VMD disease SNP functional analysis with a set of well-known *in silico* methods.

5. CONCLUSION

Current technologies and cost-effectiveness are before us now in SNP technology to expand the volume of information regarding genomic variants by several orders of magnitude in the near future, which lends itself into bioinformatics approaches to assess the potential impact of amino acid changes on gene functions. Understanding the resulting amino acid variation in proteins by sequence and structure context can assist in better understanding of phenotype–genotype relationships. Methods proposed in this work are not new to bioinformaticians. But these methods are new to the audience such as clinicians and geneticists who were not experts in the field of bioinformatics. Over the past decade, various methods were launched in the World Wide Web for the prediction of functional SNP, but the selection of the appropriate method depends on the data set. The input and output of these methods vary, but the ultimate goal is to classify deleterious or functional from neutral SNPs. Most significant is that each method has its own limitations in making their predictions. Integrations of different techniques will increase the overall prediction power of the *in silico* tools. Bioinformatics' resources do not give any information about the pathophysiology of the diseases but can be a valuable resource in filtering deleterious/functional SNPs in large data set. However, further supporting evidence is required from the traditional methods to confirm that predicted deleterious variants have a role in VMD disease processes. In this study, we have accessed the utility of existing evolutionary-based *in silico* methods for the discrimination of deleterious and neutral SNPs in the absence of protein 3D structure. We hope that our *in silico* pipeline knowledge of information will open the new possibilities in diagnostic, drug response, and therapeutic efforts and provide a useful framework for biologist investigators in choosing most appropriate functional SNPs in VMD for their study analysis.

ACKNOWLEDGMENTS

The authors take this opportunity to thank the management of Vellore Institute of Technology and Galgotias University for providing the facilities and encouragement to carry out this work. The authors have declared that no conflict of interest exists.

REFERENCES

Adzhubei, I. A., Schmidt, S., Peshkin, L., Ramensky, V. E., Gerasimova, A., & Bork, P. (2010). A method and server for predicting damaging missense mutations. *Nature Methods*, *7*, 248–249.

Amos, B., & Rolf, A. (1996). The SWISS-PROT protein sequence data bank and its new supplement TrEMBL. *Nucleic Acids Research*, *24*, 21–25.

Arbiza, L., Duchi, S., Montaner, D., Burguet, J., Pantoga-Uceda, D., Pineda-Lucena, A., et al. (2006). Selective pressures at a codon level predict deleterious mutations in human disease genes. *Journal of Molecular Biology*, *358*, 1390–1404.

Ashkenazy, H., Erez, E., Martz, E., Pupko, T., & Ben-Tal, N. (2010). ConSurf 2010: Calculating evolutionary conservation in sequence and structure of proteins and nucleic acids. *Nucleic Acids Research*, *38*, 529–533.

Balasubramanian, S., Xia, Y., Freinkman, E., & Gerstein, M. (2005). Sequence variation in G-protein-coupled receptors: Analysis of single nucleotide polymorphisms. *Nucleic Acids Research*, *33*, 1710–1721.

Best, F. (1905). Ubereine hereditare Maculaffection: Beiträge zur Vererbungslehre. *Z. Augenheilk*, *13*, 199–212.

Brendel, V., Bucher, P., Nourbakhsh, I., Blaisdell, B. E., & Karlin, S. (1992). Methods and algorithms for statistical analysis of protein sequences. *Proceedings of the National Academy of Sciences of the United States of America*, *89*, 2002–2006.

Bromberg, Y., & Rost, B. (2007). SNAP: Predict effect of non-synonymous polymorphisms on function. *Nucleic Acids Research*, *35*, 3823–3835.

Buckland, P. R. (2006). The importance and identification of regulatory polymorphisms and their mechanisms of action. *Biochimica et Biophysica Acta*, *1762*, 17–28.

Buckland, P. R., Hoogendoorn, B., Guy, C. A., Coleman, S. L., Smith, S. K., Buxbaum, J. D., et al. (2004). A high proportion of polymorphisms in the promoters of brain expressed genes influences transcriptional activity. *Biochimica et Biophysica Acta*, *1690*, 238–249.

Calabrese, R., Capriotti, E., Fariselli, P., Martelli, P. L., & Casadio, R. (2009). Functional annotations improve the predictive score of human disease-related mutations in proteins. *Human Mutation*, *30*, 1237–1244.

Capriotti, E., Calabrese, R., & Casadio, R. (2006). Predicting the insurgence of human genetic diseases associated to single point protein mutations with support vector machines and evolutionary information. *Bioinformatics*, *22*, 2729–2734.

Capriotti, E., Fariselli, P., Rossi, I., & Casadio, R. (2008). A three-state prediction of single point mutations on protein stability changes. *BMC Bioinformatics*, *9*, S6.

Cartegni, L., & Krainer, A. R. (2002). Disruption of an SF2/ASF-dependent exonic splicing enhancer in SMN2 causes spinal muscular atrophy in the absence of SMN1. *Nature Genetics*, *30*, 377–384.

Chan, P. A., Duraisamy, S., Miller, P. J., Newell, J. A., McBride, C., Bond, J. P., et al. (2007). Interpreting missense variants: Comparing computational methods in human disease genes CDKN2A, MLH1, MSH2, MECP2, and tyrosinase (TYR). *Human Mutation*, *28*, 683–693.

Cheng, J., Randall, A. Z., Sweredoski, M. J., & Baldi, P. (2005). SCRATCH: A protein structure and structural feature prediction server. *Nucleic Acids Research*, *33*, W72–W76.

Chun, S., & Fay, J. C. (2009). Identification of deleterious mutations within three human genomes. *Genome Research*, *19*, 1553–1561.

Crooks, G. E., Hon, G., Chandonia, J. M., & Brenner, S. E. (2004). WebLogo: A sequence logo generator. *Genome Research*, *14*, 1188–1190.

Edgar, R. C. (2004). MUSCLE: Multiple sequence alignment with high accuracy and high throughput. *Nucleic Acids Research*, *32*, 1792–1797.

Flicek, P., Amode, M. R., Barrell, D., Beal, K., Brent, S., & Chen, Y. (2010). Ensembl 2011. *Nucleic Acids Research*, *39*, D800–D806.

Fredman, D., Munns, G., Rios, D., Sjöholm, F., Siegfried, M., Lenhard, B., et al. (2004). HGVbase: A curated resource describing human DNA variation and phenotype relationships. *Nucleic Acids Research*, *32*, D516–D519.

Gass, J. D. (1974). A clinicopathologic study of a peculiar foveomacular dystrophy. *Transactions of the American Ophthalmological Society*, *72*, 139–156.

Greenblatt, M. S., Beaudet, J. G., Gump, J. R., Godin, K. S., Trombley, L., Koh, J., et al. (2003). Detailed computational study of p53 and p16: Using evolutionary sequence analysis and disease-associated mutations to predict the functional consequences of allelic variants. *Oncogene*, *22*, 1150–1163.

Karchin, R., Diekhans, M., Kelly, L., Thomas, D. J., Pieper, U., Eswar, N., et al. (2005). LS-SNP: Large-scale annotation of coding nonsynonymous SNPs based on multiple information sources. *Bioinformatics*, *21*, 2814–2820.

Karchin, R., Kelly, L., & Sali, A. (2005). Improving functional annotation of non-synonymous SNPs with information theory. *Pacific Symposium on Biocomputing*, *10*, 397–408.

Khan, I. A., Mort, M., Buckland, P. R., O'Donovan, M. C., Cooper, D. N., & Chuzhanova, N. A. (2006). *In silico* discrimination of single nucleotide polymorphisms and pathological mutations in human gene promoter regions by means of local DNA sequence context and regularity. *In silico Biology*, *6*, 23–34.

Kimura, M. (1983). *The neutral theory of molecular evolution*. Cambridge: Cambridge University Press.

Kjong-Van, L., & Ting, C. (2013). Exploring functional variant discovery in non-coding regions with SInBaD. *Nucleic Acids Research*, *41*, e7.

Kubo, M., Hata, J., Ninomiya, T., Matsuda, K., Yonemoto, K., Nakano, T., et al. (2007). A non-synonymous SNP in PRKCH (protein kinase Cη) increases the risk of cerebral infarction. *Nature Genetics*, *39*, 212–217.

Kumar, P., Henikoff, S., & Ng, P. C. (2009). Predicting the effects of coding non-synonymous variants on protein function using the SIFT algorithm. *Nature Protocols*, *4*, 1073–1081.

Lee, P. H., & Shatkay, H. (2008). F-SNP: Computationally predicted functional SNPs for disease association studies. *Nucleic Acids Research*, *36*, D820–D824.

Leroy, B. P., Kailasanathan, A., De Laey, J. J., Black, G. C., & Manson, F. D. (2007). Intrafamilial phenotypic variability in families with *RDS* mutations: Exclusion of ROM1 as a genetic modifier for those with retinitis pigmentosa. *The British Journal of Ophthalmology*, *91*, 89–93.

Meunier, I., Sénéchal, A., Dhaenens, C. M., Arndt, C., Puech, B., Defoort-Dhellemmes, S., et al. (2011). Systematic screening of BEST1 and PRPH2 in juvenile and adult vitelliform macular dystrophies: A rationale for molecular analysis. *Ophthalmology*, *118*, 1130–1136.

Mi, H., Muruganujan, A., & Thomas, P. D. (2013). PANTHER in 2013: Modeling the evolution of gene function, and other gene attributes, in the context of phylogenetic trees. *Nucleic Acids Research*, *41*, D377–D386.

Miller, M. P., & Kumar, S. (2001). Understanding human disease mutations through the use of interspecific genetic variation. *Human Molecular Genetics*, *10*, 2319–2328.

Mooney, S. D., & Klein, T. E. (2002). The functional importance of disease-associated mutation. *BMC Bioinformatics*, *3*, 24.

Mottagui-Tabar, S., Faghihi, M. A., Mizuno, Y., Engstrom, P. G., Lenhard, B., Wasserman, W. W., et al. (2005). Identification of functional SNPs in the 5-prime flanking sequences of human genes. *BMC Genomics*, *6*, 18.

Ng, P. C., & Henikoff, S. (2003). SIFT: Predicting amino acid changes that affect protein function. *Nucleic Acids Research*, *31*, 3812–3814.

Ng, P. C., Levy, S., Huang, J., Stockwell, T. B., Walenz, B. P., Li, K., et al. (2008). Genetic variation in an individual human exome. *PLoS Genetics*, *4*, e1000160.

Pampin, S., & Rodriguez-Rey, J. C. (2007). Functional analysis of regulatory single-nucleotide polymorphisms. *Current Opinion in Lipidology*, *18*, 194–198.

Pastinen, T., & Hudson, T. J. (2004). *Cis*-acting regulatory variation in the human genome. *Science*, *306*, 647–650.

Petrukhin, K., Koisti, M. J., Bakall, B., Li, W., Xie, G., Marknell, T., et al. (1998). Identification of the gene responsible for Best macular dystrophy. *Nature Genetics*, *19*, 241–247.

Ramensky, V., Bork, P., & Sunyaev, S. (2002). Human nonsynonymous SNPs: Server and survey. *Nucleic Acids Research*, *30*, 3894–3900.

Rishishwar, L., Varghese, N., Tyagi, E., Harvey, S. C., Jordan, I. K., & McCarty, N. A. (2012). Relating the disease mutation spectrum to the evolution of the cystic fibrosis transmembrane conductance regulator (CFTR). *PLoS One*, *7*, e42336.

Saunders, C. T., & Baker, D. (2002). Evaluation of structural and evolutionary contributions to deleterious mutation prediction. *Journal of Molecular Biology*, *322*, 891–901.

Savinkova, L. K., Ponomarenko, M. P., Ponomarenko, P. M., Drachkova, I. A., Lysova, M. V., Arshinova, T. V., et al. (2009). TATA box polymorphisms in human gene promoters and associated hereditary pathologies. *Biochemistry (Moscow)*, *74*, 117–129.

Schwarz, J. M., Rödelsperger, C., Schuelke, M., & Seelow, D. (2010). MutationTaster evaluates disease-causing potential of sequence alterations. *Nature Methods*, *7*, 575–576.

Shastry, B. S. (2003). SNPs and haplotypes: Genetics markers for disease and drug response. *International Journal of Molecular Medicine*, *11*, 379–382.

Sherry, S. T., Ward, M. H., Kholodov, M., Baker, J., Phan, L., Smigielski, E. M., et al. (2001). dbSNP: The NCBI database of genetic variation. *Nucleic Acids Research*, *29*, 308–311.

Song, X., Geng, Z., Zhu, J., Li, C., Hu, X., Bian, N., et al. (2009). Structure-function roles of four cysteine residues in the human arsenic (+3 oxidation state) methyltransferase (hAS3MT) by site-directed mutagenesis. *Chemico-Biological Interactions*, *179*, 321–328.

Stenson, P. D., Mort, M., Ball, E. V., Howells, K., Phillips, A. D., Thomas, N. S., et al. (2009). The human gene mutation database: 2008 update. *Genome Medicine*, *1*, 13.

Sun, H., Tsunenari, T., Yau, K. W., & Nathans, J. (2002). The vitelliform macular dystrophy protein defines a new family of chloride channels. *Proceedings of the National Academy of Sciences of the United States of America*, *99*, 4008–4013.

Sun, T., Zhou, Y., Yang, M., Hu, Z., Tan, W., Han, X., et al. (2008). Functional genetic variations in cytotoxic T-lymphocyte antigen 4 and susceptibility to multiple types of cancer. *Cancer Research*, *68*, 7025–7034.

Sunyaev, S., Ramensky, V., & Bork, P. (2000). Towards a structural basis of human nonsynonymous single nucleotide polymorphisms. *Trends in Genetics*, *16*, 198–200.

Sunyaev, S., Ramensky, V., Koch, I., Lathe, W., Kondrashov, A. S., & Bork, P. (2001). Prediction of deleterious human alleles. *Human Molecular Genetics*, *10*, 591–597.

Tavtigian, S. V., Deffenbaugh, A. M., Yin, L., Judkins, T., Scholl, T., Samollow, P. B., et al. (2006). Comprehensive statistical study of 452 BRCA1 missense substitutions with classification of eight recurrent substitutions as neutral. *Journal of Medical Genetics*, *43*, 295–305.

Tavtigian, S. V., Greenblatt, M. S., Lesueur, F., Byrnes, B., & IARC Unclassified Genetic Variants Working Group (2008). *In silico* analysis of missense substitutions using sequence alignment-based methods. *Human Mutation*, *29*, 1327–1336.

Teng, S., Madej, T., Panchenko, A., & Alexov, E. (2009). Modeling effects of human single nucleotide polymorphisms on protein-protein interactions. *Biophysical Journal*, *96*, 2178–2188.

The 1000 Genome Consortium, (2010). A map of human genome variation from population-scale sequencing. *Nature*, *467*, 1061–1073.

Thusberg, J., Olatubosun, A., & Vihinen, M. (2011). Performance of mutation pathogenicity prediction methods on missense variants. *Human Mutation*, *32*, 358–368.

Thusberg, J., & Vihinen, M. (2009). Pathogenic or not? And if so, then how? Studying the effects of missense mutations using bioinformatics methods. *Human Mutation*, *30*, 703–714.

Ueki, M., Fujihara, J., Takeshita, H., Kimura-Kataoka, K., Iida, R., Nakajima, T., et al. (2010). Genetic and expression analysis of all non-synonymous single nucleotide polymorphisms in the human deoxyribonuclease I-like 1 and 2 genes. *Electrophoresis*, *31*, 2063–2069.

Vitkup, D., Sander, C., & Church, G. M. (2003). The amino acid mutational spectrum of human genetic disease. *Genome Biology*, *4*, R72–R80.

Wei, Q., Wang, L., Wang, Q., Kruger, W. D., & Dunbrack, R. L. (2010). Testing computational prediction of missense mutation phenotypes: Functional characterization of 204 mutations of human cystathionine beta synthase. *Proteins*, *78*, 2058–2074.

Wrigley, J. D., Ahmed, T., Nevett, C. L., & Findlay, J. B. (2000). Peripherin/*rds* influences membrane vesicle morphology: Implications for retinopathies. *The Journal of Biological Chemistry*, *275*, 13191–13194.

Wu, J., & Jiang, R. (2013). Prediction of deleterious nonsynonymous single-nucleotide polymorphism for human diseases. *The Scientific World Journal*, *2013*, 675851.

Yoshiura, K., Kinoshita, A., Ishida, T., Ninokata, A., Ishikawa, T., Kaname, T., et al. (2006). A SNP in the ABCC11 gene is the determinant of human earwax type. *Nature Genetics*, *38*, 324–330.

Yuan, H. Y., Chiou, J. J., Tseng, W. H., Liu, C. H., Liu, C. K., Lin, Y. J., et al. (2006). FASTSNP: An always up-to-date and extendable service for SNP function analysis and prioritization. *Nucleic Acids Research*, *34*, 635–641.

CHAPTER SEVEN

Current State-of-the-Art Molecular Dynamics Methods and Applications

Dimitrios Vlachakis[*], Elena Bencurova[*,†], Nikitas Papangelopoulos[*], Sophia Kossida[*,1]

[*]Bioinformatics & Medical Informatics Team, Biomedical Research Foundation, Academy of Athens, Athens, Greece

[†]Laboratory of Biomedical Microbiology and Immunology, University of Veterinary Medicine and Pharmacy, Kosice, Slovakia

[1]Corresponding author: e-mail address: skossida@bioacademy.gr

Contents

1. Introduction 270
2. The Role of Computer Experiments in Modern Science 272
3. Brief History of Computer Molecular Simulations 274
4. Physics in MD 276
5. Algorithms for MD Simulations 278
6. Computational Complexity of MD Simulations and Methods to Increase Efficiency 281
7. High-Performance Parallel Computing 285
8. General-Purpose Computing Using Graphics Processing Units 288
9. Software for MD Simulations 290
10. Force Fields for MD 293
    10.1 AMBER 295
    10.2 CHARMM force field 302
    10.3 Consistent force field 304
    10.4 DREIDING force field 304
    10.5 GROMOS 305
    10.6 Merck molecular force field 305
    10.7 MM2/MM3 force fields 306
11. Current Limitations of MD 307
12. Conclusions 308
Acknowledgments 309
References 309

Abstract

Molecular dynamics simulations are used to describe the patterns, strength, and properties of protein behavior, drug–receptor interactions, the solvation of molecules, the

Advances in Protein Chemistry and Structural Biology, Volume 94
ISSN 1876-1623
http://dx.doi.org/10.1016/B978-0-12-800168-4.00007-X

conformational changes that a protein or molecule may undergo under various conditions, and other events that require the systematic evaluation of molecular properties in dynamic molecular systems. Only few years ago proteins were considered to be rigid body structures with very limited conformational flexibility. However, it is now clear that proteins are highly dynamic structures, the internal organization of which is the key to their 3D spatial arrangement and hence biological function. The study of protein dynamics in the lab is a very complicated, expensive, and time-consuming process. Therefore, a lot of effort and hope lies with the computers and the *in silico* study of protein structure and molecular dynamics. Herein, an effort has been made to describe the ever-evolving field of molecular dynamics, the different algorithms, and force fields that are being used as well as to provide some insight on what the near future holds for this auspicious field of computational structural biology.

1. INTRODUCTION

Molecular dynamics (MD), in a few words, refers to numerically solving the classical equations of motion for a group of atoms. In order to make this happen, we need a law that reveals how atoms interact with each other in the system (Stryer, 1995). That combined with the atomic positions can give us the associated potential energy, the forces on the atoms, and the stress on the container walls. Although such a law is generally unknown, we can use approximations that differ in accuracy and realism depending on a force field, or models constructed after performing electronic structure calculations, which can also be done at different levels of theory (Wilkinson, van Gunsteren, Weiner, & Wilkinson, 1997). Also it is necessary to have an algorithm that incorporates numerically the equations of motion for the atoms in the system. Until now many different approaches have been proposed in this direction. Of course, we cannot solve the equations of motion, unless we provide the system with initial values, such as initial positions and velocities for every atom in the system. All the abovementioned are sufficient to carry out an MD simulation.

Since MD is concerned with the motion of atoms and molecules, let us have a look at the example of the normal mode analysis, where the dynamic motion of a molecular system is evaluated from its total energy (Ding et al., 1995). According to Hook's law, the total energy of a small and simple diatomic molecule should be given by

$$E_{(r)} = \frac{1}{2}k\left(r - r^0\right)^2 \quad \text{and} \quad F(r) = -\frac{\mathrm{d}E}{\mathrm{d}r} = -k\left(r - r^0\right),$$

where E is the energy, F is the force, r is the distance, r^0 is the initial distance, and k is the Hook's constant.

Here, the forces on the atoms are estimated by the derivative of the energy (Ding et al., 1995). When the forces have been assigned, the Newton law of motion can be used to solve the molecular motion ($F = m \times a$, force = mass × acceleration). In the case of the small and simple diatomic molecule, the displacement from the equilibrium bond length (x) will be given by the formula:

$$x = r - r^{\circ} \quad \text{and} \quad \mu = \frac{m_C m_O}{m_C + m_O}$$

where μ is the effective mass of the vibrating diatomic molecule.

In MD simulations, the kinetic energy of the system will depend on the temperature of the system (Eriksson, Pitera, & Kollman, 1999). The total energy will be the sum of the kinetic and potential components of the system. The acceleration of each atom is estimated from the set of forces it accepts, under the given force field. The results generated can be used to estimate the configurational and momentum information for each atom of the system (i.e., energy and pressure). Molecular mechanics therefore can be used to further optimize a model generated by homology modeling as well as docking results, where the protein–ligand interactions can be analyzed for their stability during a specific duration of time (Potter, Kirchhoff, Carlson, & McCammon, 1999).

At this point it would be interesting to discuss about the utility of performing an MD simulation and what we can gain from it. The simulation offers an incredibly detailed approach of the real dynamics of the system we study in order to observe how distinct atoms act. The simulation is so vivid that gives you the image of moving between the atoms. As experiments do not usually provide sufficient resolution, they offer an inadequate perception of the important processes taking place at atomic and molecular level, so with MD simulations, we can have a better understanding of these processes. In addition, it is easier to configure and control the environment in which the experiment takes place, for example, temperature, pressure, and atomic configuration, through a simulation compared to an experiment.

We have made two additional assumptions apart from the basic numerical approximation involved in the integration of the equations of motion, which are fundamental in MD simulations. According to the first assumption, atoms' behavior is similar to classical entities, which means that they obey Newton's equations of motion (Hug, 2013). The accuracy of this approach is related with the system we study every time as well as the conditions of simulation. It is expected that the previously mentioned approach is unrefined for light atoms at low temperatures, but generally, it is not a bad approach. Besides liquid helium (He) and other light atoms, normally, quantum effects on atomic dynamics are rather small. In cases where quantum effects cannot be neglected, the path integral approach can be used or a relative method. The second basic assumption has to do with the modeling of how atoms interact with each other in the system. The only way to form a functional and responsible image of the atomic procedures in the system is by having representative description of those interactions. On the contrary, when gathering information about a significant class of the system such as low density gases or liquid metals, there is no necessity in high-accuracy description, a model capturing the essential features, the defining physics of a significant class is acceptable (Monticelli & Tieleman, 2013). At this point trying to be specific can have negative results in productivity and darken the general image. So it is crucial to know how in depth a description should be for a particular issue.

All the abovementioned may give the impression that MD does nothing more than solving Newton's equations for atoms and molecules. The truth is that MD is not limited in this function; it is feasible to make artificial forms of MD, which can be used in the simulation of a system according to certain demands in temperature and pressure. This would not be possible using a simple solution of the standard equations of motion. Moreover, MD provides the ability to synthesize the physical dynamics of ions with an imaginary dynamics of electronic wave functions, which results in efficient understanding of atomic dynamics from first principles (Cheng & Ivanov, 2012). In conclusion, MD has many possibilities and uses and can do more than numerically incorporating the equation of motion for atoms and molecules.

2. THE ROLE OF COMPUTER EXPERIMENTS IN MODERN SCIENCE

In the past, physical sciences were defined by an interaction between experiment and theory. From the experimental point of view, a system is subjected to measurements and provides results expressed numerically. From the theoretical point of view, a model is constructed to represent the system, usually in the form of a set of mathematical equations. The validation of the model depends on its ability to simulate the function of the system in a few selected cases, by computing a solution from the given equations. A considerable amount of simplification is usually necessary in order to eliminate all the complexities that characterize real-world problems (DeLisi, 1988).

Few years ago, theoretical models could be easily tested only in a few simple special circumstances. An example from the scientific field of condensed matter physics is a model for intermolecular forces in a specific material, which could be verified either in a diatomic molecule or in a perfect, infinite crystal. However, approximations were often required in order for the calculation to be conducted (Sadek & Munro, 1988). Unfortunately, numerous scientific problems of intense academic and practical interest fall outside the sphere of these special circumstances. The physics and chemistry of organic molecules, which include many degrees of freedom, are an example of them.

Since the 1950s, the development of high-speed computers introduced the computer experiment, a new step between experiment and theory. In a computer experiment, a model is provided by theorists, but the calculations are conducted mechanically by a computer, following a defined set of instructions, which constitute the algorithm. In this way, the complexity of a solution can be calculated before its execution. Also, more complex solutions can be carried out, and therefore more realistic systems can be modeled, leading to a better understanding of real-life experiments. The development of computer experiments changed substantially the traditional relationship between theory and experiment.

Computer simulations increased the demand for accuracy of the models. For example, an MD simulation gives us the opportunity to evaluate the melting temperature of a certain material, which has been modeled according to a certain interaction law (Durrant & McCammon, 2011). Simulation gives the opportunity to test the accuracy of the theoretical model concerning that goal, something that could not be achieved in the past. Therefore, simulation expands the modeling process, disclosing critical areas and providing suggestions to make improvements. In addition, simulation can often approximate experimental conditions, so that in some cases computer results can be compared with experimental results directly. This prospect makes simulation a potent tool, not only to comprehend and interpret the experiments but also to examine regions that are not experimentally accessible, or which would otherwise require very demanding experiments, such as under extremely high temperature or pressure (Zwier & Chong, 2010). Last but not least, computer simulations give the opportunity to implement imaginative experiments and help realize things impossible to carry out in reality, but whose outcome increases our understanding of phenomena significantly, thus giving an invaluable outlet to scientific creativity.

3. BRIEF HISTORY OF COMPUTER MOLECULAR SIMULATIONS

The simulation of the motion of molecules, which lies within the field of study of MD, provides a better comprehension of various physical phenomena that originate from molecular interactions. MD comprise the motion of many molecules as in a fluid, as well as the motion of a single large molecule made of hundreds or thousands of atoms, as a protein molecule. Therefore, as the need to determine the motion of a large number of interacting particles arises, computers seem a really essential implement. The first who dealt with these computations were Berni Alder and Tom Wainwright in the 1950s at Lawrence Livermore National Laboratory (Alder & Wainwright, 1957). In order to study the arrangement of molecules in a liquid, they used a model in which the molecules were depicted as "hard spheres" interacting like billiard balls. The simulation of the motions of 32 and 108 molecules in computations that required 10–30 h became possible with the fastest computer at that time, an IBM 704. Now we are able to perform hard sphere computations on systems consisting of over a billion particles. Another extremely important paper appeared in 1967 (Verlet, 1967). The concept of neighbor list, as well as the Verlet time integrator algorithm, which we will analyze, was introduced by Loup Verlet and its use is to calculate the phase diagram of argon and to test theories of the liquid state by computing correlation functions. Verlet's studies were based on a more realistic molecular model than the hard sphere one, known as Lennard–Jones model. Lawrence Hannon, George Lie, and Enrico Climenti also utilized the Lennard–Jones model for their study of the flow of fluids in 1986 at IBM Kingston (Hannon, Lie, & Clementi, 1986). In their study, they represented the fluid by $\sim$104 interacting molecules. This number is extremely small compared with the real number of molecules in a gram of water. However, the behavior of the flow was the same as that of a real fluid. Another object of study of MD computations is the internal motion of molecules especially proteins and nucleic acids, which constitute vital elements of biological systems. The aim of these computations is to gain a deeper perception of the behavior of these molecules in biochemical reactions. Extremely interesting are the findings about the influence of quantum effects has on the overall dynamics of proteins, which appears only at low temperatures. Consequently, although classical mechanics is adequate to model the motion of a given drug-like compound, the computational power needed for monitoring the motion of a large biomolecule is immense (Vlachakis, Koumandou, & Kossida, 2013). For instance, the simulation of the motion of a 1500 atom molecule, like a small protein, for a time interval of 10^{-7} s requires computation of 6 h when performed by a modern Intel Xeon quadcore processor.

In the article "The Molecular Dynamics of Proteins" (Karplus & McCammon, 1986), Martin Karplus and Andrew McCammon present a discovery regarding the molecule myoglobin, which could have been made only by using MD. The significance of myoglobin lies in its function as oxygen repository in biological systems. For example, whales' ability of staying under water for such a long time is due to the large amount of myoglobin in their bodies. The fact that the oxygen molecule binds to a particular site in the myoglobin molecule was known. Nevertheless, the way the binding could occur was not fully perceived yet. X-ray crystallography work has proven that large protein molecules tend to fold into compact three-dimensional structures. In the interior of such a structure is where the oxygen sites lie, in the case of myoglobin. There seemed to be no way in which an oxygen atom could enter the structure to reach the binding site. The answer to this puzzle was given by MD. X-ray crystallography provided the picture of a molecule where the average positions of the atoms in the molecule are obvious. In fact, these atoms are never still but constantly vibrate about positions of their own equilibrium. For a short period of time, a path to the site could open up, whose width would be enough for an oxygen atom, as shown by MD simulation of the internal motion of the molecule (Karplus & McCammon, 1986). In order to understand the results of an MD simulation, scientific visualization is highly essential. The millions of numbers that represent the history of the locations and velocities of the particles do not offer exactly a revealing image of the motion. It is questionable how the development of a vortex or the nature of the bending and stretching of a large molecule becomes recognizable in this mass of data? How can one gain new insights? The formation of vortices, as well as the protein bending, is literally visible to the scientists through pictures and animations, which also offer new insights into specific features of these phenomena. The computations that are responsible for monitoring the motion of a large number of interacting particles, regardless whether the particles are atoms of one molecule, or molecules of a fluid or solid body or even particles in the distinct model of a vibrating string or membrane, are pretty much the same (Wang, Shaikh, & Tajkhorshid, 2010). They involve a long sequence of time steps. At each time step, Newton's laws are applied so as to define the new atomic locations and velocities from the old locations and velocities as well as the forces. These computations are quite easy but there are plenty of them. In order to obtain accuracy, the time steps must be quite short, and as a consequence, many may be needed in order to simulate a particularly long real-time interval. In computational chemistry, one time step takes usually about one femtosecond (10^{-15} s), and the whole computation may represent approximately 10^{-8} s costing about 100 h of computer time. Each step may carry a quite extensive amount of computation. In a system comprising of N particles, the force computations may include $O(N^2)$ operations at each step. Therefore, it becomes apparent that these computations may consume a large number of machine cycles (Klepeis, Lindorff-Larsen, Dror, & Shaw, 2009). Finally, the resulting animation of the motion of the biomolecular system may also demand large amounts of processing power and computer memory.

4. PHYSICS IN MD

As mentioned in the preceding text, classical MD is the discipline that deals with the molecular properties of liquids, solids, and molecules over a given period of time. The foundational equations of MD are Newton's equations; N atoms or molecules in the simulation are treated as a point mass, and their combination computes motion of the ensemble of atoms. Additional to motion, important microscopic and macroscopic variables can be determined, such as transport coefficients, phase diagrams, and structural or conformational properties (Stryer, 1995). The physics of the model is contained in a potential energy functional for the system from which individual force equations for each atom are derived. MD can be characterized as the opposite of memory intensive computing because only vectors of atom information are stored. Instead, their simulations contain plenty of information linked to the number of atoms and time steps.

The expression of the atoms size is accomplishing in Angstroms. For configuring, the dimensional scale of atoms needs to be considered that in three dimensions, many thousands or millions of atoms must be simulated to approach even the submicron scale. The conformation of the time step in the liquids and solids is obtaining from the demand that the vibrational motion of the atoms can be accurately tracked (Schleif, 2004). The simulation of the "real" time is linked to the femtosecond unit as tens or hundreds of thousands of time steps are needed for just picoseconds. Taking into account the special requirements of the MD simulation, lot of effort and research was submitted by the scientists in order to provide accurate hardware and optimized methods.

Focusing on the parallelism of the MD computations, many considerable attempts of conveying this approach in machines were submitted in the past years. However, it is empirically proven that the message-passing model of programming for multiple-instruction/multiple-data parallel machines is based on their capability to implement all structural data and computational enhancements that are commonly exploited in MD codes on serial and vector machines (Karplus & McCammon, 2002).

There is efficient number of attempts providing algorithms for the simulation of the performance of specific kinds of classical MD. For the moment, the center of attention will be placed in algorithms, which are appropriate for a general class of MD problems that has two special characteristics. The first characteristic refers to the ability of each atom to interact only with the other atoms that are geometrically nearby. This characteristic describes the limited range of the atoms. It is used for solids and liquids to be modeled this way due to electronic screening effects or simply to avoid the computational cost of including long-range Coulombic forces. In the procedure of short-range MD, the number of atoms is scaled as *N*, in the computational effort per time step. Second characteristic in MD problems constitutes the fact that the atoms can undergo large displacements over the duration of the simulation. Generally solid, liquid, or conformational changes are the responsible for diffusion in the biological molecule. The most important variable in the MD simulation is that each atom's neighbors change during the computational progress. The aforementioned algorithms could be used for fixed-neighbor simulations (e.g., all atoms remain on lattice sites in a solid). On the other hand, following the process in a parallel machine is proved to be a demanding task to continually track the neighbors of each atom and maintain efficient $O(N)$ scaling for the overall computation (Hansson, Oostenbrink, & van Gunsteren, 2002).

Additionally, it is important for the algorithms to function efficiently in small number of atoms where *N* is chosen to be as small as possible, instead of large scale, which usually facilitates the parallelism and is accurate enough to model the desired physical effects. The ultimate achievement of the calculations consists on the fact that each time step has to be accomplished as quickly as possible. This fact is especially true in nonequilibrium MD where the macroscopic changes in the system may take significant time to evolve, requiring millions of time steps to model. Concluding is significant to analyze modes as small as a few hundred atoms but is equally important to scale parallel algorithms to larger and faster parallel machines, that is, to scale optimally with respect to N and P (the number of processors) (Elber, 1996).

5. ALGORITHMS FOR MD SIMULATIONS

In this section, we provide a brief overview of the most commonly used algorithms in MD, and we also mention some details on parallelization techniques. Simply put, MD simulations consist of computing all interactions between particles of a system. Because of the typically huge number of atoms, this problem cannot be solved analytically. To bypass this limitation, MD uses numerical methods to solve Newton's equation of motion. The first category of algorithms used in MD deals with exactly this problem: the numerical integration Newton's equation. One of the simplest and best algorithms is the Verlet integrator (Hairer, Lubich, & Wanner, 2003). It is routinely used to calculate the trajectories of interacting atoms in MD simulations, because it offers properties such as greater stability and time-reversibility that are important for physical systems. Also, it efficiently reduces local errors that can accumulate during an MD run. Another reason why Verlet is the preferred integration algorithm is that constraints between atoms are easy to implement. Imposing constraints on the rigidity and position of particles is a very important practice in MD and, as we will see later, plays an important role in designing computationally efficient simulations. The Beeman algorithm is a variation of Verlet that was designed specifically to cater for the vast number of particles in MD calculations (Beeman, 1976). The main difference is in the function it uses to calculate the particle velocities and it also comes in implicit and explicit flavors. Other integrator algorithms worth noting include the Runge–Kutta method and the leapfrog integration.

In MD simulations, it is a common practice to split the interactions between atoms in short range and long range. This is done by setting a cutoff distance, outside of which two atoms are considered to have only long-range interactions. This way, different specialized algorithms can be used to deal with each case separately, improving the overall efficiency of the simulation. The two main algorithms used for short-range contacts are cell lists (or linked lists) (Mattson & Rice, 1999) and Verlet lists. Cell lists work by dividing the simulation space into smaller cells with an edge length at least the same as the cutoff distance. Each particle belongs to exactly one cell and atoms that are found in the same cell or in adjacent cells can be considered to interact with short-range contacts such as van der Waals forces and electrostatic interactions. The adjacent cells are the primary neighbor region of each particle and the sphere with radius equal to the cutoff distance is its interaction sphere. Verlet lists are very similar to linked lists in that they are also a data structure that keeps track of the molecules that are inside the same sphere of short-range interactions. Given a particle, a Verlet list is compiled that contains all atoms that fall within the preset cutoff distance. This method can also be easily modified to be applied to Monte Carlo simulations.

To cater for long-range interactions, we need to employ other specialized algorithms. The most prominent one is Ewald summation (with its many variations), which was initially formulated for theoretical physics. It is a method for calculating electrostatic interactions in MD systems based on the Poisson summation formula, but using Fourier space instead. This approach is more efficient for long-range interactions, which can be considered to have infinite range, because of the faster convergence in Fourier space. As we have mentioned already, interactions also have a short-range part. These are easily calculated as usual (in real space) and everything is summed up. Hence the Ewald "summation" name. One important note is that this method assumes that the system under investigation is periodic. This is certainly true when, for example, crustal structures of macromolecules are simulated, but not so much in other cases. A variation of the previously mentioned method is the particle mesh Ewald algorithm (Darden, Perera, Li, & Pedersen, 1999). As in regular Ewald summation, short- and long-range interactions are treated differently and eventually everything is summed up. The difference is that to improve the performance of the calculations, fast Fourier transformation is used, which in order to be implemented, the density field must be assessed on a discrete lattice in the simulated space, restricting it in a grid (mesh) of unit cells. This restriction introduces errors in the calculations, since particles that are close together end up having limited resolution when the force field calculations are carried out. To remedy this situation, the particle–particle–particle mesh was developed, which treats the close particles differently, applying direct summation on their force fields. A less commonly used family of algorithms is hierarchical methods that include Barnes and Hut, the fast multipole method, and the cell multipole method.

Because of the way MD simulations are constructed, there is a lot of room for parallelization of the algorithms in order to take advantage of multicore processors and speed up the calculations by several orders of magnitude. An MD simulation is basically the application of the same calculations on different subsets of the whole system and its parallelization should be straightforward. In essence, each particle could be assigned to a different thread and the calculation of its interactions could be carried out completely independently. Despite the large number of algorithms that have been described for parallel MD, most of them follow one of the three basic approaches for parallelization (Plimpton, 1995). The first approach assigns to each node a subset of atoms (evenly divided between available nodes), and the node is responsible for the calculation of the forces, just for these particles. When all the nodes have finished calculating the interactions for their assigned atoms, they must each communicate with all the rest in order to update the atom positions. This step is the limiting factor of this approach, since for large-scale MD, the node communication can become very expensive in computational power and is limited by the bandwidth of the available memory, introducing a bottleneck for the whole simulation. Some methods to overcome this limitation include overlapping calculations between nodes and storing information about the whole system in every processor. The second approach is very similar to the first one, but instead of dividing atoms between nodes, a specific subset of interactions that need to be calculated are assigned to each node. This can result in efficient computation, since each core becomes specialized to run only specific calculations, but can increase memory requirements, since the whole system needs to be visible to each node, at any given time. The last approach is based on allocating a distinct spatial region of the simulated space to each thread (spatial decomposition). By intelligently selecting the size of these regions and since every node is "aware" of its position in relation to the other nodes, the update step that comes after the completion of the calculations can be restricted only to communication between neighboring nodes and not to an all-to-all exchange of information. Global communication is restricted to system-wide variables, such as temperature. This leads to greatly reduced computational requirements and minimization of memory usage. What is more, this type of parallelization fits nicely with the domain decomposition algorithms (linked lists and Verlet lists) that were discussed earlier. More specifically, if we use coarse-grained nodes, more than one cell space, from the cell or Verlet lists, can be mapped in the same processor. This further reduces the need for inter-node communication of the updated positions of the atoms that are found close to the boundaries of each node. The main issue with spatial decomposition is when the MD simulation includes large macromolecules, such as proteins. In this case, it is not trivial how to best assign parts of the protein to different nodes and it is subject. On the contrary, this area is still the subject of ongoing research. One last approach that does not follow any of the previously mentioned three principles but should be mentioned because of its sheer simplicity is task parallelization. This is parallel computing in its most basic form. Instead of devising specialized methods and algorithms, we simply run the exact same simulation, but with somewhat different starting parameters. This approach may look over simplistic but one should keep in mind that the main goal of MD simulations is to observe the interactions and trajectories of the atoms inside a system long enough to be able to make useful inferences about its behavior. By running similar simulations on different processors, essentially we divide the time needed to gather enough statistical data between the number of cores we use. So, we start with a single simulation of a system that is in equilibrium and we assign independent copies of the simulation to every available processor. Then we introduce small perturbations to the system, but each perturbation is assigned to run on a different core. The reason this works is that given enough time the early perturbations will cause the systems to leave the initial equilibrated state and diverge in completely different directions. As such, we can terminate the simulations earlier and combine the data from each, to assemble an appropriate amount of statistics for our research purposes. Another way we can approach this method, is if we start from equilibrium but instead follow the behavior of the system along opposite time paths (before and after equilibration). To summarize, even though this is a very coarse approach to parallel MD, first of all, it works and, secondly, it works with existing serial and parallel algorithms and with minimal need for customization from the user point of view.

6. COMPUTATIONAL COMPLEXITY OF MD SIMULATIONS AND METHODS TO INCREASE EFFICIENCY

As we have stated already, MD is, in essence, a computer simulation that models the physical laws that govern the movement and trajectories of atoms and molecules as they interact for a period of time in a system of interacting particles. The inherent problem with any computer simulation is the complexity of the simulated system in relation to the available computing power. Simple MD simulations can be run on any normal computer needing only limited resources, but as the complexity and size of the emulated system rises, so does the required computational resources, rendering the simulation, in more than a few cases, intractable.

In general, the most important factors that affect the complexity of an MD run are the size of the system, the time step (i.e., the time interval between sampling the system for the evaluation of the potential), and of course the total time of the simulation. The main reason that the size of the system is important is that MD needs to calculate all pair-wise interactions between atoms as a function of their coordinates inside the system. More specifically, MD simulations scale linearly as $O(t)$ (where t is the number of time steps) in regard to the length of the time simulated, if, of course, we assume that the time steps remain unchanged and each one is only dependent on the previous sampling. As such, increasing the number of time steps results in a linear increase of the total time required to complete the simulation and also of the computing resources needed. This is very important if one considers that in order to have a meaningful simulation, the total run time must be relevant to the kinetics of the natural process under investigation. For example, when researching the dynamics of a chemical reaction, there is no point in setting a total time that is shorter than the time it takes for the formation of the products of said chemical reaction. As far as the time step is concerned, it must be smaller than the shortest vibrational frequency of the system, in order to avoid the accumulation of discretization errors (Davidchack, 2010). The issue that rises here is that, for a given total simulation time, selecting shorter time steps, results in an increased number of time steps and consequently more computationally intensive simulations. Unfortunately, there is no clear-cut way for choosing the appropriate time step, since it is highly dependent on the system simulated. As such, this choice is usually left to the discretion of the researcher. Having said that, typical time steps are in the order of femtosecond (10^{-15} s) and total simulation times range from nanoseconds (10^{-9} s) to microseconds (10^{-6} s).

Since the algorithmic complexity scales linearly with time, it is not the deciding factor of the total computational complexity in an MD experiment. The most computationally intensive task is the calculation of the force field, where all interactions between atoms need to be computed. In this step, the interactions for each of the n particles in the simulation are calculated with respect to all the rest $(n-1)$ particles. This leads to a complexity that scales by $O(n^2)$. At this point we also need to mention that apart from the total number of actual particles, another deciding factor for the complexity is the method used for implementing the solvent (Bizzarri & Cannistraro, 2002). Usually, in MD experiments, we examine the behavior of a molecule (or molecules) in an aqueous environment, that is, the macromolecule is solvated inside the space of the simulated system, which is filled with solvent molecules. As such, we need to incorporate in our calculations the solvent–solvent interactions and the solvent–solute ones. This greatly increases the total number of interacting pairs and raises the computational demands. Following this approach, we get a total algorithmic complexity of an MD simulation that scales as $O(n^2) \times O(t)$.

Calculating a force field this way can be considered a bit naive and an oversimplified approach to the problem that unnecessarily increases complexity. A more efficient approach is to intelligently adjusting the length of the time steps. There are a number of methods and algorithms that impose constraints on allowable bond lengths and bond angles (Ryckaert, Ciccotti, & Berendsen, 1977; Streett, Tildesley, & Saville, 1978; Tuckerman, Berne, & Martyna, 1991). Usually, they involve an iterative procedure that, at each time step, fine-tunes the calculated particle positions to adhere to the imposed constraints. If explicit constraint forces are used, the time steps are shortened, which is the opposite of the desired effect. Thus, implicit forces are preferred when calculating the vector of a bond, especially if the available computing power is an issue. By imposing bond geometry constraints, especially in the case of long-range interactions (Tuckerman et al., 1991), sampling intervals can be extended, reducing the number of time steps. Another way to increase performance is to take into account the range of the interactions. Most short-range atom contacts have very specific, limited ranges and this can be incorporated in the simulation by imposing appropriate cutoff distances. If two atoms have a distance that falls outside the preset boundaries for a type of interaction, then it can be ignored, thus limiting the total number of interactions that need to be calculated. By denoting r, the range of the interactions and assuming that the particle density is mostly uniform inside the system, we can define a cutoff sphere for each atom that should contain approximately the same number of particles, for a given r. This way we can decrease the computational complexity to $O(n \times r)$. At this point, we can also incorporate the cell list algorithm (mentioned in section 5) by saving and managing in memory which particles belong to the primary neighbor region, allowing us to determine the particles that are located in the same sphere, without having to recalculate their position at every time step. If we skip this recalculation for a significant number of intervals, we can achieve a considerable boost in performance by further reducing the complexity to $O(n)$. On the flip side, we must be cautious because errors can accumulate, and so, it is better not to skip the recalculation altogether, but rather to set it to regular intervals of time steps. What is more, assigning to each particle its own sphere of interacting atoms increases the memory requirements. The amount of the extra memory depends on the density and the cutoff range, but for large-scale MD simulations, this can easily become an issue. One method to lower the amount of the additional memory requirements is to group together particles that are in close proximity. At any rate, computer memory is easily upgradable and much less expensive than catering for increases in computing power requirements. Furthermore, for long-range interactions, we can use well-established hierarchical methods (Barnes & Hut, 1986) that generally have a complexity of $O(n \times \log_n)$. Also, the computational power required can be further decreased by using electrostatics methods (Hockney & Eastwood, 1988; Sadus, 1999), which result in a complexity of $O(n/\log(n))$.

As we have already mentioned, one final consideration is how to implement the solvent in the system. Again, using an explicit model is more accurate but also more computationally demanding. In this case, the water molecules are considered as important as the molecules under examination (e.g., protein molecules), and all their physical properties need to be accurately described, so that the intermolecular water interactions can be calculated. There are various models that can effectively solve the solute–solvent interactions. These differ mainly in the allowed number of interacting areas, the geometric rigidity of the water molecules, and the number of available charged sites (Jorgensen, Chandrasekhar, Madura, Impey, & Klein, 1983). If an implicit implementation of the model is used, then all the parameters for the interactions that include solvent atoms are set to average values in an attempt to model a mean solvent force field (Bizzarri & Cannistraro, 2002; Fraternali & Van Gunsteren, 1996), greatly reducing the runtime. The issue that arises here is that, typically, solute–solvent interactions play an important role in the simulations, greatly affecting the solute–solute interactions. This is especially true in kinetics studies, where the details of the explicit solvent model are necessary to accurately reproduce the investigated system of solute molecules. As a result, even though the explicit model is much more computationally intensive, requiring the addition of as much as 10 times more particles in the system, it is in most cases unavoidable. To overcome this limitation, "hybrid" approaches have been developed that apply the explicit model for the solvent molecules that interact only with the part of the macromolecule that is important for the purposes of the simulation. The rest of the solvent molecules are using stochastic boundary conditions (Brooks & Karplus, 1989). As one might expect, in all the previously mentioned cases, there are large variations in complexity, mainly due to the variety of the available algorithms and their specific implementations. Since a typical MD simulation can include thousands or even millions of interacting particles, it becomes clear that when all types of interactions are taken into account, even after implementing the aforementioned adjustments for increased efficiency, the calculations quickly become extremely computationally expensive.

7. HIGH-PERFORMANCE PARALLEL COMPUTING

Previously, we analyzed the algorithmic complexity of the various MD simulations. In practice, a few simulated nanoseconds of a particle system can take actual hours or even days of CPU runtime to complete. We have also mentioned ways to dramatically improve performance using parallel MD simulations. This makes it clear that the way forward for MD is parallelism. The same rings true for high-performance computing (HPC) (Strey, 2001). Also, the level of similarities and analogies between parallel MD and parallel computing is astounding. As such, this work would not be complete without an analysis of the basic principles of parallel computing.

Parallel computing is a type of computation where different calculations related to the same problem are run concurrently. It is based on the assumption that a problem can be divided into smaller parts that can be solved in parallel, on different hardware (GOTTLIEB, 1994). Moore's law states that computing power doubles every 18 months. Up until a few years ago, the law held true because of the increases in CPU frequencies. Nowadays, frequencies are relatively stable, but Moore's law is still in effect, thanks to the increase of the number of cores in a single processor. It would not be farfetched to say that in order to increase available computing power, more cores need to be incorporated to an existing system. That is why, after all, almost all HPC is based on parallel systems. There are two main strategies for hardware parallelism: multicore and multiprocessors that are part of the same computer, or clusters and grids that interconnect multiple computers (Barney, 2012).

Just like increases in CPU frequencies did not result in corresponding increases in performance, adding more cores rarely produces the expected speed up in calculations. Optimally, adding a second core to a single-core machine should cut in half the time required to complete the same task. This is rarely the case, because not all parts of an algorithm can be parallelized and the parts that can only be expressed sequentially are the actual determinants of the total runtime. This concept is formally expressed by Amdahl's law (Gustafson, 1988), which states that the theoretical speed up of parallelization is limited by the small parts of a program that cannot be parallelized. Unfortunately, all but a few algorithms have serial parts and as a consequence, most parallel algorithms achieve an almost linear increase for the first few additional cores, which then remains constant, regardless of the number of extra cores. A more formal way is to convey this through data dependencies. The critical path of a program is the longest sequence of dependent calculations that need to be run in specific order, that is, to proceed to the next calculation, the previous one must be completed first. The time it takes for this sequence to complete is the absolute minimum time the program needs to run. Thankfully, critical paths are only a small part of most programs and the rest are independent calculations that can run in parallel and as such, benefit from multicore setups. One last important point, which also relates to parallel MD principles, is that the different parts of a program that run concurrently need at some point to communicate with each other, for example, to update the value of some variable. If the program is split into a large enough number of parallel parts and these parts need to be in constant interaction, the time expended in communicating can easily nullify any performance increases, or even extend the overall runtime. This phenomenon is called parallel slowdown.

In parallel programming, the different parts of the program run on independent "threads" of computation. Each core can cater only for one thread, or can be multithreaded. Also, threads need to be able to communicate between them, to update and exchange shared data. This makes parallel programming generally harder than the traditional serial way. Matters like concurrency, communication, and synchronization all need to be taken into account in order for bugs to be resolved early on. Lastly, depending on the frequency with which threads talk to each other, an application can be classified as having fine-grained parallelism (interthread communication many times per second), coarse-grained parallelism (few communications per second), and embarrassing parallelism (little or no communication). The last type of application is the easiest to implement in parallel, but the most common type is applications with coarse-grained parallelism. Another method to classify parallel as well as serial hardware and software is called Flynn's taxonomy. Without going into too much detail, the most commonly used type is the single-instruction, multiple-data classification, which means that the same operation is being executed repeatedly over different parts of a single large dataset. This is the type used in the case of MD, where the same interactions need to be calculated for every atom in the simulated system.

We mentioned that threads need to be able to regularly communicate between them. This is accomplished via the system memory. In parallel computers, there are two main memory architectures: shared memory and distributed memory. The term shared memory implies that all processing hardware has equal access to the same memory address and that the memory components are physically in the same space. The computer systems that implement this type of architecture provide to their various components a uniform way of accessing the memory (equal latency and bandwidth) and are called uniform memory access systems. In contrast, distributed memory provides each element with a different memory address and is usually not in the same physical space (spread among different systems), and the systems with this type of memory implement the nonuniform memory access architecture. Another type of computer memory is the cache. These are small but extremely fast memory modules that are part of the CPU chip and are used for saving temporary copies of memory addresses. This architecture can cause complications in parallel computing, since it is possible that the same memory value is stored in more than one cache, which can lead to programming bugs and errors in calculations.

Next we examine the different types of parallelism. Instruction-level parallelism occurs when parts of an executed program that can be divided into specific instructions are grouped together and run concurrently, without changing the final output of the program. Modern processors can execute more than one of these basic instructions at the same time, greatly increasing performance.

More relevant to MD type of parallelism is task parallelism, where completely different calculations are carried out on the same data, or different subsets of a single dataset. A variation of this type is data parallelism, where the exact same calculations are carried out on the same data, or different subsets of a single dataset. Both of these types are applicable to parallel MD, depending on the algorithms used.

Lastly, we give some details about the different types of parallel computers, in relation to the way the hardware implements parallelization. Multicore CPUs are the most common and broadly available parallel hardware. These CPUs contain multiple cores (processing units) on the same chip that can be used for completely independent tasks or synergistically for the execution of a single parallel program. All modern CPUs found in personal computers are multicore. An early attempt at multiple cores was Intel's Hyper-Threading technology. Even though these CPUs contained a single core, they were able to run more than one thread in a pseudo-parallel way. Symmetric multiprocessor is another type of parallel computer system that comprises matching processors that connect via shared memory and a communication bus. Because of the limited bandwidth of the bus and the contention in interrupt requests, these systems are typically limited to 32 CPUs. Despite of this drawback, they are broadly used, because they are cost-effective. The other main type of parallel computers is distributed systems. These usually consist of physically separated computer systems that are connected via a network. Each computer can have one or more multi-core CPUs and all networked computers can work in tandem. Examples of this architecture include computer clusters, grid computing that is based on computers connected via the Internet, and massively parallel processors that consist of a large number of networked processors that are placed inside a single computer. Graphics processing units (GPUs) are also examples of parallel computers, but we provide more details about them in section 8.

At this point, we should also mention the existence of specialized hardware, specifically tailored to MD calculations. MDGRAPE-3 is one such a supercomputer system that consists of 201 units of 24 custom chips, making a total of 4824 cores, with additional dual-core Xeon processors forming the basis of the system. It was designed by the RIKEN Institute of Japan, and it is especially efficient in protein structure prediction calculations. Another example is Anton, which is based in New York. It consists entirely of purpose-built, interconnect application-specific integrated circuits (ASICs). Each ASIC contains two subunits that are responsible for the calculation of different types of atom interactions. Anton is mostly used for MD simulations of biological macromolecules. Finally, we should also refer to field-programmable gate arrays. Even though these are considered general-purpose computers, they are essentially programmable hardware components that can rearrange themselves to cater for specific types of calculations, including MD simulations.

8. GENERAL-PURPOSE COMPUTING USING GRAPHICS PROCESSING UNITS

In section 7, we discussed various parallel-computing architectures in relations to MD. Most of these systems should be powerful enough to be used for even large-scale MD simulations. But they all suffer from a fundamental disadvantage: they are not cost-effective. This makes them unaffordable for any normal-sized laboratory, which simply wants to use MD for their research. This is where GPUs come in, combining affordability, expandability, and extremely high parallelization capabilities. What is more, each of the two GPU manufacturers provides easy to use programming tools, in order to quickly write efficient parallel programs that can harness the full potential of the GPU architecture. NVIDIA pioneered the general-purpose computing using graphics processing units (GPGPU) computing by offering the Compute Unified Device Architecture platform (NVIDIA, 2013) and AMD followed by releasing the Stream SDK. Also, both companies comply to the OpenCL specification, which is another tool for writing parallel programs specifically for GPUs.

The first GPUs appeared in the mid-1990s and where specialized hardware used solely for rendering 3D graphics, mainly in computer games. These early graphics cards contained only one processing unit with a host of limitations that included poor rendering quality, low processing power, and inflexible APIs for programming. But, driven by the insatiable market demand for ever-increasing fidelity in computer game graphics, these early efforts were soon replaced by powerful, multicore GPUs specializing in highly parallel computations, uniquely suited for displaying photorealistic graphics. At the same time, scientists from a wide range of disciplines, working on computationally intensive projects, started realizing that these affordable, widely available cards could be used for so much more than just computer games. This was how the first endeavors in GPGPU computing came to be. The problem was that programming on the GPUs for other purposes than 3D graphics was impossible in practice and researchers had to think in terms of polygons, adapting their algorithms accordingly. Thankfully, NVIDIA and AMD soon realized the potential of GPUs and redesigned their GPUs to be fully programmable, using their respective, specifically designed software tools, based on already established and widely used programming languages. Today, using GPUs for scientific calculations is an accelerating trend that does not show any signs of slowing down. As one might expect, the highly parallel nature of GPUs is perfectly suited for running the various parallel MD algorithms.

At this point, we should note that GPUs cannot work as stand-alone computers but must be based on a system built with traditional CPUs. Most successful algorithms make efficient usage of both types of processors, exploiting the unique advantages of each architecture. CPUs, even the latest multicore ones, have orders of magnitude less cores than high-end GPUs. They are, however, optimized for serial computing and as such can be used for the execution of the critical path of a program. On the other hand, to efficiently use the thousands of parallel cores of a GPU, a programmer can simply port only the parallel parts of an algorithm to the GPU architecture. This way, the usage of all available computing power is maximized, resulting in high-performance gains.

From an architectural point of view, GPU design is based on a scalable array of multithreaded streaming multiprocessors. This means that each one of these processors can run concurrently hundreds of threads, adding an additional layer of parallelization. The whole design is based on the single-instruction, multiple-thread (SIMT) parallel architecture. SIMT enables the GPU to divide a single task into a huge number of identical tasks, each running independently on each own thread, on the same or different core. At the same time, threads can communicate with each other and if needed, they can cooperate in completing part of the calculations. An additional advantage is the ability to create new threads on the fly, instantly adapting to changes in computing requirements of the program currently being run. This is called dynamic parallelism and is very important because it removed the need for the GPU to communicate with the CPU. This addresses the main disadvantage of GPU programming, which is the relatively slow speed of the PCIe interface that is used by the GPU to transfer data to and from the CPU, causing bottlenecks that slowed down calculations. Another important technology implemented by GPUs is automatic scalability. This means that the code is designed in a way that is not depended on some specific hardware, but can run on any number of cores, according to hardware availability. In practice, this means that, for example, to speed up calculations, a newer card can be installed, and the code will automatically take advantage of the additional processing power, without any extra effort from the user. Automatic scalability is even more important if one takes into consideration another advantage of GPU architecture, which is the ability to install more than one graphics cards in the same system. These cards can seamlessly cooperate as if all the cores were placed on a single board. In theory, the performance boost scales linearly with the number of the additional cards. NVIDIA's implementation of this capability is called scalable link interface and while AMD uses Crossfire.

To summarize, all the previously mentioned characteristics of GPUs make them uniquely suitable for running MD simulations. For example, it is possible to assign each atom of the system to a different thread, or allocate the calculation of a specific interaction to a subset of the available cores. This not only would lead to impressive performance gains but also could render feasible highly detailed simulations that were previously impossible on conventional hardware.

9. SOFTWARE FOR MD SIMULATIONS

In this section, we offer a comprehensive list of available software for running MD simulations. We list both commercial and free tools and specifically note the ones that are designed to take advantage of GPUs. We do not mention any tool in detail because we do not wish to prejudice the reader into selecting a specific tool. Our goal is to help the perspective user to correctly identify the tool that best fits his or her needs, by using the provided Table 7.1.

Table 7.1 List of available software for running MD simulations

Tool name	Availability of GPU implementation	License	Short description
Abalone	Yes	Free	Protein folding and MD simulations of biomolecules
ACEMD	Yes	Free limited version, commercial full version	Molecular dynamics using CHARMM and Amber. Optimized for CUDA
ADUN	Yes	Free (GNU GPL)	CHARMM, AMBER, and user specified force fields
AMBER	Yes	Not free	
Ascalaph Designer	Yes	Free (GNU GPL) and commercial	Molecular dynamics with GPU acceleration
CHARMM	No	Commercial	
COSMOS	No	Free (without GUI) and commercial	Hybrid QM/MM COSMOS-NMR force field
Culgi	No	Commercial	Atomistic simulations and mesoscale methods
Desmond	Yes	Free and commercial	High-performance MD with comprehensive GUI
Discovery Studio	No	Closed source/ trial version available	Comprehensive modeling and simulation suite
fold.it	No	Free	Structure prediction. Protein folding
GPIUTMD	Yes	Closed source, commercial. Demo available	General-purpose particle dynamics simulations using CUDA

Continued

Table 7.1 List of available software for running MD simulations—cont'd

Tool name	Availability of GPU implementation	License	Short description
GROMACS	Yes	Free (GNU GPLv2)	High-performance MD
GROMOS	Yes	Commercial	MD focused on biomolecules
GULP	No	Free for academic use	Molecular dynamics and lattice optimization
HOOMD-blue	Yes	Free, open source	General-purpose molecular dynamics optimized for GPUs
LAMMPS	Yes	Free, open source (GNU GPLv2)	Potentials for soft and solid-state materials and coarse-grain systems
MacroModel	No	Commercial	MD, conformational sampling, minimization. Includes comprehensive GUI
MAPS	Yes	Closed source/ trial available	Building, visualization, and analysis tools combined with access to multiple simulation engines
Materials Studio	Yes	Closed source/ trial available	Software environment for materials simulation
MedeA	No	Closed source/ trial available	Sophisticated materials property prediction, analysis, and visualization
MDynaMix	No	Free	Parallel MD
MOE	No	Commercial	Molecular operating environment
MOIL	No	Free	Includes action-based algorithms and locally enhanced sampling
ORAC	No	Free, open source	Molecular dynamics simulation program
Protein Local Optimization Program	No	Commercial	Helix, loop, and side chain optimization. Fast energy minimization
RedMD	No	Free on GNU licence	Reduced MD. Package for coarse-grained simulations

Table 7.1 List of available software for running MD simulations—cont'd

Tool name	Availability of GPU implementation	License	Short description
StruMM3D	No	Limited version free	Sophisticated 3D molecule builder and viewer with featured molecular modeling capabilites
SCIGRESS	No	Commercial	Parallel MD
TeraChem	Yes	Closed source/ trial licenses available	High-performance *ab initio* molecular dynamics, optimized for CUDA
TINKER	No	Free	Software tools for molecular design

10. FORCE FIELDS FOR MD

In the biological research, the knowledge about the behavior of the molecules is necessary. Each molecule is characterized by the individual characteristic, from the overall structure, all the interatomic bond angles, atomic radii and charge distribution, to the ability to interact with the other molecules or chemical compounds. Interactions between the molecules are characterized by several analytical functions, which are crucial for the applications of computer-based modeling. Currently, the characterization of analytical potential energy surface is one of the important steps to reveal the secrets of nucleic acid: protein interaction, ligand bindings to proteins, enzymatic reactions, characterization of interactions between molecules, or behavior of proteins in solvents. The properties between two molecules are the main problem. The main roles of force field are to reasonably describe the general properties of molecules, as a molecular geometry, conformational and stereoisometric energy, torsional barriers, torsional deformation, and energy of intermolecular interactions; to assess geometry between interacting molecules; and to evaluate the vibrational frequency and heat of formation. One of the approach to reveal all this properties is to use computer simulations or computer experiments, which providing exclusive insights into the life interaction at the molecular level. In general, there are two types of computer simulations: (i) Monte Carlo, where several molecules or ions are situated in a box and random particle is moved to different random positions (Allen & Tildesley, 1987), and (ii) MD, which are giving the information about the actual movement of molecules based on force field. Here, the computer calculated the force of each molecule and then determine the move of molecule in response to this force (Leach, 2001). Even if the general indoctrination of force field does not exist, we can say that force field is the functional combination of a mathematic formula and associated parameters resulted in analytical potential energy function—energy of the protein as a function of its atomic coordinates.

Force field represents two groups of molecular properties: (i) bonded interactions, which characterize stretching of bonds, bending of valence angels, and the rotation of dihedrals and (ii) nonbonded interactions that evaluate electrostatic data, Pauli exclusion, and dispersion. The parameters for the calculation of force field can be obtained from *ab initio*, semiempirical quantum mechanical calculations or by experimental methods such as X-ray and electron diffraction and NMR or Raman and neuron spectroscopy. Despite several protein force fields, most of them are using simple potential energy function:

(A) for bonded interactions

$$V(r) = \sum_{\text{bonds}} k_b (b - b_0)^2 + \sum_{\text{angles}} k_\theta (\theta - \theta_0)^2 + \sum_{\text{torsions}} k_{\phi[\cos(n\phi+\delta)+1]} + \sum_{\text{unbonded pairs}} \left[\frac{q_i q_j}{r_{ij}} + \frac{A_{ij}}{r_{ij}^{12}} - \frac{C_{ij}}{r_{ij}^{6}} \right]$$

(B) for nonbonded interactions

$$V(r) = \sum_{\text{nonbonded pairs}} \left(\varepsilon_{ij} \left[\left(\frac{R_{\min,ij}}{r_{ij}} \right)^{12} - 2 \times \left(\frac{R_{\min,ij}}{r_{ij}} \right)^{6} \right] + \frac{q_i q_j}{r_{ij}} \right)$$

where b is the interatomic distance, k_b and b_0 appreciate the stiffness and the equilibrium length of the bond, θ characterizes the angle formed by the two bond vectors, and the values of θ_0 and k_θ evaluate the stiffness and equilibrium geometry of the angle. The torsional potential in the equilibrium is characterized by cosine function, where ϕ is the torsional angle, δ is the phase, and n represent the dihedral potential. The last part of the equation included the electrostatic interaction, where the prefactor ε_{ij} is a parameter based on the two interacting atoms i and j, q_i and q_j described the effective charge on i and j atoms, and $R_{\min,ij}$ defines the distance at which the energy of Lennard–Jones equation is at minimum. This part also involves the bending of angles between atoms A and C (Guvench & MacKerell, 2008).

The total energy from bonded and nonbonded interaction is described as

$$E_{\text{total}} = E_{\text{bonded}} + E_{\text{nonbonded}} + E_{\text{other}}$$

where E_{other} includes the repulsive and van der Waals interactions and the Coulombic interaction.

The history of force field started in the 1960s, when the new development in the field of molecular mechanics and dynamics resulted in hunger for knowledge on deeper analysis of molecules. The first force field was developed by Aligner's group. They developed the first force field in order to study hydrocarbons, but after few years, their interest was also extended to other types of molecules, such as alcohols, sulfides, amides, or ethers. Later, the research was focused on the MD of complex system, which leads to the development of more complicated force fields. The most popular are AMBER, DREIDING, CHARMM, UFF, GROMS, OPLS, and COMPASS, which differ in several properties (Table 7.2). For example, DREIDING and UFF contain parameters of all atoms of the periodic table; CHARMM, AMBER, and GROMS are used mostly by the simulations of biomolecules; and the condensed matter is detected by OPLS and COMPASS. In general, there are two groups of force fields based on their energy expression: first-generation (class I force fields; AMBER, GROMS, CHARMM, and OPLS) and second-generation (class II force fields; COMPASS, UFF, MM2, MM3, MM4, consistent force field (CFF), and Merck molecular force field (MMFF)). The development of the force field is still continues, and still new versions are available.

10.1. AMBER

The AMBER (Assisted Model Building with Energy Refinement) includes several classical molecular mechanics force fields, which describes the structural and dynamic properties of biomolecules, canonical and noncanonical nucleic acids, and proteins in water, as well as complex conformational changes such as the A→B transition in duplex and triplex DNA and simulation of DNA in extremely conditions (Cheatham & Kollman, 1996; Miller & Kollman, 1997; Rueda, Kalko, Luque, & Orozco, 2003). The new versions allow also the study of polymers and small molecules; however, in general, there are the limitations of parameter that reduce the application

Table 7.2 List of the force field in molecular dynamics

Force field	Type of force field	Target molecules	Properties	Limitation	References
Amber94	All-atom	Proteins and nucleic acids	Partial charges are dictionary based; compatible with RESP and AM1-BCC charges. Polar hydrogens have zero van der Waals radii	Small organic molecules	Pearlman et al. (1995)
CHARMM22	All-atom	Proteins and heme group	Partial charges are based on bond-charge increments that reproduce the original dictionary charges	Small organic molecules or DNA	MacKerell et al. (1998)
DREIDING	All-atoms	Proteins and small molecules	General force and geometry parameters are based on simple hybridization, all bonds distances are derived from atomic radii		Mayo, Olafson, & Goddard (1990)
Engh–Huber	United atom	Crystallographic refinement of proteins	Explicit hydrogens are required for polar atoms (N and O). Partial charges are based on bond-charge increments that reproduce the original dictionary charges	Small organic molecules nor DNA	Engh & Huber (1991)

GAFF	All-atom	Small molecules, proteins, and nucleic acid	Compatible to the AMBER force field, including parameters for almost all the organic molecules made of C, N, O, H, S, P, F, Cl, Br, and I	Nonorganic molecules	Wang, Wolf, Caldwell, Kollman, & Case (2004)
GROMOS	United atoms	Nucleic acids, proteins, lipids, carbohydrates, and drug-like organic compounds	Force field parameters are set for biomolecules in aqueous, apolar, and gas phase; van der Waals parameters are derived from calculation of the crystal structures of hydrocarbons and calculations on amino acids using nonbonded cutoff radii		Jorgensen, Maxwell, & Tirado-Rives (1996)
MMFF94	All-atom	Small organic molecules	Partial charges are based on bond-charge increments. Suitable for use with generalized Born solvation models. Conjugated nitrogens are less tetrahedral than with MMFF94		Halgren (1999)
OPLS-AA	All-atom	Proteins and small organic molecules	Partial charges are based on bond-charge increments that reproduce the original dictionary charges. Polar hydrogens have zero van der Waals radii		Kony, Damm, Stoll, & Van Gunsteren (2002)

Continued

Table 7.2 List of the force field in molecular dynamics—cont'd

Force field	Type of force field	Target molecules	Properties	Limitation	References
PEF95SAC	All-atom	Carbohydrate	Partial charges are based on bond-charge increments that reproduce the original dictionary charges		Fabricius, Engelsen, & Rasmussen (1997)
UNIVERSAL (UFF)	All-atom	Proteins, nucleic acids	Contain parameter for all atoms in the periodic table		Rappé, Casewit, Colwell, Goddard Iii, & Skiff (1992)
SYBYL	All-atom	Nucleic acids	Force field for the calculation of internal geometries and conformational energies. Does not contain electrostatic term	Study of condensed-phase properties	Clark, Cramer, & Van Opdenbosch (1989)

of AMBER force fields within the small organic molecules. AMBER force field use the Leach approach, where C atom at the junction between a six- and a five-membered ring is assigned an atom type that differ from the C atom in an isolated five-membered ring such as histidine, which in turn differ from the atom type of a carbon atom in a benzene ring (Leach, 2001). The most significant difference between AMBER and other force fields is in employing the united atom representation, which means that an all-atom representation in nonpolar hydrogen atoms is not represented explicitly but is amalgamated into the description of the heavy atoms to which they are bonded. Furthermore, AMBER includes a hydrogen-bond term that increases the value of the hydrogen-bond energy derived from dipole–dipole interaction of the donor and acceptor groups. Second differentiation is the using of general torsion parameters, which depend exclusively upon the atom types of the two atoms that include the central bond, not the atom types of the terminal atoms as in the other force fields. The torsion terms are improper to maintain the stereochemistry at chiral centers (Leach, 2001).

The equation of the AMBER force field is very similar to the equation of the general force field and includes all properties, bonds, angles, torsions, electrostatic interactions, and van der Waals:

$$V = \sum_{\text{bonds}} K_2(b - b_0)^2 + \sum_{\text{angles}} k_\theta\left(\theta - \theta_{\text{eq}}\right)^2 + \sum_{\text{torsion}} \frac{V_n}{2}(1 + \cos[n\phi - \gamma]) + \sum_{\text{nonbonded}} \frac{A_{ij}}{R_{ij}^{12}} - \frac{B_{ij}}{R_{ij}^{6}} + \sum_{\text{nonbonded}} \frac{q_i q_j}{\varepsilon R_{ij}}$$

The most used versions of AMBER force fields are ff94 (Cornell et al., 1995), ff99SB (Hornak et al., 2006), ff03 (Duan et al., 2003), general AMBER force field (GAFF) (Wang et al., 2004), and others.

The AMBER94 is a simple force field based on the Weiner force field that is used for the study of organic molecules, peptides, and nucleic acids. The force field is parameterized with simple compounds such as ethane, but for special parameters, more complex particles should be added. AMBER94 is also known as "helix-friendly." The extended version of ABMER94 is AMBER96 where the torsional model is improved, which leads to better agreement between molecular mechanical and quantum mechanical energetics. AMBER99 is the third generation of AMBER, which provides the parameters for amino acids and nucleic acids. Also, a new optimization program *parmscan* is used, which optimizes the exploration of parameter space in a rational manner. AMBER-03 provides the new quantum mechanical calculations for the solvent model, not for the gas phase as previous versions do, and also each amino acid is allowed unique main-chain charges (Cornell et al., 1995; Duan et al., 2003).

All previously mentioned force fields are focused mostly on the amino acids and the nucleic acids. However, carbohydrates, oligosaccharides, and polysaccharides play important function in the interaction in the host–pathogen interactions, cell–cell recognition, or posttranslation modification (Bencurova, Mlynarcik, & Bhide, 2011; Delacroix, Bonnet, Courty, Postel, & Van Nhien, 2013; Muramatsu, 2000). For the determination of the structural properties of carbohydrates and lipids, GLYCAM force fields were designed. The first GLYCAM_93 version has been evaluated against quantum mechanical calculation data and chemometric analysis from 20 second-generation carbohydrate force fields (Pérez et al., 1998; Rudd et al., 1995). The second version, GLYCAM2000, was also able to characterize the experimental solution data for disaccharides, oligosaccharides, and oligosaccharide–protein complexes (Kirschner & Woods, 2001), while the current version, GLYCAM06, is suitable for all biomolecules without some limitations (Kirschner et al., 2008; Tessier, Demarco, Yongye, & Woods, 2008). The list of components including GLYCAM is depicted in Table 7.3. GLYCAM employs the AMBER force field with PARM94 parameter set for van der Waals parameter. The B3LYP/6-31++G(2d,2p)//HF/6-31G* level was selected as a reference for valence quantum calculation and to derive quantum mechanical molecular electrostatic potential. ESP fitting is used to assign the partial atomic charge of each atom in the molecule. In the current version, GLYCOM06, 1-4 electrostatic or nonbonded scaling factors are removed, which balanced these interactions leading to an ability to predict rotamer populations of carbohydrates. Moreover, GLYCAM utilizes a common set of terms of α- and β-carbohydrate anomers, which provide a means of predicting the relative energies of interconverting ring forms (Kirschner et al., 2008). Gather from the features of GLYCAM force field, we can mention it as a valuable tool for the structural study of carbohydrates in live organisms, post-translation modification of proteins, protein–carbohydrate interaction, and the protein–membrane complexes.

GAFF is an extension of AMBER force field that also allows the characterization of pharmaceutical molecules. The main advantage of GAFF is its atomic spectrum consisting of 33 basic atom types and 22 special atom types, which allow to characterize also organic molecules composed of C, N, O,

Table 7.3 List of compounds included in MMFF94 and GLYCAM06 force field core parameterizations

Force field	Compounds	References
MMFF94	Alkenes (conjugated, nonconjugated), alkanes aromatic hydrocarbons, five- and six-membered heteroaromatic, alcohols (vinylic, allylic), phenols, ethers (vinylic), 1,2-diols, amines (allylic, hydroxyl, N-oxides), imines, amidines, guanidines, 1,2-diamines, amides, dipeptides, ureas, imides, aldehydes and ketones (α,β-unsaturated, allylic), ketals, acetals, hemiketals, hemiacetals, carboxylic acids and esters (α,β-unsaturated, vinylic), dicarboxylic acids, β-ketoacids, β-hydroxyesters, thiols, sulfides, disulfides, 1,2-dithiols, halides, pyridine, imidazole cations, carboxylate anions, α,β-unsaturated-carboxylate anions	Halgren (1996)
GLYCAM06	Hydrocarbons, alcohols, ethers, amides, esters, carboxylates, molecules with simple ring system, cyclic carbohydrates, lipids, lipid bilayers, phospholipids, glycolipids, triacylglycerols, sphingolipids (cerebrosides, gangliosides, sphingomyelin), steroids (cholesterol), glycerophospholipids (choline, ethanolamines, glycerols, inositols)	Kirschner et al. (2008) and Tessier et al. (2008)

H, S, P, F, Cl, Br, and I. The atomic spectrum cover large chemical space and all type of the bond length and angles parameters, also torsional angle parameters are calculated with empiric rules. Parameters for all combinations of atoms type are determined algorithmically for each input molecule based on topology of the bond and its geometry (Wang et al., 2004). Parameterization is based on Parm99 approach for the nonbonded pairs, and van der Waals parameter used restrained electrostatic potential fit model at HF/6-31G*, which was chosen for its physical picture and direct implementation scheme (Cornell, Cieplak, Bayly, & Kollmann, 1993; Junmei Wang, Cieplak, & Kollman, 2000). In summary, GAFF is a useful molecular mechanical tool parameterized for the organic molecules, mainly for the drug designing and protein–ligand and DNA–ligand simulations.

10.2. CHARMM force field

Chemistry at HARvard Macromolecular Mechanics (CHARMM) force field is used to perform the calculation of interaction and conformation energies, local minima, geometry of the particles, time-dependent dynamic behavior, barriers to rotation, vibrational frequencies, and free energy of the wide range or particles, from the small molecules to large biological macromolecules, including lipids and carbohydrates. CHARMM uses two approaches for the molecular simulation and modeling, (i) classical energy functions, including empirical and semiempirical functions, and (ii) quantum mechanical energy functions, which are semiempirical or *ab initio*. Similar to AMBER, there are several versions of CHARMM. The first version was published in 1983 (Brooks, Bruccoleri, Olafson, Swaminathan, & Karplus, 1983), but during the years, the new platforms were developed. The first version of CHARMM force field included static structure, energy comparisons, time series, correlation functions of molecular dynamic trajectories, and statistical data. In 1985, the new version of CHARMM19 was developed with improved parameters, where hydrogen atoms bonded to nitrogen and oxygen were explicitly represented, while hydrogen atoms bonded to carbon or sulfur were treated as part of extended atoms. CHARMM19 was developed on gas-phase simulations; however, it is often used in conjugation with a distance-dependent dielectric constant as a rough continuum solvation model (MacKerell et al., 1998).

CHARMM belongs to class II of force fields and also includes MMFF and consisted force field. The parameterization of CHARMM allows to study the organic compounds for the pharmaceutical interest and also to use alternative methods for the treatment of the nonbonded interactions. These methods act to smooth the transition in the energy and force at the cutoff distance to reduce the errors in studied region and can be used in electrostatic and Lennard–Jones interactions in the energy shifting and switching (Brooks et al., 1983). The importance of the force shift/switch methods is in the determination of the trajectories of the atoms and mostly for the highly charged biomolecules.

CHARMM force field includes the internal parameters, such as length of the bond, bond angle, Urey–Bradley, torsions, and improper torsional term, which reproduce the crystal structures, infrared and Raman spectroscopic data, and *ab initio* calculations, as well as interaction parameters including electrostatic and van der Waals terms. Those are used to fit HF/6-31G* *ab initio* energy of interaction and geometry for the water molecules that are bonded to polar sites of the model particles. This involves a series of supermolecular calculations of the model compound, such as formamide or *N*-methylacetamide and a single water molecule at each of several interaction sites.

Moreover, CHARMM also includes experimental condensed-phase properties or heats of vaporization and molecular volumes, which gives the satisfied clarification of the behavior of biomolecules in extreme conditions (MacKerell et al., 1998).

CHARMM22 uses a flexible and wide-ranging energy functional form:

$$V = \sum_{\text{bonds}} k_b (b - b_0)^2 + \sum_{\text{angles}} k_\theta (\theta - \theta_0)^2 + \sum_{\text{torsions}} k_\phi \left[1 + \cos\left(n_\phi - \delta\right)\right]$$
$$+ \sum_{\text{impropers}} k_\omega (\omega - \omega_0)^2 + \sum_{\text{Urey-Bradley}} k_\mu (\mu - \mu_0)^2$$
$$+ \sum_{\text{nonbonded}} \in \left(\left[\left(\frac{R_{\min,ij}}{r_{ij}} \right)^{12} - 2 \times \left(\frac{R_{\min,ij}}{r_{ij}} \right)^{6} \right] + \frac{q_i q_j}{s r_{ij}} \right)$$

where k_b is the constant value for bond force, $b - b_0$ is the l form equilibrium that the atom has moved, k_θ is the constant for angle force, the value of $\theta - \theta_0$ specifies the angle from equilibrium between three bonded atoms, k_ϕ is the constant for torsion force, n is the multiplicity of the function, ϕ is the torsion angle, and δ is the phase shift. The fourth term of the equation characterizes the improper that is out of plane bending. Here, k_ω is the force constant and $\omega - \omega_0$ define the out of plane angle. In the Urey–Bradley part (cross-term accounting for angle bending using 1,3-nonbonded interactions), k_μ is the respective force constant and $\mu - \mu_0$ is the distance between the 1,3-atoms in the harmonic potential. In the last part, the term of nonbonded interactions between pairs of atoms was described in the preceding text. For the deep characterization of intermolecular and intramolecular energetics, the dielectric constant ϵ is set as equal to unity in this potential energy function, mainly for the peptide groups (MacKerell et al., 1998). As a water model, the modified TIP3P model is used and nonbonded interaction terms include all atoms separated by three or more covalent bonds. In CHARMM, there is not used general scaling of the electrostatic or Lennard–Jones interactions for the atoms, which are separated by three bonds. Also, no explicit hydrogen-bond terms are included, according to Coulomb and Lennard–Jones terms. For the vibrational calculations, *ab initio* HF/6-31G(d) values are used, which are further analyzed by

MOLVIB program (MacKerell et al., 1998). Taken together, TIP3P model and HF/6-31G provide a balanced interaction between solute–water and water–water energies.

10.3. Consistent force field

The CFF family was developed in 1994 by Halgren and Maple. These force fields have cross-term improvements and are derived from *ab initio* methods. In overall, the CFF was the first force field that was using exclusively the data from *ab initio* quantum mechanical calculations, which were performed on structures slanted from equilibrium to the expected calculations on structures at equilibrium. CFF simulates the structural and dynamical properties of peptides and proteins using quartic polynomials for bond stretching and angle bending. For the definition or torsions, the CFF is using three-term Fourier expansion and the van der Waals interactions are denoted by inverse 9th-power term for repulsive behavior (Maple et al., 1994).

10.4. DREIDING force field

DREIDING is a force field using general force constants and geometry parameters based on simple hybridization. The distances between the bonds are derived from atomic radii. The parameters are not generated automatically, but the parameters are derived by a rule-based approach. DREIDING uses only one force constant for each bond, angle, and inversion and six values for torsional barrier. All possible combinations of atoms are available, and also, the new atoms can be added easily. Atom types in the DREIDING force field are denoted by naming several characters, the elemental symbol (e.g., C for carbon), and hybridization state (1 = linear, sp^1; 2 = trigonal, sp^2; 3 = tetrahedral, sp^3; and R = an sp^2 atom involved in resonance), and the number of implicit hydrogen atoms and the last character define other characteristics, for example, H_A denotes a hydrogen atom that is capable of forming a hydrogen bond. The main interest is on the atoms of the columns B, C, N, O, and F of the periodic table, and other elements—Na, Ca, Zn, and Fe—were added for the successful simulation of biological system. DREIDING force field also allows rapid reflection of the new structures and complexes and also renewal of the predicted structures for a particular system, and both allow prediction of structures and dynamics of organic, biological, and inorganic molecules, as well as prediction of novel combination of elements (Mayo et al., 1990). DREIDING is available in two versions, the DREIDING force field and DREIDING II force field.

10.5. GROMOS

GROningen Molecular Simulation (GROMOS) force field was designed for broad range of biomolecules including proteins, nucleic acids, lipids, carbohydrates in various phases. Currently, a few forms of GROMOS are available. The first version, GROMOS96, used 43A1 parameter set, which means that this approach used 1 parameter set with 43 atom types in condensed phase. Later versions were designed with 53A5 and 53A6 parameters. The latest version is also able to, very trustfully, reproduce the interactions between different media, which is widely used, for example, in membrane protein simulations. The data for GROMOS force field are obtained not from the *ab initio* but from the crystallographic data and atomic polarizabilities. The advantage of the latest version is that it brings the importance on reproduction of solvation-free enthalpies of amino acid analogs in cyclohexane, which is usually problematic in other types of classical force fields (Oostenbrink, Villa, Mark, & Van Gunsteren, 2004) (Scott et al., 1999).

10.6. Merck molecular force field

MMFF was published for the first time in 1996 as MMFF94 and depends fully on the computational data. MMFF94 uses the high quality of computational data and broader range for the study of the molecular properties of nucleic acids and proteins, mainly in the ligand–receptor relationship in the bounded and isolated form. The strong emphasis is placed on the molecular geometry, conformational energies, and intermolecular energies that are necessary to preserve the right conformation of the ligand or receptor upon binding. Moreover, the wrong conformer and energy calculation can lead to the wrong estimate of the energy cost required for the ligand–receptor interaction. The approach of MMFF94 does not impose the significant importance on the vibrational frequencies or spectroscopic precision. The authors suggest that it is unlikely to affect the differential binding energy of macromolecules only by splitting of high-frequency models for the angle bending or bond stretching (Halgren, 1996).

Based on the application fields of MMFF—study of biomolecules, pharmaceutical compounds, and chemical mixtures, the requirements on the data used in the parameterization of MMFF are relatively tricky. First, force field must include all the organic structural types within in the Merck Index and Fine Chemical Directory (Blake, 2004). The list of components included in MMFF94 is in Table 7.3. Second, the data cannot be obtained from the experiments like in AMBER or CHARMM because of the complicated calculation of conformational energies and energies of interacting molecules and also because of time-dependent mining of those data. Thus, MMFF uses *ab initio* data, although the experimental data are also using to confirm their accuracy.

The pattern of MMFF94 is based on HF/6-31G* calculations to characterize the quantum mechanical energy surface used to derive the quantum mechanical force field, prototype of CFF93. MMFF94 employs the program PROBE, which derives force constants for bond stretching and angle bending from data on the curvature on the quantum mechanical surface contained in the HF/6-30G* second derivatives (Cheng, Best, Merz, & Reynolds, 2000; Halgren, 1996). At the first step, the van der Waals and electrostatic representation, enhanced by 10%, are chosen from the scaled HF/6-31G*40 molecular dipole moments. The second step is to adjust the key parameters of van der Waals and electrostatic interactions obtained from HF/6-31G calculations, and in the last step, torsion terms are derived to fir the *ab initio* gas-phase conformational data (Halgren, 1996). This approach allows proper balance between solvents, solutes, and solvent–solute interactions, which is crucial for accurately describing energetics of ligand–receptor binding in aqueous solution.

MMFF94 is intended for the studies of MD, and the current versions are suitable also for energy-minimization studies. This version offers unique functional for describing the van der Waals interaction and employs the new rules that combine a systematic correlation of van der Waals parameters with experimental data describing the interaction between small molecules and rare-gases atoms. Moreover, the core parameterization includes the *ab initio* calculations of approximately 3000 molecular structures evaluated with different levels and also provides well-defined parameters for more than 20 chemical families, which makes it a suitable tool for studying the molecular interaction in biological systems.

10.7. MM2/MM3 force fields

The group of MM force fields is used for solving the structural and dynamic properties of small molecules. They are typical with special characterization of the eight types of carbon atom: sp^3, sp^2, sp, carbonyl, cyclopropane, radical cyclopropene, and carbonium ion. The parameterization of MM2/MM3 fits to values, which were obtained from the electron diffraction, which affords mean distances between atoms averaged over vibrational movement (Leach, 2001). The potential of the bond-stretching potential is represented by the Hooke law to approximate the Morse curve, but in the MM2, the Hooke law is expanded only by quadratic and cube term, which arranges extremely long bond in some of the experiments. However, MM3 uses the cubic contribution only in necessary case, which eliminates the enormous length of the bonds. Furthermore, in contrast of other force field, MM2/MM3 assigns dipoles to the bonds in the molecules; thus, the electrostatic energy is given by a sum of dipole–dipole interaction energies, which are useful for the molecules with formal charge and which require charge–charge and charge–dipole terms for the energy expression (Allinger, 1977; Lii & Allinger, 1991; Lii et al., 1989).

11. CURRENT LIMITATIONS OF MD

It would be unfair to complete these basic notes about MD without discussing their limitations. The issue of time scales is the most obvious limitation, with scales ranging from a few picoseconds up to a few nanoseconds at most, depending on the level at which the system is modeled (first principles or empirical force field). Straightforward MD can get you nowhere in cases like slow-diffusion problems in solids, dynamics of glassy or polymer systems, or protein folding because many processes of chemical and physical interest happen over time scales, which can be many orders of magnitude larger than this. Hyperdynamics (Voter, 1997), temperature-accelerated dynamics, and the parallel-replica method are some techniques of Voter and others that try to address this problem (Sørensen & Voter, 2000; Voter, 1998).

Calculating the energy and necessary forces for performing MD simulations is one of the causes that the time scale that can be covered is limited. It has been reported that custom designed neural networks prediction capability for the energetics and forces of a given system can be trained by feeding the neural network with a enough data that have been obtained from simulations (Behler & Parrinello, 2007). With this method, we reduce the cost of performing accurate simulations, with result to extend the length of time scales accessible. Even so, this technique is still quite new and its abilities must be proved.

MD is similar sampling technique with MC, but it also provides dynamical information, with extra cost of forces' calculation. We must be careful with the cases of systems with complicated energy landscapes, as it is more problematic to sample them adequately and it also takes too long for the exploration of configuration space by the dynamics. An improvement of this problem is proposed by Parrinello and coworkers, a method known as metadynamics (Laio, Bernard, Chiarotti, Scandolo, & Tosatti, 2000). In metadynamics, a simulation trajectory is constructed to eliminate the chance of revisiting the same part of the 3D conformational space twice. So, we gradually fill up the potential energy minima with facilitation of the system's escape from such trapping funnels and improvement of the configuration space sampling (Martonák et al., 2005).

The issue of varying length scales is another challenge for MD, where in many systems it is covered by the phenomena under observation. The most typical example of that is crack propagation. In crack propagation some material is stressed to its breaking point when a crack tip will form. Near this crack tip, chemical bonds are broken, and atoms are strongly rearranged. Away from this crack tip, the material can be deformed, without the break of bonds and even more away from tip, the atomic position deviate from a perfect crystal's positions. To model systems for this, we need large simulation cells, which can grow into more than six orders of magnitude figures. To deal with these problems effectively, different length scales at different levels of theory must be treated. For example, in a very small region surrounding the crack tip, a quantum mechanical algorithm will be applied rather than the universal Newton force field that is being applied to the full system. Then, for large-length scales, we match this at some point with a continuum mechanics description. Similar is the case of enzymatic reactions.

12. CONCLUSIONS

Summing up, MD is a really useful tool in the hands of the modern scientist computational expert. A lot of challenges are lying ahead in the field of biomolecular simulation in general and MD in particular. These are new, interesting, upgradeable, and scalable fields that bear a lot of promise for the future of computational basic research. They respond to the new issues that are being raised all the time, where, with the recent help of quantum physics and nanotechnology and late breakthroughs in the field of structural biology, they strive toward the deepening of our knowledge in the fields of physics, biochemistry, and medicine. There is a great hope for the further understanding of numerous unresolved mechanisms in nature and the development of novel strategies, discovery of new drugs, and cures for many lethal diseases, such as viral infections, autoimmunity, and even cancer.

ACKNOWLEDGMENTS

This study was supported by the European Union (European Social Fund—ESF), received Greek national funds through the Operational Program "Education and Lifelong Learning" of the National Strategic Reference Framework (NSRF)—Research Funding Program: Thales, the Investing in Knowledge Society through the European Social Fund, COST-Action FA1002, and ITMS26220220185.

REFERENCES

Alder, B., & Wainwright, T. (1957). Phase transition for a hard sphere system. *Journal of Chemical Physics*, *27*(5), 1208–1209.

Allen, M. P., & Tildesley, D. J. (1987). *Computer simulation of liquids*. Oxford, England, New York: Clarendon Press, Oxford University Press.

Allinger, N. L. (1977). Conformational analysis. 130. MM2. A hydrocarbon force field utilizing V1 and V2 torsional terms. *Journal of the American Chemical Society*, *99*(25), 8127–8134.

Barnes, J., & Hut, P. (1986). A hierarchical *O* (*N* log *N*) force-calculation algorithm. *Nature 324*(4), 446–449. http://dx.doi.org/10.1038/324446a0.

Barney, B. (2012). *Introduction to parallel computing*. California: Lawrence Livermore National Laboratory.

Beeman, D. (1976). Some multistep methods for use in molecular dynamics calculations. *Journal of Computational Physics*, *20*(2), 130–139.

Behler, J., & Parrinello, M. (2007). Generalized neural-network representation of high-dimensional potential-energy surfaces. *Physical Review Letters*, *98*(14), 146401.

Bencurova, E., Mlynarcik, P., & Bhide, M. (2011). An insight into the ligand-receptor interactions involved in the translocation of pathogens across blood-brain barrier. *FEMS Immunology and Medical Microbiology*, *63*(3), 297–318. http://dx.doi.org/10.1111/j.1574-695X.2011.00867.x.

Bizzarri, A. R., & Cannistraro, S. (2002). Molecular dynamics of water at the protein-solvent interface. *The Journal of Physical Chemistry. B*, *106*(26), 6617–6633.

Blake, M. R. (2004). The Merck Index. *Library Journal*, *129*(6), 131–132.

Brooks, B. R., Bruccoleri, R. E., Olafson, B. D., Swaminathan, S., & Karplus, M. (1983). CHARMM: A program for macromolecular energy, minimization, and dynamics calculations. *Journal of Computational Chemistry*, *4*(2), 187–217.

Brooks, C. L., & Karplus, M. (1989). Solvent effects on protein motion and protein effects on solvent motion: Dynamics of the active site region of lysozyme. *Journal of Molecular Biology*, *208*(1), 159–181.

Cheatham, T. E., 3rd., & Kollman, P. A. (1996). Observation of the A-DNA to B-DNA transition during unrestrained molecular dynamics in aqueous solution. *Journal of Molecular Biology*, *259*(3), 434–444.

Cheng, A., Best, S. A., Merz, K. M., Jr., & Reynolds, C. H. (2000). GB/SA water model for the Merck molecular force field (MMFF). *Journal of Molecular Graphics and Modelling*, *18*(3), 273–282.

Cheng, X., & Ivanov, I. (2012). Molecular dynamics. *Methods in Molecular Biology*, *929*, 243–285. http://dx.doi.org/10.1007/978-1-62703-050-2_11.

Clark, M., Cramer, R. D., & Van Opdenbosch, N. (1989). Validation of the general purpose Tripos 5.2 force field. *Journal of Computational Chemistry*, *10*(8), 982–1012.

Cornell, W. D., Cieplak, P., Bayly, C. I., Gould, I. R., Merz, K. M., Ferguson, D. M., et al. (1995). A second generation force field for the simulation of proteins, nucleic acids, and organic molecules. *Journal of the American Chemical Society*, *117*(19), 5179–5197.

Cornell, W. D., Cieplak, P., Bayly, C. I., & Kollmann, P. A. (1993). Application of RESP charges to calculate conformational energies, hydrogen bond energies, and free energies of solvation. *Journal of the American Chemical Society, 115*(21), 9620–9631.

Darden, T., Perera, L., Li, L., & Pedersen, L. (1999). New tricks for modelers from the crystallography toolkit: The particle mesh Ewald algorithm and its use in nucleic acid simulations. *Structure,* 7(3), R55–R60.

Davidchack, R. L. (2010). Discretization errors in molecular dynamics simulations with deterministic and stochastic thermostats. *Journal of Computational Physics, 229*(24), 9323–9346.

Delacroix, S., Bonnet, J. P., Courty, M., Postel, D., & Van Nhien, A. N. (2013). Glycosylation mediated-BAIL in aqueous solution. *Carbohydrate Research, 381,* 12–18. http://dx.doi.org/10.1016/j.carres.2013.08.009.

DeLisi, C. (1988). Computers in molecular biology: Current applications and emerging trends. *Science, 240*(4848), 47–52.

Ding, J., Das, K., Moereels, H., Koymans, L., Andries, K., Janssen, P. A., et al. (1995). Structure of HIV-1 RT/TIBO R 86183 complex reveals similarity in the binding of diverse nonnucleoside inhibitors. *Nature Structural and Molecular Biology, 2*(5), 407–415.

Duan, Y., Wu, C., Chowdhury, S., Lee, M. C., Xiong, G., Zhang, W., et al. (2003). A point-charge force field for molecular mechanics simulations of proteins based on condensed-phase quantum mechanical calculations. *Journal of Computational Chemistry, 24*(16), 1999–2012. http://dx.doi.org/10.1002/jcc.10349.

Durrant, J. D., & McCammon, J. A. (2011). Molecular dynamics simulations and drug discovery. *BMC Biology, 9,* 71. http://dx.doi.org/10.1186/1741-7007-9-71.

Elber, R. (1996). Novel methods for molecular dynamics simulations. *Current Opinion in Structural Biology, 6*(2), 232–235.

Engh, R. A., & Huber, R. (1991). Accurate bond and angle parameters for X-ray protein structure refinement. *Acta Crystallographica Section A, 47,* 392–400.

Eriksson, M. A., Pitera, J., & Kollman, P. A. (1999). Prediction of the binding free energies of new TIBO-like HIV-1 reverse transcriptase inhibitors using a combination of PROFEC, PB/SA, CMC/MD, and free energy calculations. *Journal of Medicinal Chemistry, 42*(5), 868–881.

Fabricius, J., Engelsen, S. B., & Rasmussen, K. (1997). The Consistent Force Field. 5. PEF95SAC: Optimized potential energy function for alcohols and carbohydrates. *Journal of Carbohydrate Chemistry, 16,* 751–772.

Fraternali, F., & Van Gunsteren, W. (1996). An efficient mean solvation force model for use in molecular dynamics simulations of proteins in aqueous solution. *Journal of Molecular Biology, 256*(5), 939–948.

Gottlieb, A. A. (1994). *Highly parallel computing.* Boston, MA: Addison-Wesley Longman.

Gustafson, J. L. (1988). Reevaluating Amdahl's law. *Communications of the ACM, 31*(5), 532–533.

Guvench, O., & MacKerell Jr., A. D. (2008). Comparison of protein force fields for molecular dynamics simulations. *Molecular modeling of proteins* (pp. 63–88). Totowa, New Jersey: Humana Press.

Hairer, E., Lubich, C., & Wanner, G. (2003). Geometric numerical integration illustrated by the Stormer–Verlet method. *Acta Numerica, 12*(12), 399–450.

Halgren, T. A. (1996). Merck molecular force field. I. Basis, form, scope, parameterization, and performance of MMFF94. *Journal of Computational Chemistry, 17*(5–6), 490–519. http://dx.doi.org/10.1002/(SICI)1096-987X(199604)17:5/6<490::AID-JCC1>3.0.CO;2-P.

Hannon, L., Lie, G., & Clementi, E. (1986). Molecular dynamics simulation of flow past a plate. *Journal of Scientific Computing, 1*(2), 145–150.

Hansson, T., Oostenbrink, C., & van Gunsteren, W. (2002). Molecular dynamics simulations. *Current Opinion in Structural Biology, 12*(2), 190–196.

Hockney, R. W., & Eastwood, J. W. (1988). Particle–particle–particle-mesh (P3M) algorithms. *Computer simulation using particles* (pp. 267–304). Boca Raton, Florida: CRC Press.

Hornak, V., Abel, R., Okur, A., Strockbine, B., Roitberg, A., & Simmerling, C. (2006). Comparison of multiple Amber force fields and development of improved protein backbone parameters. *Proteins, 65*(3), 712–725. http://dx.doi.org/10.1002/prot.21123.

Hug, S. (2013). Classical molecular dynamics in a nutshell. *Methods in Molecular Biology, 924*, 127–152. http://dx.doi.org/10.1007/978-1-62703-017-5_6.

Jorgensen, W. L., Chandrasekhar, J., Madura, J. D., Impey, R. W., & Klein, M. L. (1983). Comparison of simple potential functions for simulating liquid water. *Journal of Chemical Physics, 79*, 926.

Jorgensen, W. L., Maxwell, D. S., & Tirado-Rives, J. (1996). Development and testing of the OPLS all-atom force field on conformational energetics and properties of organic liquids. *Journal of the American Chemical Society, 118*, 11225–11236.

Karplus, M., & McCammon, J. A. (1986). The dynamics of proteins. *Scientific American, 254*(4), 42–51.

Karplus, M., & McCammon, J. A. (2002). Molecular dynamics simulations of biomolecules. *Nature Structural Biology, 9*(9), 646–652. http://dx.doi.org/10.1038/nsb0902-646.

Kirschner, K. N., & Woods, R. J. (2001). Solvent interactions determine carbohydrate conformation. *Proceedings of the National Academy of Sciences of the United States of America, 98*(19), 10541–10545.

Kirschner, K. N., Yongye, A. B., Tschampel, S. M., Gonzalez-Outeirino, J., Daniels, C. R., Foley, B. L., et al. (2008). GLYCAM06: A generalizable biomolecular force field. Carbohydrates. *Journal of Computational Chemistry, 29*(4), 622–655. http://dx.doi.org/10.1002/jcc.20820.

Klepeis, J. L., Lindorff-Larsen, K., Dror, R. O., & Shaw, D. E. (2009). Long-timescale molecular dynamics simulations of protein structure and function. *Current Opinion in Structural Biology, 19*(2), 120–127. http://dx.doi.org/10.1016/j.sbi.2009.03.004.

Kony, D., Damm, W., Stoll, S., & Van Gunsteren, W. F. (2002). An improved OPLS-AA force field for carbohydrates. *Journal of Computational Chemistry, 23*(15), 1416–1429. http://dx.doi.org/10.1002/jcc.10139.

Laio, A., Bernard, S., Chiarotti, G., Scandolo, S., & Tosatti, E. (2000). Physics of iron at Earth's core conditions. *Science, 287*(5455), 1027–1030.

Leach, A. R. (2001). *Molecular modelling: Principles and applications* (2nd ed.). Harlow, Essex, England: Prentice Hall.

Lii, J. H., & Allinger, N. L. (1991). The MM3 force field for amides, polypeptides and proteins. *Journal of Computational Chemistry, 12*(2), 186–199.

Lii, J. H., Gallion, S., Bender, C., Wikström, H., Allinger, N. L., Flurchick, K. M., et al. (1989). Molecular mechanics (MM2) calculations on peptides and on the protein Crambin using the CYBER 205. *Journal of Computational Chemistry, 10*(4), 503–513.

MacKerell, A. D., Jr., Bashford, D., Bellott, M., Dunbrack, R. L., Jr., Evanseck, J. D., Field, M. J., et al. (1998). All-atom empirical potential for molecular modeling and dynamics studies of proteins. *The Journal of Physical Chemistry B, 102*(18), 3586–3616.

Maple, J., Hwang, M. J., Stockfisch, T. P., Dinur, U., Waldman, M., Ewig, C. S., et al. (1994). Derivation of class II force fields. I. Methodology and quantum force field for the alkyl functional group and alkane molecules. *Journal of Computational Chemistry, 15*(2), 162–182.

Martoňák, R., Laio, A., Bernasconi, M., Ceriani, C., Raiteri, P., & Parrinello, M. (2005). Simulation of structural phase transitions by metadynamics. *Zeitschrift für Kristallographie, 220*, 489, arXiv, preprint cond-mat/0411559.

Mattson, W., & Rice, B. M. (1999). Near-neighbor calculations using a modified cell-linked list method. *Computer Physics Communications, 119*(2), 135–148.

Mayo, S. L., Olafson, B. D., & Goddard, W. A. (1990). DREIDING: A generic force field for molecular simulations. *Journal of Physical Chemistry*, *94*(26), 8897–8909.

Miller, J. L., & Kollman, P. A. (1997). Observation of an A-DNA to B-DNA transition in a nonhelical nucleic acid hairpin molecule using molecular dynamics. *Biophysical Journal*, *73*(5), 2702–2710. http://dx.doi.org/10.1016/S0006-3495(97)78298-5.

Monticelli, L., & Tieleman, D. P. (2013). Force fields for classical molecular dynamics. *Methods in Molecular Biology*, *924*, 197–213. http://dx.doi.org/10.1007/978-1-62703-017-5_8.

Muramatsu, T. (2000). Protein-bound carbohydrates on cell-surface as targets of recognition: An odyssey in understanding them. *Glycoconjugate Journal*, *17*(7–9), 577–595.

NVIDIA (July 19, 2013). CUDA C Programming Guide. Retrieved November 25, 2013.

Oostenbrink, C., Villa, A., Mark, A. E., & Van Gunsteren, W. F. (2004). A biomolecular force field based on the free enthalpy of hydration and solvation: The GROMOS force-field parameter sets 53A5 and 53A6. *Journal of Computational Chemistry*, *25*(13), 1656–1676.

Pearlman, D. A., Case, D. A., Caldwell, J. W., Ross, W. S., Cheatham III, T. E., DeBolt, S., et al., (1995). AMBER, a package of computer programs for applying molecular mechanics, normal mode analysis, molecular dynamics and free energy calculations to simulate the structural and energetic properties of molecules. *Computer Physics Communications, 91*, 1–41.

Pérez, S., Imberty, A., Engelsen, S. B., Gruza, J., Mazeau, K., Jimenez-Barbero, J., et al. (1998). A comparison and chemometric analysis of several molecular mechanics force fields and parameter sets applied to carbohydrates. *Carbohydrate Research*, *314*(3), 141–155.

Plimpton, S. (1995). Fast parallel algorithms for short-range molecular dynamics. *Journal of Computational Physics*, *117*(1), 1–19.

Potter, M. J., Kirchhoff, P. D., Carlson, H. A., & McCammon, J. A. (1999). Molecular dynamics of cryptophane and its complexes with tetramethylammonium and neopentane using a continuum solvent model. *Journal of Computational Chemistry*, *20*(9), 956–970.

Rappé, A. K., Casewit, C. J., Colwell, K., Goddard Iii, W., & Skiff, W. (1992). UFF, a full periodic table force field for molecular mechanics and molecular dynamics simulations. *Journal of the American Chemical Society*, *114*(25), 10024–10035.

Rudd, P. M., Woods, R. J., Wormald, M. R., Opdenakker, G., Downing, A. K., Campbell, I. D., et al. (1995). The effects of variable glycosylation on the functional activities of ribonuclease, plasminogen and tissue plasminogen activator. *Biochimica et Biophysica Acta*, *1248*(1), 1–10.

Rueda, M., Kalko, S. G., Luque, F. J., & Orozco, M. (2003). The structure and dynamics of DNA in the gas phase. *Journal of the American Chemical Society*, *125*(26), 8007–8014. http://dx.doi.org/10.1021/ja0300564.

Ryckaert, J.-P., Ciccotti, G., & Berendsen, H. J. (1977). Numerical integration of the Cartesian equations of motion of a system with constraints: Molecular dynamics of *n*-alkanes. *Journal of Computational Physics*, *23*(3), 327–341.

Sadek, M., & Munro, S. (1988). Comparative review of molecular modelling software for personal computers. *Journal of Computer-Aided Molecular Design*, *2*(2), 81–90.

Sadus, R. J. (1999). Particle–particle and particle-mesh (PPPM) methods. *Molecular simulation of fluids* (pp. 162–169). Amsterdam: Elsevier Science.

Schleif, R. (2004). Modeling and studying proteins with molecular dynamics. *Methods in Enzymology*, *383*, 28–47. http://dx.doi.org/10.1016/S0076-6879(04)83002-7.

Scott, W. R., Hünenberger, P. H., Tironi, I. G., Mark, A. E., Billeter, S. R., Fennen, J., et al. (1999). The GROMOS biomolecular simulation program package. *The Journal of Physical Chemistry A*, *103*(19), 3596–3607.

Sørensen, M. R., & Voter, A. F. (2000). Temperature-accelerated dynamics for simulation of infrequent events. *Journal of Chemical Physics*, *112*, 9599.

Streett, W., Tildesley, D., & Saville, G. (1978). Multiple time-step methods in molecular dynamics. *Molecular Physics*, *35*(3), 639–648.

Strey, A. (2001). High performance computing. In *International encyclopedia of the social & behavioral sciences* (pp. 6693–6697). Oxford, United Kingdom: Elsevier Science Ltd.

Stryer, L. (1995). *Biochemistry*. New York: W.H. Freeman.

Tessier, M. B., Demarco, M. L., Yongye, A. B., & Woods, R. J. (2008). Extension of the GLYCAM06 biomolecular force field to lipids, lipid bilayers and glycolipids. *Molecular Simulation*, *34*(4), 349–363. http://dx.doi.org/10.1080/08927020701710890.

Tuckerman, M. E., Berne, B. J., & Martyna, G. J. (1991). Molecular dynamics algorithm for multiple time scales: Systems with long range forces. *Journal of Chemical Physics*, *94*, 6811.

Verlet, L. (1967). Computer "experiments" on classical fluids. I. Thermodynamical properties of Lennard-Jones molecules. *Physical Review*, *159*(1), 98–103.

Vlachakis, D., Koumandou, V. L., & Kossida, S. (2013). A holistic evolutionary and structural study of flaviviridae provides insights into the function and inhibition of HCV helicase. *PeerJ*, *1*, e74. http://dx.doi.org/10.7717/peerj.74.

Voter, A. F. (1997). Hyperdynamics: Accelerated molecular dynamics of infrequent events. *Physical Review Letters*, *78*(20), 3908.

Voter, A. F. (1998). Parallel replica method for dynamics of infrequent events. *Physical Review B*, *57*(22), R13985.

Wang, J., Cieplak, P., & Kollman, P. A. (2000). How well does a restrained electrostatic potential (RESP) model perform in calculating conformational energies of organic and biological molecules? *Journal of Computational Chemistry*, *21*(12), 1049–1074.

Wang, Y., Shaikh, S. A., & Tajkhorshid, E. (2010). Exploring transmembrane diffusion pathways with molecular dynamics. *Physiology (Bethesda)*, *25*(3), 142–154. http://dx.doi.org/10.1152/physiol.00046.2009.

Wang, J., Wolf, R. M., Caldwell, J. W., Kollman, P. A., & Case, D. A. (2004). Development and testing of a general amber force field. *Journal of Computational Chemistry*, *25*(9), 1157–1174. http://dx.doi.org/10.1002/jcc.20035.

Wilkinson, T., van Gunsteren, W. F., Weiner, P. K., & Wilkinson, A. (1997). *Computer simulation of biomolecular systems: Theoretical and experimental applications*, Vol. 3. New York City, USA: Springer.

Zwier, M. C., & Chong, L. T. (2010). Reaching biological timescales with all-atom molecular dynamics simulations. *Current Opinion in Pharmacology*, *10*(6), 745–752. http://dx.doi.org/10.1016/j.coph.2010.09.008.

CHAPTER EIGHT

Intrinsically Disordered Proteins—Relation to General Model Expressing the Active Role of the Water Environment

Barbara Kalinowska[*,†], **Mateusz Banach**[*,†], **Leszek Konieczny**[‡], **Damian Marchewka**[*,†], **Irena Roterman**[*,1]

[*]Department of Bioinformatics and Telemedicine, Medical College, Jagiellonian University, Krakow, Poland
[†]Faculty of Physics, Astronomy and Applied Computer Science - Jagiellonian University, Krakow, Poland
[‡]Chair of Medical Biochemistry, Medical College, Jagiellonian University, Krakow, Poland
[1]Corresponding author: e-mail address: myroterm@cyf-kr.edu.pl

Contents

1. Introduction 316
2. Definition of the Fuzzy Oil Drop Model 317
   2.1 Observed hydrophobic density distribution 318
   2.2 What is the expected hydrophobic core structure? 319
   2.3 Do real-world proteins actually follow the presented distribution? 319
   2.4 Two ways to determine *O/T* and *O/R* values 322
   2.5 Internal force field 323
3. Results 324
   3.1 Summary of results 324
   3.2 Selected examples 325
4. Discussion 339
5. Conclusions 343
Acknowledgments 344
References 344

Abstract

This work discusses the role of unstructured polypeptide chain fragments in shaping the protein's hydrophobic core. Based on the "fuzzy oil drop" model, which assumes an idealized distribution of hydrophobicity density described by the 3D Gaussian, we can determine which fragments make up the core and pinpoint residues whose location conflicts with theoretical predictions. We show that the structural influence of the water environment determines the positions of disordered fragments, leading to the formation of a hydrophobic core overlaid by a hydrophilic mantle. This phenomenon is further described by studying selected proteins which are known to be unstable and contain intrinsically disordered fragments. Their properties are established quantitatively,

Advances in Protein Chemistry and Structural Biology, Volume 94
ISSN 1876-1623
http://dx.doi.org/10.1016/B978-0-12-800168-4.00008-1

explaining the causative relation between the protein's structure and function and facilitating further comparative analyses of various structural models.

1. INTRODUCTION

The existence of loosely packed polypeptide chain fragments suggests mechanisms which counter the natural tendency of proteins to fold. Generally, protein folding usually results in a tightly packed, stable structure which can be described in terms of secondary and supersecondary motifs. While the basic question remains the same (i.e., "how are such structures generated?"), an equally interesting problem is to explain the formation of structures dominated by disordered regions.

The above-mentioned phenomenon is related to 2 of 14 problems (specifically, nos. 8 and 9) which—despite 50 years of research, aided by the dynamic growth of bioinformatics—have so far eluded solution. These challenges are summarized in Dill and MacCallum (2012):

> *We know little about the ensembles and functions of intrinsically disordered proteins, even though nearly half of all eukaryotic proteins contain large disordered regions. This is sometimes called the "protein unfolding problem" or "unstructural biology."*

In an overview of to-date research into the properties of disordered fragments (Uversky & Dunker, 2010), the authors present a list of phenomena closely related to the presence of such fragments (citation—selected points):

> *5. Increased interaction (surface) area per residue;*
> *6. A single disordered region may bind to several structurally diverse partners;*
> *7. Many distinct (structured) proteins may bind a single disordered region;*
> *8. Intrinsic disorder provides ability to overcome steric restrictions, enabling larger interaction surfaces in protein-protein and protein-ligand complexes than those obtained with rigid partners;*
> *9. Unstructured regions fold to specific bound conformations according to the template provided by structured partners;*
> *14. The possibility of overlapping binding sites due to extended linear conformation.*

This is why the problem of disordered proteins becomes the focus of attention of many researchers. Information on known disordered proteins can be found in a specialized database called DisProt (http://www.disprot.org; Vucetic et al., 2005). The database lists proteins (or fragments thereof) whose native form lacks a stable 3D representation. Its content is derived from published experimental data confirming the unstructured nature of such proteins (or fragments) (Abramavicius & Mukamel, 2004; Asplund, Zanni, & Hochstrasser, 2000; Baiz, Peng, Reppert, Jones, & Tokmakoff, 2012). In addition, DisProt contains information on the biological profile of disordered fragments, methods of detecting such fragments, and links to other databases (Sickmeier et al., 2007).

Our analysis in this chapter focuses on a set of proteins (retrieved from the DisProt database on March 15, 2013) whose crystal structures can be found in PDB. Proteins for which PDB data do not provide 3D structure of the unstructured fragments (where DisProt fragments are unambiguously references) have been excluded from analysis.

Along with a presentation of the causes and effects of the presence of disordered fragments in proteins, an important issue raised in this work concerns the lack of a generalized model describing the relation between the protein and its water environment. It seems that interaction with water plays a pivotal role in ensuring that the polypeptide chain folds in a correct fashion. The presence of water also determines which fragments should remain stable and which ones may retain dynamic properties. This is why our discussion of disordered fragments is presented in the context of a formal model expressing the influence of the water environment upon the structure and properties of proteins in living organisms.

The abbreviation DisProt is used in this chapter as identification of intrinsically disordered fragments (identified in DisProt database).

2. DEFINITION OF THE FUZZY OIL DROP MODEL

The presence and role of the hydrophobic core in tertiary structural stabilization is well described in biochemistry handbooks, although no accurate *in silico* model has so far been proposed. Hydrophobic interactions are usually accounted for by structural prediction software, for example, in Levitt's model (Levitt, 1976), where such interactions are modeled in a pairwise fashion, referring to individual protein atoms and individual water molecules (two- or three-atom models) (Urbic & Dill, 2010). In such models, interaction between the protein and its water environment is described and measured in terms of electrostatic components and Leonard-Jones potentials in pairwise system. As a result, hydrophilic residues preferentially aggregate on the protein's surface, while hydrophobic residues are expected to remain buried, forming a hydrophobic core (Biancardi,

Cammi, Cappelli, Mennucci, & Tomasi, 2012; Kauzmann, 1959; Murphy et al., 2012; Priyakumar, 2012).

In this work, the influence of the water environment, leading to the expected distribution of hydrophobicity (i.e., hydrophobicity density gradient, peaking near the center of the protein body and reaching near-zero values on its surface), is modeled on a global scale, that is, by considering the protein as a whole. The formation of a hydrophobic core appears to be directly related to interactions between polar residues and the water environment, isolating the remaining hydrophobic residues. In the "fuzzy oil drop" model, the notion of a "hydrophobic core" refers to a set of properties which describe the entire molecule, including the aggregation of hydrophobic residues near its center as well as the existence of a hydrophilic "mantle" which remains in contact with water (Yang, Jiao, & Li, 2012). The role of this mantle is to stabilize the core and thus both elements need to be considered as part of a hydrophobic core as a whole.

The "fuzzy oil drop" model is based on an idealized distribution of hydrophobicity density represented by a 3D Gaussian. Actual (observed) hydrophobicity density—dependent on the placement of each residue in a properly folded chain—can be obtained by tracing interactions between pairs of residues and ascribing to each amino a value which reflects its affinity for water. Quantitative comparison of both distributions (theoretical and observed) enables us to determine whether a given protein conforms to the model and contains a well-defined hydrophobic core.

2.1. Observed hydrophobic density distribution

If we assume that hydrophobicity density distribution within the protein molecule results from interactions between side chains, each of which is represented by its so-called effective atom (located at the geometric center of the side chain), then the force of such interactions is given by Levitt's formula (Levitt, 1976):

$$\widetilde{\mathrm{Ho}}_j = \frac{1}{\widetilde{\mathrm{Ho}}_{\mathrm{sum}}} \sum_{i=1}^{N} \left(H_i^r + H_j^r \right) \begin{cases} \left[1 - \frac{1}{2}\left(7\left(\frac{r_{ij}}{c}\right)^2 - 9\left(\frac{r_{ij}}{c}\right)^4 + 5\left(\frac{r_{ij}}{c}\right)^6 - \left(\frac{r_{ij}}{c}\right)^8 \right) \right] & \text{for}\, r_{ij} \le c \\ 0 \quad \text{for}\, r_{ij} > c \end{cases}$$

N is the number of amino acids in the protein, $\widetilde{H}_i^r$ expresses the hydrophobicity of the ith residue, r_{ij} expresses the distance between two interacting residues (jth effective atom and ith effective atom), while c expresses the cutoff distance for hydrophobic interactions, which is taken as 9.0 Å (following Levitt, 1976). The $\widetilde{\mathrm{Ho}}_{\mathrm{sum}}$ coefficient, representing the aggregate sum of all components, is needed to normalize the distribution.

2.2. What is the expected hydrophobic core structure?

The resulting empirical distribution can be compared to a corresponding idealized distribution (Konieczny, Brylinski, & Roterman, 2006). We expect to find the greatest hydrophobicity at the center of the molecule, with hydrophobicity values decreasing along with distance from the center, approaching values close to 0 on the surface. This kind of distribution can be approximated by the 3D Gaussian:

$$\widetilde{\mathrm{Ht}}_j = \frac{1}{\widetilde{\mathrm{Ht}}_{\mathrm{sum}}} \exp\left(\frac{-\left(x_j-\bar{x}\right)^2}{2\sigma_x^2}\right) \exp\left(\frac{-\left(y_j-\bar{y}\right)^2}{2\sigma_y^2}\right) \exp\left(\frac{-\left(z_j-\bar{z}\right)^2}{2\sigma_z^2}\right)$$

The above formula expresses the distribution of probability in an ellipsoid capsule whose dimensions are determined by the values of σ parameters calculated for each of the three cardinal directions. If we place the molecule in such a way that the 3D Gaussian completely encapsulates it (fine tuning the σ values as needed), the value of the function will express the expected (theoretical) distribution of hydrophobicity throughout the protein body increased by additional 9 Å cutoff distance to include also the possible space available for hydrophobic interaction. The superscribed $\bar{x}$, $\bar{y}$, and $\bar{z}$ parameters reflect the placement of the center of the ellipsoid—if this center coincides with the origin of the coordinate system, then all three values are equal to 0. Values of σ are calculated as 1/3 of the greatest distance between an effective atom belonging to the molecule and the origin of the system, once the molecule has been oriented in such a way that its greatest breadth coincides with a system axis (for each axis separately).

The $1/\mathrm{Ht}_{\mathrm{sum}}$ and $1/\mathrm{Ho}_{\mathrm{sum}}$ coefficients ensure normalization of both distributions (empirical and theoretical), enabling meaningful comparisons.

2.3. Do real-world proteins actually follow the presented distribution?

The answer to this question requires quantitative analysis of the similarities/differences between both distributions based on Kullback–Leibler's entropy criterion (Kullback & Leibler, 1951):

$$D_{\mathrm{KL}}\left(p|p^0\right) = \sum_{i=1}^{N} p_i \log_2\left(p_i/p_i^0\right)$$

The value of D_{KL} expresses the distance between the empirical (p) and target (p^0) distributions. In our case, the target distribution is supplied by the 3D Gaussian. According to the definition, D_{KL} is a measure of entropy and thus cannot be interpreted on its own. This is why an independent separate target distribution is required—one in which the aggregation of hydrophobicity near the center of the molecule is absent. In this so-called unified distribution, each residue is assigned a hydrophobicity density value of $1/N$, where N is the number of residues in the polypeptide chain.

To simplify matters, we can introduce the following notation:

$$O/T = \sum_{i=1}^{N} O_i \log_2 O_i/T_i$$

$$O/R = \sum_{i=1}^{N} O_i \log_2 O_i/R_i$$

where O/T is the difference (distance) between the observed and theoretical distributions, while O/R is the corresponding difference (distance) between the observed distribution and a distribution in which each residue carries the same hydrophobicity density value (the hydrophobic core absent).

Comparing O/T and O/R profiles reveals the "closeness" between the observed and theoretical distributions for a given protein. A binary predicate can be adopted at this stage: $O/T < O/R$ indicates the presence of a hydrophobic core. Quantitative analysis is also possible, leading to a ranked list of proteins which expresses their adherence to the idealized core model. This measure of "closeness" can be expressed by the following formula:

$$\mathrm{RD} = \frac{O/T}{O/T + O/R} \tag{8.1}$$

RD stands for *relative distance* between the observed and theoretical distributions. Clearly, the lower the value of RD, the more closely a given protein approximates the theoretical optimum. Figure 8.1 provides a graphical depiction of this relationship.

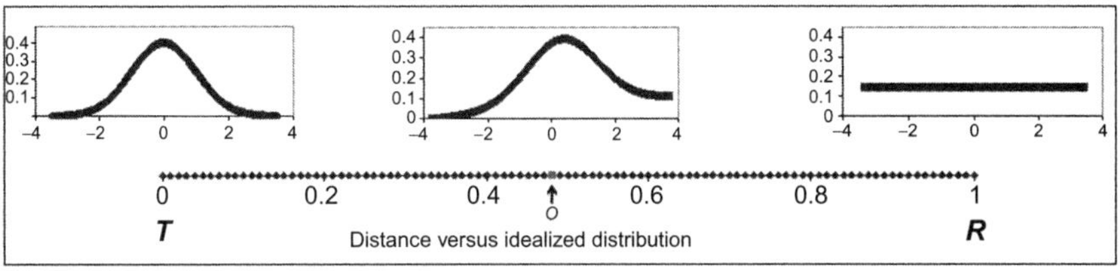

Figure 8.1 Graphical representation of the theoretical versus observed distribution, shown using a linear scale. The horizontal axis expresses the relative distance (RD; cf. Eq. (8.1)) between the idealized 3D Gaussian (leftmost image) and the actual hydrophobicity profile. The right-hand image presents a distribution which does not include a hydrophobic core. The pink dot tagged "O" represents the placement of an arbitrary empirical (observed) distribution which is shown in the central image. (For interpretation of the references to color in this figure legend, the reader is referred to the online version of this chapter.)

We can also suspect that the status of some residues may have been affected by external factors, such as interaction with ligands or other proteins (Marchewka, Jurkowski, Banach, & Roterman, 2013).

As already mentioned, the *O*/*T* versus *O*/*R* relation is the primary method of determining whether the structure in question contains a well-ordered hydrophobic core. Table 8.3 lists *O*/*T* and *O*/*R* values for each structural element (complex, chain, or domain) separately. It also provides information regarding the influence of external chains, ligands, or ions. Assessing this influence requires us to calculate *O*/*T* and *O*/*R* values with and without residues involved in external interactions (as it is assumed that the presence of a ligand may distort hydrophobicity density distribution in the target protein) (Brylinski, Konieczny, & Roterman, 2007).

This, in turn, means that *T*, *O*, and *R* values need to be normalized following elimination of ligand-binding residues, while other coefficients of the 3D Gaussian remain the same. For each structural unit, the Gaussian is computed separately. Determining *O*/*T* and *O*/*R* for fragments (parts) of a structural unit calls for normalization of *T* and *O* values and subsequent calculation of *O*/*T* and *O*/*R* which express the participation of the eliminated fragment in shaping the unit's hydrophobic core. Values obtained for units from which active residues have been eliminated reflect the degree to which such "external" activity (ligand binding, complexation, SS bonds, etc.) distorts the core. However, the groups of proteins belonging to downhill proteins appear to represent the structure of hydrophobic core highly accordant with the idealized one (Banach, Prymula, Jurkowski, Konieczny, & Roterman, 2012).

2.4. Two ways to determine *O/T* and *O/R* values

The values of D_{KL}, both for O/T and O/R, can be calculated for each structural unit separately.

Possible units include individual domains, entire chains, and protein complexes. Depending on the size of the unit, the encapsulating "drop" must be appropriately selected by adjusting its σ coefficients (σ_x, σ_y, σ_z). The O/T versus O/R relation (or, correspondingly, the value of DR) identifies the status of each unit with respect to the theoretical hydrophobicity density distribution.

Another way to calculate O/T and O/R (as well as RD), used when trying to identify the structural role of each fragment, is as follows.

A specific fragment (for instance, a fragment listed in DisProt) is selected from the hydrophobicity density distribution profile. T and R values representing this fragment are removed from the profile, and the remaining values are again normalized (ensuring that the aggregate total T and R remains equal to 1). For this new chain, O/T_F and O/R_F values are calculated, along with RD_F. The F subscript indicates that we are dealing with a truncated molecule (fragmented). If, following removal of this fragment, the inequality between O/T and O/R flips (e.g., $O/T > O/R$ while $O/T_F < O/R_F$), we can conclude that the excised fragment is indeed responsible for the discordance between the given structural unit and its corresponding theoretical representation.

Similar analysis can be performed for the selected fragment itself. Following normalization, T and O values are fed into O/T_{FR} and O/R_{FR} (as well as RD_{FR}) formulae to determine the hydrophobicity density distribution status of the fragment (FR—fragment). This status can be compared to the corresponding status of the entire structural unit—such as a complex (if one exists), a chain, or a domain (if the given complex and/or chain can be subdivided into domains). Applying this algorithm to each fragment listed in the DisProt database yields information regarding its status with respect to the "fuzzy oil drop" model.

Identification of residues engaged in ligand binding or protein complexation was performed using the PDBSum criteria depending on the distance between interacting molecules (Laskowski, 2009).

The analysis of protein structures in respect to "fuzzy oil drop" revealed that the presence of cavity (ligand binding, enzymatic cavity, protein–protein interaction area) significantly influences the fashion of the protein body (Banach, Konieczny, & Roterman, 2012a, 2012b, 2013;

Prymula, Jadczyk, & Roterman, 2011). This is why the calculation of *O/T* and *O/R* for proteins under consideration was performed also for molecules with residues engaged in intermolecular interaction eliminated from calculation.

2.5. Internal force field

The calculation of internal interaction in protein molecule was performed to make possible comparison between hydrophobicity density distribution and nonbonding interaction density distribution (Marchewka, Banach, & Roterman, 2011).

The following procedure was carried out:

1. the structure of the protein (as listed in PDB) was subject to energy minimization (EM) in order to eliminate steric clashes (e.g., resulting from inclusion of hydrogen atoms, which are not present in the protein's crystal form);
2. for each amino acid residue (and its constituent atoms), interactions with remaining part of the molecule were calculated;
3. separate computations were carried out for electrostatic and van der Waals interactions;
4. the Gromacs program was used to perform the energy optimization and crystal relaxation procedure. All EM calculations were conducted with the use of Gromacs software package v4.0.3 and Gromos96 43a1 force field (Berendsen, Postma, van Gunsteren, & Hermans, 1981; Berendsen, van der Spoel, & van Drunen, 1995; Lin & van Gunsteren, 2013; Lindahl, Hess, & van der Spoel, 2001; van der Spoel et al., 2005, 1995). The grouping option was used to tag each residue as a separate "group" interacting with the rest of the protein body. The interaction between each residue and the rest of molecule was performed to attribute the local interaction of each residue with its local surrounding and concentrated in the position of effective atoms to make possible the comparison with the hydrophobicity density distribution. The electrostatic and vdW interaction was taken under consideration. The electrostatic and vdW interaction density was normalized to make applicability of Kullback–Leibler distance entropy calculation possible and comparable to hydrophobicity density distribution in the protein molecule under consideration (Banach, Marchewka, Piwowar, & Roterman, 2012; Banach, Prymula, et al., 2012).

3. RESULTS

The calculation methods as introduced above when applied to the set of DisProt proteins deliver the general overview of different characteristics of structural units and intrinsically disordered fragments as well.

3.1. Summary of results

The summary results classifying particular structural units (chains and domains) as accordant or discordant in respect to "fuzzy oil drop" model is given in Table 8.1. This classification suggests the presence of hydrophobic core as defined in the model. The results of this analysis are given in Table 8.1.

Table 8.1 Summary of DisProt fragments as they appear in appropriate structural unit, the status

Chain accordant Domain accordant	Chain discordant Domain accordant	Chain discordant Domain discordant
2CV4(2–8), 1ECF(472–492), 3GZP(67–78), (125–139), 1ECF (471–492), 2B76(91–158), 1GUA (104–107), 1OLG_A(319–323), 1OS2(110–116), 1RJ7_A (312–317), 1RJ8_A(312–315), 1RRP_A(60–62), 2BZS (228–231), 1L0I(17–36), 1DDS (9–24), (63–72), (116–132), (142–150), 1CWX_A(1–45), 2KOG_A(55–76), 2BRZ(38–45), 2JV4(24–45), 1SS3_A(34–50), 1JK3(110–116), (146–157), 1RXR(169–189), (172–176), (178–187), (181–187), (202–206), 2NLN(42–71), (82–109), 1YYZ (204–222), 1RG7(16–22), 1L0H (258–267), 1KAO(60–63), 1LXL (28–80), 1KAO(62–63), 1AA9 (57–64), (58–66), 1CRD(58–66), 121P(59–72), 1GHZ(183–188), (224–227), 1SVA_1(15–89), 1SRY(258–267), 1TBA(11–17)	2IS2 (495–564) 2WB0 (401–406) 1RRP_A (108–109) 1FAQ_A (136–138) 1FAQ_A (185–187)	3C66(80–105), 1GME(1–42), 1ECF(73–84), 1SVA_1(90–107), (297–301), (302–330), (331–341), (342–362), 1CRD(30–38), 1HPW_A(35–36), 1EOT_A (1–8), 1HPW_A(35–36), 2JU4 (74–87), 1G2S_A(1–9), 1HRT_I (50–65), 1JSU_C(22–34), 1OLG_A(357–360), 1AA9_A (31–39), 1AA9_A(31–39), 1OS2_A(146–157), 2BZS_A (1–4), 2PTL_A(1–17), 1KAO (45–55), (32–36), 1YDV(66–80), (165–178), 2KOG_A(1–35), (36–54), (77–88)

Table 8.1 lists the results obtained for each fragment from the DisProt database. Several cases merit further interpretation. For instance, 2KOG is a membrane protein whose shape is far from globular. It seems that the "fuzzy oil drop" model does not adequately reflect its structural properties. Proteins labeled 2PTL and 1GME are globular but have an outstretched "arm" which causes them to diverge from the model. Many other proteins, such as 1TEW (as well as the small 1EOT protein with two SS bonds), are rich in disulfide bonds which impose additional structural constraints and distort the shape of the molecule in relation to the model.

As an example showing the applied methodology, the protein 1QO9 was arbitrarily selected (Harel et al., 2000).

The analysis of 1QO9 protein—a hydrolase (E.C.3.1.1.7)—acetylcholinesterase from *Drosophila melanogaster* in complex with two inhibitor proteins is presented in details.

The status of 1QO9 chain can be represented by its RD value, which is equal to 0.63. The DisProt fragment (142–173), described in the database as "flexible linkers/spacers," has an RD value of 0.74.

Figure 8.2 (bottom) presents the O/T and O/R profiles calculated using a 20 aa open reading frame. As can be seen, only the beginning and the end of the chain (as well as a single fragment near position 350) satisfy $O/T < O/R$. The DisProt fragment, tagged light blue, diverges from the theoretical hydrophobicity density distribution model, with the exception of its C-terminal fragment and some short intermediate sequences.

The localization of DisProt fragment, which occupies rather central position in the protein body of 1QO9 molecule, is shown in Fig. 8.3 in 3D presentation.

Do any other proteins follow the idealized hydrophobicity density distribution?

As a matter of fact—yes—such proteins are relatively easy to identify in PDB. The answer to this question depends, however, on the structural unit for which accordance is measured. The presented work is an attempt at addressing this question in the scope of DisProt fragments. Our study set contains accordant proteins, as well as proteins in which disorganized fragments diverge from the model. As a result, we have decided to perform further analysis for each group separately.

3.2. Selected examples

Individual proteins were selected to present the different status of particular proteins. The proteins with many structural and functional profiles are aimed to verify the applicability of "fuzzy oil drop" model for the DisProt analysis.

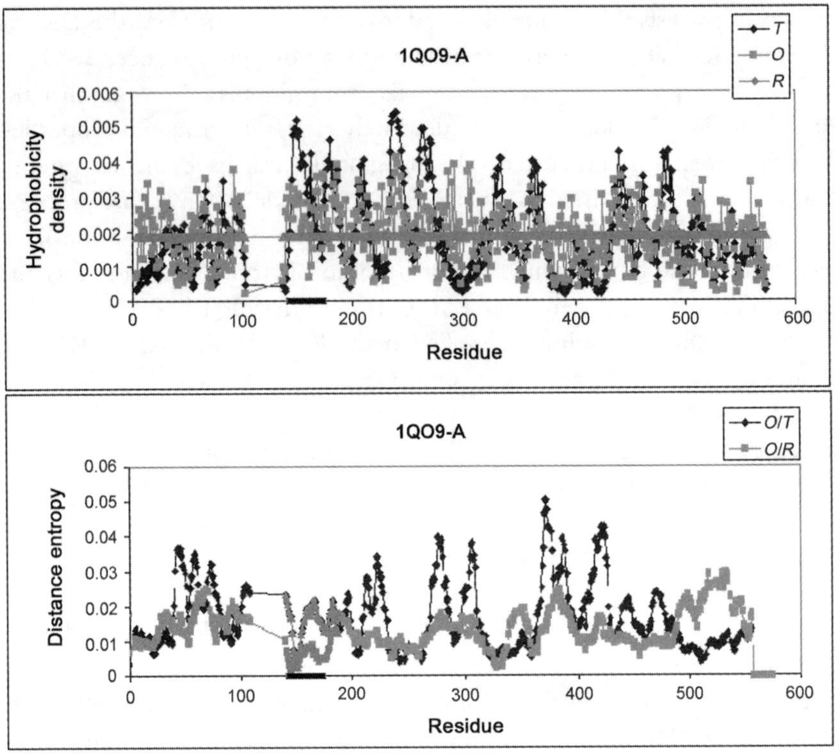

Figure 8.2 Visual representation of the 1QO9-A chain profile. Top diagram: values of *T* (dark blue—rhombs), *O* (pink squares), and *R* (light blue—continuous line) showing to what extend the observed *O* distribution resembles *R* or *T* distribution. The black bar on the horizontal axis identifies the disordered fragment (according to DisProt database). Bottom diagram: *O*/*T* and *O*/*R* profiles calculated for fragments using a 20 aa open reading frame. Light blue marks (and black fragment of *X*-axis) correspond to the DisProt fragment. Both diagrams indicate high similarity between the observed distribution (*O*) and the random distribution (*R*), which also applies to the DisProt fragment. (For interpretation of the references to color in this figure legend, the reader is referred to the online version of this chapter.)

3.2.1 Varied hydrophobicity distribution status of DisProt fragments—2JU4

2JU4 is a gamma subunit domain cgmp phosphodiesterase (retinal rod rhodopsin-sensitive cgmp 3′,5′-cyclic phosphodiesterase subunit gamma of *Bos taurus* cattle—retina; Song et al., 2008). The domain under consideration does not contain enzymatically active residues. In addition, the crystal structure of this protein contains a number of ligands (six to be exact—three monomers, two dimers, and a trimer designated RCY).

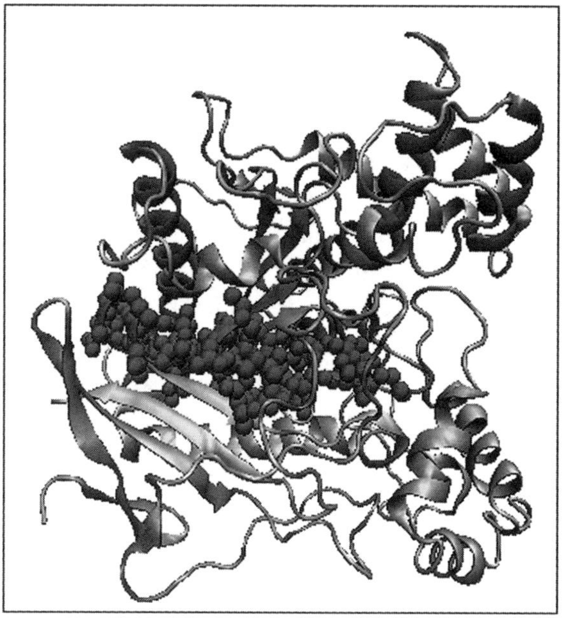

Figure 8.3 The 3D presentation of 1QO9 with DisProt fragment distinguished in red (balls). (For interpretation of the references to color in this figure legend, the reader is referred to the online version of this chapter.)

The retinal phosphodiesterase (PDE6) inhibitory gamma subunit (PDEgamma) plays a central role in vertebrate phototransduction by alternate interactions with the catalytic alphabeta subunits of PDE6 and the alpha subunit of transducin (alpha(t)). In-depth, analysis of its structure (using NMR) suggests a high degree of intrinsic disorder. NMR scans also point to high structural variability of the 24–45 and 74–87 fragments (listed in the DisProt database). This is further confirmed by analysis of 100 candidate structures for 2JU4 in PDB.

The structure of 2JU4, from the point of view of Φ and Ψ angles (Ramachandran plot), does not indicate the presence of ordered secondary structural fragments (see Ramachandran plot in PDBSum database). Structural analysis of this domain based on the "fuzzy oil drop" model reveals the presence of a hydrophobic core—the domain satisfies $O/T < O/R$ (0.216 and 0.244, respectively).

The intrinsically disordered fragments listed in DisProt (24–45 and 74–87) possess RD values of 0.49 and 0.54, respectively. Analysis of the 1–23 fragment, which is only loosely integrated with the rest of the molecule, produces an RD value of 0.53, which means that this fragment does not contribute to hydrophobic core formation. Following elimination of ligand-binding residues, the remainder of the chain has an RD value of 0.50 (Fig. 8.4).

Based on *O*/*T* and *O*/*R* analysis, the entirety of the polypeptide chain (84 amino acids) seems to exhibit an ordered distribution of hydrophobicity density, approximating the 3D Gaussian. Differences in the status of disordered fragments (24–45—accordant; 74–87—discordant) may indicate variable stability. As one of these fragments—along with the molecule as a whole—exhibits accordance with the model, it can be assumed that 2JU4 does indeed contain a hydrophobic core, as predicted by the "fuzzy oil drop" model. Regarding structural variability, DisProt mentions that the "function (of this domain) arises via a disorder to order transition," which seems to be driven by hydrophobic interactions.

Plotting the distribution of hydrophobicity density in 2JU4 reveals greater accordance between the theoretical and observed values for the 24–45 fragment than for the 74–87 fragment.

Figure 8.5 shows the 3D presentation of 2JU4, highlighting the status of intrinsically disordered fragments. Shades of gray indicate the concentration of hydrophobicity, which is greatest near the center of the ellipsoid and close to zero on its surface. Highlighted fragments also illustrate the "fuzzy oil drop" model, which not only assumes the existence of a hydrophobic core

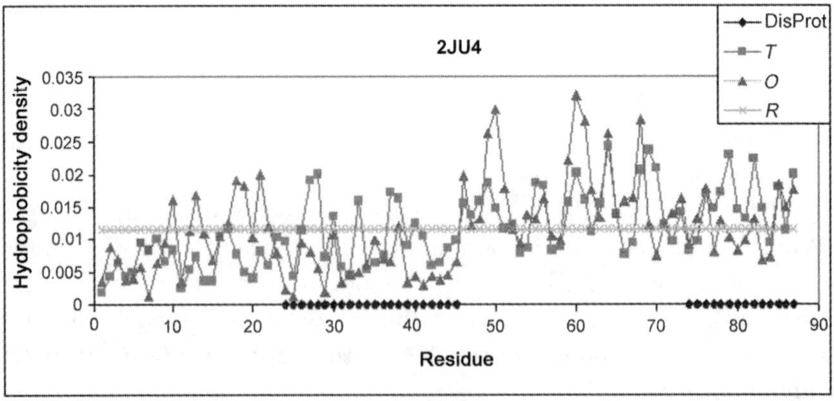

Figure 8.4 Hydrophobicity density distribution in 2JU4: *T*, theoretical (pink squares); *O*, observed (gray triangles); *R*, unified (no hydrophobic core—horizontal line). Dark blue marks on the horizontal axis denote intrinsically disordered fragments. (For interpretation of the references to color in this figure legend, the reader is referred to the online version of this chapter.)

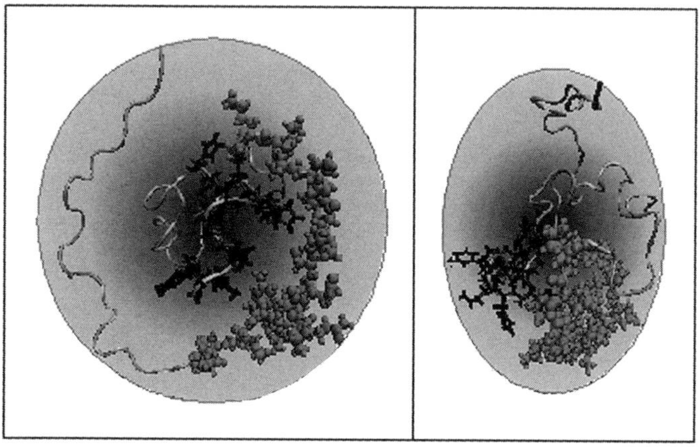

Figure 8.5 3D representation of 2JU4 in relation to the "fuzzy oil drop" model (different perspectives). The red fragment (balls) is structurally consistent with the "fuzzy oil drop" model, while the black (sticks) fragment diverges from it. Grayscale saturation increases along with distance from the surface, corresponding to the hydrophobicity density gradient which peaks near the center of the ellipsoid. The ellipses visualize encapsulation of the molecule according to the 3D Gaussian. (For interpretation of the references to color in this figure legend, the reader is referred to the online version of this chapter.)

but also predicts a hydrophobicity gradient through which peripheral fragments are capable of shielding the core from entropically disadvantageous contact with water. In this sense, the presented fragments are also consistent with the model.

Highly dynamic fragments which undergo significant structural modifications as a result of their function are also capable of reverting to their original ordered form (this is especially true of the 24–45 fragment—red balls in Fig. 8.5).

3.2.2 Sample discordant protein: 1YDV

The 1YDV homodimer represents an interesting study case. It is a triosephosphate isomerase E.C. 5.3.1.1. (*source*: *Plasmodium falciparum*—malaria parasite; Velanker et al., 1997). Each monomer contains two fragments which have been identified as intrinsically disordered. Of particular note is the presence of three enzymatically active residues in one of these fragments.

Neither the dimer itself nor any of its disordered fragments (as given by DisProt) exhibit accordance with "fuzzy oil drop" model. However, such

Table 8.2 RD values for structural units in the 1YDV homodimer

Structural unit	Complex	Chain	DisProt1 66–80	DisProt2 165–178
COMPLEX	0.69	0.69	0.60	0.75
NO P–P	0.66	0.66	**0.48**	–
NO ENZYM.	0.69	0.69	–	0.69
CHAIN		**0.45**	0.63	0.67
NO P–P		**0.40**	0.53	–
NO ENZYM.		**0.45**	–	0.50

DisProt lists two intrinsically disordered fragments identified in this chain (identification according to DisProt database). Values printed in boldface indicate accordance with the hydrophobicity density distribution model.

accordance is observed for a single chain (treated as a structural unit) (Table 8.2).

While interpreting the results shown in Table 8.2, it should be noted that only the 1YDV chain itself—as a structural unit—remains accordant with the assumed model. Neither DisProt fragment exhibits such accordance, regardless of the structural unit in which it is analyzed. The structure of a single isolated chain represents the structure with hydrophobic core (according to "fuzzy oil drop" model).

Figure 8.6 visualizes the localization of DisProt fragments in 1YDV which appear to be localized in the protein–protein interface (homodimer).

Comparison of RD values reveals the likely sequence of events involved in forming the presented homodimer. It seems that each chain folds on its own, reaching a hydrophobicity density distribution which is consistent with the model. Elimination of residues involved in enzymatic activity or protein–protein interaction does not alter this behavior.

DisProt fragments, treated as elements of the complex as well as of individual chains, do not match the expected hydrophobicity distribution. Regarding DisProt1, elimination of residues involved in protein–protein interactions renders this fragment accordant with the model. However, DisProt2 remains discordant even when all residues involved in p–p interactions and enzymatic activity are eliminated.

3.2.3 2TPI as an example of a protein of highly differentiated structure

To further illustrate practical use of the "fuzzy oil drop" model, we will refer to the protein complex composed of E.C.3.4.21.4 hydrolase—Trypsin

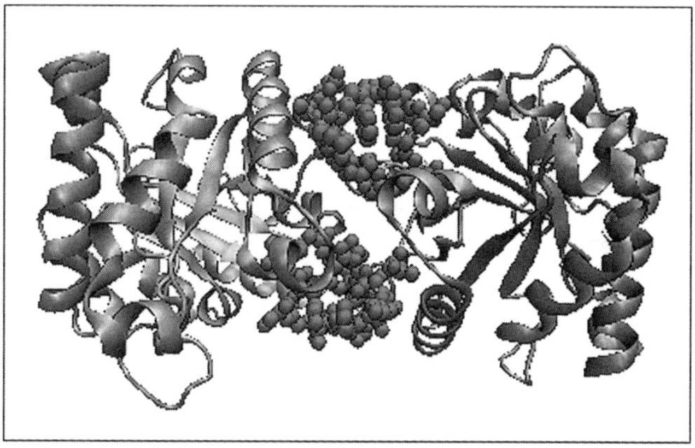

Figure 8.6 The 3D presentation of 1YDV dimer with DisProt fragments distinguished by red balls. The position in the protein–protein interface is visualized. (For interpretation of the references to color in this figure legend, the reader is referred to the online version of this chapter.)

(Z chain) and its inhibitor (I chain) (PDB ID: 2TPI—Walter et al., 1982). This protein is a very interesting study subject in the context of the relation between its secondary structural domains (two of which can be distinguished within the Z chain which facilitates enzymatic activity) and the hydrophobic core structure. It should be noted that 2TPI is a complex (meaning that the influence of the complexed protein can be studied) contains disulfide bonds (which can affect the core in measurable ways), includes a ligand (i.e., the ILE–VAL dipeptide), contains a mercury ion (likely not associated with biological activity), and, finally, contains disordered fragments in its Z and I chains.

The "fuzzy oil drop" model can be applied to various structural units: complex, chains, or domains in 2TPI. For each of these entities, a separate "drop" is defined by establishing its volume and location. Additionally, the model permits quantitative analysis of the involvement of each fragment in shaping the common hydrophobic core. When determining the input of each fragment (or even of individual residues), a separate "drop" is not necessary—rather, the residues corresponding to the fragment in question are eliminated from *O*, *T*, and *R* calculations, and the resulting profile is renormalized. Computing O/T and O/R yields the status of the remainder of the initial molecule. If the inequality flips (from $O/T > O/R$ to $O/T < O/R$), we can surmise that the eliminated residues cause the chain to diverge from the model. The presented case study involves a trypsin inhibitor complex designated 2TPI. Its Z chain is known to contain an enzymatic active site (E.C.3.4.21.4—Trypsin in preferential cleavage reaction: Arg-|-Xaa, Lys-|-Xaa), while the I chain is a basic protease inhibitor (aprotinin). The structure of the resulting complex enables us to study various factors which affect hydrophobicity density distribution. Our analysis focuses on two chains, one of which (labeled Z) comprises two domains: domain 1 (19–27 and 121–233) and domain 2 (28–120 and 233–245). The remaining chain (I) consists of a single domain.

Each of external factors (ligand binding, protein–protein interaction, enzymatic activity, and SS bonds) can be studied separately to determine its influence upon the final structure of 2TPI.

2TPI is listed in the database of protein disorder (http://www.disprot.org—accessed April 20, 2013). Its Z chain includes four disordered fragments: 18–27, 137–147, 19–190, and 209–216. The shared DisProt identifier of these fragments is DP00728.

Regardless of the DisProt classification, our analysis of disordered structures also covers the following three fragments in the I chain: 8–18, 24–28, and 35–47. These fragments have been selected on the basis of subjective visual assessment.

The status of each disordered fragment in relation to the "fuzzy oil drop" model has been computed for each structural unit (complex, chain, or domain) separately. Results—specifically, the relation between O/T and O/R for individual units—are listed in Table 8.1.

3.2.3.1 Complex

The complex as a whole does not appear to contain a shared hydrophobic core (as evidenced by the $O/T > O/R$ relation) (Table 8.3).

The Z chain in the complex has been identified as having significant influence upon the formation of a common hydrophobic core. Its domain 2 is consistent with the theoretical hydrophobicity distribution model. Three out of four disordered fragments also appear consistent with the model (applied to the complex as a whole). Regarding chain I, its disordered fragments seem to match the expected distribution of hydrophobicity throughout the complex.

One disordered fragment in chain Z diverges from the model.

Eliminating residues involved in external interactions (ligand binding, inhibitor complexation, enzymatic activity, disulfide bonds) does not change the status of the complex, implying that such residues do not affect the common hydrophobic core.

Table 8.3 *O/T* and *O/R* values calculated for each fragment under the assumption that the structural unit (in the sense of the "fuzzy oil drop" model) is the entire complex of 2TPI

Struc/Func	**Residues excluded** ***Residues present***	***O/T***	***O/R***	**RD**
COMPLEX		0.199	0.154	0.56
NO LIGAND	Z: 19, 142–144, 156–159, 187–189, 194,221	0.206	0.159	0.56
NO ENZYM.	Z: 57, 102, 193, 195, 196	0.199	0.154	0.56
NO P–P	Z: 39–42, 57, 151, 189–193, 195, 214–216, 226 I: 11–19, 36–39	0.198	0.156	0.56
NO ION	Z: 145, 146, 191, 220	0.202	0.153	0.57
NO SS bonds	Z: 22;157,42;58,128;232,136;201,168;182 I: 5;55,14;38,30;51	0.180	0.142	0.56
NO L-P-P-I	Z: 19, 142–144, 156–159, 187–189, 194,221 39–42, 57, 151, 189–193, 195, 214–216, 226 145, 146, 191, 220 I: 5;55,14;38,30;51	0.209	0.158	0.57
NO L-P-P-I-SS	Z: 19, 142–144, 156–159, 187–189, 194,221 39–42, 57, 151, 189–193, 195, 214–216, 226 145, 146, 191, 220 22;157,42;58,128;232,136;201,168;182 I: 5;55,14;38,30;51	0.183	0.149	0.55
NO DisProt1	19–27	0.213	0.158	0.57
CHAIN Z	*19–245*	**0.135**	**0.141**	**0.49**
CHAIN I	*2–58*	0.460	0.209	0.69
DOMAIN 1—Z	*19–27, 121–233*	0.138	0.136	0.50
DOMAIN 2—Z	*28–120, 233–245*	**0.130**	**0.141**	**0.48**
Z—DisProt 1	*19–27*	0.093	0.078	0.54
Z—DisProt 2	*137–147*	**0.021**	**0.083**	**0.20**
Z—DisProt 3	*179–190*	**0.149**	**0.177**	**0.46**
Z—DisProt 4	*209–216*	**0.021**	**0.039**	**0.35**
I—Disorder	*8–18*	**0.089**	**0.193**	**0.31**

Continued

Table 8.3 *O/T* and *O/R* values calculated for each fragment under the assumption that the structural unit (in the sense of the "fuzzy oil drop" model) is the entire complex of 2TPI—cont'd

Struc/Func	Residues excluded *Residues present*	O/T	O/R	RD
I—Disorder	*24–28*	**0.125**	**0.153**	**0.45**
I—Disorder	*35–47*	**0.157**	**0.345**	**0.31**

Rows labeled "NO *XX*" list *O/T* and *O/R* values for a common "oil drop" encapsulating the entire complex, without residues involved in activity *XX*.
Chains, domains, and fragments labeled "DisProt" or "Disorder" are characterized by *O/T* and *O/R* values which reflect their contribution to the common hydrophobic core for the entire complex.
Values listed in boldface represent hydrophobicity density distribution accordant with the theoretical model.
It should be noted that the discussed fragments in chain I (disordered) are selected on the basis of subjective visual analysis (not present in the DisProt database).
The numbers of residues in italics represent residues present in calculation.

3.2.3.2 Chains

Chain Z treated as a separate structural unit satisfies *O/T*> *O/R*, which—according to the model—indicates the lack of a hydrophobic core. In contrast, when analyzed as part of the complex, both Z chain domains remain accordant with the model (Table 8.4). Elimination of residues involved in enzymatic activity and inhibitor complexation renders the remainder of the chain accordant with the model. Other types of interactions (with ligands and ions, as well as participation in disulfide bonds) do not result in significant deformations within chain Z.

Regarding chain Z, three out of four disordered fragments assume conformations consistent with the expected hydrophobicity density gradient. Similar to the entire complex, the first disordered fragment remains discordant.

Chain I: The I chain, treated as a separate structural unit, also does not appear to contain a well-ordered hydrophobic core since it satisfies *O/T*> *O/R* (note, however, that eliminating residues involved in enzyme complexation renders the chain accordant with the model).

From the three fragments arbitrarily deemed disordered, only one exhibits the expected hydrophobicity density gradient (as long as the I chain is treated as a separate structural unit).

3.2.3.3 Domains

The summary characteristics of domains present in 2TPI are shown in Table 8.5 together with the calculations of domains deprived of residues

Table 8.4 *O/T* and *O/R* values calculated for each chain of 2TPI individually

Struc/Func	**Residues excluded** *Residues present*	*O/T*	*O/R*	**RD**
CHAIN Z	*19–245*	0.141	0.139	0.50
NO LIGAND	19, 138, 140, 142, 143, 144, 156, 157, 158, 187, 188, 189, 194, 221	0.144	0.144	0.50
NO ENZYMATIC	57, 102, 193, 195,196	**0.138**	**0.140**	**0.49**
NO P–P	39–42, 57, 97, 99, 151, 189–195 214–216, 226	**0.137**	**0.142**	**0.49**
NO ION	145, 146, 191, 220	0.142	0.139	0.50
NO SS bonds	22;157,42;58,128;232,136;201, 168;182	0.137	0.131	0.51
DOMAIN 1	*19–27, 121–232*	**0.121**	**0.146**	**0.45**
DOMAIN 2	*28–120, 233–245*	**0.111**	**0.146**	**0.43**
NO–DisProt	All residues of domain 2 with residues recognized as DisProt eliminated	0.142	0.141	0.50
DisProt 1	*19–27*	0.087	0.078	0.53
DisProt 2	*137–147*	**0.060**	**0.083**	**0.42**
DisProt 3	*179–190*	**0.152**	**0.183**	**0.45**
DisProt 4	*209–216*	**0.021**	**0.056**	**0.27**
CHAIN I	*2–58*	0.254	0.216	0.54
NO P–P	9, 11–19, 36–39	**0.207**	**0.214**	**0.49**
NO SS	5;55,14;38,30;51	0.231	0.187	0.55
Disorder 1	*8–18*	0.409	0.238	0.63
Disorder 2	*24–28*	**0.099**	**0.154**	**0.39**
Disorder 3	*35–47*	0.487	0.339	0.59

Rows labeled "NO *XX*" list *O/T* and *O/R* values for an "oil drop" encapsulating the chain, without residues involved in activity *XX*.
Chains, domains, and fragments labeled "DisProt" or "Disorder" are characterized by *O/T* and *O/R* values which reflect their contribution to the common hydrophobic core for the given chain.
Values listed in boldface represent hydrophobicity density distribution accordant with the theoretical model.

engaged in chain–chain interaction, ligand binding, enzymatic activity, and ion binding. The influence of Cys engaged in SS bonds formation is also discussed (Table 8.5).

Analysis of individual domains (where each domain is treated as a separate structural unit for the purposes of drop encapsulation) in the Z chain reveals high concentration of hydrophobicity density near the center of each domain (treated as the unit for drop definition). Hydrophobicity density decreases along with distance from the center, reaching near-zero values on the surface, as implied by the $O/T < O/R$ relationship. Elimination of residues responsible for external interactions does not alter the overall status of the domain (in fact, it produces a distribution which matches theoretical values even more closely). Figure 8.7 provides a graphical representation of the 2TPI complex.

3.2.3.4 Suggested order of events in the formation of the complex

By analyzing the status of individual structural units (complex, chain, or domain), we can speculate several stages in the formation of the 2TPI complex (enzyme + inhibitor). Assuming that the polypeptide chain—when placed in an aqueous environment—exhibits a natural tendency to form a hydrophobic core shielded by a hydrophilic mantle, we propose that:

1. The structures which emerge first are those characterized by structural stability owing to the presence of a hydrophobic core. The structural status of three (out of four) disordered fragments seems to be accordant with the status of the entire domain. Elimination of residues involved in external interactions does not affect this status.
2. The second stage involves formation (folding) of the protein chain, with participation of individual domains. This produces a specific catalytic pocket, as the participating domains do not significantly alter their shape and therefore, do not form a shared hydrophobic core. The role of disordered fragments does not change at this stage.
3. Targeted external interactions lead to creation of an enzyme–inhibitor complex. This joint structure also does not contain a shared hydrophobic core.

The role of disulfide bonds is noteworthy, as their elimination does not affect hydrophobicity density distribution in 2TPI.

The DisProt1 fragment does not follow the theoretical hydrophobicity density distribution model in any structural unit. It appears to be strongly

Table 8.5 *O/T* and *O/R* values calculated for each domain individually

Unit	Struc/Func	**Residues excluded** *Residues present*	*O/T*	*O/R*	RD
DOMAIN 1	CHAIN (Z)	*19–27, 121–233*	**0.121**	**0.146**	**0.45**
	NO LIGAND	19, 138, 142–144, 156–158, 187–189, 194, 221	**0.115**	**0.150**	**0.43**
	NO ENZYMATIC	193, 195, 196	**0.124**	**0.149**	**0.45**
	NO P–P	151, 189–195, 214–216, 226	**0.107**	**0.154**	**0.41**
	NO ION	145, 146, 191, 220	**0.125**	**0.144**	**0.46**
	NO SS bonds	22:157,128;232,136;201,168:182	**0.116**	**0.134**	**0.46**
	NODisProt		**0.124**	**0.147**	**0.46**
	DisProt 1	*19–27*	0.224	0.159	0.58
	DisProt 2	*137–147*	**0.037**	**0.095**	**0.28**
	DisProt 3	*179–190*	**0.131**	**0.211**	**0.38**
	DisProt 4	*209–216*	**0.044**	**0.055**	**0.44**
DOMAIN 2	CHAIN (Z)	*28–120, 233–245*	**0.111**	**0.146**	**0.43**
	NO ENZYMATIC	57, 102	**0.112**	**0.148**	**0.43**
	NO P–P	36–42, 57, 97, 99	**0.110**	**0.145**	**0.43**
	NO SS bonds	42:58	**0.114**	**0.142**	**0.44**

Rows labeled "NO *XX*" list *O/T* and *O/R* values for an "oil drop" encapsulating the domain, without residues involved in activity *XX*. Chains, domains, and fragments labeled "DisProt" (identified according to DisProt database) or "Disorder" (subjectively recognized) are characterized by *O/T* and *O/R* values which reflect their contribution to the common hydrophobic core for the given domain.

Values listed in boldface represent hydrophobicity density distribution accordant with the theoretical model.

The residues present in calculation are given in italics.

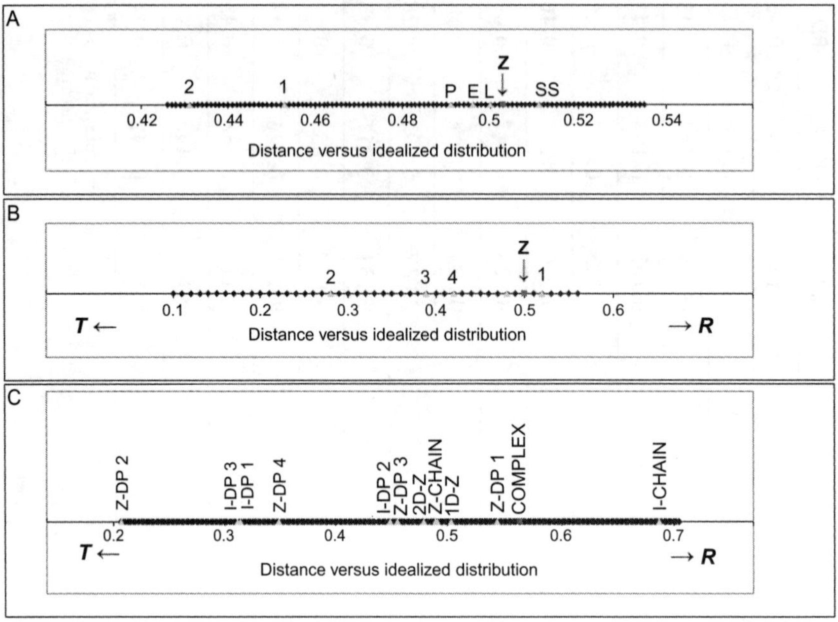

Figure 8.7 Status of individual structural units in the 2TPI complex. (A) The Z chain (pink dot on the axis) and its structure following elimination of residues responsible for ligand binding (L), enzymatic activity (E), complexation of the I chain (P), and forming disulfide bonds (SS), as well as domains 1 and 2, respectively. (B) Participation of disordered fragments in generating a distribution accordant with the model (numbering matches Table 8.5). The Z chain has been highlighted by a pink dot. (C) Participation of structural fragments in generating the "oil drop" for the entire complex. Labels—chain name followed by DP indicates a disordered fragment, as listed in Table 8.5. Numbers followed by D indicate Z chain domains. Equation ((8.1)) has been applied to construct the summary chart presented in this figure. (For interpretation of the references to color in this figure legend, the reader is referred to the online version of this chapter.)

implicated in structural changes associated with the protein's biological role, as stated in the DisProt database (based on experimental data).

3.2.3.5 Internal force field

The hydrophobic core structure (which is understood as an area characterized by higher than average hydrophobicity density) could potentially result from dense packing of residues near the center of the molecule. In order to determine the relationship between residue packing and hydrophobicity density distribution, we will refer to other types of internal interactions, specifically, electrostatic and van der Waals forces.

Calculating O/T and O/R values provides a basis upon which to determine the distribution of internal force fields within the molecule. All electrostatic and van der Waals interactions were normalized (ensuring that their aggregate sum was equal to 1.0 in each case). O/T and O/R values were established in a similar manner to hydrophobicity density computations. The 2TPI molecule was used as a representative case study.

Neither electrostatic nor van der Waals interactions are accordant with the theoretical model for any structural unit. This, however, should come as no surprise as the hydrophobicity density distribution in 2TPI also diverges from theoretical expectations. Regarding hydrophobicity density, the Z chain exhibits a well-defined hydrophobic core, whereas no such core can be distinguished in the scope of nonbinding interactions. High RD values (0.72 and 0.85 for electrostatic and van der Waals forces, respectively) suggest that neither interaction is significantly concentrated in the center of the domain.

Similar results of electrostatic and van der Waals density calculations were obtained for a larger group of proteins, including proteins which contain a "fuzzy oil drop"-compliant hydrophobic core. This suggests that the "fuzzy oil drop" model applies exclusively to hydrophobicity density distribution. The observation may be of practical importance for *in silico* structure prediction algorithms. While pairwise optimization does not treat any part of the molecule preferentially (and thus tends to produce uniform distributions of the optimized quantity), hydrophobic interactions instead require a holistic model which differentiates parts of the protein body with respect to their placement.

4. DISCUSSION

Can a protein contain a hydrophobic core which is in perfect accordance with the theoretical profile, with all hydrophobic residues concentrated near the center and all hydrophilic residues exposed on the surface?

Such a protein would possess two key characteristics: it would be highly water soluble (which is somewhat expected) but also incapable of interacting with other types of molecules, given its high preference for contact with water. Analysis of numerous proteins reveals that most of them depart from the idealized hydrophobicity density distribution. Areas where such departures are concentrated often mediate the protein's biological activity. For example, ligand-binding pockets are typically associated with hydrophobicity deficiencies (Brylinski, Kochanczyk, Broniatowska, & Roterman, 2007; Brylinski, Prymula, et al., 2007) while protein complexation sites frequently exhibit excess hydrophobicity. As a result, distortions in the hydrophobicity density distribution throughout the protein body appear to be a critical factor in ensuring that the protein remains biologically active—whether dissolved in water or confined to a hydrophobic environment (e.g., cellular membranes).

Do any proteins follow the theoretical model with high accuracy?

Yes, several such proteins can be found in databases. This group includes antifreeze (class II) proteins, which counteract growth of ice crystals and prevent them from damaging cellular structures. The biological role of antifreeze proteins requires them to be water soluble. In addition, they should not form clusters and instead need to be uniformly distributed in the aqueous environment, which explains their adherence to the "fuzzy oil drop" model (Banach, Prymula, et al., 2012).

Another group which also exhibits high accordance with the theoretical model comprises the so-called fast-folding (downhill) proteins (Roterman, Konieczny, Jurkowski, Prymula, & Banach, 2011). In their case, high reversibility of folding (as determined experimentally) indicates that the structural influence of water is sufficient to guide the process in such a way as to ensure aggregation of hydrophobic residues near the center of the protein with simultaneous exposure of hydrophilic residues on the surface.

The role of structural fragments identified as disordered seems to be related to the goal of stabilizing the molecule (or complex) as a whole. They should therefore be studied in the context of the structure in which they participate. Algorithms which attempt to assemble the protein's native form from individual structural "building blocks" are accurate only in selected cases, as determined by the CASP initiative (Dill & MacCallum, 2012). "Bottom-up" models which seek minimal free energy conformations for short fragments have since been abandoned in favor of more holistic energy optimization approaches.

The "fuzzy oil drop" model interprets the molecule as a whole, rather than a sum of its (locally optimized) parts. According to the model, optimization of hydrophobic interactions should not be performed in a pairwise fashion (in contrast to electrostatic and van der Waals interactions). Instead, hydrophobic forces should act upon the molecule as a whole, leading to formation of a hydrophobic core. In fact, pairwise optimization usually results in a static (unified) distribution of the optimized quantity as it implicitly assumes that each part of the molecule tends to a local optimum regardless of its placement. In the "fuzzy oil drop" model, the concept of a "hydrophobic core" goes beyond aggregation of hydrophobic molecules at the center of the molecule. In order to remain stable, the core must be shielded by a layer which prevents contact with water. This layer is characterized by a hydrophobicity gradient, resulting in near-zero hydrophobicity on the surface. The interplay between the core and its hydrophilic mantle ensures stability in a water environment. However, few molecules are expected to follow the idealized hydrophobicity density profile with perfect accuracy—such molecules would exhibit perfect solubility (owing to the unbroken mantle) but would also be unable to interact with any external molecules such as ligands, ions, substrates, or other proteins. It seems that by reaching an equilibrium between perfect adherence to the model and localized discrepancies, the protein can remain both soluble and capable of interactions. Such localized discrepancies express the link between hydrophobicity density distribution and the protein's biological function.

Variations in the status of polypeptide chain fragments point to their participation in stabilization (and perhaps also function) of the target protein, requiring dynamic adaptation to changing environmental conditions (such as the presence of a ligand). This model is further detailed in Hartman et al. (2013), where the authors propose certain assumptions with regard to disordered fragments.

Analysis of fragments characterized by loose packing can be likened to tracing the "history" of the folding process. If the proposed model is correct and accurately reflects the tendency of the protein to develop a hydrophobic core as a result of interactions with water, we can determine the sequence of events involved in the folding process by studying the adherence of specific structural units to theoretical predictions. The role of disordered fragments in this process, as well as in structural stabilization of individual structural units (complexes, chains, or domains), seems clear.

The presence of a water environment—which is ubiquitous in biological systems—seems to play an active part in shaping protein structures. This work proposes a method of analyzing the effects of this environment with regard to individual fragments of the polypeptide chain which undergoes folding. The "fuzzy oil drop" model reflects the global influence of water upon the function of proteins, forcing us to rethink the way in which protein structure is defined. Traditional approaches view the protein as a geometric shape whose form can be described in terms of secondary and supersecondary structural motifs. By adopting additional criteria (not related to geometry), it becomes evident that the distribution of hydrophobicity density may substantially influence the protein in ways unrelated to its secondary, tertiary, or quaternary structure (Shanmugham et al., 2012; Vugmeyster et al., 2011; Wood et al., 2013; Zhang et al., 2012).

As already mentioned, a protein chain exhibiting perfect adherence to the "fuzzy oil drop" model (as described by the 3D Gaussian) would be incapable of interacting with any molecules other than water. Local deformations in the hydrophobicity density distribution facilitate contact with ligands, substrates, and other complexation targets.

The protein's native structure, associated with its intended biological activity, is determined by juxtaposing two processes: optimization of internal interactions (producing secondary and supersecondary structural motifs) and environmental effects which lead to internalization of hydrophobic residues, along with exposure of hydrophilic residues on the surface. Both processes are somewhat contradictory: internal free energy optimization may counteract the entropic effects of water. Achieving a proper balance of these forces (expressed by ionic potentials or pH values) is necessary if the protein is to assume a stable and active form. For this reason, models based solely on internal free energy optimization do not produce adequate results, that is, their predictions are a poor match for experimentally determined structures.

Unfolding (denaturation)—especially if reversible—is not a direct consequence of the denaturing agent (such as urea) acting upon the protein body. Instead, the agent modifies the structural properties of water and alters the way in which the polypeptide chain interacts with its environment. Once the agent is removed, the environment reverts to a state which promotes protein folding as predicted by the "fuzzy oil drop" model.

The presence of other environmental factors may also affect the final distribution of hydrophobicity density throughout the protein body, ensuring high biological specificity. The same mechanism acts upon parts of the polypeptide chain, which—in addition to highly ordered fragments (described by secondary structural motifs)—also contain disordered fragments. The presented example suggests that lack of secondary structural ordering may be a direct consequence of another form of ordering, related to global hydrophobicity density distribution. Such "disordered" fragments, therefore, reflect the balance between local (internal free energy) and global (fuzzy oil drop) optimization.

The two-step model (first one—solely backbone dependent, second—environmental dependent)—proposed to simulate the protein folding *in silico* (Roterman et al., 2011) seems to be supported by experimental observation reported in Chung and Tokmakoff (2008).

5. CONCLUSIONS

Analysis of fragments listed in the DisProt database (http://www.disprot.org) suggests that their involvement in shaping the hydrophobic core in the protein (complex, chain, or domain) depends on their status vis-a-vis the "fuzzy oil drop" model. If the given structural unit is not accordant with theoretical predictions, the corresponding DisProt fragment usually also diverges from the model. It should also be noted that accordance depends critically upon selection of the structural unit. Tables 8.3–8.5 with results concerning the 2TPI list individual fragments of a sample protein along with their status within various structural units. For example—if the fragment is accordant within an individual domain but not within the entire chain, the given domain probably underwent folding on its own. Chains composed of distinct domains usually do not possess unified hydrophobic cores.

Plotting RD values for various structural units appears to accurately reflect the relations between these units and may enable a holistic interpretation of the folding process. Accordance of DisProt fragments with the "fuzzy oil drop" model may be related to the stability of crystal structures; it therefore appears that structures derived from NMR measurements are better suited to such studies.

DisProt fragments listed as discordant in Table 8.1 usually belong to proteins which, as a whole, do not possess regular hydrophobic cores. Such proteins are generally less stable than those in which a unified hydrophobic core (as predicted by the "fuzzy oil drop" model) can be found.

Synergy between disordered regions and structured domains increases the functional versatility of proteins and strengthens their interaction networks (Babu, Kriwacki, & Pappu, 2012; Mészáros, Dosztányi, Magyar, & Simon, 2014; Mészáros, Dosztányi, & Simon, 2012).

Structural accordance of DisProt fragments may be correlated with the protein's ability to revert to a stable conformation following function-related deformations. Proteins in which neither DisProt fragments nor any structural units (domains, chains, or complexes) conform to the model remain something of a mystery (at least in terms of their structural preferences).

ACKNOWLEDGMENTS

This work was made possible by the Jagiellonian University Medical College Grant No. K/ZDS/001531. We would also like to thank Piotr Nowakowski and Anna Zaremba-Śmietańska for their technical and editorial assistance.

REFERENCES

Abramavicius, D., & Mukamel, S. (2004). Many-body approaches for simulating coherent nonlinear spectroscopies of electronic and vibrational excitons. *Chemical Reviews*, *104*, 2073–2098.

Asplund, M. C., Zanni, M. T., & Hochstrasser, R. M. (2000). Two-dimensional infrared spectroscopy of peptides by phase-controlled femtosecond vibrational photon echoes. *Proceedings of the National Academy of Sciences of the United States of America*, *97*, 8219–8224.

Babu, M. M., Kriwacki, R. W., & Pappu, R. V. (2012). Versatility from protein disorder. *Science*, *337*, 1460–1461.

Baiz, C. R., Peng, C. S., Reppert, M. E., Jones, K. C., & Tokmakoff, A. (2012). Coherent two-dimensional infrared spectroscopy: Quantitative analysis of protein secondary structure in solution. *Analyst*, *137*, 1793–1799.

Banach, M., Konieczny, L., & Roterman, I. (2012a). Ligand-binding-site recognition. In I. Roterman-Konieczna (Ed.), *Protein folding in silico* (pp. 78–94). Oxford, Cambridge, Philadelphia, New Dehli: Woodhead Publishing.

Banach, M., Konieczny, L., & Roterman, I. (2012b). Use of the "fuzzy oil drop" model to identify the complexation area in protein homodimers. In I. Roterman-Konieczna (Ed.), *Protein folding in silico* (pp. 95–122). Oxford, Cambridge, Philadelphia, New Dehli: Woodhead Publishing.

Banach, M., Konieczny, L., & Roterman, I. (2013). Can the structure of hydrophobic core determine the complexation area? In Irena Roterman-Konieczna (Ed.), *Identification of ligand binding site and protein–protein interaction area* (pp. 41–54). Dordrecht, Heidelberg, New York, London: Springer.

Banach, M., Marchewka, D., Piwowar, M., & Roterman, I. (2012). The divergence entropy characterizing the internal force field in proteins. In I. Roterman-Konieczna (Ed.), *Protein folding in silico* (pp. 55–78). Oxford, Cambridge, Philadelphia, New Dehli: Woodhead Publishing.

Banach, M., Prymula, K., Jurkowski, W., Konieczny, L., & Roterman, I. (2012). Fuzzy oil drop model to interpret the structure of antifreeze proteins and their mutants. *Journal of Molecular Modeling*, *18*(1), 229–237.

Berendsen, H. J. C., Postma, J. P. M., van Gunsteren, W. F., & Hermans, J. (1981). Interaction models for water in relation to protein hydration. In B. Pullman (Ed.), *Intermolecular forces* (pp. 331–342). Dordrecht: Reidel Publishing Company.

Berendsen, H. J., van der Spoel, D., & van Drunen, R. (1995). GROMACS: A message-passing parallel molecular dynamics implementation. *Computational Physics Communication*, *91*, 43–56.

Biancardi, A., Cammi, R., Cappelli, C., Mennucci, B., & Tomasi, J. (2012). Modelling vibrational coupling in DNA oligomers: A computational strategy combining QM and continuum solvation models. *Theoretical Chemistry Accounts*, *131*, 1157.

Brylinski, M., Kochanczyk, M., Broniatowska, E., & Roterman, I. (2007). Localization of ligand binding site in proteins identified in silico. *Journal of Molecular Modelling*, *13*, 665–675.

Brylinski, M., Konieczny, L., & Roterman, I. (2007). Is the protein folding an aim-oriented process? Human haemoglobin as example. *International Journal of Bioinformatics Research and Application*, *3*(2), 234–260.

Brylinski, M., Prymula, K., Jurkowski, W., Kochańczyk, M., Stawowczyk, E., Konieczny, L., et al. (2007). Prediction of functional sites based on the fuzzy oil drop model. *PLoS Computational Biology*, *3*, e94.

Chung, H. S., & Tokmakoff, A. (2008). Temperature-dependent downhill unfolding of ubiquitin. I. Nanosecond-to-milisecond resolved nonlinear infrared spectroscopy. *Proteins*, *72*, 474–487.

Dill, K. A., & MacCallum, J. L. (2012). The protein-folding problem, 50 years on. *Science*, *338*, 1042–1046.

Harel, M., Kryger, G., Rosenberry, T. L., Mallender, W. D., Lewis, T., Fletcher, R. J., et al. (2000). Three-dimensional structures of Drosophila melanogaster acetylcholinesterase and of its complexes with two potent inhibitors. *Protein Science: A Publication of the Protein Society*, *9*, 1063–1072.

Hartman, E., Wang, Z., Zhang, Q., Roy, K., Chanfreau, G., & Feigon, J. (2013). Intrinsic dynamics of an extended hydrophobic core in the S. cerevisiae RNase III dsRBD contributes to recognition of specific RNA binding sites. *Journal of Molecular Biology*, *425*(3), 546–562. http://www.disprot.org.

Kauzmann, W. (1959). Some factors in the interpretation of protein denaturation. *Advances in Protein Chemistry*, *14*, 1–63.

Konieczny, L., Brylinski, M., & Roterman, I. (2006). Gauss function based model of hydrophobicity density in proteins. *In Silico Biology*, *6*(1–2), 15–22.

Kullback, S., & Leibler, R. A. (1951). On information and sufficiency. *Annals of Mathematical Statistics*, *22*, 79–86.

Laskowski, R. A. (2009). PDBsum new things. *Nucleic Acids Research*, *37*, D355–D359.

Levitt, M. (1976). A simplifed representation of protein conformations for rapid simulation of protein folding. *Journal of Molecular Biology*, *104*, 59–107.

Lin, Z., & van Gunsteren, W. F. (2013). On the choice of a reference state for one-step perturbation calculations between polar and nonpolar molecules in a polar environment. *Journal of Computational Chemistry*, *34*(5), 387–393.

Lindahl, E., Hess, B., & van der Spoel, D. (2001). GROMACS 3.0: A package for molecular simulation and trajectory analysis. *Journal of Molecular Modeling*, *7*, 306–317.

Marchewka, D., Banach, M., & Roterman, I. (2011). Internal force field in proteins seen by divergence entropy. *Bioinformation*, *6*(8), 300–302.

Marchewka, D., Jurkowski, W., Banach, M., & Roterman, I. (2013). Prediction of protein-protein binding interfaces. In I. Roterman-Konieczna (Ed.), *Identification of ligand binding site and protein–protein interaction area* (pp. 105–134). Dordrecht, Heidelberg, New York, London: Springer Verlag.

Mészáros, B., Dosztányi, Z., Magyar, C., & Simon, I. (2014). Bioinformatical approaches to unstructured/disordered proteins and their interactions. In A. Liwo (Ed.), *Computational methods to study the structure and dynamics of biomolecules and biomolecular processes: from bioinformatics to molecular quantum mechanics. Springer Series in Bio-/Neuroinformatics*, Vol. 1, Springer.

Mészáros, B., Dosztányi, Z., & Simon, I. (2012). Disordered binding regions and linear motifs—Bridging the gap between two models of molecular recognition. *PLoS One*, 7(10), e46829.

Murphy, G. S., Mills, J. L., Miley, M. J., Machius, M., Szyperski, T., & Kuhlman, B. (2012). Increasing sequence diversity with flexible backbone protein design: The complete redesign of a protein hydrophobic core. *Structure*, *20*(6), 1086–1096.

Priyakumar, U. D. (2012). Role of hydrophobic core on the thermal stability of proteins-molecular dynamics simulations on a single point mutant of sso7d. *Journal of Biomolecular Structure and Dynamics*, *29*(5), 1–11.

Prymula, K., Jadczyk, T., & Roterman, I. (2011). Catalytic residues in hydrolases: Analysis of methods designed for ligand-binding site prediction. *Journal of Computer-Aided Molecular Design*, *25*(2), 117–133.

Roterman, I., Konieczny, L., Jurkowski, W., Prymula, K., & Banach, M. (2011). Two-intermediate model to characterize the structure of fast-folding proteins. *Journal of Theoretical Biology, 283*(1), 60–70.

Shanmugham, A., Bakayan, A., Völler, P., Grosveld, J., Lill, H., & Bollen, Y. J. (2012). The hydrophobic core of twin-arginine signal sequences orchestrates specific binding to Tat-pathway related chaperones. *PLoS One*, 7(3), e34159.

Sickmeier, M., Hamilton, J. A., LeGall, T., Vacic, V., Cortese, M. S., Tantos, A., et al. (2007). DisProt: The database of disordered proteins. *Nucleic Acids Research, 35*(Database issue), D786–D793, Epub 2006 Dec 1.

Song, J., Guo, L. W., Muradov, H., Artemyev, N. O., Ruoho, A. E., & Markley, J. L. (2008). Intrinsically disordered gamma-subunit of cGMP phosphodiesterase encodes functionally relevant transient secondary and tertiary structure. *Proceedings of the National Academy of Sciences of the United States of America, 105*, 1505–1510.

Urbic, T., & Dill, K. A. (2010). A statistical mechanical theory for a two-dimensional model of water. *Journal of Chemical Physics, 132*, 224507.

Uversky, V. N., & Dunker, A. K. (2010). Understanding protein non-folding. *Biochimica et Biophysica Acta, 1804*, 1231–1264.

van der Spoel, D., Lindahl, E., Hess, B., Groenhof, G., Mark, A. E., & Berendsen, H. J. (2005). GROMACS: Fast, flexible, and free. *Journal of Computational Chemistry, 26*, 1701–1718.

van der Spoel, D., Lindahl, E., Hess, B., van Buuren, A.R., Apol, E., & Berendsen, H.J. (1995) Gromacs User Manual version 3.3.

Velanker, S. S., Ray, S. S., Gokhale, R. S., Suma, S., Balaram, H., Balaram, P., et al. (1997). Triosephosphate isomerase from Plasmodium falciparum: The crystal structure provides insights into antimalarial drug design. *Structure, 5*(6), 751–761.

Vucetic, S., Obradovic, Z., Vacic, V., Radivojac, P., Peng, K., Iakoucheva, L. M., et al. (2005). DisProt: A database of protein disorder. *Bioinformatics, 21*(1), 137–140.

Vugmeyster, L., Ostrovsky, D., Khadjinova, A., Ellden, J., Hoatson, G. L., & Vold, R. L. (2011). Slow motions in the hydrophobic core of chicken villin headpiece subdomain and their contributions to configurational entropy and heat capacity from solid-state deuteron NMR measurements. *Biochemistry, 50*(49), 10637–10646.

Walter, J., Steigemann, W., Singh, T. P., Bartunik, H., Bode, W., & Huber, R. (1982). On the disordered activation domain in trypsinogen. Chemical labelling and low-temperature crystallography. *Acta Crystallographica Section B, 38*, 1462–1472.

Wood, K., Gallat, F. X., Otten, R., van Heel, A. J., Lethier, M., van Eijck, L., et al. (2013). Protein surface and core dynamics show concerted hydration-dependent activation. *Angewandte Chemie International Edition in English, 52*(2), 665–668.

Yang, H., Jiao, X., & Li, S. (2012). Hydrophobic core-hydrophilic shell-structured catalysts: A general strategy for improving the reaction rate in water. *Chemical Communications (Cambridge, England), 48*(91), 11217–11219.

Zhang, X., Tan, Y., Zhao, R., Chu, B., Tan, C., & Jiang, Y. (2012). Site-directed mutagenesis study of the Ile140 in conserved hydrophobic core of Bcl-x(L). *Protein and Peptide Letters, 19*(9), 991–996.

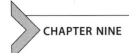

CHAPTER NINE

Conformational Elasticity can Facilitate TALE–DNA Recognition

Hongxing Lei[*,†,1], Jiya Sun[*,‡], Enoch P. Baldwin[§], David J. Segal[¶], Yong Duan[†,1]

[*]CAS Key Laboratory of Genome Sciences and Information, Beijing Institute of Genomics, Chinese Academy of Sciences, Beijing, China
[†]UC Davis Genome Center and Department of Biomedical Engineering, One Shields Avenue, Davis, California, USA
[‡]University of Chinese Academy of Sciences, Beijing, China
[§]Department of Molecular and Cellular Biology, University of California, Davis, California, USA
[¶]Genome Center and Department of Biochemistry and Molecular Medicine, University of California, Davis, California, USA
[1]Corresponding authors: e-mail address: leihx@big.ac.cn; duan@ucdavis.edu

Contents

1. Introduction 348
2. Methods 350
   2.1 MD simulations of TALE 350
   2.2 Evaluation of binding free energy for different RVDs 351
3. Results 352
   3.1 Elastic motion of the ligand-free TALE 352
   3.2 Evaluation of the binding free energy between RVDs and bases 357
4. Discussion 360
   4.1 Low free energy barrier implied from the high elasticity 360
   4.2 Technical considerations for the binding free energy evaluation 361
   4.3 Concluding remarks 362
Acknowledgments 363
References 363

Abstract

Sequence-programmable transcription activator-like effector (TALE) proteins have emerged as a highly efficient tool for genome engineering. Recent crystal structures depict a transition between an open unbound solenoid and more compact DNA-bound solenoid formed by the 34 amino acid repeats. How TALEs switch conformation between these two forms without substantial energetic compensation, and how the repeat-variable di-residues (RVDs) discriminate between the cognate base and other bases still remain unclear. Computational analysis on these two aspects of TALE–DNA interaction mechanism has been conducted in order to achieve a better understanding of the energetics. High elasticity was observed in the molecular dynamics simulations of DNA-free TALE structure that started from the bound conformation where it

Advances in Protein Chemistry and Structural Biology, Volume 94
ISSN 1876-1623
http://dx.doi.org/10.1016/B978-0-12-800168-4.00009-3

sampled a wide range of conformations including the experimentally determined apo and bound conformations. This elastic feature was also observed in the simulations starting from the apo form which suggests low free energy barrier between the two conformations and small compensation required upon binding. To analyze binding specificity, we performed free energy calculations of various combinations of RVDs and bases using Poisson–Boltzmann surface area (PBSA) and other approaches. The PBSA calculations indicated that the native RVD–base structures had lower binding free energy than mismatched structures for most of the RVDs examined. Our theoretical analyses provided new insight on the dynamics and energetics of TALE–DNA binding mechanism.

1. INTRODUCTION

Transcription activator-like effectors (TALEs) are sequence-programmable transcription factors derived from bacterial plant pathogens. They have garnished wide attention in recent years due to their modular design consisting of highly similar repeats. Each repeat can recognize one base by the repeat-variable di-residues (RVDs) with well-documented specificity, including NI (Asn-Ile) to A, HD (His-Asp) to C, NH (Asn-His) to G, and NG (Asn-Gly) to T (Boch et al., 2009; Moscou & Bogdanove, 2009). This simple recognition code as well as the low toxicity has led to its fast-developing applications in diverse fields (Bedell et al., 2012; Sanjana et al., 2012; Tremblay, Chapdelaine, Coulombe, & Rousseau, 2012). For instance, the precise targeting of genomic loci in numerous species has been demonstrated using engineered TALE nucleases (Perez-Pinera, Ousterout, & Gersbach, 2012). Engineered TALE transcription factors and recombinases have also been described (Boch et al., 2009; Mercer, Gaj, Fuller, & Barbas, 2012).

However, our understanding of the mechanism by which TALE proteins interact with DNA and achieve such high specificity lags far behind our ability to use them as successful tools. For example, the recent structures of several TALE–DNA complexes have been determined by X-ray crystallography (Deng et al., 2012; Mak, Bradley, Cernadas, Bogdanove, & Stoddard, 2012). Based on these structures, each repeat consists of two helical segments connected by a short loop that contains the RVD sequences. Surprisingly, only the second residue of the RVD contributes directly to the base recognition, while the first residue mainly contributes to the C-terminal capping of the first helix (Deng et al., 2012; Mak et al., 2012). Although specific hydrogen bonding and other interactions have been observed from the X-ray structures, the available data do not provide a quantitative explanation for the apparent high specificity imparted by the RVDs. Structural data of mismatched RVD–base pairings are presently lacking. Another interesting finding is that although the apo- and DNA-bound forms share the same helical architecture, the bound TALE is much more compact (Deng et al., 2012). Specifically, while both contain 11 repeats per turn, the pitch changes from 60 to 35 Å per turn upon binding, accompanied by subtle repacking at the repeat interfaces. These two distinct conformations have been observed in independent X-ray structures (Gao, Wu, Chai, & Han, 2012; Mak et al., 2012). However, the mechanism by which the ligand-free TALE switches from the apo form to the bound form upon DNA binding is not yet revealed. This is a critical issue because it would require large compensation upon binding if there is a significant free energy barrier separating these two conformations.

Semiquantitative experiments have investigated the binding specificity of RVDs. Using a reporter assay, Cong, Zhou, Kuo, Cunniff, and Zhang (2012) interrogated the binding specificity of 23 RVDs, which confirmed the specific recognition of NI to A, HD to C, NN to G/A, and NG to T and discovered highly specific recognition of NH to G. They further evaluated the binding free energy and found that NH–G binding was 0.86 kcal/mol more favorable than NN–G binding. Streubel, Blucher, Landgraf, and Boch (2012) examined the specificity and efficiency of 14 RVDs also using a reporter assay and various TALE constructs. HD and NN were identified as strong RVDs, while NG, NI, NK, and N* were scored as weak RVDs (* indicates the absence of the second RVD residue). In addition, NH displayed higher specificity to G than did NN, while NS, NT, and HN displayed recognition to both A and G. Our more recent quantitative study, using DNA electrophoretic mobility shift assays with highly purified TALE proteins, showed the relative RVD affinity in the order $NG > HD \sim NN \gg NI > NK$, with each repeat contributing an average of 1–4 kJ/mol to binding free energy (Meckler et al., 2013). The discrepancies with the cellular measurements underscore the need for more quantitative measurements *in vitro* and *in silico* in order to probe the physical basis and mechanisms of TALE–DNA binding. Despite the great importance of TALEs, a comprehensive investigation of the binding specificity by free energy calculation has yet to be reported, partly due to the challenge of evaluating protein–DNA interaction energies.

In this work, we investigated the dynamics and energetics of TALE–DNA binding mechanism through computational analyses. First, we conducted molecular dynamics (MD) simulations to investigate the conformational elasticity of TALE. Our MD simulations started with both bound and free forms where consistent features were observed. Second, we applied Poisson–Boltzmann surface area (PBSA) (Kollman et al., 2000) calculations to evaluate binding free energies between RVDs and bases. This physics-based approach was compared with two empirical approaches, namely, Rosetta (Leaver-Fay et al., 2011) and DDNA3 (Zhao, Yang, & Zhou, 2010). Here, we report insights gained from our computational analyses.

2. METHODS

2.1. MD simulations of TALE

The AMBER (version 12) software package (Case et al., 2012) and FF03 force field (Duan et al., 2003) were used for the MD simulations. The initial TALE coordinates for our MD simulations were extracted from X-ray crystallographic structures of the free apo (PDB code: 3V6P) and bound (PDB code: 3V6T) forms of dHAX3 (Deng et al., 2012). In order to allow room for substantial movement, large water boxes were used with minimum 27 Å from the protein or complex surface to the solvent wall, resulting in 128,774 atoms for the bound system and 127,978 atoms for the free system. The systems were neutralized by adding Na^+ and Cl^- to the systems using the tleap program in AMBERTOOLS. Short minimization (500 steps, steepest decent) and equilibration (500 ps, NPT, constant pressure, and temperature) with positional restraints on TALE were performed to bring the solvated systems to normal pressure (1 atm.) and room temperature (300°K). For the bound system, standard production run was performed for 50 ns with triple replicates using different random seeds at the beginning of the replicate simulations. For the free system extracted from the bound structure after the removal of the DNA, a standard production run was performed for 250 ns with triple replicates, again using different random seeds at the beginning of the replicate simulations. Briefly, the production simulations were conducted at NVT mode (constant volume and temperature, $T=300$ K). No positional restraints were applied in the production run. Temperature was controlled by using Berendsen's thermostat with a coupling constant of 2.0 ps. SHAKE was applied to constrain all bonds connecting hydrogen atoms. The particle-mesh Ewald method was used to treat long-range electrostatic interaction under periodic boundary condition. The cutoff distance for short-range nonbonded interaction was 10 Å, while the long-range van der Waals (VDW) interaction was treated by a uniform density approximation. To reduce computation, nonbonded forces were calculated using the two-step RESPA approach. To eliminate the "block of ice" problem, we reset the translation and rotation of the center of mass every 500 steps. Coordinates were saved every 10 ps, resulting in 5000 snapshots for each bound system and 25,000 snapshots for each free system. In addition, simulations were also performed on free system starting from the apo structure with a set of triple simulations conducted for 50 ns each using the same protocol. The simulations were performed on NVIDIA GPU using the GPU version of pmemd (Gotz et al., 2012). Each 10 ns of the simulations required about 50 h.

2.2. Evaluation of binding free energy for different RVDs

Our template system for binding energy calculation was extracted from the DNA-bound dHAX3 X-ray structure (PDB code: 3V6T) (Deng et al., 2012). There are 11.5 repeats in the original structure ("0.5" refers to the most C-terminal DNA binding repeat). We extracted repeats 7–11 as the template for permutation and kept the original DNA intact. The RVDs for repeats 7–11 were NS-NG-HD-NG-HD in the X-ray structure. We fixed all the RVDs except for the central repeat 9, which was mutated to 15 other RVDs, NG, NN, NH, NK, NI, NS, NT, NP, ND, N* (* indicates the absence of the second RVD residue), NA, HG, HN, HS, and HT. For each of these 16 RVDs (including the original HD), all 4 possible base pairs at the recognition site were also constructed by the tleap program in AMBERTOOLS, resulting in a total of 64 systems. During the *in silico* mutagenesis, only the backbone atoms of the mutated amino acid or base were kept and the side chain or base atoms were automatically generated by tleap. In order to relax the mutation site, a short MD run (500 ps, NVT) with explicit solvent was performed, preceded by energy minimization (500 steps, steepest decent) and equilibration (500 ps, NPT) according to the standard AMBER protocol. During the entire process of the simulations, all atoms were fixed by positional restraints with harmonic forces (5 kcal/mol/Å^2) except for the central RVD loop and the central base pair at the mutation site. From each MD simulation, 100 snapshots were saved for energy evaluation. We conducted three types of binding free energy calculations. PBSA was performed with the AMBER package following the standard protocol. The solvation-free energies were calculated for the structures obtained from the simulations. The average binding free energy of the 100 snapshots was reported directly by the PBSA calculation. Rosetta (version 3.4) software (www.rosettacommons.org) was installed following the instruction. The 100 snapshots were converted to individual PDB files by the ptraj program in AMBERTOOLS and submitted to Rosetta for energy calculation, and the free energies were averaged for each of the 64 complexes. The scoring option for the RosettaDNA module was used to calculate the binding energy. No further structural optimization was applied prior to energy calculation. The binary code of DDNA3 was downloaded (http://sparks.informatics.iupui.edu/yueyang/DFIRE/ddna3-service). No options for structure optimization were available in DDNA3. Similar to the Rosetta calculation, DDNA3 binding free energies of the individual 100 snapshots were evaluated and then averaged.

3. RESULTS

3.1. Elastic motion of the ligand-free TALE

One of the main goals of this study was to examine the elasticity of ligand-free TALE and how it may contribute to DNA binding. As a reference and a validation of the simulation protocols, we first conducted simulations on the bound TALE–DNA complex. Within the 50-ns MD simulations with explicit solvent, the three independent trajectories displayed similar stable features with fluctuations around a 3-Å backbone RMSD compared to the starting structure (Fig. 9.1). This fluctuation was considerably smaller

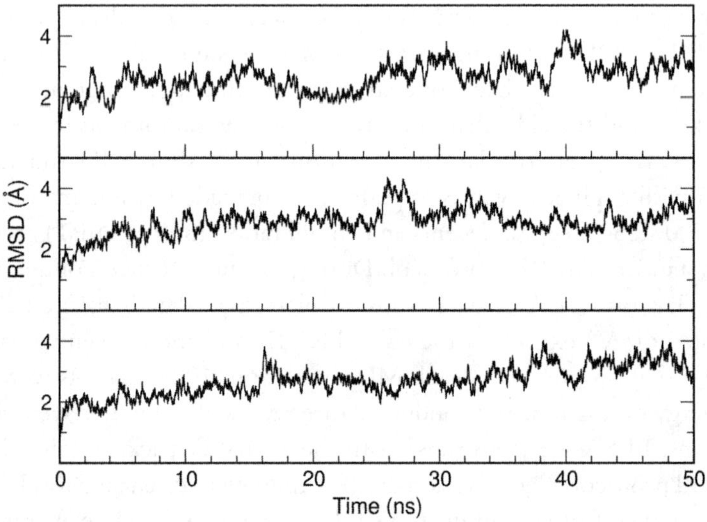

Figure 9.1 RMSD profiles from the three 50-ns MD simulations with the TALE–DNA complex (PDB code: 3V6T, the complete system).

when compared to the simulations with the free TALE which will be described later. The results indicated that the bound TALE conformation in the presence of the DNA is in a free energy minimum as would be expected from the experimental X-ray structure. According to previous experimental studies (Deng et al., 2012), two major forces contribute to the favorable interaction between TALE and DNA. These include the nonspecific contribution from the interaction between Lys16/Gln17 near the RVD loop and the DNA backbone phosphate group, and the specific contribution from the RVD–base interactions. The combination of the two favorable forces led to the free energy minimum observed in our simulation. In addition, this stability test provided a validation for the suitability of the simulation protocol used in this study.

In contrast to the simulations in the presence of the DNA, the simulations with ligand-free TALE all displayed high elasticity. Two sets of MD simulations were performed in the absence of DNA: one started from the apo form (PDB code: 3V6P) and the other from the bound form (PDB code: 3V6T) with the DNA removed. All simulations were performed with explicit solvent. Large water boxes were used in anticipation of substantial elastic movement. The backbone RMSDs from these two simulations are shown in Figs. 9.2 and 9.3, respectively. The RMSDs were calculated relative to both the apo (black trace) and bound (green trace) forms in both cases.

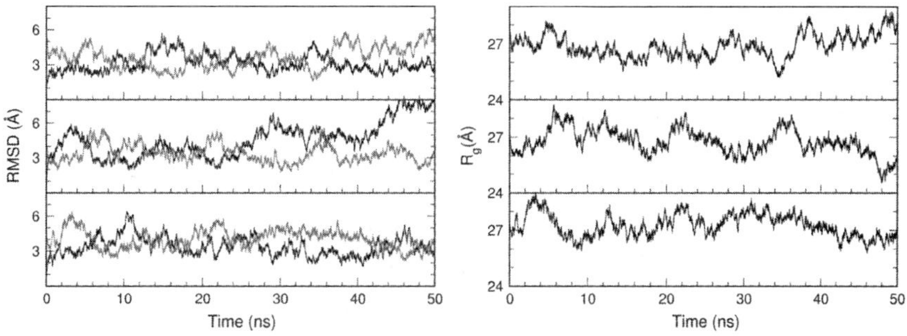

Figure 9.2 The profiles of RMSD (left) and radius of gyration (Rg, right) from the three 50-ns MD simulations with the ligand-free TALE starting from the apo structure (PDB code: 3V6P). In the RMSD profiles, the RMSDs against the apo structure are shown in black and the RMSDs against the bound structure are shown in green. (For interpretation of the references to color in this figure legend, the reader is referred to the online version of this chapter.)

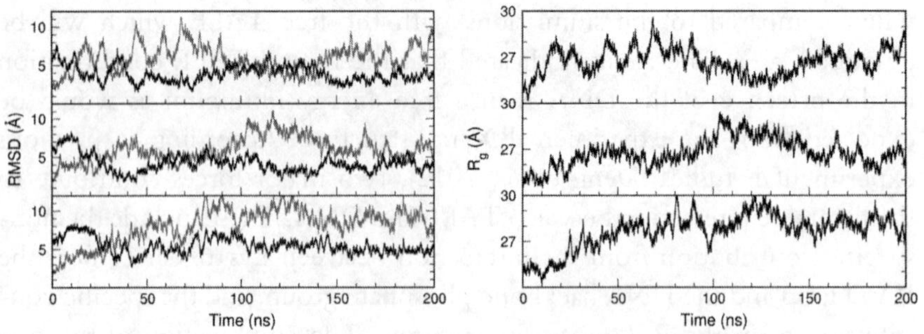

Figure 9.3 The profiles of RMSD (left) and Rg (right) from the three 200-ns MD simulations with the ligand-free TALE starting from the bound structure (PDB code: 3V6T, DNA removed). In the RMSD profiles, the RMSDs against the bound structure are shown in green and the RMSDs against the apo structure are shown in black. (For interpretation of the references to color in this figure legend, the reader is referred to the online version of this chapter.)

A consistent picture emerged from these simulations was the constant oscillation of TALE although these simulations started from two different conformations. This is clearly illustrated by the fact that RMSDs relative to apo (black trace) and bound (green trace) forms both exhibit large degree of fluctuations. These two RMSD profiles also moved generally in the opposite directions. For example, when the RMSD relative to the apo structure went up to 6 Å, the RMSD relative to the bound structure decreased to below 3 Å and vice versa. Thus, TALE is quite elastic in the absence of DNA. More importantly, TALE oscillates in a wide range encompassing the apo and bound forms.

Since the apo-form TALE is an experimentally determined structure, we expected it to be reasonably stable in the simulation. Indeed, the backbone RMSD was fluctuating around 3 Å in two of the three 50-ns simulation trajectories started from the apo form (Fig. 9.2, left, black trace). However, we still observed considerable fluctuations as shown by RMSD and radius of gyration (Rg) (Fig. 9.2, RMSD, left, and Rg, right). In one of these three trajectories (Fig. 9.2, top panel), although the apo basin around 3 Å backbone RMSD was the predominant conformation, it moved away to 5–6 Å a few times, and for several short periods, it was close to the bound form (green trace). In another trajectory (Fig. 9.2, bottom panel), it displayed high fluctuation in the first half, but the apo basin was heavily sampled in the second half. Transient sampling close to the bound form was also observed in this trajectory. In the third trajectory (Fig. 9.2, middle panel), it moved away from the apo basin near 25 ns and stayed away during the second half of the 50-ns simulation. Sampling close to the bound form was also observed in the second half of the simulation. The substantial elastic motion could also be seen from the Rg profiles. The Rg of the apo form was near 27 Å according to the initial values at the beginning of the three simulations. It fluctuated between 24.5 and 29 Å during the simulations, with the lower boundary close to the bound form and the upper boundary more extended than the apo form.

With the removal of the DNA, the bound conformation was expected to be less stable in the simulation. Indeed, in all three trajectories that started from the bound form, TALE moved away from the bound conformation within 50 ns (Fig. 9.3, left, green trace). Another interesting observation was the transient sampling back to the bound form in two trajectories (Fig. 9.3, top and middle). This elastic motion can also be seen from the Rg profiles that fluctuated between 24 and 32 Å during the simulations (Fig. 9.3, right). It should be noted that these Rg values are not directly comparable to the Rg values in Fig. 9.2 because the bound structure (PDB code: 3V6T) had a longer chain than the apo structure (PDB code: 3V6P). Nonetheless, we still observed the similar features that the TALE repeat structures sampled a wide range of conformations including the apo and bound forms.

To further illustrate the substantial elastic motion, we selected three representative snapshots based on the closeness to the apo or bound forms (Fig. 9.4, red color) from the trajectory shown in the top panel of

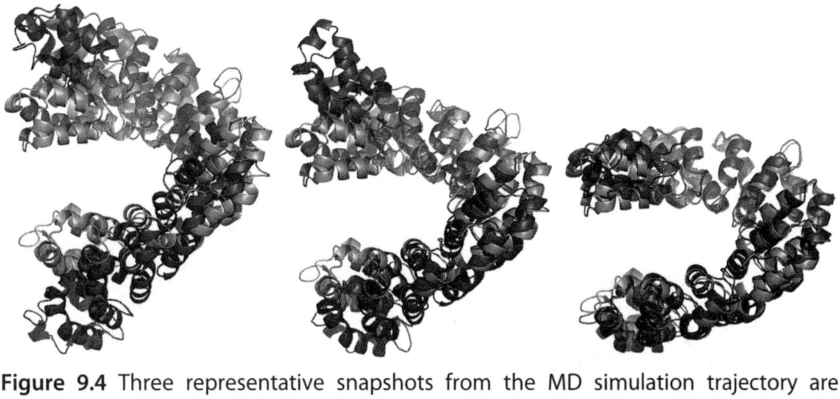

Figure 9.4 Three representative snapshots from the MD simulation trajectory are shown in the top panel of Fig. 9.3 (left, 68.25 ns, highly extended; middle, 73.95 ns, close to the apo form; and right, 130.29 ns, close to the bound form). The structures from the simulation are shown in red and the reference bound structure is shown in green. (For interpretation of the references to color in this figure legend, the reader is referred to the online version of this chapter.)

Fig. 9.3 and compared them against the starting bound form (Fig. 9.4, green color). At 68.25 ns (Fig. 9.4, left), it adopted a rather extended conformation with significant deviation from the starting bound structure (11.08 Å backbone RMSD to the bound form, Rg=29.4 Å). At 73.95 ns (Fig. 9.4, middle), it reached a conformation very close to the apo form (backbone RMSD=1.56 Å to the apo form, Rg=27.1 Å). At 130.29 ns (Fig. 9.4, right), however, it transiently moved back to the compact conformation very close to the starting structure (2.08 Å backbone RMSD to the bound form, Rg=24.9 Å).

In summary, we observed consistent features in the DNA-free simulations started from both the apo and bound forms: (1) the DNA-free TALE is highly dynamic with constant elastic movement; (2) the apo form is closer to the energy minima than the bound form as demonstrated by the RMSD profiles; and (3) the bound form can be transiently reached during the elastic motions.

The conformational sampling can be further illustrated using a three-dimensional contour map (Fig. 9.5). To construct this map, sampling data from the three trajectories shown in Fig. 9.3 (started with the bound form, DNA removed) were merged. Sampling over a large conformational space is evident from this map. The RMSD to the bound TALE varied mostly from 2 to 12 Å, while the RMSD to the apo TALE varied mostly from 1.5 to 8 Å. The most heavily sampled region was within RMSD 3–4.5 Å to the apo form and 5–7 Å to the bound form. We have also performed clustering

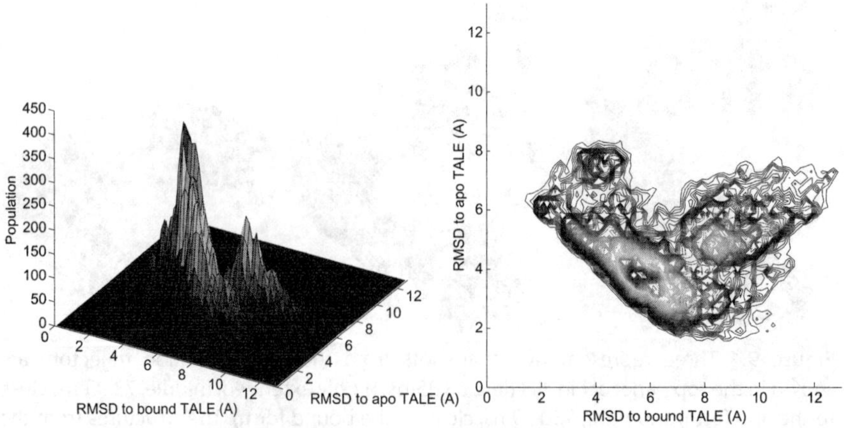

Figure 9.5 Conformational sampling of the ligand-free TALE from the three MD simulations are shown in Fig. 9.3. (For color version of this figure, the reader is referred to the online version of this chapter.)

analysis for all the conformations on the map. The top five clusters comprise ~75% of the conformations, and the representative structures for the top five clusters are shown on the map. In brief, clusters #1 and #4 are close to the apo form (RMSD_apo = 3.51 and 2.32 Å, population = 33.9% and 8.0%, respectively) for a combined ~42% population. Cluster #5 is close to the bound form (RMSD_bound = 3.01 Å, population = 5.2%). The other two clusters have intermediate conformations (RMSD_apo = 4.76 Å, RMSD_bound = 8.72 Å, and population = 17.5% for cluster #2; RMSD_apo = 4.77 Å, RMSD_bound = 3.95 Å, and population = 10.5% for cluster #3).

Overall, the sampling was biased toward the apo form in these simulations, even though they all started from the bound structure. Since the RMSD to the bound form also reflected the compactness of the TALE, the map illustrated the broad conformational sampling with large variation in the compactness. The DNA-free TALE constantly underwent oscillation movement with center close to the apo form, while the bound form can be reached during the oscillation process.

3.2. Evaluation of the binding free energy between RVDs and bases

Another critical component of the binding mechanism is the specific recognition of RVDs and bases. In order to understand the energetic contribution to the binding specificity, we evaluated the binding free energy between RVDs and bases using three different methods, PBSA, Rosetta, and DDNA3. A minimal local environment was included in the calculations with a five-repeat segment from the high resolution X-ray structure (PDB code: 3V6T, repeats 7–11) and in silico mutations were performed on the central repeat (repeat #9). Such a minimal environment helps to reduce the uncertainty associated with inevitable fluctuation due to the remaining parts. Furthermore, during the relaxation of the structure, only the central RVD loop and the central base pair were allowed full flexibility, whereas all other atoms were restrained by harmonic forces. We attempted other protocols and found that DNA has the tendency to untwist when simulations exceed 10 ns. Thus, to reduce the influence of the inevitable approximation in simulations including both parameterization and limited sampling, it was necessary to keep the TALE–DNA close to the experimental structures. We evaluated 16 RVDs (Table 9.1) for which observed experimental binding preferences are described in the literature (Boch et al., 2009; Cong et al., 2012; Streubel et al., 2012). All 64 possible RVD–base pair

Table 9.1 An overall comparison of the performance for binding free energy evaluation by PBSA, VDW, Rosetta, and DDNA3

RVD–base	PBSA	VDW	DDNA3	Rosetta
NI–A	1	2	X	X
NS–A	1	2	2	X
NK–G	1	X	X	1
NH–G	X	2	X	1
NN–G	2	1	X	1
NG–T	2	X	X	X
HG–T	2	X	1	1
HD–C	1	X	X	X
HN–AG	A	A	T(x)	G
NT–AG	A	A	C(x)	C(x)
NP–ACT	T	C	A	G(x)
N*–CT	T	T	C	T
HT–AG	C(x)	G	A	G
NA–CT	C	C	A(x)	T
ND–C	1	2	1	2
HS–A	1	2	2	X

Note: For RVDs with single preference, the energy ranking of the base is shown as 1 (lowest energy), 2 (second lowest energy), or X (others). For RVDs, which recognize multiple bases, the base with lowest energy is shown ("x" stands for wrong energy ranking). The second preferred base for NN is A. NS also recognizes other bases. For comparison, the performance by VDW (van der Waals) is also shown.

combinations were evaluated and, for each, the average binding free energy of 100 relaxed complex structures was calculated by the three methods.

A summary of the energy evaluation by the three methods is shown in Table 9.1 (more detailed free energy values can be found in Supplementary Table 1 (see http://dx.doi.org/10.1016/B978-0-12-800168-4.00009-3). Overall, the PBSA energies exhibited better correlations with experimental observations. For the 10 RVDs with single base preferences, PBSA had 6 ranked at No. 1 and 3 ranked at No. 2, while Rosetta had 4 ranked at No. 1 and 1 ranked at No. 2, and DDNA3 only had 2 ranked at No. 1 and 2 ranked at No. 2. For the six RVDs, which recognize multiple bases, PBSA had incorrect ranking for only one RVD, while Rosetta had two

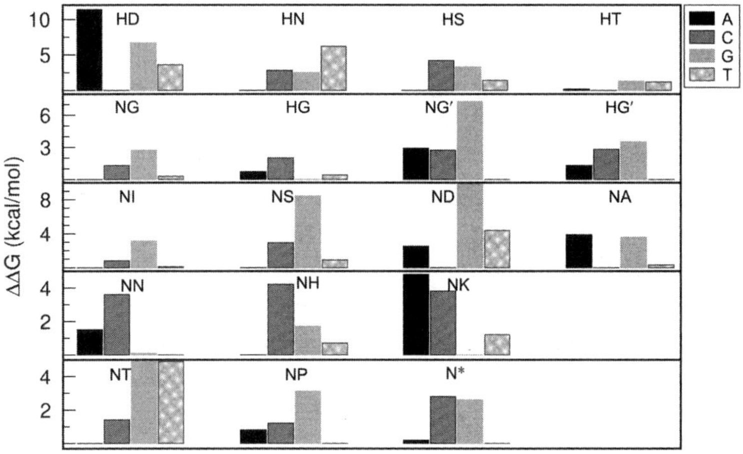

Figure 9.6 Binding free energy evaluation of 16 RVDs and all 4 possible bases for each RVD by PBSA. For each RVD, the lowest binding free energy was set to zero, while others were assigned to positive energy based on the energy difference. For NG and HG, a second template with NG at the central RVD of the original structure was used for energy evaluation (shown as NG′ and HG′). (For color version of this figure, the reader is referred to the online version of this chapter.)

RVDs incorrectly ranked, and DDNA3 had three RVDs incorrectly ranked. A potential problem with Rosetta was that it showed preference over either G or T for all but 2 of the 16 RVDs examined.

The results of the PBSA energy evaluation are summarized in Fig. 9.6. For HT, NG, HG, and NN, the native recognition was only 0.2–0.6 kcal/mol away from the lowest binding free energy. The consistency of PBSA with experimental findings prompted us to further dissect the energy components of PBSA. The ranking performance by VDW is shown in Table 9.1. It is evident that the ranking by VDW is much less satisfactory than the total PBSA energy. Similarly, none of the other energy components demonstrated better correlation with experimental observation than the total PBSA energy (data not shown). Therefore, the specific recognition of TALE arises from the combination of VDW, electrostatics, and solvation-free energy, not dominated by any of the individual terms.

In order to examine the effect of the native conformation on energy evaluation, we reconducted energy evaluations for RVDs NG and HG using another template that consisted of repeats 6–10 with NG as the central RVD (repeat #8) in the original structure (PDB code: 3V6T). It is evident that both NG and HG had clear preference over T using this template (NG′ and HG′ in Fig. 9.6). This suggests that the NG–T interaction was more optimized in the X-ray structure than our constructed structure by mutation. Another interesting insight regarding NG–T interaction can be gained from the energy evaluation. Based on the X-ray structure, it has been hypothesized that specific recognition of NG to T was likely due to the exclusion mechanism, that is, that NG can accommodate the thymine 5-methyl group and other RVD side chains would be expected to clash with the group. However, in our *in silico*-constructed TALE systems, the RVD interactions with T at the recognition site were all well tolerated; there were no visible clashes in any of the structures even with restraints on most of the atoms; and the T recognition was not the least favorable interaction for most of the RVDs examined (Fig. 9.6), suggesting that NG–T recognition may not be due to the exclusion mechanism.

4. DISCUSSION

4.1. Low free energy barrier implied from the high elasticity

The observation of two distinctive TALE conformations at the apo and bound states from crystallography data prompted us to conduct a computational analysis of the conformational space for the DNA-free TALE. The MD simulations starting from both the apo and bound forms demonstrated consistent features. Although the apo conformation was more favorable, the DNA-free TALE was highly elastic. A wide range of conformations were sampled in the simulations, and some were significantly more extended than the apo form while others were more compact. The bound conformation was also transiently sampled in the DNA-free simulations, and the overall feature was the constant oscillation with center close to the apo conformation. Together with the more favorable binding free energy for the specific RVD–base recognition, the high elasticity may help us to dissect the energetics in TALE–DNA interactions. The cylindrical TALE–DNA complex structure requires the wrapping of TALE around DNA major groove, which can be difficult without the high elasticity observed in our simulation. The ability of apo TALE to reach the bound conformation implies a low free energy barrier separating the bound and unbound conformations. Favorable interactions with the DNA backbone as described earlier can help TALE to overcome this small free energy barrier. Since TALE does not bind to a random DNA sequence, this nonspecific TALE–DNA interaction is likely in the similar scale as the free energy barrier between the two TALE states. The overall favorable binding free energy likely comes entirely from the specific RVD–base interaction including the neighbor effect. Therefore, it is critical to quantitatively determine the binding energies of RVD–base interactions. We have also attempted *ab initio* TALE–DNA binding simulation. However, the preliminary test showed that the timescale for binding is far beyond our reach. Therefore, more details regarding the initial binding process cannot be revealed from the simulation.

4.2. Technical considerations for the binding free energy evaluation

Due to the errors in parameterization and difficulty in evaluating entropy, accurate and quantitative free energy calculation has been a major challenge in the field of computational biology. Not surprisingly, the evaluation of binding free energy for RVD–base recognition in this study turned out to be technically challenging. We tested several alternative strategies to perform the analysis. Since extended simulations can provide extensive conformation sampling, we first attempted longer simulations to allow the structures to relax to their bound states. However, extended relaxation of the central repeat or the whole five-repeat segment without restraints led to significantly distorted DNA structure with the DNA clearly untwisted. Calculations using those simulated structures had notably worse correlation with experimental RVD specificities, and in many cases, yielded values close to random ranking for all three energy evaluation methods (data not shown). This implies inherent problems in the underlying simulation parameters, in particular, the parameter set representing DNA because notable distortion of DNA conformation was observed consistently in the simulations without restraints. Good correlation with experimental observation was obtained only when stringent restraints were applied (Table 9.1 and Fig. 9.6). The limited conformational sampling during the short MD simulations (500 ps) ensured that the simulation sampled the local minimum only and retained the critical features of experimental structures. Clearly, much work is needed to improve the simulation parameters. Nevertheless, it is encouraging that such difficulty can be partially circumvented.

The selection of template was also critical in this study. We have conducted full analyses on these 16 RVDs using 2 different templates, one with repeats 7–11 (template #1) and the other with repeats 6–10 (template #2). Although template #2 gave better results for NG and HG (NG′ and HG′ in Fig. 9.6), the overall performance was less satisfactory for all three methods (data not shown). One of the potential problems with template #2 was the side-chain interaction among neighboring RVDs, which was weak with template #1 because the flanking RVDs were NG on both sides of the central repeat. Again, this suggests deficiency in the force field.

Since the RVD side-chain orientations are critical for the favorable RVD–base interaction, we also used the side-chain orientations from the X-ray structures of the same or similar RVDs whenever possible, whereas the direct assignment of side-chain orientation by AMBER tleap led to less satisfactory correlations (data not shown).

The empirical methods Rosetta and DDNA3 had less satisfactory performance compared to PBSA even though the experimental protein–DNA interactions were not included in the parameterization process of PBSA. Although PBSA has not been extensively tested for protein–DNA interactions, the results from this study suggest that PBSA might be the better choice for understanding the energetics of TALE–DNA interactions. The lessons learned from this study shall be carefully considered in future computational studies on TALE–DNA binding mechanism. However, we note that this result does not necessarily diminish the usefulness of Rosetta, DDNA3, or other empirical methods. For example, structure refinement may lead to better ranking in Rosetta, which was not tested in this work. Given the increasing availability of DNA–protein complex structures, these methods are expected to improve over time.

4.3. Concluding remarks

In this work, we conducted computational analyses on the conformational elasticity and specific recognition of TALEs. Novel insights regarding the binding mechanism were gained from the MD simulations of the DNA-free TALE. While the DNA-bound TALE structure was relatively stable, the DNA-free TALE underwent significant and reversible conformational transition in the simulations irrespective of the starting conformation. This spring-like motion may be a critical part of the binding mechanism for TALE–DNA interactions. The PBSA binding free energy calculation was validated by the result that the native pairing of RVD and base was favored compared to the mismatched pairings, and showed better consistency than the empirical approaches including Rosetta and DDNA3. An additional insight from the free energy evaluation is the proposition that NG to T recognition is not due to exclusion of other larger side chains by the base as suggested by many, since all the substitutions examined were well tolerated in the simulations. Based on the computational analyses on these two aspects, we propose that the high elasticity of DNA-free TALE leads to low free energy barrier between the apo and bound states, which requires only small compensation from the nonspecific TALE–DNA interaction upon binding. Therefore, the binding affinity may come entirely from the specific RVD–base interaction.

ACKNOWLEDGMENTS

We would like to thank Dr. Yuedong Yang for helping us solve some technical issues with DDNA3. This work was supported by Research Grants from NIH (GM79383 to Y. D.; GM097073 to D. J. S.) and MOST (Grant 2014CB964901 to H. L.). The authors have no conflicts of interest to declare.

REFERENCES

Bedell, V. M., Wang, Y., Campbell, J. M., Poshusta, T. L., Starker, C. G., Krug, R. G., 2nd., et al. (2012). In vivo genome editing using a high-efficiency TALEN system. *Nature*, *491*, 114–118.

Boch, J., Scholze, H., Schornack, S., Landgraf, A., Hahn, S., Kay, S., et al. (2009). Breaking the code of DNA binding specificity of TAL-type III effectors. *Science*, *326*, 1509–1512.

Case, D. A., Darden, T. A., Cheatham, T. E., III, Simmerling, C. L., Wang, J., Duke, R. E., Luo, R., et al. (2012). *AMBER 12*. San Francisco: University of California.

Cong, L., Zhou, R., Kuo, Y. C., Cunniff, M., & Zhang, F. (2012). Comprehensive interrogation of natural TALE DNA-binding modules and transcriptional repressor domains. *Nature Communications*, *3*, 968.

Deng, D., Yan, C., Pan, X., Mahfouz, M., Wang, J., Zhu, J. K., et al. (2012). Structural basis for sequence-specific recognition of DNA by TAL effectors. *Science*, *335*, 720–723.

Duan, Y., Wu, C., Chowdhury, S., Lee, M. C., Xiong, G., Zhang, W., et al. (2003). A point-charge force field for molecular mechanics simulations of proteins based on condensed-phase quantum mechanical calculations. *Journal of Computational Chemistry*, *24*, 1999–2012.

Gao, H., Wu, X., Chai, J., & Han, Z. (2012). Crystal structure of a TALE protein reveals an extended N-terminal DNA binding region. *Cell Research*, *22*, 1716–1720.

Gotz, A. W., Williamson, M. J., Xu, D., Poole, D., Le Grand, S., & Walker, R. C. (2012). Routine microsecond molecular dynamics simulations with AMBER on GPUs. 1. Generalized born. *Journal of Chemical Theory and Computation*, *8*, 1542–1555.

Kollman, P. A., Massova, I., Reyes, C., Kuhn, B., Huo, S., Chong, L., et al. (2000). Calculating structures and free energies of complex molecules: Combining molecular mechanics and continuum models. *Accounts of Chemical Research*, *33*, 889–897.

Leaver-Fay, A., Tyka, M., Lewis, S. M., Lange, O. F., Thompson, J., Jacak, R., et al. (2011). ROSETTA3: An object-oriented software suite for the simulation and design of macromolecules. *Methods in Enzymology*, *487*, 545–574.

Mak, A. N., Bradley, P., Cernadas, R. A., Bogdanove, A. J., & Stoddard, B. L. (2012). The crystal structure of TAL effector PthXo1 bound to its DNA target. *Science*, *335*, 716–719.

Meckler, J. F., Bhakta, M. S., Kim, M.-S., Ovadia, R., Habrian, C. H., Zykovich, A., et al. (2013). Quantitative analysis of TALE-DNA interactions suggests polarity effects. *Nucleic Acids Research*, *41*(7), 4118–4128.

Mercer, A. C., Gaj, T., Fuller, R. P., & Barbas, C. F., 3rd. (2012). Chimeric TALE recombinases with programmable DNA sequence specificity. *Nucleic Acids Research*, *40*, 11163–11172.

Moscou, M. J., & Bogdanove, A. J. (2009). A simple cipher governs DNA recognition by TAL effectors. *Science*, *326*, 1501.

Perez-Pinera, P., Ousterout, D. G., & Gersbach, C. A. (2012). Advances in targeted genome editing. *Current Opinion in Chemical Biology*, *16*, 268–277.

Sanjana, N. E., Cong, L., Zhou, Y., Cunniff, M. M., Feng, G., & Zhang, F. (2012). A transcription activator-like effector toolbox for genome engineering. *Nature Protocols*, 7, 171–192.

Streubel, J., Blucher, C., Landgraf, A., & Boch, J. (2012). TAL effector RVD specificities and efficiencies. *Nature Biotechnology*, *30*, 593–595.

Tremblay, J. P., Chapdelaine, P., Coulombe, Z., & Rousseau, J. (2012). Transcription activator-like effector proteins induce the expression of the frataxin gene. *Human Gene Therapy*, *23*, 883–890.

Zhao, H., Yang, Y., & Zhou, Y. (2010). Structure-based prediction of DNA-binding proteins by structural alignment and a volume-fraction corrected DFIRE-based energy function. *Bioinformatics*, *26*, 1857–1863.

CHAPTER TEN

Computational Approaches and Resources in Single Amino Acid Substitutions Analysis Toward Clinical Research

C. George Priya Doss[*,1], **Chiranjib Chakraborty**[†], **Vaishnavi Narayan**[‡], **D. Thirumal Kumar**[*]

[*]Medical Biotechnology Division, School of Biosciences and Technology, VIT University, Vellore, Tamil Nadu, India
[†]Department of Bio-Informatics, School of Computer and Information Sciences, Galgotias University, Greater Noida, Uttar Pradesh, India
[‡]BioMolecules & Genetics Division, School of Biosciences and Technology, VIT University, Vellore, Tamil Nadu, India
[1]Corresponding author: e-mail address: georgecp77@yahoo.co.in; georgepriyadoss@vit.ac.in

Contents

1. Introduction 366
2. Computational Methods in SAP Analysis 371
3. Database Resources for SAPs 374
4. Molecular Phenotypic Effect Analysis 375
5. Sequence Information Analysis 376
6. Computational Methods for Structure Determination 377
7. Docking 380
   7.1 Force-field scoring functions 390
   7.2 Empirical free-energy scoring functions 390
   7.3 Knowledge-based scoring functions 390
   7.4 Consensus-based scoring functions 391
8. Types of Docking 391
   8.1 Advantages of docking 392
   8.2 Limitations of docking 393
9. Molecular Dynamics 393
10. Concluding Remarks 396
Acknowledgments 399
References 399

Abstract

Single amino acid substitutions (SAPs) belong to a class of SNPs in the coding region, which alter the protein function during the translation process. Storage of more information regarding SAPs in public databases will soon become a major hurdle in

Advances in Protein Chemistry and Structural Biology, Volume 94
ISSN 1876-1623
http://dx.doi.org/10.1016/B978-0-12-800168-4.00010-X

characterizing the functional SAPs. In such a demanding era, biology has to rely on bioinformatics, which can work its way through to solve the problems at hand by cutting huge amount of time and resources that are otherwise wasted. Here, we describe an overview of the existing repositories of variant databases and computational methods in predicting the effects of functional SAPs on protein stability, structure, function, drug response, and protein dynamics. This chapter will inspire many biologists with a greater promise in identifying the functional SAPs at the structural level, thereby understanding the molecular effects that are critical for personalized medicine diagnosis, prognosis, and treatment for diseases.

1. INTRODUCTION

Technological advances in high-throughput research have modernized the whole field of biology and medicine with the introduction of terms like genomics, proteomics, pharmacogenomics, and epigenomics. The completion of the Human Genome Project in 2003 (International Human Genome Sequencing Consortium, 2004) and HapMap project in 2007 (Frazer et al., 2007; The International HapMap Consortium, 2003) followed by initiation of 1000 Genomes Project (The 1000 Genomes Project Consortium, 2010) and the Exome Sequencing Project has led to deposition of large volume of genetic variation information in public databases. Numerous efforts were underway in understanding the effects of genetic variation between the individuals and consequences in phenotypic variation and disease susceptibility (De Baets et al., 2012; Taillon-Miller, Gu, Li, Hillier, & Kwok, 1998). The DNA variation consists of insertions, deletions, copy number variations, and single nucleotide substitutions (SNPs). Change in a single nucleotide base from any one of the four nucleotides (A, T, G, and C) to another one is termed as single nucleotide substitution and is found to be the most common DNA variation (ENCODE Project Consortium, 2012). SNP alleles are created either by transition (C/T or G/A) or transversion (C/T, A/G, C/A, or T/G) substitutions. All of these transition and transversion events appear to be more or less similar in occurrence, except for the extreme overabundance of the C to T transition. Over 70% of all the SNPs found in the human genome involve a C to T transition (Kimura, 1980). So far 11 million SNPs have been cataloged; among them 7 million SNPs are designated as common variants occurring with a minor allele frequency above 5%, while the remaining SNPs with minor allele frequency below 5% are designated as rare variants

(Frazer, Murray, Schork, & Topol, 2009; Raychaudhuri, 2011). SNPs are not only designated as markers for constructing genetic maps but also have the potential to direct functional polymorphic variants that are involved in monogenic and complex disorders such as diabetes, cardiovascular diseases, and cancers. Understanding the involvement of functional SNPs might shed some light in disease susceptibility to monogenic and complex disorders and also help in designing more effective treatments to individuals by monitoring adverse drug effects. SNPs are classified based on the location within coding sequences of genes, in noncoding regions of genes, or in the intergenic regions between genes (Risch, 2000). Maximum numbers of SNPs are found in noncoding regions without any biological function of a protein (silent), though they may affect gene expression or splicing. An SNP in which nucleotide substitution leads to no change in amino acid sequence is termed as synonymous (silent mutation), whereas substitution of nucleotide that leads to alteration in the amino acid sequence is defined as nonsynonymous (missense or "nonsense mutations") also called as single amino acid substitutions (SAPs) (Mooney, 2005; Stenson et al., 2003). SNPs within coding or regulatory region of a gene are of biological significance (Pastinen, Ge, & Hudson, 2006). Tennessen et al. (2012) estimated the occurrence of 13,000 exonic SNPs per person, of which 58% are nonsynonymous. Among the class of SNPs, SAPs are of broad research interest due to their accountability in causing half of the known gene lesions responsible for human inherited diseases (Krawczak et al., 2000; Stenson et al., 2003). Therefore, these SAPs are classified as deleterious ones, which have an impact on protein function, thereby leading to dramatic phenotypic change (Sunyaev et al., 2001; Wang & Moult, 2001; Yue & Moult, 2006). However, majority of the SAPs were hypothesized to be neutral or tolerant SAPs, which do not contribute to any phenotype (Masso & Vaisman, 2010; Ng & Henikoff, 2006; Shastry, 2006a, 2006b). Differentiation of deleterious from neutral or tolerant SAPs is very essential in characterizing the genetic basis and pathogenesis of human disease in medical genetics, thereby able to access individual susceptibility to disease (Dimmic, Sunyaev, & Bustamante, 2005). SAPs affect the functional roles of proteins in signal transduction of visual, hormonal, and other stimulants (Dryja et al., 1990; Smith et al., 1994) in gene regulation by altering DNA and transcription factor binding (Barroso et al., 1999), and in maintaining the structural integrity of cells and tissues (Thomas et al., 1999). SAPs inactivate functional sites of enzymes or alter splice sites and thereby form defective gene products (Jaruzelska et al., 1995; Yoshida, Huang, & Ikawa, 1984).

SAPs may affect drug–receptor or drug–enzyme interactions by inducing structural change in receptors or active target-enzyme sites (Bonnardeaux et al., 1994; Erdin, Ward, Venner, & Lichtarge, 2010; Rignall et al., 2002; Ung, Lu, & McCammon, 2006; Vatsis, Martell, & Weber, 1991), ion channels (Wang et al., 1996), and proteins involved in the detoxification pathways (Hassett, Aicher, Sidhu, & Omiecinski, 1994). Furthermore, SAPs may destabilize proteins, or reduce protein solubility (Proia & Neufeld, 1982), and also have functional effects on transcriptional regulation, by affecting transcription factor binding sites in the promoter or intronic enhancer regions (Prokunina & Alarcon-Riquelme, 2004), or alternatively splicing regulation by disrupting exonic splicing enhancers or silencers (Cartegni & Krainer, 2002).

To understand the mechanism of phenotypic variations due to SAPs, it is important to measure the structural consequences due to change in amino acid residue. A well-known classical example is sickle-cell anemia, studied by Sir John Kendrew 55 years ago, which results from the substitution of V instead of E in sixth position of the beta chain of hemoglobin reducing the solubility of the deoxygenated form of hemoglobin markedly (Stryer, 1995). Several studies have illustrated the importance of SAPs in affecting cellular function in the variety of ways. It includes occurrence of SAPs in the active sites (Stevanin et al., 2004; Yamada et al., 2006) or surrounding amino acid residue involved in ligand binding or amino acid residue involved in contact with surrounding proteins will alter the function of the protein. When an SAP occurs near the active site, it might alter the characteristic of the catalytic groups (Koukouritaki et al., 2007; Takamiya, Seta, Tanaka, & Ishida, 2002; Zhang, Norris, Schwartz, & Alexov, 2011; Zhang, Wang, et al., 2010). This will alter the kinetic properties (optimum cellular environment) such as pH, temperature, and salt concentration (Alexov, 2004; Fujiwara et al., 2000). Furthermore, these SAPs can affect the protein stability (Dobson, 2003; Gromiha, Oobatake, Kono, Uedaira, & Sarai, 1999; Ode et al., 2007; Shirley, Stanssens, Hahn, & Pace, 1992; Wang & Moult, 2001), protein flexibility (Karplus & Kuriyan, 2005; Song et al., 2005; Tang & Dill, 1998; Young, Gonfloni, Superti-Furga, Roux, & Kuriyan, 2001), protein folding (Dobson, 2003; Thomas, Qu, & Pedersen, 1995), solvent accessibility (Gromiha et al., 1999; Karchin, Diekhans, et al., 2005; Karchin, Kelly, & Sali, 2005; Kleina & Miller, 1990; Rennell, Bouvier, Hardy, & Poteete, 1991; Rose & Wolfenden, 1993; Stitziel et al., 2003), secondary structure elements (Chasman & Adams, 2001; Ferrer-Costa, Orozco, & de la Cruz, 2002; Gromiha & Ponnuswamy, 1993;

Saunders & Baker, 2002), protein aggregation (Board, Pierce, & Coggan, 1990; Keage et al., 2009; Valerio et al., 2005; Wong, Fritz, & Frishman, 2005), protein–protein interaction (Akhavan et al., 2005; Dixit, Torkamani, Schork, & Verkhivker, 2009; Hardt & Laine, 2004; Jones et al., 2007; Ma, Elkayam, Wolfson, & Nussinov, 2003; Ortiz, Light, Maki, & Assa-Munt, 1999; Ozbabacan, Gursoy, Keskin, & Nussinov, 2010; Rignall et al., 2002; Teng, Madej, Panchenko, & Alexov, 2009; van Wijk, Rijksen, Huizinga, Nieuwenhuis, & van Solinge, 2003; Zhang et al., 2011), protein–DNA interaction (Elles & Uhlenbeck, 2008; Venkatesan et al., 2007; Wright & Lim, 2007), subcellular localization (Boulling et al., 2007; Castella et al., 2011; Hanemann, D'Urso, Gabreëls-Festen, & Müller, 2000; Kim, Hyrc, et al., 2011; Kim, Kim, et al., 2011; Laurila & Vihinen, 2009; Moosawi & Mohabatkar, 2009), protein expression (Boulling et al., 2007; Hanemann et al., 2000), and posttranslational modifications (Grasbon-Frodl et al., 2004; Radivojac et al., 2008; Ryu et al., 2009; Thomas et al., 2004; Tolkacheva et al., 2001; Vazquez, 2000; Vogt et al., 2007). These mounting studies imply the varying functional role of SAPs, which can have a large effect on an organism or species. It is assumed that SAPs in the protein sequences that are observed among living organisms have survived natural selection. Population genetic studies describe that a significant fraction of functional SAPs was present in the highly conserved regions. Residues that evolve under strong selective pressure are found to be significantly associated with human diseases (Arbiza et al., 2006). Disease-causing or deleterious mutations are most likely to correspond to evolutionarily conserved positions in protein sequence due to their functional importance (Tavtigian et al., 2006; Thusberg & Vihinen, 2009). Generally, functional consequences of SAPs fall into two types, namely, disease-associated (deleterious) and benign (no observable phenotypic effect) (Bao & Cui, 2006). The researches of structural and evolutionary features that discriminate the two classes of SAPs have many important applications. First, such features will help to identify disease-associated SAPs from the majority of benign SAPs and to reveal the molecular background of genetic diseases (Karchin, Kelly, et al., 2005). Second, such features will help to determine crucial residues and to elucidate the sequence–structure–function paradigms for individual proteins (Murphy, Barrantes-Reynolds, Kocherlakota, Bond, & Greenblatt, 2004; Wang & Moult, 2003). Finally, such features can be used to guide the selection of target sites in artificial mutagenesis experiments (Damborsky, Prokop, & Koca, 2001). Importantly, SAPs result in altered protein products, which might lead to change in drug–target phenotypes and thereby cause dysfunction of drugs. Moreover, SAPs may produce altered effects in drug transporters, drug-metabolizing enzymes, and drug–target proteins (Ingelman-Sundberg, Sim, Gomez, & Rodriguez-Antona, 2007; Tomalik-Scharte, Lazar, Fuhr, & Kirchheiner, 2008; Zhou et al., 2009), which results in variability of patient–drug responses. To address this, gaining a detailed understanding of the effect of genetic variants on patient–drug response and underlying mechanism is a key part in the establishment of personalized medicine (Fernald, Capriotti, Daneshjou, Karczewski, & Altman, 2011; Rodriguez-Casado, 2012).

Knowledge of a protein's three-dimensional (3D) structure is not only used for energy calculations but also necessary for a full understanding of a mutational effect on its functionality (Capriotti & Altman, 2011), drug–target interaction, and the relationships between mutations and drug response (Lahti, Tang, Capriotti, Liu, & Altman, 2012; Rodriguez-Casado, 2012; Weigelt, 2010). To understand the structure–function relationship, it is necessary to map a mutation onto known protein 3D structure, which acts as a powerful tool in revealing the mechanistic explanation of their effects on function. Proteins with mutations do not always have 3D structures that are solved and deposited in Protein Data Bank (PDB). Therefore, it is necessary to construct 3D models using homology modeling by locating the variation in 3D. This acts as a powerful tool to reveal what kind of adverse effects a mutation can have on protein. This process of detailed structural analysis of protein–drug interactions was not always feasible in the past, but advances in structural genomics have resulted in an explosion of high-resolution structures of known and potential drug–target proteins.

The last decade has witnessed a drastic increase in genomic information, especially SNPs in public databases and lends itself to an informatics approach. Bioinformatics, especially computational molecular biology, is playing a vital role in extracting knowledge from the vast amount of genomic information generated by different genomics technologies. As a result, various computational resources were developed to aid identification and characterization of the functional SAPs and to study the impact of SAPs in protein, patient–drug response, and current therapeutic targets. In this chapter, we intend to provide an overview of the existing computational methods in SAP analysis available on the World Wide Web. Summing up, the techniques presented in this chapter will build a bridge between computational methods and clinicians toward personalized medicine in tailoring new treatment strategies.

2. COMPUTATIONAL METHODS IN SAP ANALYSIS

With the ever-increasing influx of high-throughput technologies and the pronounced ascendency of the Human Genome Project (HGP), the scientific communities of disease research and drug design have witnessed a paradigm shift toward single amino acid substitutions (SAP). With a successfully completed genome, genome-wide association studies were initiated using SAPs as markers to study disease–gene associations. The commonality of SAPs in a genome makes them suitable detection points for studying disease susceptibility (Curtis, North, & Sham, 2001). The last decade has witnessed extraordinary advances in experimental and computational technologies to identify, characterize, and differentiate pathogenic SAPs from neutral ones. Traditional methods find a unique place in disease diagnosis and treatment of an individual. Employing traditional methods for SAPs analysis will consume precious time, require increased labor, and thus often turn detrimental to the study itself. Computational techniques with cutting edge software and innovative algorithms are fast improving the arduous task of analyzing genetic polymorphism data. This will serve as a powerful screening process to single out potentially deleterious SAPs from the whole stack and examine it as a possible drug–target through stability studies (Mah & Chia, 2007). Increased deliberation in this regard led to the generation of numerous algorithms and methods to provide an edge to the study of SAPs. Many computational methods were developed to predict whether an SAP is deleterious to the structure or the function of the gene and will, therefore, lead to disease. These predictions of SAP are of three classes: (1) predict the effect on protein function, (2) predict the effect based on the pathogenicity, and (3) predict the effect on protein structure stability. Researchers have taken many input features such as sequence-based properties, physical properties of the wild-type and mutant amino acids, protein structural properties (solvent accessibility, location within beta strands or active sites, and participation in disulphide bridges), and evolutionary properties derived from a phylogeny or sequence alignment to predict an SAP as deleterious/disease/pathogenic/intolerable/nonneutral or neutral/tolerable. To classify whether an SAP will be tolerable, a training set is usually constructed of mutations known to be deleterious. For example, these training sets can be derived from saturation mutagenesis experiments, where the mutation severity is determined in activity assays (Cai et al., 2004;

Chasman & Adams, 2001; Krishnan & Westhead, 2003; Ng & Henikoff, 2001; Saunders & Baker, 2002), multiple sequence alignments where tolerance to mutation is derived from evolutionary analyses of sequence positions (Sunyaev et al., 2001), or known deleterious human mutations. In this chapter, we have provided a brief classification and description of some well-known reliable SAP prediction methods available to this day. For the accurate description of the process of SAP analysis, we have provided a pictorial representation of the entire process in Fig. 10.1.

The computational prediction methods can be divided into four most important families of algorithms: machine-learning methods, empirical rule-based methods, physics-based models, and evolutionary theory-based models to classify SAPs. Machine-learning approaches develop classification models that automatically learn from the training data, extract patterns from complex data, and make predictions of new cases. Random forests (Bao, Zhou, & Cui, 2005; Li et al., 2009; Mathe et al., 2006), neural networks (Bromberg & Rost, 2007; Ferrer-Costa et al., 2005; Linding, Russell, Neduva, & Gibson, 2003; Yang, Thomson, McNeil, & Esnouf, 2005), Decision Trees (Yuan et al., 2006), support vector machines (Calabrese, Capriotti, Fariselli, Martelli, & Casadio, 2009; Capriotti & Altman, 2011; Capriotti et al., 2013; Capriotti, Fariselli, & Casadio, 2005; Capriotti, Fariselli, Rossi, & Casadio, 2008; Karchin, Kelly, et al., 2005; Mathe et al., 2006; Parthiban, Gromiha, Hoppe, & Schomburg, 2007; Tian et al., 2007; Yue & Moult, 2006), naive Bayes approach (Adzhubei et al., 2010; Schwarz, Rodelsperger, Schuelke, & Seelow, 2010), hidden Markov models (Mi, Muruganujan, & Thomas, 2013; Shihab et al., 2013), and rule-based methods (Kumar, Henikoff, & Ng, 2009; Ramensky, Bork, & Sunyaev, 2002; Reva, Antipin, & Sander, 2011; Tavtigian et al., 2006; Zhou & Zhou, 2002) are among the most widely used machine-learning methods for SAP analysis. Few meta-analysis suite tools are also available, which combine the prediction information from the abovementioned prediction methods (Jegga, Gowrisankar, Chen, & Aronow, 2007; Lee & Shatkay, 2008; Olatubosun, Valiaho, Harkonen, Thusberg, & Vihinen, 2012; Schaefer, Meier, Rost, & Bromberg, 2012; Wang, Ronaghi, Chong, & Lee, 2011) and SAP prioritization tools (Conde et al., 2004; Freimuth, Stormo, & McLeod, 2005; Lee & Shatkay, 2008; Wjst, 2004; Xu et al., 2005). The abovementioned methods use several different inputs such as NCBI GI number OR RefSeq ID, wild-type protein FASTA sequences, and wild and new residue after mutation (single-letter amino acid code) for making their predictions. In order to quantify the destabilization

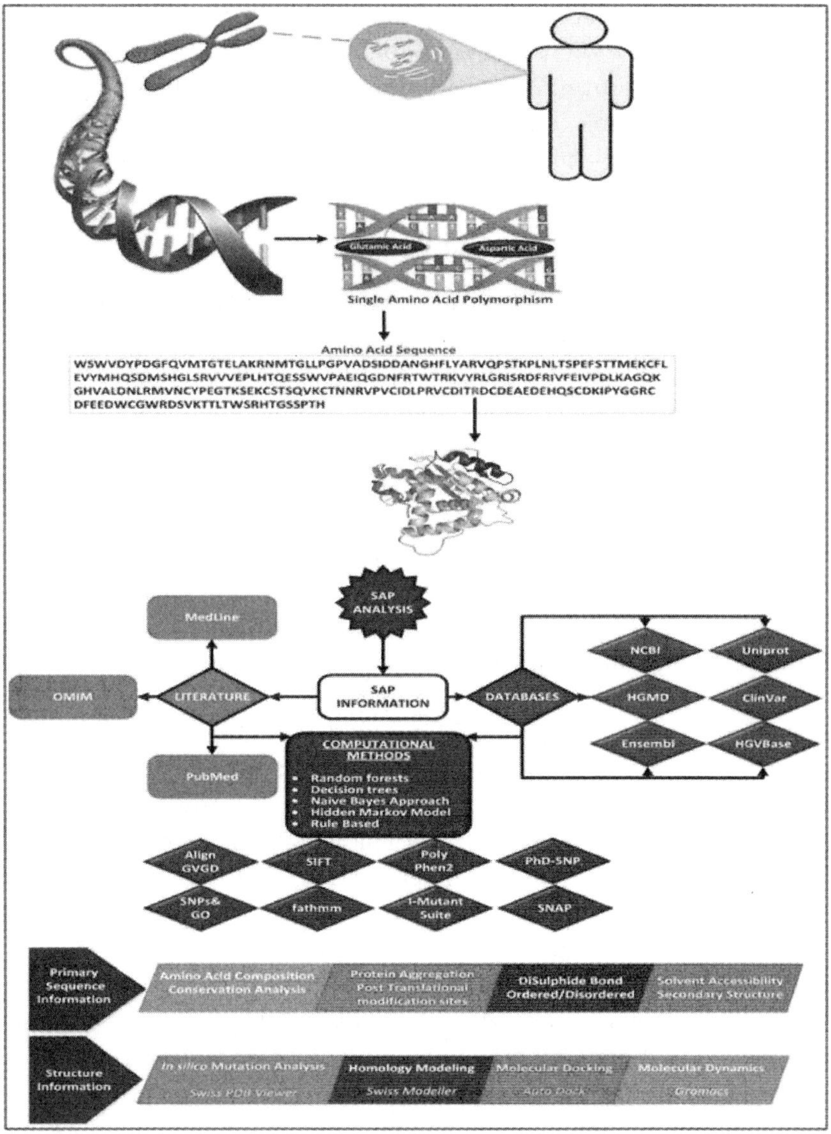

Figure 10.1 Computational pipeline in Single Amino Acid Substitutions analysis. (For color version of this figure, the reader is referred to the online version of this chapter.)

effects of SAPs, the protein stability change upon mutations can be evaluated by calculating the difference in folding free energy change between wild-type and mutant proteins ($\Delta\Delta G$ or ddG) without performing any experimental results (Capriotti et al., 2005; Chen, Lin, & Chu, 2013; Cheng,

Randall, & Baldi, 2006; Dehouck et al., 2009; Dosztanyi, Magyar, Tusnady, & Simon, 2003; Magyar, Gromiha, Pujadas, Tusnády, & Simon, 2005; Masso & Vaisman, 2008; Parthiban, Gromiha, & Schomburg, 2006). SAP prediction tools can be classified for better understanding into sequence-, structure-, and sequence- and structure-based methods.

3. DATABASE RESOURCES FOR SAPs

As of November 7, 2013, about 62,676,337 SNPs of Homo sapiens were identified and deposited in the major repository National Center for Biotechnology Information (NCBI) database (Sherry et al., 2001) (http://www.ncbi.nlm.nih.gov/projects/SNP/snp_summary.cgi?view+summary=view+summary&build_id=138). In addition, there are now a few other extensive databases that provide information about the DNA variations. This includes UniProt database (Apweiler et al., 2010), ClinVar database (Maglott et al., 2013), human genome variation database, HGVBase (Fredman et al., 2002), Human Gene Mutation Database (HGMD) (Stenson et al., 2012), Online Mendelian Inheritance in Man (OMIM) (Amberger, Bocchini, & Hamosh, 2011), and MutDB (Mooney & Altman, 2003). Repositories like dbSNP and UniProt contain information about the experimentally proved SAPs, but a few of these are annotated with respect to function. OMIM contains disease-related literature information; HGMD contains disease variants with one or more references in the literature, and ClinVar database contains clinically significant variant information. The availability of a comprehensive SNP catalog offers the possibility of identifying many disease loci and eventually pinpointing functionally important variants in which the nucleotide change alters the function or expression of a gene that directly influences a disease outcome. The study of the distribution of SNPs particularly in different populations is also valuable for investigating molecular events that underlie the evolution, namely, genetic drift, mutation, recombination, and selection. These illustrate important changes in human history, for example, tracing the origin of populations and their migrations. In addition, disease-specific databases were initiated to specifically collect variant information pertaining to one disease (Basu, Kollu, & Banerjee-Basu, 2009; caBIG Strategic Planning Workspace (caBIG), 2007; George Priya Doss, Nagasundaram, Srajan, & Chiranjib, 2012; Ingman & Gyllensten, 2006; Nuytemans, Theuns, Cruts, & Van Broeckhoven, 2010; Ruiz-Pesini et al., 2007) and in defining the molecular basis of the disease (Giardine et al., 2007; Kawabata, Ota, & Nishikawa,

1999) and structural stability (Gromiha et al., 1999). In addition, visualization tools were developed to analyze the effect of SAP on protein structure along with conservation and physicochemical properties (Chang & Fujita, 2001; Han et al., 2006; Luu, Rusu, Walter, Linard, et al., 2012; Luu, Rusu, Walter, Ripp, et al., 2012; Uzun, Leslin, Abyzov, & Ilyin, 2007; Venselaar, Te Beek, Kuipers, Hekkelman, & Vriend, 2010). Several researchers have reviewed in detail about the computational methods and also compared the performance of each computational method over the other existing methods, which are available online (Castellana & Mazza, 2013; Frousios, Iliopoulos, Schlitt, & Simpson, 2013; George Priya Doss, Rajasekaran, Arjun, & Sethumadhavan, 2010; George Priya Doss & Sethumadhavan, 2009a, 2009b; George Priya Doss et al., 2008; Gnad, Baucom, Mukhyala, Manning, & Zhang, 2013; Gray, Kukurba, & Kumar, 2012; Khan & Vihinen, 2010; Peterson, Doughty, & Kann, 2013; Reumers, Schymkowitz, & Rousseau, 2009; Thusberg, Olatubosun, & Vihinen, 2011; Tiffin, Okpechi, Perez-Iratxeta, Andrade-Navarro, & Ramesar, 2008; Wang, Eickholt, & Cheng, 2010; Wang, Sun, Akutsu, & Song, 2013).

4. MOLECULAR PHENOTYPIC EFFECT ANALYSIS

SAPs can alter the biophysical properties of amino acid residues such as size, charge, hydrogen bonding, hydrophobic contacts, disulfide bonds, van der Waals, and electrostatic interactions at critical folding positions, which will have an impact on residue contacts and thereby lead to loss of protein stability as well as folding, flexibility, and aggregation of the protein (Betz, 1993; Dill, Ozkan, Weikl, Chodera, & Voelz, 2007; Eriksson et al., 1992; Horovitz, Serrano, Avron, Bycroft, & Fersht, 1990; Pace et al., 2011). Most often the change in protein stability results in increased propensity for protein aggregation. As a consequence, SAPs can have a considerable effect in the solubility and aggregation propensity of a protein (Karplus & Kuriyan, 2005; Keage et al., 2009; Valerio et al., 2005; Wong et al., 2005). Several computational algorithms have been proposed to predict the aggregation-nucleating sequences in proteins using either sequence-based or structural bioinformatics tools (Conchillo-Sole et al., 2007; De Baets et al., 2012; Fernandez-Escamilla, Rousseau, Schymkowitz, & Serrano, 2004; Garbuzynskiy, Lobanov, & Galzitskaya, 2010; Maurer-Stroh et al., 2010; Trovato, Seno, & Tosatto, 2007; Tsolis, Papandreou, Iconomidou, & Hamodrakas, 2013; Van Durme et al., 2009). SAPs also introduce disorder predisposition in a target protein, which may affect the protein conformation, increase the flexibility, and lead to alterations in function. SAP effects on structural disorders can be analyzed by providing a sequence as the input (Cheng, Randall, Sweredoski, & Baldi, 2005; Cheng, Sweredoski, & Baldi, 2005; Dosztányi, CsizmLok, Tompa, & Simon, 2005; Galzitskaya, Garbuzynskiy, & Lobanov, 2006; Ishida & Kinoshita, 2007, 2008; Linding, Russell, Neduva, & Gibson, 2003; Prilusky et al., 2005; Sickmeier et al., 2007; Vullo, Bortolami, Pollastri, & Tosatto, 2006). There are significant numbers of reports that explain the involvement of missense mutations in posttranslational target sites leading to diseases (Grasbon-Frodl et al., 2004; Radivojac et al., 2008; Vogt et al., 2007). Posttranslation modification sites (PTMs) are implicated in many cellular processes and have a vital role in regulating the functional and structural properties of protein (Walsh, 2006). Different PTMs of known protein like phosphorylation, glycosylation, methylation, acetylation, and sumoylation can be analyzed by various computational methods (Blom, Sicheritz-Ponten, Gupta, Gammeltoft, & Brunak, 2004; Chang et al., 2009; Gupta et al., 1999; Huang, Lee, Tseng, & Horng, 2005; Kiemer, Bendtsen, & Blom, 2005).

5. SEQUENCE INFORMATION ANALYSIS

Protein primary sequence provides the most direct and readily available information regarding the clues for functional mutation sites that can be extracted from the amino acid sequence in cases where no structural information is available. Population genetic studies describe that a significant fraction of functional SAPs was present in a conserved region. Residues that evolve under strong selective pressure are found to be significantly associated with human diseases (Arbiza et al., 2006). The importance of residue for maintaining the structure and function of a protein can usually be inferred from how conserved it appears in a multiple sequence alignment of that protein and its homologues. A comparative analysis of amino acid conservation from multiple species by protein sequence alignments gives an indication of which amino acid residues are truly conserved and which of them represent localized evolution. It is assumed that SAPs in the protein sequences that are observed among living organisms have survived natural selection. Disease causing, or deleterious mutations are most likely to correspond to evolutionarily conserved positions in protein sequence due to their functional importance (Tavtigian et al., 2006; Thusberg & Vihinen, 2009). Several methods were made available online predict the conservation analysis of multiple sequence alignments (Ashkenazy, Erez, Martz, Pupko, & Ben-Tal, 2010; Berezin et al., 2004; Gu & Vander Velden, 2002; Pupko, Bell, Mayrose, Glaser, & Ben-Tal, 2002; Siepel et al., 2005). ConSurf (Ashkenazy et al., 2010) is a tool based on comparative analysis of amino acid conservation from multiple species by protein sequence alignments and provides an indication of which amino acid residues are truly conserved and which of them represent localized evolution. There is extensive research indicating the function of substituted amino acids in causing diseases (Dobson, Munroe, Caulfield, & Saqi, 2006; Khan & Vihinen, 2007; Vitkup, Sander, & Church, 2003) and Trp and Cys residues in determining protein stability (Arbiza et al., 2006). Few studies have illustrated the importance of Cys residues in a protein sequence since most of the protein foldings are dependent on disulfide bonds (Song, Geng, et al., 2009; Song, Lim, & Tong, 2009). SNP-associated residue changes to or from Cys will likely destabilize the protein structure. Composition of each amino acid can be calculated by statistical analysis of protein sequences (Brendel, Bucher, Nourbakhsh, Blais-dell, & Karlin, 1992; Cheng et al., 2006). Solvent accessibility, considered as a discriminating feature in disease-associated SAPs, tends to occur at buried sites; benign substitutions tend to occur at solvent accessible sites (Ferrer-Costa et al., 2002; Sunyaev et al., 2001). Solvent accessibility from an exposed to buried state could be considered functionally significant in the mutant protein at the structural level (Chen & Zhou, 2005), which can be accessed by ACCpro (Pollastri, Baldi, Fariselli, & Casadio, 2002), WHAT IF (Vriend, 1990) and WESA (Chen & Zhou, 2005). Secondary structure elements can be analyzed using NetSurfP-1.1 (Petersen, Petersen, Andersen, Nielsen, & Lundegaard, 2009), Jpred (Cole, Barber, & Barton, 2008), YASPIN (Lin, Simossis, Taylor, & Heringa, 2005), STRIDE (Heinig & Frishman, 2004), DSSP (Kabsch & Sander, 1983), and SSPro (Cheng, Randall, et al., 2005; Cheng, Sweredoski, et al., 2005).

6. COMPUTATIONAL METHODS FOR STRUCTURE DETERMINATION

As of November 2013, PDB contains 95280 entries (Bernstein et al., 1997) of experimentally solved structures, which includes multiple structures of the same protein, while UniProtKB/Swiss-Prot contains 541561 sequence entries (Apweiler et al., 2004) and NCBI RefSeq database contains 33,139,144 protein entries (Pruitt, Tatusova, Brown, & Maglott, 2012).

Due to the advent of cost-effective high-throughput gene sequencing technologies, the number of sequence entries in the aforementioned databases is increasing. Furthermore, the number of solved structure determination will tend to increase; the number of newly discovered sequences grows much faster than the number of structures solved (Levitt, 2007). The protein structures are solved by X-ray crystallography, nuclear magnetic resonance (NMR), and high-resolution molecular microscopy (EM). These methods of structural determination are limited by cost, time consumption, and requirement of specialized instruments, which leads to a large gap between the solved structures and available protein sequences in the databases. Due to this, the application of computational-based methods in 3D structure predictions has increased and also become a valuable resource in defining protein function (Hermann et al., 2007) and studying the impact of mutation at structural level (George Priya Doss, Chakraborty, Rajith, & Nagasundaram, 2013; Kosinski, Hinrichsen, Bujnicki, Friedhoff, & Plotz, 2010) and drug discovery (Liu, Tang, & Capriotti, 2011). Existing computation methods for structure determination fall into two categories (Zhang, 2008a, 2008b): templates-based comparative (or homology) and threading methods, which utilize structures of known homologous proteins as starting templates (Kolinski, Rotkiewicz, Ilkowski, & Skolnick, 1999; Rost, Fariselli, & Casadio, 1996), and free modeling methods (*de novo* and *ab initio*), which apply the principles of physical chemistry in protein folding, often in combination with efficient fragment searching techniques (Jothi, 2012; Lesk, 1997; Zemla, Venclovas, Reinhardt, Fidelis, & Hubbard, 1997). These computational structure prediction methods were discussed in detail (Kryshtafovych & Fidelis, 2009; Pierri, Parisi, & Porcelli, 2010; Werner, Morris, Dastmalchi, & Church, 2012; Zhang, 2009). In this section, we have discussed the steps followed in homology modeling. In template-based methods, the tertiary structure of an unknown protein can be modeled using a known 3D structure of protein with the homologous sequence (homology modeling), while, in fold recognition, the protein structure was modeled based on the proteins with known structures having the same fold but no homology to the proteins with known structure (Daga, Patel, & Doerksen, 2010; Martí-Renom et al., 2000; Qu, Swanson, Day, & Tsai, 2009). Homology modeling uses only sequence similarity, whereas fold recognition uses both structure and sequence relationship. Homology modeling consists of four major steps: (a) template identification, (b) alignment of target sequence with template structures, (c) model building, and (d) model evaluation.

Template identification is one of the most important steps in homology modeling, which is performed by searching the target sequence in databases such as PDB, which includes solved structures. The commonly used searched methods BLAST (Altschul, Gish, Miller, Myers, & Lipman, 1990), FASTA (Lipman & Pearson, 1985; Pearson & Lipman, 1988), PSI-BLAST (Altschul et al., 1997), HHSearch (Koehl & Delarue, 1994), HHpred (Söding, Biegert, & Lupas, 2005), and Phyre (Kelley & Sternberg, 2009) provide a ranking of templates along with "*E*-value" alignment scores. A good candidate template is selected based on the *E* value equal to zero, with the highest similarity and a template of solved structures by X-ray crystallography. A sequence identity cutoff of 30% is considered as the standard threshold in homology modeling. In the case of low identity, Doolittle (1986) formulated three rules for template selection: sequences longer than 100 amino acids with <25% identities (with gaps) are probably related; sequences with 15–25% identity might be related ("twilight zone") but need additional statistical analyses to help establish this with confidence; sequences with 15% identity are most likely not related. Choosing multiple templates can improve the quality of the model when compared to use a single template (Fernandez-Fuentes, Rai, Madrid-Aliste, Fajardo, & Fiser, 2007; Kosinski, Tkaczuk, Kasprzak, & Bujnicki, 2008; Wallner, Lindahl, & Elofsson, 2008). Once a final template is selected, target-template alignment should be performed by using unique pair-wise or multiple sequence alignment tools such as Clustal Omega (Sievers et al., 2011), T-Coffee (Notredame, Higgins, & Heringa, 2000), ClustalW2 (Larkin et al., 2007), 3DCoffee (O'Sullivan, Suhre, Abergel, Higgins, & Notredame, 2004), and Muscle (Edgar, 2004). Tress, Jones, and Valencia (2003) proposed that inclusion of regions sharing highest sequence similarity along with common motifs can be considered as correctly aligned because they tend to be evolutionarily conserved. Additional sequence information, localization of hydrophobic regions, secondary structure elements, and disulphide bonds are considered to improve alignment. The next step in homology modeling involves model building, where a 3D structure model is built based on the given target-template alignment and template structures. Nowadays, this procedure has become fully automated. Basically, model building methods are grouped as follows (Wallner & Elofsson, 2005; Xiang, 2006): rigid-body assembly methods, which build a model from the structurally conserved regions of the template that align to the target sequence, like 3D-JIGSAW (Bates, Kelley, MacCallum, & Sternberg, 2001), BUILDER (Koehl & Delarue, 1995), and SWISS-MODEL (Arnold, Bordoli, Kopp, & Schwede, 2006);

segment matching methods like SegMod/ENDCAD (Levitt, 1992); spatial restraint methods like MODELLER (Eswar et al., 2006); and artificial evolution methods like NEST (Petrey et al., 2003). A spatial restraint method utilizes satisfying restraints such as bond lengths and angles, van der Waals contact distances, and dihedral angles to map onto the target-template structure alignments. Studies comparing the model building methods have rated MODELLER as the best among the existing model building methods (Dalton & Jackson, 2007; Wallner & Elofsson, 2005). Model evaluation remains as the fundamental and most important step in homology modeling and in defining whether the model created is of good quality. The error in homology structure comes from the side chains and loops. For this, many approaches including hybrid methods have been proposed to rectify these errors (Arnold et al., 2006; Das & Baker, 2008; Deane & Blundell, 2001; Fernandez-Fuentes, Zhai, & Fiser, 2006; Holm & Sander, 1992; Hwang & Liao, 1995; Koehl & Delarue, 1994; Krivov, Shapovalov, & Dunbrack, 2009; Lee & Subbiah, 1991; Liang, Zheng, Zhang, & Standley, 2011; Rohl, Strauss, Misura, & Baker, 2004; Samudrala & Moult, 1998; Sippl, 1993; Xiang, Soto, & Honig, 2002; Xu & Berger, 2006). Additionally, many model quality assessment programs are made available, which define a scoring function that is capable of discriminating good and bad models (Bowie, Lüthy, & Eisenberg, 1991; Davis et al., 2007; Hooft, Vriend, Sander, & Abola, 1996; Laskowski, MacArthur, Moss, & Thornton, 1993; Lüthy, Bowie, & Eisenberg, 1992; Sippl, 1993; Wiederstein & Sippl, 2007). Threading methods can be used in place of homology modeling, when template structures share less than ~30% sequence identity with the target sequence with evolutionary relationship (Miller, Jones, & Thornton, 1996; Xu, Li, Kim, & Xu, 2003). The structures that are not modeled by homology modeling and threading can be performed using free modeling servers (Das et al., 2007; Jayaram et al., 2006; Kim, Chivian, & Baker, 2004; Kinch et al., 2011; Rohl et al., 2004; Wang, Yang, Li, Liu, & Zhou, 2010; Zhang, 2008a, 2008b; Zhang, Wang, et al., 2010). The abovementioned structure prediction methods have their own advantages and disadvantages, and their consistency varies for each structural problem. Choosing appropriate methods depends mainly on the availability of a suitable template and computational resources.

7. DOCKING

All the drugs presently available in the market have gone through years of clinical research and drug trials (Jorgensen, 2004). Even a single disease can have many options for a drug. It is estimated that a typical drug-discovery cycle, from lead identification to clinical trials, can take 14 years with a cost of 800 million US dollars before it can be sold in the market to the general public. It takes the scientists even longer to finally select that one "potential" drug from among the many other options. Rapid advancements in the fields of genomics, proteomics, and biotechnology have fuelled the drug-discovery process. This constant expansion has led to the approval of 18 drugs for human use by the US Food and Drug Administration, with approximately four acting on novel target structures (Rask-Andersen, Almen, & Schioth, 2011). In spite of this success, an "efficacy–effectiveness gap" exists and is, ultimately, the result of variability in patient–drug responses. Also, marketed drugs exhibit limited efficacy of about 30–60% (Sadee & Dai, 2005; Wilkinson, 2005) and may also exhibit drug toxicity (Wilke & Dolan, 2011) toward the patient. Drug discovery and drug research have contributed more to the progress in the field of medicine than any other scientific factor. Many studies have highlighted that drug discovery and designing is a complex, highly expensive, risky, and quite cumbersome *process* (Congreve, Murray, & Blundell, 2005; DiMasi & Grabowski, 2007; DiMasi, Grabowski, & Vernon, 2004; DiMasi, Hansen, Grabowski, & Lasagna, 1991; Kolb & Sharpless, 2003). The drug-discovery process aids in designing a candidate drug or lead compound, which binds with the target-specific protein whose function is thought to be essential for the disease phenotype. The traditional drug-discovery process involves three major steps: target identification and validation, lead identification, and lead optimization. This process is expensive and takes a very long time in identifying the drug. Modern drug discovery involves application of computational methods in an efficient manner to predict drug–target interaction, stability, and activity. Bioinformatics, through various tools and techniques, can eliminate the numerous drug choices to just 10 drugs or even fewer. Virtual screening is the screening of many compounds that might dock with a particular target macromolecule and lead to the formulation of a potential drug (Taboureau, Baell, Fernández-Recio, & Villoutreix, 2012). It is a high-throughput screening of millions of compound databases in the hopes of finding a unique compound or a drug that can replace an existing drug or that can shed light on diseases with no drugs. All the compounds that are screened do not necessarily exist (Lavecchia & Di Giovanni, 2013). Millions of compounds can be screened easily without having to spend much time, effort, or money. Virtual screening can be divided into two broad categories: ligand-based and structure-based. There are stringent rules that have to be followed in virtual screening. Many *in silico* tools can be used to design libraries of compounds with drug-like properties (Villoutreix et al., 2007). These are predominantly biophysical properties based on empirical rules. A well-known example is Lipinski's "rule of five" (Lipinski, 2000), which states that a compound is likely to be "nondrug-like" if it has more than 5 hydrogen bond donors and more than 10 hydrogen bond acceptors; molecular mass is greater than 500 and lipophilicity is above 5. New pharmacokinetic data found in rats have caused this rule to be revisited (Ridder, Wang, de Vlieg, & Wagener, 2011). Many related rules have been subsequently modified and proposed as the "rule of three" (Rees, Congreve, Murray, & Carr, 2004), which defines fragment properties with an average molecular weight $\leq$300 Da, a Clog $P \leq 3$, the number of hydrogen bond donors $\leq$3, the number of hydrogen bond acceptors $\leq$3, and the number of rotatable bonds <3. Recently, Pfizer's "rule of 3/75" has been described, which states that compounds with a calculated partition coefficient (Clog *P*) of <3 and topological polar surface area (TPSA) >75 have the best chances of being well tolerated from a safety perspective *in vivo* (Hughes et al., 2008).

There are various new drugs whose development was heavily influenced by computational methods and screening strategies (Durrant & McCammon, 2011). One such important example is the HIV protease inhibitor. Most drugs are ligand–protein complexes, where the ligand enhances the function of the protein, which in turn helps it to fight against the disease. The other alternative is when the ligand helps in downregulating the expression of a particular protein it binds to. Both the scenarios depend on how well and easily the ligand and protein bind to each other. Such results of ligand–protein binding can be obtained by performing molecular docking (Gulati, Cheng, & Bates, 2013). There are many sources for the structure of the protein and drugs or ligands separately (Pak & Wang, 2000). There are not much data for protein and drug complexes.

There are three major advances in the field of drug designing (Rao & Srinivas, 2011):

- The first one is the conformational modeling of all small molecules, ligands, macromolecules, and their complexes; these are molecules that are potential drug candidates.
- The next is to determine their physical, chemical, and biological properties called property modeling.
- The last is to optimize the chemical, physical, and biological properties or molecular design.

Docking is not the only process that is important. Much before docking, the search and retrieval of sequences or 3D structures of target proteins are more important. Some of the databases for retrieval of sequences include NCBI, GenBank, and UniProt. Drug banks have formed an integral part of today's drug discovery and designing. A drug bank is a unique database that consolidates information on drugs. Drug banks combine the sequence, structure, and pathway information of drugs or target drug molecules with chemical, physical, biological, pharmacological, and pharmaceutical information (Wishart, 2007). Several other databases also exist with known 3D structures of potential drug–targets, ligands, diseases, and their associated pathways.

Molecular docking is an invaluable tool in the field of molecular biology, computational structural biology, computer-aided drug designing, and pharmacogenomics. Docking also plays a vital role in virtual screening of huge libraries. The subsequent results are ranked accordingly, and structural hypothesis of how the ligands inhibit/activate the target macromolecule can be deduced (Azam & Abbasi, 2013). This proves to be an invaluable part in lead optimization of drugs. Most importantly, docking provides certain information that is difficult to deduce through conventional experimental methods. The docking procedure fits two molecules together, protein and ligand in 3D space. Their binding complementarity is then evaluated, and the results are displayed with the best fit "scores." This method is widely used in hit identification and lead optimization of drugs. Docking, simply put, tries to find the best "fit" between two molecules. It can be a protein–ligand docking or a protein–protein docking or a nucleic acid–macromolecule docking. Docking aims at finding out if two molecules can interact. If they do interact, docking attempts to find out the orientation in which the interactions are at its maximum and where the binding energy is at its minimum (Gold, 2007). In a ligand–protein docking, the main goal is to predict the predominant binding pockets; most effective docking softwares perform this by searching high-dimensional spaces efficiently and thoroughly. The protein used for docking purpose should have previously solved 3D structure or a constructed structure by homology modeling (Alonso, Bliznyuk, & Gready, 2006). The two main aspects considered for molecular docking include accurate structural modeling and rendition of active binding site(s). Basically, docking consists of four main steps: (a) preparation of ligand and the receptor, (b) identification of active sites, (c) generation of putative complexes, and (d) evaluation of the complexes by scoring.

The initial step in docking is to identify or define the binding sites by using experimental information derived from the mutagenesis or cross-linking studies or from homologous structures whose binding site is already known. Any molecule, whether a protein or a ligand, may undergo slight structural or conformational changes after binding with another molecule. This makes binding site analysis slightly difficult. Structure-based algorithms are used for binding site analysis (Lahti et al., 2012) (Ghersi & Sanchez, 2011). For the prediction of binding sites, numerous softwares are available, which are categorized into structural similarity approach, geometric approach, and energy-based approach and docking. In the next step of docking, the ligand pose has to be predicted in defined pocket. Several various molecular docking algorithms (Table 10.1) are now available that can fit or "dock" small molecules like ligands into pockets of macromolecules like proteins or sometimes DNA, with different scoring and search algorithms (Abagyan, Totrov, & Kuznetsov, 1994; B-Rao, Subramanian, & Sharma, 2009; Claussen, Buning, Rarey, & Lengauer, 2001; Ewing, Makino, Skillman, & Kuntz, 2001; Friesner et al., 2004; Goodsell, Morris, & Olson, 1996; Jones, Willett, Glen, Leach, & Taylor, 1997; McGann, Almond, Nicholls, Grant, & Brown, 2003; Morris et al., 1998; Rarey, Kramer, Lengauer, & Klebe, 1996), which can predict in an accurate and fast manner. These algorithms determine all possible optimal conformations for a given complex (protein–protein and protein–ligand) in an environment, where each conformation is linked with a final score. In addition, each algorithm calculates the energy of all the resulting conformations of each individual interaction (Sushma & Suresh, 2012). Most docking softwares use scoring techniques that correctly rank the docking conformations (McConkey et al., 2002). Scoring is an important component of docking. The work is futile if there are many conformations but no ranking system. When docking is performed, it is very important to have the right conformation of the docked molecule and also to have each individual conformation to be correctly ranked. This helps to identify the most probable biological conformations (Kollman, 1993). The scoring functions should be able to differentiate between different orientations of the same receptor and ligand. The scoring functions are mathematical functions that assign a value based on the strength of the interaction between the two docked molecules. Each docked conformation is scored for best fit (Zsoldos, Reid, Simon, Sadjad, & Johnson, 2007). This scoring process is repeated the number of times defined by the user or according to the maximum iterations supported by the program. Scoring functions predict factors like van der

Table 10.1 List of computational methods employed in docking analysis

Database name	**Resource**
Drug banks	
Therapeutic Target Database (Chen et al., 2002)	http://bidd.nus.edu.sg/group/cjttd/
DrugBank (Knox et al., 2011)	http://www.drugbank.ca/
PubChem (Bolton, Wang, Thiessen, & Bryant, 2008; Wang et al., 2009)	http://pubchem.ncbi.nlm.nih.gov/
Binding MOAD (Hu, Benson, Smith, Lerner, & Carlson, 2005)	http://bindingmoad.org/
PDBbind (Wang, Fang, Lu, & Wang, 2004; Wang, Wolf, Caldwell, Kollman, & Case, 2004)	http://sw16.im.med.umich.edu/databases/pdbbind/index.jsp
PDTD (Gao et al., 2008)	http://www.dddc.ac.cn/pdtd/
DGIdb (Griffith et al., 2013)	http://dgidb.genome.wustl.edu/
TDR Targets (Magariños et al., 2012)	http://tdrtargets.org/
SuperDrug (Goede, Dunkel, Mester, Frommel, & Preissner, 2005)	http://bioinf.charite.de/superdrug/
ChemBank (Seiler et al., 2008)	http://chembank.broadinstitute.org/
BindingDB (Liu, Lin, Wen, Jorissen, & Gilson, 2007)	http://www.bindingdb.org/bind/index.jsp
CancerDR (Kumar et al., 2013)	http://crdd.osdd.net/raghava/cancerdr/
Binding site prediction tools	
CASTp (Dundas et al., 2006)	http://sts.bioengr.uic.edu/castp/
LIGSITE (Hendlich, Rippmann, & Barnickel, 1997; Huang & Schroeder, 2006)	http://projects.biotec.tu-dresden.de/pocket/
SURFNET (Laskowski, 1995)	http://www.ebi.ac.uk/thornton-srv/software/SURFNET/
SMAP-WS (Ren, Xie, Li, & Bourne, 2010)	http://nbcr-222.ucsd.edu/smap_ws/
PocketPicker (Weisel, Proschak, & Schneider, 2007)	http://gecco.org.chemie.uni-frankfurt.de/pocketpicker/
FINDSITE (Brylinski & Skolnick, 2008)	http://cssb.biology.gatech.edu/findsite

Continued

Table 10.1 List of computational methods employed in docking analysis—cont'd

Database name	Resource
PBBinder (Hooft et al., 1996)	http://160.80.35.80/PDBinder/
PDBSite (Ivanisenko, Grigorovich, & Kolchanov, 2000)	http://wwwmgs.bionet.nsc.ru/mgs/gnw/pdbsite/
LigAsite (Dessailly, Lensink, Orengo, & Wodak, 2008)	http://www.bigre.ulb.ac.be/Users/benoit/LigASite/index.php?home
3DLigandSite (Wass et al., 2010)	http://www.sbg.bio.ic.ac.uk/~3dligandsite/
PocketAnnotate (Anand, Yeturu, & Chandra, 2012)	http://proline.biochem.iisc.ernet.in/pocketannotate/reference.php
Active Site Prediction (Singh, Biswas, & Jayaram, 2011)	http://www.scfbio-iitd.res.in/dock/ActiveSite_new.jsp
Docking softwares	
AutoDock4.2 (Morris et al., 2009)	http://autodock.scripps.edu
PatchDock (Schneidman-Duhovny et al., 2005a)	http://bioinfo3d.cs.tau.ac.il/PatchDock
ClusPro (Comeau, Gatchell, Vajda, & Camacho, 2004)	http://cluspro.bu.eduhttp://nrc.bu.edu/cluster
DockingServer (Bikadi & Hazai, 2009)	http://www.dockingserver.com
DOCK 6.6 (Brozell et al., 2012)	http://dock.compbio.ucsf.edu
3DLigandSite (Wass, Kelley, & Sternberg, 2010)	http://www.sbg.bio.ic.ac.uk/~3dligandsite
@TOME (Pons & Labesse, 2009)	http://atome.cbs.cnrs.fr/AT2/meta.html
AutoDock Vina (Trott & Olson, 2010)	http://vina.scripps.edu
BSP-SLIM (Lee & Zhang, 2012)	http://zhanglab.ccmb.med.umich.edu/BSP-SLIM
FiberDock—Flexible induced-fit backbone refinement in molecular docking (Mashiach, Nussinov, & Wolfson, 2009)	http://bioinfo3d.cs.tau.ac.il/FiberDock
GEMDOCK—Generic evolutionary method for molecular docking (Yang & Chen, 2004)	http://gemdock.life.nctu.edu.tw/dock

Table 10.1 List of computational methods employed in docking analysis—cont'd

Database name	Resource
Hex (Ghoorah, Devignes, Smaïl-Tabbone, & Ritchie, 2013)	http://hex.loria.fr
idTarget (Wang et al., 2012)	http://idtarget.rcas.sinica.edu.tw
iGEMDOCK (Yang & Chen, 2004)	http://gemdock.life.nctu.edu.tw/dock/igemdock.php
iScreen (Tsai, Chang, & Chen, 2011)	http://iscreen.cmu.edu.tw
ParDOCK (Gupta, Gandhimathi, Sharma, & Jayaram, 2007)	http://www.scfbio-iitd.res.in/dock/pardock.jsp
Surflex-Dock (Jain, 2003)	http://www.tripos.com/index.php?family=modules,SimplePage,&page=Surflex_Dock
AuPosSOM (Mantsyzov, Bouvier, Evrard-Todeschi, & Bertho, 2012)	https://www.biomedicale.univ-paris5.fr/aupossom
BetaDock (Kim, Kim, et al., 2011)	http://voronoi.hanyang.ac.kr/software.htm
DOCK Blaster (Irwin et al., 2009)	http://blaster.docking.org
eHiTS—Electronic high-throughput screening (Zsoldos, Reid, Simon, Sadjad, & Johnson, 2006)	http://www.simbiosys.ca/ehits
FITTED—Flexibility induced through targeted evolutionary description (De Cesco et al., 2012)	http://fitted.ca/index.php?option=com_content&task=view&id=50&Itemid=40
Fleksy (Nabuurs, Wagener, & de Vlieg, 2007)	http://www.cmbi.ru.nl/software/fleksy
FlexX (Sousa, Fernandes, & Ramos, 2006)	http://www.biosolveit.de/flexx
FLIPDock (Zhao & Sanner, 2007)	http://flipdock.scripps.edu/what-is-flipdock
FRED—Fast exhaustive docking (McConkey, Sobolev, & Edelman, 2002)	http://www.eyesopen.com/docs/oedocking/current/html/fred.html
GlamDock (Tietze & Apostolakis, 2007)	http://www.chil2.de/Glamdock.html

Continued

Table 10.1 List of computational methods employed in docking analysis—cont'd

Database name	Resource
GOLD (Verdonk, Cole, Hartshorn, Murray, & Taylor, 2003)	http://www.ccdc.cam.ac.uk/products/life_sciences/gold
GPCRautomodel (Launay et al., 2012)	http://genome.jouy.inra.fr/GPCRautomdl/cgi-bin/welcome.pl
GRAMM-X (Tovchigrechko & Vakser, 2006)	http://vakser.bioinformatics.ku.edu/resources/gramm/grammx
HADDOCK—High ambiguity-driven biomolecular docking (Dominguez, Boelens, & Bonvin, 2003)	http://www.nmr.chem.uu.nl/haddock
HomDock (Marialke, Tietze, & Apostolakis, 2008)	http://www.chil2.de/HomDock.html
ICM-Docking (Fernandez-Recio, Totrov, & Abagyan, 2003)	http://www.molsoft.com/docking.html
kinDOCK—A ligand transposition server (Martin, Catherinot, & Labesse, 2006)	http://abcis.cbs.cnrs.fr/LIGBASE_SERV_WEB/PHP/kindock.php
Lead Finder (Novikov et al., 2012)	http://www.moltech.ru
MVD—Molegro Virtual Docker (Thomsen & Christensen, 2006)	http://www.molegro.com/mvd-product.php
ParaDocks—Parallel Docking Suite (Muegge, 2006)	http://www.paradocks.org
PLANTS—Protein–Ligand ANT System (Korb, Stützle, & Exner, 2006)	http://www.tcd.uni-konstanz.de/research/plants.php
Rosetta FlexPepDock—High-resolution modeling of peptide–protein interactions (London, Raveh, Cohen, Fathi, & Schueler-Furman, 2011)	http://flexpepdock.furmanlab.cs.huji.ac.il/index.php
RosettaLigand (Hirst, Alexander, McHaourab, & Meiler, 2011)	http://www.rosettacommons.org/software
SwissDock (Grosdidier, Zoete, & Michielin, 2011)	http://www.swissdock.ch/docking
SymmDock—Prediction of complexes with Cn symmetry by geometry-based docking (Schneidman-Duhovny et al., 2005b)	http://bioinfo3d.cs.tau.ac.il/SymmDock

Table 10.1 List of computational methods employed in docking analysis—cont'd

Database name	Resource
TarFisDock—Target fishing dock (Li et al., 2006)	http://www.dddc.ac.cn/tarfisdock
VEGA ZZ (Pedretti, Villa, & Vistoli, 2004)	http://www.vegazz.net
VLifeDock	http://www.vlifesciences.com/products/VLifeMDS/VLifeDock.php
Visualization tools	
BALL View (Moll, Hildebrandt, Lenhof, & Kohlbacher, 2006)	http://www.ball-project.org/
Visual Molecular Dynamics (Humphrey, Dalke, & Schulten, 1996)	http://www.ks.uiuc.edu/Research/vmd/
UCSF Chimera (Pettersen et al., 2004)	http://www.cgl.ucsf.edu/chimera/
PyMOL (The PyMOL Molecular Graphics System, Version 1.2r3pre, Schrödinger, LLC)	http://pymol.org/
RasMol (Sayle & White, 1995)	http://www.rasmol.org/

Waals force, binding energy, the number of hydrogen bonds formed between the docked receptor and ligand, and other types of intermolecular interactions in the final docked molecule. Scoring only decides the best conformations for further research or for the purpose of creating a new drug (Alonso et al., 2006). Currently, scoring functions are utilized by docking softwares in one of two ways. In the first approach, the docked molecule, for example, the ligand–protein conformations, is ranked completely by the scoring functions (Huang, Grinter, & Zou, 2010). The search algorithms then modify this arrangement a little. These rearranged conformations are again ranked by the scoring functions. The second approach uses a two-stage scoring function. First, the reduced function is used to direct the search, and then more a vigorous function is used to rank the resulting structures. There are drawbacks with this directed approach. These directed methods make assumptions about the energy hyper surface, often omitting computationally expensive terms such as electrostatics, and considering only a few types of interactions such as hydrogen bonds. Such algorithms are, therefore, directed to areas of importance as determined by the reduced scoring function (Taylor, Jewsbury, & Essex, 2002). There are many algorithms used for docking, which include search algorithm, molecular dynamics, Monte Carlo, genetic algorithms, fragment-based methods, point complementary methods, distance geometry methods, and systemic searches. It is important to note that many docking technologies incorporate multiple or blended approaches into their techniques. In rigid-body docking, ligand and target protein are rigid, whereas in flexible docking, the flexibility of the ligand and/or the target protein (receptor) is considered. Rigid-body docking is fast but does not consider induced fit and is cheaper in comparison to flexible docking. Major types of scoring functions are listed in the succeeding text.

7.1. Force-field scoring functions

The force-field scoring function decides the binding energy by calculating the sum on nonbonding interactions such as van der Waals, electrostatics, bond stretching/bending/torsional forces, and entropy contributions (Taylor et al., 2002). The electrostatic force is calculated by Coulombic formulation; electrostatic potential energy is represented as a pair-wise summation of Coulombic interactions (Kitchen, Decornez, Furr, & Bajorath, 2004), and van der Waals terms are described by a Lennard-Jones potential function. Changing the potential increases or decreases the acceptance threshold of the score between the protein and the ligand. Changing the potential also determines the proximity of the ligand to the protein (Meng, Zhang, Mezei, & Cui, 2011).

7.2. Empirical free-energy scoring functions

In an empirical-based scoring function, the elements such as hydrogen bond, ionic interaction, the hydrophobic effect, and binding entropy are taken individually. These components are then individually multiplied by a coefficient to give an individual score. All the individual scores are then summed up to get a final score (Campbell, Gold, Jackson, & Westhead, 2003).

7.3. Knowledge-based scoring functions

Knowledge-based scoring functions use statistical analysis of the ligand–protein interactions. The function is based on the theory that if there is a high interaction between a particular protein and a ligand, then the probability of combining them together or occurring together is very high (Meng et al., 2011).

7.4. Consensus-based scoring functions

Scoring functions are prone to errors. The consensus scoring function is more recent trend. This scoring function combines the scores of all the scoring programs and gives one final score in order to minimize the errors and to find the perfect ligand for a given target macromolecule (Meng et al., 2011). There is a potential limitation to this method; sometimes instead of the error being minimized, there can be amplification in the calculation errors, which can void the balance of this scoring function aims at (Kitchen et al., 2004).

8. TYPES OF DOCKING

Molecular docking is also referred to as small molecular docking. Molecular docking is a study of how two or more molecular structures, for instance, drug and catalyst or macromolecule receptor, match along to be a perfect fit (Gane & Dean, 2000). Binding orientation of small-molecule drug candidates to their macromolecular targets predicts the affinity and activity of a given small molecule (Hakes, Lovell, Oliver, & Robertson, 2007).

Protein–protein docking is a simple procedure, which involves docking of two protein molecules without any need of experimental measurement. Flexible and rigid docking is followed in this type of docking (Ehrlich & Wadey, 2003). Shape complementarity is the most essential ingredient of the scoring functions for protein–protein docking (Chen & Weng, 2003). The steady rise in the number of protein structures elucidated has boosted the number of protein–protein docking studies, and intensive research is being carried out in the field. Many proteins that remain rigid after forming a complex can also be docked (Hakes et al., 2007).

Protein–ligand docking is the most commonly used docking technique. It predicts the position of a ligand when it is bound to its receptor molecule, in this case, a protein. The ligand might act as an inhibitor or a promoter. Large libraries of ligands are scanned to choose potential drug candidates (Smith, Engdahl, Dunbar, & Carlson, 2012).

AutoDock is a molecular docking suite consisting of automated docking tools. AutoDock consists of two main programs: AutoDock and AutoGrid. AutoDock docks the two molecules according to the grid, which is precalculated and set by AutoGrid. AutoDock is considered one of the best programs when it comes to docking and virtual screening (Park, Lee, & Lee, 2006). This section will give a brief overview of the steps followed in AutoDock 4.2 (Fig. 10.2). Various possible problems must be resolved

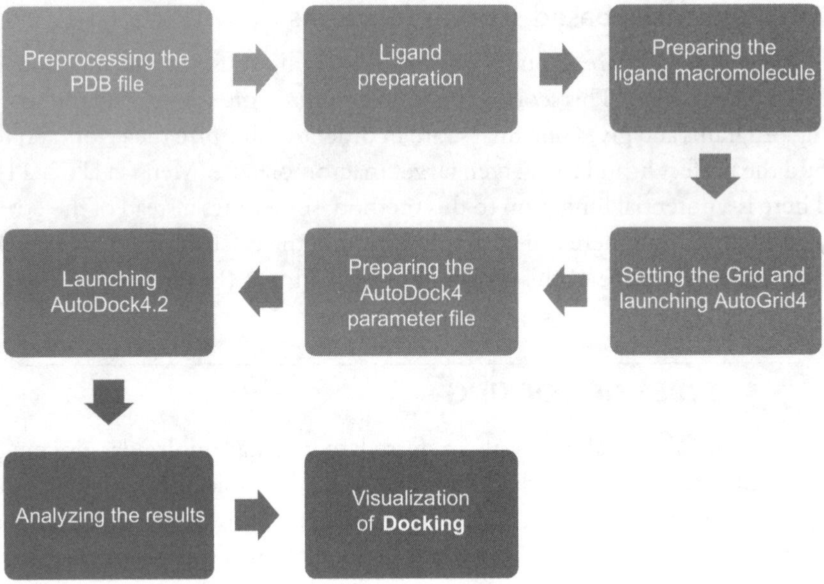

Figure 10.2 Steps in followed in docking analysis. (For color version of this figure, the reader is referred to the online version of this chapter.)

before a protein can be used for AutoDock. This includes missing atoms, chain breaks, and alternate locations. Potential energy grids are used by different docking programs. These grids represent the energy calculations, and in their most basic form, the grid stashes two types of potentials: the electrostatic and the van der Waals force. The grid was formulated so that the information about the receptor's energy contributions could be stored on grid points. This allowed the necessity of it being read only during ligand scoring. More options can be explored in AutoDock, and the options may vary depending on the complexes that are being docked and also the complexity of the problem in hand.

8.1. Advantages of docking

- The application of docking in a targeted drug-delivery system is a huge benefit. One can study the size, shape, charge distribution, polarity, hydrogen bonding, and hydrophobic interactions of both ligand (drug) and receptor (target site).
- Molecular docking helps in the identification of target sites of the ligand and the receptor molecule.

- Docking also helps in understanding of different enzymes and their mechanism of action.
- The "scoring" feature in docking helps in selecting the best fit or the best drug from an array of options.
- Not everything can be proved experimentally as traditional experimental methods for drug discovery take a long time. Molecular docking helps in moving the process of computer-aided drug designing faster and also provides every conformation possible based on the receptor and ligand molecule.
- Docking has a huge advantage when it comes to the study of protein interactions.
- There are millions of compounds, ligands, drugs, and receptors, the 3D structure of which has been crystallized. Virtual screening of these compounds can be made.

8.2. Limitations of docking

- In protein–small-molecule docking, there can be problems in the receptor structure. A reliable resolution value for small-molecule docking is below 1.2 Å (Gohlke & Klebe, 2002), while most crystallographic structures have a resolution between 1.5 and 2.5 Å. Increasing the use of homology models in docking should be looked at with care as they have even poorer resolution (Mihasan, 2010). Most applications accept and yield good results for structures below 2.2 Å. All the same, care should be taken while picking a structure.
- The scoring functions used in docking, almost all of them, do not take into account the role played by covalently bound inhibitors or ions (Mihasan, 2012).
- The methodology and research in protein–protein docking have to be greatly increased as the success in this field is greatly hampered by many false positives and false negatives (Moreira, Fernandes, & Ramos, 2010).

9. MOLECULAR DYNAMICS

In the late 1950s and the early 1960s, Alder, Wainwright, and Rahman first developed molecular dynamics (MD) to understand the atomic movement of liquids (Maginn & Elliott, 2010). With the advancement of computer science, MD has consequently become a precious and powerful tool in many domains. Since the 1970s, MD is widely applied to study the structure and dynamics of the complexity of chemical components of life, especially proteins or nucleic acids. Computer-aided drug design has come into more focus due to the development of computer science. Improvements in computer science and algorithm have played a vital role in the development of next-generation tools in computer-aided drug design. Currently, MD is used for simulations of biomolecular systems comprising many thousands of atoms within a very short span of time such as nanoseconds (Plimpton, 1995). Therefore, this method is used to solve a number of biological problems including drug discovery and development (Yang, 2010). However, in this process, more computational resources are involved. Conventional methods for the experimental determination of protein structures have relied on X-ray crystallographic and NMR techniques that have been considered to decide the structure of a protein with high precision. However, dynamic properties of molecules, especially proteins and DNA, are connected with backbone and side-chain moieties. MD enables us to understand the crucial determinants of many aspects of protein actions especially stability, folding, and function (Benkovic & Hammes-Schiffer, 2003; Eisenmesser, Bosco, Akke, & Kern, 2002; Frauenfelder, Sligar, & Wolynes, 1991; Karplus & McCammon, 2002; Rasmussen, Stock, Ringe, & Petsko, 1992; Wong & McCammon, 2003). Such dynamic processes can be understood by a variety of different techniques, particularly detecting NMR through relaxation phenomena (Kay, 1998). During drug-discovery process, crystallographic studies and NMR techniques are very important in understanding the role of protein–ligand binding as well as protein dynamics. It also helps in understanding the flexibility and motion of the proteins in the ligand-binding assay, where emerging computational methods have made the process simpler. During the prediction of protein motions, the calculations of the quantum-mechanical motion are needed, but it is difficult to explain the chemical reactions and motions in large molecular systems. MD helps to describe the chemical reactions and motion of proteins (Carloni, Rothlisberger, & Parrinello, 2002). This method helps to understand the interactions of protein and ligand or protein–drug interactions, which have multifaceted use in drug discovery and development (Utesch, Daminelli, & Mroginski, 2011).

According to quantum mechanics, physical quantities are represented by averages over microscopic conformations of the system, which are distributed in harmony with a particular statistical assembly (Weaire & Aste, 2010). Newtonian dynamics involves the conservation of energy and molecular dynamic trajectories, which can present a number of arrangements distributed according to the microcanonical ensemble. This is why physical quantity can be calculated through MD with the help of the arithmetic average over instantaneous values of that quantity obtained from the trajectories (Tuckerman, Yarne, Samuelson, Hughes, & Martyna, 2000). In the limit of infinite, simulation time of the true value of the measured thermodynamical properties. *In vitro*, the quality of sampling and accuracy of the interatomic potentials used in simulations are always limited. In fact, the quality of sampling may not be proper, especially for processes of the timescale larger than typical molecular dynamic simulations, and caution should be exerted when drawing conclusions from such computer experiments (Phillips et al., 2005). MD simulations illustrate the physical movement of atoms and molecules as they interact over time. This is accomplished by the potential energy function, also so-called force fields, used in most of the standard molecular simulation programs for biological systems. It takes the form of the summation of different additive terms that correspond to bond distance stretching (E_{bonds}), bond-angle bending (E_{angles}), bond dihedral or torsion angle ($E_{\text{dihedrals}}$), van der Waals potential (E_{vdw}), and electrostatic potential (E_{elect}). The first three terms are considered to be the intramolecular bonding interactions, and each term involves a multitude of atoms connected by chemical bonds. The other two terms represent the nonbonded interactions between atoms. The most common force fields are OPLS-AA (Jorgensen, Maxwell, & Tirado-Rives, 1996), CHARMM (MacKerell et al., 1998), GROMOS (Christen et al., 2005), and AMBER (Wang, Fang, et al., 2004; Wang, Wolf, et al., 2004), which can perform energy calculations, energy minimization, and dynamic calculations. These tools are regularly used in biomolecular simulation and principally vary as per their parameters. However, these tools usually give similar results. It is very interesting that the forces acting on every one of the system atoms are calculated; the site of these atoms is stimulated according to Newton's laws of motion, and therefore, Newtonian dynamics is applicable. The simulation time is then higher, often by only 1 or 2 quadrillionths of a second, and in this way, the procedure is replicated millions of times on average. Since several calculations are necessary, MD simulations are executed on computer clusters or supercomputer systems, which run hundreds of parallel processors at a time to calculate the process (Herschbach, 1987). Some well-accepted simulation software packages are AMBER (Case et al., 2005), CHARMM (MacKerell et al., 1998), and NAMD (Kalé et al., 1999; Phillips et al., 2005), which apply their default force fields. In the current era, pharmacogenomics is a prominent field, and drug discovery is moving in this direction. There are many studies where MD is applied in studying the conformational space accessible to proteins and protein–ligand interactions and refined experimental or modeled protein structures (Schlick, Collepardo-Guevara, Halvorsen, Jung, & Xiao, 2011). Modeled protein structures (Fan et al., 2009) reveal transient binding sites (Ivetac & McCammon, 2010), examine the stability and strength of docked protein–ligand conformations (B-Rao et al., 2009), aid drug discovery (Salsbury, 2010), and explore altered drug binding profiles of protein variants (Shan et al., 2011). In addition to this, MD simulations were carried out on both the native and mutant proteins to show its flexibility and effects on the protein (George Priya Doss, Nagasundaram, Chakraborty, Chen, & Zhu, 2013; George Priya Doss, Rajith, & Chakraborty, 2013; George Priya Doss, Rajith, Rajasekaran, et al., 2013; John et al., 2013; Miteva, Brugge, Rosing, Nicolaes, & Villoutreix, 2004; Steen, Miteva, Villoutreix, Yamazaki, & Dahlback, 2003; Witham, Takano, Schwartz, & Alexov, 2011; Zhang, Teng, Wang, Schwartz, & Alexov, 2010), stability on protein–protein interactions (George Priya Doss & Nagasundaram, 2013), and protein–ligand (George Priya Doss, Rajith, Chakraborty, Balaji, et al., 2013; Nagasundaram & George Priya Doss, 2013) and protein–DNA interactions (George Priya Doss & Nagasundaram, 2012). There are studies showing agreement between computational and experimental measurements of macromolecular dynamics (Bruschweiler & Showalter, 2007; LaConte, Voelz, Nelson, & Thomas, 2002; Markwick et al., 2010; Peter et al., 2003). Combining this information, MD analysis has the potential to be a vital resource in elucidating the molecular effects of mutations and variable drug responses in the context of target protein structures with genetic variants. Here, molecular dynamics has an advantage over experimental methods. Along with the molecular dynamic power, advances in computer technology and algorithm design will definitely act as a driving force in computer-aided drug design in the development of novel pharmacological drugs.

10. CONCLUDING REMARKS

The current world is fast paced; the number of diseases crippling the human race is on the rise. This has caused a never before seen urgency for results for drugs to combat the various diseases that plague the human race. In such a demanding era, science has to rely on bioinformatics, an emerging field that has helped to cut down on the huge amount of time and resources that are otherwise wasted. Single-nucleotide polymorphisms are the most common single amino acid substitutions found in the human genome.

There exist quite a few traditional methods of SNP analysis. Few examples of such experimental methods include DNA sequencing, capillary electrophoresis, single-strand conformation polymorphism, and restriction fragment length polymorphism. These traditional methods have many limitations. A few of these methods require a large sample of DNA, require expensive equipments, and also face difficulty in handling large sequences and are also not practically easy in the process of complicated drug designing. The completion of the Human Genome Project has changed the each core of bioinformatics. With a plethora of data available at the click of a mouse, and very little time in obtaining results, it is impossible to rely only on traditional experimental methods. Drug designing in the late twentieth century has been on the rise. This called for the integration of computational methods as they help in relaying whether the compound or a ligand of which is being screened as a potential drug can be an asset or liability to a particular research. Quality of information can also be improved with the use of computational methods. Computers can integrate results from both *in vivo* and *in vitro* studies from many different labs. This information database can be used by scientists and researchers from all over the world to come up with better and faster results, and newer technologies. Computational methods now allow scientists to cut down on animal testing, as the data provided are accurate, and further testing can be done with computer modeling. This saves wastage of chemicals and also reduces animal suffering.

Personalized medicine is another upcoming field; the scientific community is looking at now. It is common knowledge that a drug that is effective for one person may not be as effective or may even be harmful to another individual. All this is caused by the tiniest changes in our genomes, which make us different from one another (SNPs). Identifying significant SAPs that produce clinically relevant phenotype is important to provide personalized diagnosis and treatment. Personalized medicine would require sequencing each person's genome, analyzing the SNPs, and then formulating which drug would be best suited for a particular individual. Even though computational methods give a faster access to the solution, they are still only predictions. These methods can only predict which SAPs are deleterious and particular disease phenotypes. These methods cannot provide insight how the disruption happens on a molecular scale. Hence, bridging this gap between computational and clinical methods proves to be a challenging task for personalized and precision medicine. There are also limitations to abovementioned computational methods. All the data produced might not be useful, if the results or observations made during experiments vary.

Most of the methods used in *in silico* still have gaping holes that are yet to be filled. More research is required to improve the quality of results derived from such computational research. Existing SAP prediction methods, homology modeling methods, docking, and molecular dynamics softwares have different algorithms and varying protocols. The scoring algorithms of *in silico* tools have to be looked into, as even now, researchers have to rely on multiple scoring algorithms and manually cross-check to see if the results hold true. Therefore, only computational methods cannot be used to validate a drug. Most of the computational prediction methods available online were benchmarked by the curators with their known datasets and were shown to perform well. There are mounting studies that compare the prediction scores from a set of methods and the results bring to light that no single method can be rated as the best predictor. The results obtained from these indicate that a combination of different methods with sequence and structure information may provide a wider coverage and accurate prediction in the study of SAPs to be either deleterious or neutral. Basically, computational methods used different algorithms to predict the impact of deleterious variants, and therefore, the outcome may differ for each tool. However, the positive predictions overlap in all the computational disease prediction methods, which show that they have a high possibility to behave in a similar fashion. The variation in their prediction scores might be due to the difference in features utilized by the methods or the trained datasets.

Computational research has to be complemented with traditional experimental methods. Computational methods help in narrowing down the possible choices of drugs or target molecules, and the experimental biologist can take the given output and check how a drug or any compound will work in a biological system. It has been more than a decade, since the completion of the Human Genome Project. This has helped us to gain insight into the human genome. For instance, the recent ENCODE project has helped to bring before us the biochemical functions of almost 80% of the human genome. In spite of this, a revolution in the medical field is yet to be seen. Cost and time reduction methods can sequence the human genome points to a near future where DNA sequencing would be routinely used in medical practices. A human genome sequence generates a large amount of data, and additional data will be generated in the case of sequencing carried out for personalized medicines. There is a need for next-generation computers and faster networks that can easily handle and process many terabytes of data. When clinical practices will be amalgamated with computational power in the future, the systems should be able to handle the influx of a large amount of data from all the patients. The computers must also be able to take on each task based on their nature of urgency. Moreover, prioritizing the disease genes can also be incorporated. All these issues need to be dealt with so that the populace receives the best medical attention. The promise of personalized medicine and precision medicine will be made possible by the computational methods discussed in this chapter and will be a useful resource for researchers looking to widen their research scope.

ACKNOWLEDGMENTS

The authors take this opportunity to thank the management of VIT and Galgotias University for providing the facilities and encouragement to carry out this work.

REFERENCES

1000 Genomes Project Consortium, Abecasis, G. R., Altshuler, D., Auton, A., Brooks, L. D., Durbin, R. M., Gibbs, R. A., et al. (2010). A map of human genome variation from population-scale sequencing. *Nature*, *467*(7319), 1061–1073.

Abagyan, R., Totrov, M., & Kuznetsov, D. (1994). ICM-A new method for protein modeling and design: Applications to docking and structure prediction from the distorted native conformation. *Journal of Computational Chemistry*, *15*, 488–506.

Adzhubei, I. A., Schmidt, S., Peshkin, L., Ramensky, V. E., Gerasimova, A., Bork, P., et al. (2010). A method and server for predicting damaging missense mutations. *Nature Methods*, *7*, 248–249.

Akhavan, S., Miteva, M. A., Villoutreix, B. O., Venisse, L., Peyvandi, F., Mannucci, P. M., et al. (2005). A critical role for Gly25 in the B chain of human thrombin. *Journal of Thrombosis and Haemostasis*, *3*, 139–145.

Alexov, E. (2004). Numerical calculations of the pH of maximal protein stability: The effect of the sequence composition and three-dimensional structure. *European Journal of Biochemistry*, *271*(1), 173–185.

Alonso, H., Bliznyuk, A. A., & Gready, J. E. (2006). Combining docking and molecular dynamic simulations in drug design. *Medicinal Research Reviews*, *26*(5), 531–568.

Altschul, S. F., Gish, W., Miller, W., Myers, E. W., & Lipman, D. J. (1990). Basic local alignment search tool. *Journal of Molecular Biology*, *215*, 403–410.

Altschul, S. F., Madden, T. L., Schäffer, A. A., Zhang, J., Zhang, Z., Miller, W., et al. (1997). Gapped BLAST and PSI-BLAST: A new generation of protein database search programs. *Nucleic Acids Research*, *25*(17), 3389–3402.

Amberger, J., Bocchini, C., & Hamosh, A. (2011). A new face and new challenges for Online Mendelian Inheritance in Man (OMIM(R)). *Human Mutation*, *32*, 564–567.

Anand, P., Yeturu, K., & Chandra, N. (2012). PocketAnnotate: Towards site-based function annotation. *Nucleic Acids Research*, *40*, W400–W408.

Apweiler, R., Bairoch, A., Wu, C. H., Barker, W. C., Boeckmann, B., Ferro, S., et al. (2004). UniProt: The Universal Protein knowledgebase. *Nucleic Acids Research*, *32*, D115–D119.

Apweiler, R., Martin, M. J., O'Donovan, C., Magrane, M., Alam-Faruque, Y., Antunes, R., et al. (2010). The universal protein resource (UniProt) in 2010. *Nucleic Acids Research*, *38*, D142–D148.

Arbiza, L., Duchi, S., Montaner, D., Burguet, J., Pantoja-Uceda, D., Pineda-Lucena, A., et al. (2006). Selective pressures at a codon level predict deleterious mutations in human disease genes. *Journal of Molecular Biology*, *358*, 1390–1404.

Arnold, K., Bordoli, L., Kopp, J., & Schwede, T. (2006). The SWISS-MODEL workspace: A web based environment for protein structure homology modeling. *Bioinformatics*, *22*, 195–201.

Ashkenazy, H., Erez, E., Martz, E., Pupko, T., & Ben-Tal, N. (2010). ConSurf 2010: Calculating evolutionary conservation in sequence and structure of proteins and nucleic acids. *Nucleic Acids Research*, *38*, W529–W533.

Azam, S. S., & Abbasi, S. W. (2013). Molecular docking studies for the identification of novel melatoninergic inhibitors for acetylserotonin-O-methyltransferase using different docking routines. *Theoretical Biology and Medical Modelling*, *10*(1), 63.

Bao, L., & Cui, Y. (2006). Functional impacts of non-synonymous single nucleotide polymorphisms: Selective constraint and structural environments. *FEBS Letters*, *580*, 1231–1234.

Bao, L., Zhou, M., & Cui, Y. (2005). nsSNPAnalyzer: Identifying disease-associated nonsynonymous single nucleotide polymorphisms. *Nucleic Acids Research*, *33*, W480–W482.

Barroso, I., Gurnell, M., Crowley, V. E., Agostini, M., Schwabe, J. W., Soos, M. A., et al. (1999). Dominant negative mutations in human PPAR gamma associated with severe insulin resistance, diabetes mellitus and hypertension. *Nature*, *402*, 880–883.

Basu, S. N., Kollu, R., & Banerjee-Basu, S. (2009). AutDB: A gene reference resource for autism research. *Nucleic Acids Research*, *37*, D832–D836.

Bates, P. A., Kelley, L. A., MacCallum, R. M., & Sternberg, M. J. E. (2001). Enhancement of protein modeling by human intervention in applying the automatic programs 3D-JIGSAW and 3D-PSSM. *Proteins*, *45*, 39–46.

Benkovic, S. J., & Hammes-Schiffer, S. (2003). A perspective on enzyme catalysis. *Science*, *301*, 1196–1202.

Berezin, C., Glaserm, F., Rosenberg, J., Paz, I., Pupko, T., Fariselli, P., et al. (2004). ConSeq: The identification of functionally and structurally important residues in protein sequences. *Bioinformatics*, *20*, 1322–1324.

Bernstein, F. C., Koetzle, T. F., Williams, G. J., Meyer, E. F., Jr., Brice, M. D., Rodgers, J. R., et al. (1977). The Protein Data Bank: A computer-based archival file for macromolecular structures. *Journal of Molecular Biology*, *112*, 535–542.

Betz, S. F. (1993). Disulfide bonds and the stability of globular proteins. *Protein Science*, *2*(10), 1551–1558.

Bikadi, Z., & Hazai, E. (2009). Application of the PM6 semi-empirical method to modeling proteins enhances docking accuracy of AutoDock. *Journal of Cheminformatics*, *1*, 15.

Blom, N., Sicheritz-Ponten, T., Gupta, R., Gammeltoft, S., & Brunak, S. (2004). Prediction of post-translational glycosylation and phosphorylation of proteins from the amino acid sequence. *Proteomics*, *4*(6), 1633–1649.

Board, P. G., Pierce, K., & Coggan, M. (1990). Expression of functional coagulation factor XIII in Escherichia coli. *Thrombosis and Haemostasis*, *63*(2), 235–240.

Bolton, E., Wang, Y., Thiessen, P. A., & Bryant, S. H. (2008). PubChem: Integrated platform of small molecules and biological activities. *Annual Reports in Computational Chemistry*, *4*, 217–241.

Bonnardeaux, A., Davies, E., Jeunemaitre, X., Féry, I., Charru, A., Clauser, E., et al. (1994). Angiotensin II type 1 receptor gene polymorphisms in human essential hypertension. *Hypertension*, *24*, 63–69.

Boulling, A., le Marechal, C., Trouve, P., Raguenes, O., Chen, J. M., & Ferec, C. (2007). Functional analysis of pancreatitis associated missense mutations in the pancreatic secretory trypsin inhibitor (SPINK1) gene. *European Journal of Human Genetics*, *15*(9), 936–942.

Bowie, J. U., Lüthy, R., & Eisenberg, D. (1991). A method to identify protein sequences that fold into a known three-dimensional structure. *Science, 253*, 164–170.

B-Rao, C., Subramanian, J., & Sharma, S. D. (2009). Managing protein flexibility in docking and its applications. *Drug Discovery Today, 14*, 394–400.

Brendel, V., Bucher, P., Nourbakhsh, I., Blais-dell, B. E., & Karlin, S. (1992). Methods and algorithms for statistical analysis of protein sequences. *Proceedings of the National Academy of Sciences of the United States of America, 89*, 2002–2006.

Bromberg, Y., & Rost, B. (2007). SNAP: Predict effect of non-synonymous polymorphisms on function. *Nucleic Acids Research, 35*, 3823–3835.

Brozell, S. R., Mukherjee, S., Balius, T. E., Roe, D. R., Case, D. A., & Rizzo, R. C. (2012). Evaluation of DOCK 6 as a pose generation and database enrichment tool. *Journal of Computer-Aided Molecular Design, 26*(6), 749–773.

Bruschweiler, R., & Showalter, S. A. (2007). Validation of molecular dynamics simulations of biomolecules using NMR spin relaxation as benchmarks: Application to the AMBER99SB force field. *Journal of Chemical Theory and Computation, 3*, 961–975.

Brylinski, M., & Skolnick, J. (2008). A threading-based method (FINDSITE) for ligand-binding site prediction and functional annotation. *Proceedings of the National Academy of Sciences of the United States of America, 105*, 129–134.

caBIG Strategic Planning Workspace, (2007). The Cancer Biomedical Informatics Grid (caBIG): Infrastructure and applications for a worldwide research community. *Studies in Health Technology and Informatics, 129*, 330–334.

Cai, Z., Tsung, E. F., Marinescu, V. D., Ramoni, M. F., Riva, A., & Kohane, I. S. (2004). Bayesian approach to discovering pathogenic SNPs in conserved protein domains. *Human Mutation, 24*, 178–184.

Calabrese, R., Capriotti, E., Fariselli, P., Martelli, P. L., & Casadio, R. (2009). Functional annotations improve the predictive score of human disease-related mutations in proteins. *Human Mutation, 30*, 1237–1244.

Campbell, S. J., Gold, N. D., Jackson, R. M., & Westhead, D. R. (2003). Ligand binding: Functional site location, similarity and docking. *Current Opinion in Structural Biology, 13*(3), 389–395.

Capriotti, E., & Altman, R. B. (2011). Improving the prediction of disease-related variants using protein three-dimensional structure. *BMC Bioinformatics, 12*(4), S3.

Capriotti, E., Calabrese, R., Fariselli, P., Martelli, P. L., Altman, R. B., & Casadio, R. (2013). WS-SNPs&GO: A web server for predicting the deleterious effect of human protein variants using functional annotation. *BMC Genomics, 14*, S6.

Capriotti, E., Fariselli, P., & Casadio, R. (2005). I-Mutant2.0: Predicting stability changes upon mutation from the protein sequence or structure. *Nucleic Acids Research, 33*, W306–W310.

Capriotti, E., Fariselli, P., Rossi, I., & Casadio, R. (2008). A three-state prediction of single point mutations on protein stability changes. *BMC Bioinformatics, 9*(2), S6.

Carloni, P., Rothlisberger, U., & Parrinello, M. (2002). The role and perspective of ab initio molecular dynamics in the study of biological systems. *Accounts of Chemical Research, 35*(6), 455–464.

Cartegni, L., & Krainer, A. R. (2002). Disruption of an SF2/ASF-dependent exonic splicing enhancer in SMN2 causes spinal muscular atrophy in the absence of SMN1. *Nature Genetics, 30*, 377–384.

Case, D. A., Cheatham, T. E., 3rd, Darden, T., Gohlke, H., Luo, R., Merz, K. M., Jr., et al. (2005). The AMBER biomolecular simulation programs. *Journal of Computational Chemistry, 26*, 1668–1688.

Castella, M., Pujol, R., Callén, E., Trujillo, J. P., Casado, J. A., Gille, H., et al. (2011). Origin, functional role, and clinical impact of Fanconi anemia FANCA mutations. *Blood, 117*, 3759–3769.

Castellana, S., & Mazza, T. (2013). Congruency in the prediction of pathogenic missense mutations: State-of-the-art web-based tools. *Briefings in Bioinformatics*, *14*, 448–459.

Chang, H., & Fujita, T. (2001). PicSNP: A browsable catalog of nonsynonymous single nucleotide polymorphisms in the human genome. *Biochemical and Biophysical Research Communications*, *287*, 288–291.

Chang, W. C., Lee, T. Y., Shien, D. M., Hsu, J. B., Horng, J. T., Hsu, P. C., et al. (2009). Incorporating support vector machine for identifying protein tyrosine sulfation sites. *Journal of Computational Chemistry*, *30*(15), 2526–2537.

Chasman, D., & Adams, R. M. (2001). Predicting the functional consequences of non-synonymous single nucleotide polymorphisms: Structure-based assessment of amino acid variation. *Journal of Molecular Biology*, *307*, 683–706.

Chen, X., Ji, Z. L., & Chen, Y. Z. (2002). TTD: Therapeutic target database. *Nucleic Acids Research*, *30*, 412–415.

Chen, C. W., Lin, J., & Chu, Y. W. (2013). iStable: Off-the-shelf predictor integration for predicting protein stability changes. *BMC Bioinformatics*, *14*(2), S5.

Chen, R., & Weng, Z. (2003). A novel shape complementarity scoring function for protein-protein docking. *Proteins: Structure, Function, and Bioinformatics*, *51*(3), 397–408.

Chen, H.-L., & Zhou, H.-X. (2005). Prediction of solvent accessibility and sites of deleterious mutations from protein sequence. *Nucleic Acids Research*, *33*, 3193–3199.

Cheng, J., Randall, A., & Baldi, P. (2006). Prediction of protein stability changes for single-site mutations using support vector machines. *Proteins*, *62*(4), 1125–1132.

Cheng, J., Randall, A. Z., Sweredoski, M. J., & Baldi, P. (2005). SCRATCH: A protein structure and structural feature prediction server. *Nucleic Acids Research*, *33*, W72–W76.

Cheng, J., Sweredoski, M., & Baldi, P. (2005). Accurate prediction of protein disordered regions by mining protein structure data. *Data Mining and Knowledge Discovery*, *11*(3), 213–222.

Christen, M., Hünenberger, P. H., Bakowies, D., Baron, R., Bürgi, R., Geerke, D. P., et al. (2005). The GROMOS software for biomolecular simulation: GROMOS05. *Journal of Computational Chemistry*, *26*, 1719–1751.

Claussen, H., Buning, C., Rarey, M., & Lengauer, T. (2001). FlexE: Efficient molecular docking considering protein structure variations. *Journal of Molecular Biology*, *308*, 377–395.

Cole, C., Barber, J. D., & Barton, G. J. (2008). The Jpred 3 secondary structure prediction server. *Nucleic Acids Research*, *35*(2), W197–W201.

Comeau, S. R., Gatchell, D. W., Vajda, S., & Camacho, C. J. (2004). ClusPro: A fully automated algorithm for protein-protein docking. *Nucleic Acids Research*, *32*(2), W96–W99.

Conchillo-Sole, O., de Groot, N. S., Aviles, F. X., Vendrell, J., Daura, X., & Ventura, S. (2007). AGGRESCAN: A server for the prediction and evaluation of "hot spots" of aggregation in polypeptides. *BMC Bioinformatics*, *8*, 65.

Conde, L., Vaquerizas, J. M., Santoyo, J., Al-Shahrour, F., Ruiz-Llorente, S., Robledo, M., et al. (2004). PupaSNP Finder: A web tool for finding SNPs with putative effect at transcriptional level. *Nucleic Acids Research*, *32*, W242–W248.

Congreve, M., Murray, C. W., & Blundell, T. L. (2005). Structural biology and drug discovery. *Drug Discovery Today*, *10*, 895–907.

Curtis, D., North, B. V., & Sham, P. C. (2001). Use of an artificial neural network to detect association between a disease and multiple marker genotypes. *Annals of Human Genetics*, *65*(Pt. 1), 95–107.

Daga, P. R., Patel, R. Y., & Doerksen, R. J. (2010). Template-based protein modeling: Recent methodological advances. *Current Topics in Medicinal Chemistry*, *10*, 84–94.

Dalton, J. A., & Jackson, R. M. (2007). An evaluation of automated homology modelling methods at low target template sequence similarity. *Bioinformatics*, *23*(15), 1901–1908.

Dambosky, J., Prokop, M., & Koca, J. (2001). TRITON: Graphic software for rational engineering of enzymes. *Trends in Biochemical Sciences*, *26*, 71–73.
Das, R., & Baker, D. (2008). Macromolecular modeling with Rosetta. *Annual Review of Biochemistry*, *77*, 363–382.
Das, R., Qian, B., Raman, S., Vernon, R., Thompson, J., Bradley, P., et al. (2007). Structure prediction for CASP7 targets using extensive all-atom refinement with Rosetta@home. *Proteins*, *69*(8), 118–128.
Davis, I. W., Leaver-Fay, A., Chen, V. B., Block, J. N., Kapral, G. J., Wang, X., et al. (2007). MolProbity: All-atom contacts and structure validation for proteins and nucleic acids. *Nucleic Acids Research*, *35*, W375–W383.
De Baets, G., Van Durme, J., Reumers, J., Maurer-Stroh, S., Vanhee, P., Dopazo, J., et al. (2012). SNPeffect 4.0: On-line prediction of molecular and structural effects of protein-coding variants. *Nucleic Acids Research*, *40*, D935–D939.
De Cesco, S., Deslandes, S., Therrien, E., Levan, D., Cueto, M., Schmidt, R., et al. (2012). Virtual screening and computational optimization for the discovery of covalent prolyl oligopeptidase inhibitors with activity in human cells. *Journal of Medicinal Chemistry*, *55*, 6306–6315.
Deane, C. M., & Blundell, T. L. (2001). CODA: A combined algorithm for predicting the structurally variable regions of protein models. *Protein Science*, *10*, 599–612.
Dehouck, Y., Grosfils, A., Folch, B., Gilis, D., Bogaerts, P., & Rooman, M. (2009). Fast and accurate predictions of protein stability changes upon mutations using statistical potentials and neural networks: PoPMuSiC-2.0. *Bioinformatics*, *25*, 2537–2543.
Dessailly, B. H., Lensink, M. F., Orengo, C. A., & Wodak, S. J. (2008). LigASite—A database of biologically relevant binding sites in proteins with known apo-structures. *Nucleic Acids Research*, *36*(1), D667–D673.
Dill, K. A., Ozkan, S. B., Weikl, T. R., Chodera, J. D., & Voelz, V. A. (2007). The protein folding problem: When will it be solved? *Current Opinion in Structural Biology*, *17*(3), 342–346.
DiMasi, J. A., & Grabowski, H. G. (2007). The cost of biopharmaceutical R&D: Is biotech different? *Managerial and Decision Economics*, *28*, 285–291.
DiMasi, J. A., Grabowski, H. G., & Vernon, J. (2004). R&D costs and returns by therapeutic category. *Drug Information Journal*, *38*, 211–223.
DiMasi, J. A., Hansen, R. W., Grabowski, H. G., & Lasagna, L. (1991). Cost of innovation in the pharmaceutical industry. *Journal of Health Economics*, *10*, 107–142.
Dimmic, M. W., Sunyaev, S., & Bustamante, C. D. (2005). Inferring SNP function using evolutionary, structural, and computational methods. *Pacific Symposium on Biocomputing*, *10*, 382–384.
Dixit, A., Torkamani, A., Schork, N. J., & Verkhivker, G. (2009). Computational modeling of structurally conserved cancer mutations in the RET and MET kinases: The impact on protein structure, dynamics, and stability. *Biophysical Journal*, *96*(3), 858–874.
Dobson, C. M. (2003). Protein folding and misfolding. *Nature*, *426*(6968), 884–890.
Dobson, R. J., Munroe, P. B., Caulfield, M. J., & Saqi, M. A. S. (2006). Predicting deleterious nsSNPs: An analysis of sequence and structural attributes. *BMC Bioinformatics*, 7(1), 217.
Dominguez, C., Boelens, R., & Bonvin, A. M. (2003). HADDOCK: A protein-protein docking approach based on biochemical and/or biophysical information. *Journal of the American Chemical Society*, *125*, 1731–1737.
Doolittle, R. F. (1986). *Of Urfs and Orfs: A Primer on how to Analyze Derived Amino Acid.* (University Science Books) ISBN 0-935702-54-7.
Dosztányi, Z., CsizmLok, V., Tompa, P., & Simon, I. (2005). IUPred: Web server for the prediction of intrinsically unstructured regions of proteins based on estimated energy content. *Bioinformatics*, *21*(16), 3433–3434.

Dosztanyi, Z., Magyar, C., Tusnady, G. E., & Simon, I. (2003). SCide: Identification of stabilization centers in proteins. *Bioinformatics*, *19*(7), 899–900.

Dryja, T. P., McGee, T. L., Hahn, L. B., Cowley, G. S., Olsson, J. E., Reichel, E., et al. (1990). Mutations within the rhodopsin gene in patients with autosomal dominant retinitis pigmentosa. *The New England Journal of Medicine*, *323*(19), 1302–1307.

Dundas, J., Ouyang, Z., Tseng, J., Binkowski, A., Turpaz, Y., & Liang, J. (2006). CASTp: Computed atlas of surface topography of proteins with structural and topographical mapping of functionally annotated residues. *Nucleic Acids Research*, *34*, W116–W118.

Durrant, J., & McCammon, J. A. (2011). Molecular dynamics simulations and drug discovery. *BMC Biology*, *9*(1), 71.

Edgar, R. C. (2004). MUSCLE: Multiple sequence alignment with high accuracy and high throughput. *Nucleic Acids Research*, *32*(5), 1792–1797.

Ehrlich, L. P., & Wadey, R. C. (2003). Protein–protein docking. *Reviews in Computational Chemistry*, *17*, 61.

Eisenmesser, E. Z., Bosco, D. A., Akke, M., & Kern, D. (2002). Enzyme dynamics during catalysis. *Science*, *295*, 1520–1523.

Elles, L. M. S., & Uhlenbeck, O. C. (2008). Mutation of the arginine finger in the active site of Escherichia coli DbpA abolishes ATPase and helicase activity and confers a dominant slow growth phenotype. *Nucleic Acids Research*, *36*(1), 41–50.

ENCODE Project Consortium, Bernstein, B. E., Birney, E., Dunham, I., Green, E. D., Gunter, C., & Snyder, M. (2012). An integrated encyclopedia of DNA elements in the human genome. *Nature*, *489*(7414), 57–74.

Erdin, S., Ward, R. M., Venner, E., & Lichtarge, O. (2010). Evolutionary trace annotation of protein function in the structural proteome. *Journal of Molecular Biology*, *396*(5), 1451–1473.

Eriksson, A. E., Baase, W. A., Zhang, X. J., Heinz, D. W., Blaber, M., Baldwin, E. P., et al. (1992). Response of a protein structure to cavity creating mutations and its relation to the hydrophobic effect. *Science*, *255*, 178–183.

Eswar, N., Webb, B., Marti-Renom, M. A., Madhusudhan, M. S., Eramian, D., Shen, M. Y., et al. (2006). Comparative protein structure modeling using MODELLER. *Current Protocols in Bioinformatics*, Chapter 5, Unit 5.6.

Ewing, T. J., Makino, S., Skillman, A. G., & Kuntz, I. D. (2001). DOCK 4.0: Search strategies for automated molecular docking of flexible molecule databases. *Journal of Computer-Aided Molecular Design*, *15*, 411–428.

Fan, H., Irwin, J. J., Webb, B. M., Klebe, G., Shoichet, B. K., & Sali, A. (2009). Molecular docking screens using comparative models of proteins. *Journal of Chemical Information and Modeling*, *49*, 2512–2527.

Fernald, G. H., Capriotti, E., Daneshjou, R., Karczewski, K. J., & Altman, R. B. (2011). Bioinformatics challenges for personalized medicine. *Bioinformatics*, *27*, 1741–1748.

Fernandez-Escamilla, A. M., Rousseau, F., Schymkowitz, J., & Serrano, L. (2004). Prediction of sequence-dependent and mutational effects on the aggregation of peptides and proteins. *Nature Biotechnology*, *22*, 1302–1306.

Fernandez-Fuentes, N., Rai, B. K., Madrid-Aliste, C. J., Fajardo, J. E., & Fiser, A. (2007). Comparative protein structure modeling by combining multiple templates and optimizing sequence-to-structure alignments. *Bioinformatics*, *23*, 2558–2565.

Fernandez-Fuentes, N., Zhai, J., & Fiser, A. (2006). ArchPRED: A template based loop structure prediction server. *Nucleic Acids Research*, *34*, W173–W176.

Fernandez-Recio, J., Totrov, M., & Abagyan, R. (2003). ICM-DISCO docking by global energy optimization with fully flexible side-chains. *Proteins*, *52*, 113–117.

Ferrer-Costa, C., Gelpi, J. L., Zamakola, L., Parraga, I., de la Cruz, X., & Orozco, M. (2005). PMUT: A web-based tool for the annotation of pathological mutations on proteins. *Bioinformatics*, *21*, 3176–3178.

Ferrer-Costa, C., Orozco, M., & de la Cruz, X. (2002). Characterization of disease-associated single amino acid polymorphisms in terms of sequence and structure properties. *Journal of Molecular Biology, 315*, 771–786.

Frauenfelder, H., Sligar, S. G., & Wolynes, P. G. (1991). The energy landscapes and motions on proteins. *Science, 254*, 1598–1603.

Frazer, K. A., Ballinger, D. G., Cox, D. R., Hinds, D. A., Stuve, L. L., et al. (2007). A second generation human haplotype map of over 3.1 million SNPs. *Nature, 449*, 851–861.

Frazer, K. A., Murray, S. S., Schork, N. J., & Topol, E. J. (2009). Human genetic variation and its contribution to complex traits. *Nature Reviews. Genetics, 10*, 241–251.

Fredman, D., Siegfried, M., Yuan, Y. P., Bork, P., Lehväslaiho, H., & Brookes, A. J. (2002). HGVbase: A human sequence variation database emphasizing data quality and a broad spectrum of data sources. *Nucleic Acids Research, 30*(1), 387–391.

Freimuth, R. R., Stormo, G. D., & McLeod, H. L. (2005). PolyMAPr: Programs for polymorphism database mining, annotation, and functional analysis. *Human Mutation, 25*, 110–117.

Friesner, R. A., Banks, J. L., Murphy, R. B., Halgren, T. A., Klicic, J. J., Mainz, D. T., et al. (2004). Glide: A new approach for rapid, accurate docking and scoring. 1. Method and assessment of docking accuracy. *Journal of Medicinal Chemistry, 47*, 1739–1749.

Frousios, K., Iliopoulos, C. S., Schlitt, T., & Simpson, M. A. (2013). Predicting the functional consequences of non-synonymous DNA sequence variants—Evaluation of bioinformatics tools and development of a consensus strategy. *Genomics, 102*(4), 223–228.

Fujiwara, H., Tatsumi, K. I., Tanaka, S., Kimura, M., Nose, O., & Amino, N. (2000). A novel V59E missense mutation in the sodium iodide symporter gene in a family with iodide transport defect. *Thyroid, 10*(6), 471–474.

Galzitskaya, O. V., Garbuzynskiy, S. O., & Lobanov, M. Y. (2006). FoldUnfold: Web server for the prediction of disordered regions in protein chain. *Bioinformatics, 22*(23), 2948–2949.

Gane, P. J., & Dean, P. M. (2000). Recent advances in structure-based rational drug design. *Current Opinion in Structural Biology, 10*, 401–404.

Gao, Z., Li, H., Zhang, H., Liu, X., Kang, L., Luo, X., et al. (2008). PDTD: A web-accessible protein database for drug target identification. *BMC Bioinformatics, 19*(9), 104.

Garbuzynskiy, S. O., Lobanov, M. Y., & Galzitskaya, O. V. (2010). FoldAmyloid: A method of prediction of amyloidogenic regions from protein sequence. *Bioinformatics, 26*, 326–332.

George Priya Doss, C., Chakraborty, C., Rajith, B., & Nagasundaram, N. (2013). In silico discrimination of nsSNPs in hTERT gene by means of local DNA sequence context and regularity. *Journal of Molecular Modeling, 19*(9), 3517–3527.

George Priya Doss, C., & Nagasundaram, N. (2012). Investigating the structural impacts of I64T and P311S mutations in APE1-DNA complex: A molecular dynamics approach. *PLoS One*, 7(2), e31677.

George Priya Doss, C., & Nagasundaram, N. (2013). Molecular docking and molecular dynamics study on the effect of ERCC1 deleterious polymorphisms in ERCC1-XPF heterodimer. *Applied Biochemistry and Biotechnology*, http://dx.doi.org/10.1007/s12010-013-0592-5. [Epub ahead of print].

George Priya Doss, C., Nagasundaram, N., Chakraborty, C., Chen, L., & Zhu, H. (2013). Extrapolating the effect of deleterious nsSNPs in the binding adaptability of flavopiridol with CDK7 protein: A molecular dynamics approach. *Human Genomics*, 7, 10.

George Priya Doss, C., Nagasundaram, N., Srajan, J., & Chiranjib, C. (2012). LSHGD: A database for human leprosy susceptible genes. *Genomics, 100*(3), 162–166.

George Priya Doss, C., Rajasekaran, R., Arjun, P., & Sethumadhavan, R. (2010). Prioritization of candidate SNPs in colon cancer using bioinformatics tools: An alternative approach for a cancer biologist. *Interdisciplinary Sciences, 2*(4), 320–346.

George Priya Doss, C., Rajith, B., & Chakraborty, C. (2013). Predicting the impact of deleterious mutations in the protein kinase domain of FGFR2 in the context of function, structure, and pathogenesis—A bioinformatics approach. *Applied Biochemistry and Biotechnology*, *170*(8), 1853–1870.

George Priya Doss, C., Rajith, B., Chakraborty, C., Balaji, V., Magesh, R., Menon, S., et al. (2013). In silico profiling and structural insights of missense mutations in RET protein kinase domain by molecular dynamics and docking approach. *Molecular BioSystems*, http://dx.doi.org/10.1039/C3MB70427K.

George Priya Doss, C., Rajith, B., Rajasekaran, R., Srajan, J., Nagasundaram, N., & Debajyoti, C. (2013). In silico analysis of prion protein mutants: A comparative study by molecular dynamics approach. *Cell Biochemistry and Biophysics*, *67*(3), 1307–1318.

George Priya Doss, C., & Sethumadhavan, R. (2009a). Investigation on the role of nsSNPs in HNPCC genes—A bioinformatics approach. *Journal of Biomedical Science*, *24*(16), 42.

George Priya Doss, C., & Sethumadhavan, R. (2009b). Impact of single nucleotide polymorphisms in HBB gene causing haemoglobinopathies: In silico analysis. *New Biotechnology*, *25*(4), 214–219.

George Priya Doss, C., Sudandiradoss, C., Rajasekaran, R., Choudhury, P., Sinha, P., Hota, P., et al. (2008). Applications of computational algorithm tools to identify functional SNPs. *Functional & Integrative Genomics*, *8*(4), 309–316.

Ghersi, D., & Sanchez, R. (2011). Beyond structural genomics: Computational approaches for the identification of ligand binding sites in protein structures. *Journal of Structural and Functional Genomics*, *12*, 109–117.

Ghoorah, A. W., Devignes, M. D., Smaïl-Tabbone, M., & Ritchie, D. W. (2013). Protein docking using case-based reasoning. *Proteins,* http://dx.doi.org/10.1002/prot.24433.

Giardine, B., Riemer, C., Hefferon, T., Thomas, D., Hsu, F., Zielenski, J., et al. (2007). PhenCode: Connecting ENCODE data with mutations and phenotype. *Human Mutation*, *28*, 554–562.

Gnad, F., Baucom, A., Mukhyala, K., Manning, G., & Zhang, Z. (2013). Assessment of computational methods for predicting the effects of missense mutations in human cancers. *BMC Genomics*, *14*(3), S7.

Goede, A., Dunkel, M., Mester, N., Frommel, C., & Preissner, R. (2005). SuperDrug: A conformational drug database. *Bioinformatics*, *21*(9), 1751–1753.

Gohlke, H., & Klebe, G. (2002). Approaches to the description and prediction of the binding affinity of small-molecule ligands to macromolecular receptors. *Angewandte Chemie (International Ed. in English)*, *41*, 2644–2676.

Gold, A. K. (2007). Cyber infrastructure, data, and libraries, part 2: Libraries and the data challenge: Roles and actions for libraries. *Office of the Dean (Library)*, *17*.

Goodsell, D. S., Morris, G. M., & Olson, A. J. (1996). Automated docking of flexible ligands: Applications of AutoDock. *Journal of Molecular Recognition*, *9*, 1–5.

Grasbon-Frodl, E., Lorenz, H., Mann, U., Nitsch, R. M., Windl, O., & Kretzschmar, H. A. (2004). Loss of glycosylation associated with the T183A mutation in human prion disease. *Acta Neuropathologica*, *108*, 476–484.

Gray, V. E., Kukurba, K. R., & Kumar, S. (2012). Performance of computational tools in evaluating the functional impact of laboratory-induced amino acid mutations. *Bioinformatics*, *28*(16), 2093–2096.

Griffith, M., Griffith, O. L., Coffman, A. C., Weible, J. V., McMichael, J. F., Spies, N. C., et al. (2013). DGIdb: Mining the druggable genome. *Nature Methods*, *10*(12), 1209–1210.

Gromiha, M. M., Oobatake, M., Kono, H., Uedaira, H., & Sarai, A. (1999). Role of structural and sequence information in the prediction of protein stability changes: Comparison between buried and partially buried mutations. *Protein Engineering*, *12*, 549–555.

Gromiha, M. M., & Ponnuswamy, P. K. (1993). Prediction of transmembrane beta-strands from hydrophobic characteristics of proteins. *International Journal of Peptide and Protein Research*, *42*(5), 420–431.

Grosdidier, A., Zoete, V., & Michielin, O. (2011). SwissDock, a protein-small molecule docking web service based on EADock DSS. *Nucleic Acids Research*, *39*, W270–W277.

Gu, X., & Vander Velden, K. (2002). DIVERGE: Phylogeny-based analysis for functional-structural divergence of a protein family. *Bioinformatics*, *18*, 500–501.

Gulati, S., Cheng, T. M. K., & Bates, P. A. (2013). Cancer networks and beyond: Interpreting mutations using the human interactome and protein structure. *Seminars in Cancer cell Biology*, *23*(4), 219–226.

Gupta, A., Gandhimathi, A., Sharma, P., & Jayaram, B. (2007). ParDOCK: An all atom energy based Monte Carlo docking protocol for protein-ligand complexes. *Protein and Peptide Letters*, *14*(7), 632–646.

Gupta, R., Jung, E., Gooley, A. A., Williams, K. L., Brunak, S., & Hansen, J. (1999). Scanning the available Dictyostelium discoideum proteome for O-linked GlcNAc glycosylation sites using neural networks. *Glycobiology*, *9*(10), 1009–1022.

Hakes, L., Lovell, S. C., Oliver, S. G., & Robertson, D. L. (2007). Specificity in protein interactions and its relationship with sequence diversity and coevolution. *Proceedings of the National Academy of Sciences of the United States of America*, *104*(19), 7999–8004.

Han, A., Kang, H. J., Cho, Y., Lee, S., Kim, Y. J., & Gong, S. (2006). SNP@Domain: A web resource of single nucleotide polymorphisms (SNPs) within protein domain structures and sequences. *Nucleic Acids Research*, *34*, W642–W644.

Hanemann, C. O., D'Urso, D., Gabreëls-Festen, A. A., & Müller, H. W. (2000). Mutation-dependent alteration in cellular distribution of peripheral myelin protein 22 in nerve biopsies from Charcot-Marie-Tooth type 1A. *Brain*, *123*(5), 1001–1006.

Hardt, M., & Laine, R. A. (2004). Mutation of active site residues in the chitin-binding domain ChBDChiA1 from chitinase A1 of Bacillus circulans alters substrate specificity: Use of a green fluorescent protein binding assay. *Archives of Biochemistry and Biophysics*, *426*(2), 286–297.

Hassett, C., Aicher, L., Sidhu, J. S., & Omiecinski, C. J. (1994). Human microsomal epoxide hydrolase: Genetic polymorphism and functional expression in vitro of amino acid variants. *Human Molecular Genetics*, *3*, 421–428.

Heinig, M., & Frishman, D. (2004). STRIDE: A web server for secondary structure assignment from known atomic coordinates of proteins. *Nucleic Acids Research*, *32*, W500–W502.

Hendlich, M., Rippmann, F., & Barnickel, G. (1997). LIGSITE: Automatic and efficient detection of potential small molecule-binding sites in proteins. *Journal of Molecular Graphics & Modelling*, *15*, 359–389.

Hermann, J. C., Marti-Arbona, R., Fedorov, A. A., Fedorov, E., Almo, S. C., Shoichet, B. K., et al. (2007). Structure-based activity prediction for an enzyme of unknown function. *Nature*, *448*, 775–779.

Herschbach, D. R. (1987). Molecular dynamics of elementary chemical reactions (Nobel lecture). *Angewandte Chemie (International Ed. in English)*, *26*(12), 1221–1243.

Hirst, S. J., Alexander, N., McHaourab, H. S., & Meiler, J. (2011). RosettaEPR: An integrated tool for protein structure determination from sparse EPR data. *Journal of Structural Biology*, *173*, 506–514.

Holm, L., & Sander, C. (1992). Fast and simple Monte Carlo algorithm for side-chain optimization in proteins: Application to model. *Proteins: Structure, Function, and Genetics*, *14*, 213–223.

Hooft, R. W., Vriend, G., Sander, C., & Abola, E. E. (1996). Errors in protein structures. *Nature*, *381*, 272.

Horovitz, A., Serrano, L., Avron, B., Bycroft, M., & Fersht, A. R. (1990). Strength and cooperativity of contributions of surface salt bridges to protein stability. *Journal of Molecular Biology*, *216*, 103–144.

Hu, L., Benson, M. L., Smith, R. D., Lerner, M. G., & Carlson, H. A. (2005). Binding MOAD (Mother Of All Databases). *Proteins*, *60*, 333–340.

Huang, S. Y., Grinter, S. Z., & Zou, X. (2010). Scoring functions and their evaluation methods for protein–ligand docking: Recent advances and future directions. *Physical Chemistry Chemical Physics*, *12*(40), 12899–12908.

Huang, H. D., Lee, T. Y., Tseng, S. W., & Horng, J. T. (2005). KinasePhos: A web tool for identifying protein kinase-specific phosphorylation sites. *Nucleic Acids Research*, *33*, W226–W229.

Huang, B., & Schroeder, M. (2006). LIGSITEcsc: Predicting ligand binding sites using the Connolly surface and degree of conservation. *BMC Structural Biology*, *6*, 19.

Hughes, J. D., Blagg, J., Price, D. A., Bailey, S., Decrescenzo, G. A., Devraj, R. V., et al. (2008). Physiochemical drug properties associated with in vivo toxicological outcomes. *Bioorganic & Medicinal Chemistry Letters*, *18*(17), 4872–4875.

Humphrey, W., Dalke, A., & Schulten, K. (1996). VMD—Visual molecular dynamics. *Journal of Molecular Graphics*, *14*, 33–38.

Hwang, J., & Liao, W. (1995). Side-chain by neural networks and simulated annealing optimization. *Protein Engineering*, *8*(4), 363–370.

Ingelman-Sundberg, M., Sim, S. C., Gomez, A., & Rodriguez-Antona, C. (2007). Influence of cytochrome P450 polymorphisms on drug therapies: Pharmacogenetic, pharmacoepigenetic and clinical aspects. *Pharmacology & Therapeutics*, *116*, 496–526.

Ingman, M., & Gyllensten, U. (2006). mtDB: Human Mitochondrial Genome Database, a resource for population genetics and medical sciences. *Nucleic Acids Research*, *34*, D749–D751.

International HapMap Consortium. (2003). The International HapMap Project. *Nature*, *426*, 789–796.

International Human Genome Sequencing Consortium. (2004). Finishing the euchromatic sequence of the human genome. *Nature*, *431*, 931–945.

Irwin, J. J., Shoichet, B. K., Mysinger, M. M., Huang, N., Colizzi, F., Wassam, P., et al. (2009). Automated docking screens: A feasibility study. *Journal of Medicinal Chemistry*, *52*, 5712–5720.

Ishida, T., & Kinoshita, K. (2007). PrDOS: Prediction of disordered protein regions from amino acid sequence. *Nucleic Acids Research*, *35*, W460–W464.

Ishida, T., & Kinoshita, K. (2008). Prediction of disordered regions in proteins based on the meta approach. *Bioinformatics*, *24*(11), 1344–1348.

Ivanisenko, V. A., Grigorovich, D. A., & Kolchanov, N. A. (2000). PDBSite: A database on biologically active sites and their spatial surroundings in proteins with known tertiary structure. In *The Second International Conference on Bioinformatics of Genome Regulation and Structure (BGRS'2000), Novosibirsk, Russia, 2* (pp. 171–174). .

Ivetac, A., & McCammon, J. A. (2010). Mapping the druggable allosteric space of G-protein coupled receptors: A fragment-based molecular dynamics approach. *Chemical Biology & Drug Design*, *76*, 201–217.

Jain, A. N. (2003). Surflex:? Fully automatic flexible molecular docking using a molecular similarity-based search engine. *Journal of Medicinal Chemistry*, *46*(4), 499–511.

Jaruzelska, J., Abadie, V., d'Aubenton-Carafa, Y., Brody, E., Munnich, A., & Marie, J. (1995). In vitro splicing deficiency induced by a C to T mutation at position-3 in the intron 10 acceptor site of the phenylalanine hydroxylase gene in a patient with phenylketonuria. *Journal of Biological Chemistry*, *270*, 20370–20375.

Jayaram, B., Bhushan, K., Shenoy, S. R., Narang, P., Bose, S., Agrawal, P., et al. (2006). Bhageerath: An energy based web enabled computer software suite for limiting the search space of tertiary structures of small globular proteins. *Nucleic Acids Research*, *34*(21), 6195–6204.

Jegga, A. G., Gowrisankar, S., Chen, J., & Aronow, B. J. (2007). PolyDoms: A whole genome database for the identification of nonsynonymous coding SNPs with the potential to impact disease. *Nucleic Acids Research*, *35*, D700–D706.

John, A. M., C, G. P., Ebenazer, A., Seshadri, M. S., Nair, A., Rajaratnam, S., et al. (2013). P. Arg82Leu von Hippel-Lindau (VHL) gene mutation among three members of a family with familial bilateral pheochromocytoma in India: Molecular analysis and in silico characterization. *PLoS One*, *8*(4), e61908.

Jones, R., Ruas, M., Gregory, F., Moulin, S., Delia, D., Manoukian, S., et al. (2007). CDKN2A mutation in familial melanoma that abrogates binding of p16INK4a to CDK4 but not CDK6. *Cancer Research*, *67*, 9134–9141.

Jones, G., Willett, P., Glen, R. C., Leach, A. R., & Taylor, R. (1997). Development and validation of a genetic algorithm for flexible docking. *Journal of Molecular Biology*, *267*, 727–748.

Jorgensen, W. L. (2004). The many roles of computation in drug discovery. *Science*, *303*(5665), 1813–1818.

Jorgensen, W. L., Maxwell, D. S., & Tirado-Rives, J. (1996). Development and testing of the OPLS all-atom force field on conformational energetics and properties of organic liquids. *Journal of the American Chemical Society*, *118*(11), 225–236.

Jothi, A. (2012). Principles, challenges and advances in ab initio protein structure prediction. *Protein and Peptide Letters*, *19*(11), 1194–1204.

Kabsch, W., & Sander, C. (1983). Dictionary of protein secondary structure: Pattern recognition of hydrogen-bonded and geometrical features. *Biopolymers*, *22*(12), 2577–2637.

Kalé, L., Skeel, R., Bhandarkar, M., Brunner, R., Gursoy, A., Krawetz, N., et al. (1999). NAMD2: Greater scalability for parallel molecular dynamics. *Journal of Computational Physics*, *151*, 283–312.

Karchin, R., Diekhans, M., Kelly, L., Thomas, D. J., Pieper, U., Eswar, N., et al. (2005). LS-SNP: Large-scale annotation of coding non-synonymous SNPs based on multiple information sources. *Bioinformatics*, *21*, 2814–2820.

Karchin, R., Kelly, L., & Sali, A. (2005). Improving functional annotation of non-synonymous SNPs with information theory. *Pacific Symposium on Biocomputing*, *10*, 397–408.

Karplus, M., & Kuriyan, J. (2005). Molecular dynamics and protein function. *Proceedings of the National Academy of Sciences of the United States of America*, *102*(19), 6679–6685.

Karplus, M., & McCammon, J. A. (2002). Molecular dynamics simulations of biomolecules. *Nature Structural and Molecular Biology*, *9*, 646–652.

Kawabata, T., Ota, M., & Nishikawa, K. (1999). The protein mutant database. *Nucleic Acids Research*, *27*, 355–357.

Kay, L. E. (1998). Protein dynamics from NMR. *Nature Structural and Molecular Biology*, *5*, 513–517.

Keage, H. A., Carare, R. O., Friedland, R. P., Ince, P. G., Love, S., Nicoll, J. A., et al. (2009). Population studies of sporadic cerebral amyloid angiopathy and dementia: A systematic review. *BMC Neurology*, *9*, 3.

Kelley, L. A., & Sternberg, M. J. E. (2009). Protein structure prediction on the Web: A case study using the Phyre server. *Nature Protocols*, *4*, 363–371.

Khan, S., & Vihinen, M. (2007). Spectrum of disease-causing mutations in protein secondary structures. *BMC Structural Biology*, 7(1), 1–18.

Khan, S., & Vihinen, M. (2010). Performance of protein stability predictors. *Human Mutation*, *31*(6), 675–684.

Kiemer, L., Bendtsen, J. D., & Blom, N. (2005). NetAcet: Prediction of N-terminal acetylation sites. *Bioinformatics*, *21*(7), 1269–1270.

Kim, D. E., Chivian, D., & Baker, D. (2004). Protein structure prediction and analysis using the Robetta server. *Nucleic Acids Research*, *32*, W526–W531.

Kim, E., Hyrc, K. L., Speck, J., Salles, F. T., Lundberg, Y. W., Goldberg, M. P., et al. (2011). Missense mutations in Otopetrin 1 affect subcellular localization and inhibition of purinergic signaling in vestibular supporting cells. *Molecular and Cellular Neurosciences*, *41*, 655–661.

Kim, D. S., Kim, C. M., Won, C. I., Kim, J. K., Ryu, J., Cho, Y., et al. (2011). BetaDock: Shape-priority docking method based on beta-complex. *Journal of Biomolecular Structure & Dynamics*, *29*(1), 219–242.

Kimura, M. (1980). A simple method for estimating evolutionary rates of base substitutions through comparative studies of nucleotide sequences. *Journal of Molecular Evolution*, *16*, 111–120.

Kinch, L., Yong Shi, S., Cong, Q., Cheng, H., Liao, Y., & Grishin, N. V. (2011). CASP9 assessment of free modeling target predictions. *Proteins*, *79*(10), 59–73.

Kitchen, D. B., Decornez, H., Furr, J. R., & Bajorath, J. (2004). Docking and scoring in virtual screening for drug discovery: Methods and applications. *Nature Reviews. Drug Discovery*, *3*(11), 935–949.

Kleina, L. G., & Miller, J. H. (1990). Genetic studies of the lac repressor. XIII. Extensive amino acid replacements generated by the use of natural and synthetic nonsense suppressors. *Journal of Molecular Biology*, *212*, 295–318.

Knox, C., Law, V., Jewison, T., Liu, P., Ly, S., Frolkis, A., et al. (2011). DrugBank 3.0: A comprehensive resource for 'Omics' research on drugs. *Nucleic Acids Research*, *39*, D1035–D1041.

Koehl, P., & Delarue, M. (1994). Application of a self-consistent mean field theory to predict protein side-chains conformation and estimate their conformational entropy. *Journal of Molecular Biology*, *239*(2), 249–275.

Koehl, P., & Delarue, M. (1995). A self consistent mean field approach to simultaneous gap closure and side-chain positioning in homology modeling. *Nature Structural Biology*, *2*, 163–170.

Kolb, H. C., & Sharpless, K. B. (2003). The growing impact of click chemistry on drug discovery. *Drug Discovery Today*, *8*(24), 1128–1137.

Kolinski, A., Rotkiewicz, P., Ilkowski, B., & Skolnick, J. (1999). A method for the improvement of threading-based protein models. *Proteins*, *37*(4), 592–610.

Kollman, P. (1993). Free energy calculations: Applications to chemical and biochemical phenomena. *Chemical Reviews*, *93*(7), 2395–2417.

Korb, O., Stützle, T., & Exner, T. E. (2006). PLANTS: Application of ant colony optimization to structure-based drug design. *Lecture Notes in Computer Science*, *4150*, 247–258.

Kosinski, J., Hinrichsen, I., Bujnicki, J. M., Friedhoff, P., & Plotz, G. (2010). Identification of Lynch syndrome mutations in the MLH1-PMS2 interface that disturb dimerization and mismatch repair. *Human Mutation*, *31*, 975–982.

Kosinski, J., Tkaczuk, K. L., Kasprzak, J. M., & Bujnicki, J. M. (2008). Template based prediction of three-dimensional protein structures: Fold recognition and comparative modeling. In J. M. Bujnicki (Ed.), *Prediction of protein structures, functions, and interactions*. Chichester: John Wiley & Sons, Ltd.

Koukouritaki, S. B., Poch, M. T., Henderson, M. C., Siddens, L. K., Krueger, S. K., VanDyke, J. E., et al. (2007). Identification and functional analysis of common human flavin-containing monooxygenase 3 genetic variants. *The Journal of Pharmacology and Experimental Therapeutics*, *320*, 266–273.

Krawczak, M., Ball, E. V., Fenton, I., Stenson, P. D., Abeysinghe, S., Thomas, N., et al. (2000). Human gene mutation database—A biomedical information and research resource. *Human Mutation*, *15*, 45–51.

Krishnan, V. G., & Westhead, D. R. (2003). A comparative study of machine-learning methods to predict the effects of single nucleotide polymorphisms on protein function. *Bioinformatics*, *19*(17), 2199–2209.

Krivov, G., Shapovalov, M., & Dunbrack, L. D., Jr. (2009). Improved prediction of protein side-chain conformations with SCWRL4. *Proteins: Structure, Function, and Bioinformatics*, 77, 778–795.

Kryshtafovych, A., & Fidelis, K. (2009). Protein structure prediction and model quality assessment. *Drug Discovery Today*, *14*, 386–393.

Kumar, R., Chaudhary, K., Gupta, S., Singh, H., Kumar, S., Gautam, A., et al. (2013). CancerDR: Cancer drug resistance database. *Scientific Reports*, *3*, 1445.

Kumar, P., Henikoff, S., & Ng, P. C. (2009). Predicting the effects of coding non-synonymous variants on protein function using the SIFT algorithm. *Nature Protocols*, *4*(7), 1073–1081.

LaConte, L. E. W., Voelz, V. A., Nelson, W. D., & Thomas, D. D. (2002). Molecular dynamics simulation of site-directed spin labeling: Experimental validation in muscle fibers. *Biophysical Journal*, *83*(4), 1854–1866.

Lahti, J. L., Tang, G. W., Capriotti, E., Liu, T., & Altman, R. B. (2012). Bioinformatics and variability in drug response: A protein structural perspective. *Journal of the Royal Society, Interface*, *9*(72), 1409–1437.

Larkin, M. A., Blackshields, G., Brown, N. P., Chenna, R., McGettigan, P. A., McWilliam, H., et al. (2007). Clustal W and Clustal X version 2.0. *Bioinformatics*, *23*(21), 2947–2948.

Laskowski, R. A. (1995). SURFNET: A program for visualizing molecular surfaces, cavities, and intermolecular interactions. *Journal of Molecular Graphics*, *13*, 323–328.

Laskowski, R. A., MacArthur, M. W., Moss, D. S., & Thornton, J. M. (1993). PROCHECK: A program to check the stereochemical quality of protein structures. *Journal of Applied Crystallography*, *26*, 283–291.

Launay, G., Téletchéa, S., Wade, F., Pajot-Augy, E., Gibrat, J. F., & Sanz, G. (2012). Automatic modeling of mammalian olfactory receptors and docking of odorants. *Protein Engineering, Design & Selection*, *25*(8), 377–386.

Laurila, K., & Vihinen, M. (2009). Prediction of disease-related mutations affecting protein localization. *BMC Genomics*, *10*, 122.

Lavecchia, A., & Di Giovanni, C. (2013). Virtual screening strategies in drug discovery: A critical review. *Current Medicinal Chemistry*, *20*(23), 2839–2860.

Lee, P. H., & Shatkay, H. (2008). F-SNP: Computationally predicted functional SNPs for disease association studies. *Nucleic Acids Research*, *36*, D820–D824.

Lee, C., & Subbiah, S. (1991). Prediction of protein side-chain conformation by packing optimization. *Journal of Molecular Biology*, *213*, 373–388.

Lee, H. S., & Zhang, Y. (2012). BSP-SLIM: A blind low-resolution ligand-protein docking approach using theoretically predicted protein structures. *Proteins*, *80*, 93–110.

Lesk, A. M. (1997). CASP2: Report on ab initio predictions. *Proteins: Structure, Function, and Genetics*, *1*, 151–166.

Levitt, M. (1992). Accurate modeling of protein conformation by automatic segment matching. *Journal of Molecular Biology*, *226*, 507–533.

Levitt, M. (2007). Growth of novel protein structural data. *Proceedings of the National Academy of Sciences of the United States of America*, *104*, 3183–3188.

Li, H., Gao, Z., Kang, L., Zhang, H., Yang, K., Yu, K., et al. (2006). TarFisDock: A web server for identifying drug targets with docking approach. *Nucleic Acids Research*, *34*, W219–W224.

Li, B., Krishnan, V. G., Mort, M. E., Xin, F., Kamati, K. K., Cooper, D. N., et al. (2009). Automated inference of molecular mechanisms of disease from amino acid substitutions. *Bioinformatics*, *25*, 2744–2750.

Liang, S., Zheng, D., Zhang, C., & Standley, D. M. (2011). Fast and accurate prediction of protein side-chain conformations. *Bioinformatics*, *27*(20), 2913–2914.

Lin, K., Simossis, V. A., Taylor, W. R., & Heringa, J. (2005). A simple and fast secondary structure prediction algorithm using hidden neural networks. *Bioinformatics*, *21*(2), 152–159.

Linding, R., Jensen, L. J., Diella, F., Bork, P., Gibson, T. J., & Russell, R. B. (2003). Protein disorder prediction implications for structural proteomics. *Structure*, *11*, 1453–1459.

Linding, R., Russell, R. B., Neduva, V., & Gibson, T. J. (2003). GlobPlot: Exploring protein sequences for globularity and disorder. *Nucleic Acids Research*, *31*(13), 3701–3708.

Lipinski, C. A. (2000). Drug-like properties and the causes of poor solubility and poor permeability. *Journal of Pharmacological and Toxicological Methods*, *44*(1), 235–249.

Lipman, D., & Pearson, W. (1985). Rapid and sensitive protein similarity searches. *Science*, *227*(4693), 1435–1441.

Liu, T., Lin, Y., Wen, X., Jorissen, R. N., & Gilson, M. K. (2007). BindingDB: A web-accessible database of experimentally determined protein-ligand binding affinities. *Nucleic Acids Research*, *35*, D198–D201.

Liu, T., Tang, G. W., & Capriotti, E. (2011). Comparative modeling: The state of the art and protein drug target structure prediction. *Combinatorial Chemistry & High Throughput Screening*, *14*(6), 532–547.

London, N., Raveh, B., Cohen, E., Fathi, G., & Schueler-Furman, O. (2011). Rosetta FlexPepDock web server—High resolution modeling of peptide-protein interactions. *Nucleic Acids Research*, *9*, W249–W253.

Lüthy, R., Bowie, J. U., & Eisenberg, D. (1992). Assessment of protein models with three-dimensional profiles. *Nature*, *356*, 83–85.

Luu, T. D., Rusu, A., Walter, V., Linard, B., Poidevin, L., Ripp, R., et al. (2012). KD4v: Comprehensible knowledge discovery system for missense variant. *Nucleic Acids Research*, *40*, W71–W75.

Luu, T. D., Rusu, A. M., Walter, V., Ripp, R., Moulinier, L., Muller, J., et al. (2012). MSV3d: Database of human missense variants mapped to 3D protein structure. *Database (Oxford)*, *2012*, bas018.

Ma, B., Elkayam, T., Wolfson, H., & Nussinov, R. (2003). Protein-protein interactions: Structurally conserved residues distinguish between binding sites and exposed protein surfaces. *Proceedings of the National Academy of Sciences of the United States of America*, *100*, 5772–5777.

MacKerell, A. D., Bashford, D., Bellott, M., Evanseck, R. L. J. D., Field, M. J., Fischer, S., et al. (1998). All-atom empirical potential for molecular modeling and dynamics studies of proteins. *The Journal of Physical Chemistry. B*, *102*, 3586–3616.

Magariños, M. P., Carmona, S. J., Crowther, G. J., Ralph, S. A., Roos, D. S., Shanmugam, D., et al. (2012). TDR targets: A chemogenomics resource for neglected diseases. *Nucleic Acids Research*, *40*, D1118–D1127.

Maginn, E. J., & Elliott, J. R. (2010). Historical perspective and current outlook for molecular dynamics as a chemical engineering tool. *Industrial & Engineering Chemistry Research*, *49*(7), 3059–3078.

Maglott, D., Chitipiralla, S., Church, D., Feolo, M., Garner, J., Jang, W., et al. (2013). ClinVar. NAR Molecular Biology Database Collection entry number 1570.

Magyar, C., Gromiha, M. M., Pujadas, G., Tusnády, G. E., & Simon, I. (2005). SRide: A server for identifying stabilizing residues in proteins. *Nucleic Acids Research*, *33*, W303–W305.

Mah, J. T. L., & Chia, K. S. (2007). A gentle introduction to SNP analysis: Resources and tools. *Journal of Bioinformatics and Computational Biology*, *5*, 1123–1138.

Mantsyzov, A. B., Bouvier, G., Evrard-Todeschi, N., & Bertho, G. (2012). Contact-based ligand-clustering approach for the identification of active compounds in virtual screening. *Advances and Applications in Bioinformatics and Chemistry*, *5*, 61–79.

Marialke, J., Tietze, S., & Apostolakis, J. (2008). Similarity based docking. *Journal of Chemical Information and Modeling*, *48*(1), 186–196.

Markwick, P. R. L., Cervantes, C. F., Abel, B. L., Komives, E. A., Blackledge, M., & McCammon, J. A. (2010). Enhanced conformational space sampling improves the prediction of chemical shifts in proteins. *Journal of the American Chemical Society*, *132*, 1220–1221.

Martin, L., Catherinot, V., & Labesse, G. (2006). kinDOCK: A tool for comparative docking of protein kinase ligands. *Nucleic Acids Research*, *34*(2), W325–W329.

Martí-Renom, M. A., Stuart, A. C., Fiser, A., Sánchez, R., Melo, F., & Sali, A. (2000). Comparative protein structure modeling of genes and genomes. *Annual Review of Biophysics and Biomolecular Structure*, *29*, 291–325.

Mashiach, E., Nussinov, R., & Wolfson, H. J. (2009). FiberDock: Flexible induced-fit backbone refinement in molecular docking. *Proteins*, *78*(6), 1503–1519.

Masso, M., & Vaisman, I. I. (2008). Accurate prediction of stability changes in protein mutants by combining machine learning with structure based computational mutagenesis. *Bioinformatics*, *24*(18), 2002–2009.

Masso, M., & Vaisman, I. I. (2010). Knowledge-based computational mutagenesis for predicting the disease potential of human non-synonymous single-nucleotide polymorphisms. *Journal of Theoretical Biology*, *266*(4), 560–568.

Mathe, E., Olivier, M., Kato, S., Ishioka, C., Hainaut, P., & Tavtigian, S. V. (2006). Computational approaches for predicting the biological effect of p53 missense mutations: A comparison of three sequence analysis based methods. *Nucleic Acids Research*, *34*, 1317–1325.

Maurer-Stroh, S., Debulpaep, M., Kuemmerer, N., Lopez de la Paz, M., Martins, I. C., Reumers, J., et al. (2010). Exploring the sequence determinants of amyloid structure using position-specific scoring matrices. *Nature Methods*, 7, 237–242.

McConkey, B. J., Sobolev, V., & Edelman, M. (2002). The performance of current methods in ligand-protein docking. *Current Science*, *83*(7), 845–856.

McGann, M. R., Almond, H. R., Nicholls, A., Grant, J. A., & Brown, F. K. (2003). Gaussian docking functions. *Biopolymers*, *68*, 76–90.

Meng, X. Y., Zhang, H. X., Mezei, M., & Cui, M. (2011). Molecular docking: A powerful approach for structure-based drug discovery. *Current Computer-Aided Drug Design*, 7(2), 146.

Mi, H., Muruganujan, A., & Thomas, P. D. (2013). PANTHER in 2013: Modeling the evolution of gene function, and other gene attributes, in the context of phylogenetic trees. *Nucleic Acids Research*, *41*, D377–D386. http://dx.doi.org/10.1093/nar/gks1118.

Mihasan, M. (2010). Basic protein structure prediction for the biologist: A review. *Archives of Biological Sciences*, *62*, 857–871.

Mihasan, M. (2012). What in silico molecular docking can do for the 'bench-working biologists'. *Journal of Biosciences*, *37*(1), 1089–1095.

Miller, R. T., Jones, D. T., & Thornton, J. M. (1996). Protein fold recognition by sequence threading: Tools and assessment techniques. *The FASEB Journal*, *10*, 171–178.

Miteva, M. A., Brugge, J. M., Rosing, J., Nicolaes, G. A. F., & Villoutreix, B. O. (2004). Theoretical and experimental study of the D2194G mutation in the C2 domain of coagulation factor V. *Biophysical Journal*, *86*(1), 488–498.

Moll, A., Hildebrandt, A., Lenhof, H. P., & Kohlbacher, O. (2006). BALLView: a tool for research and education in molecular modeling. *Bioinformatics*, *22*(3), 365–366.

Mooney, S. (2005). Bioinformatics approaches and resources for single nucleotide polymorphism functional analysis. *Briefings in Bioinformatics*, *6*, 44–56.

Mooney, S. D., & Altman, R. B. (2003). MutDB: Annotating human variation with functionally relevant data. *Bioinformatics*, *19*(14), 1858–1860.

Moosawi, F., & Mohabatkar, H. (2009). Computer-assisted analysis of subcellular localization signals and post-translational modifications of human prion proteins. *Journal of Biomedical Science and Engineering*, *2*, 70–75.

Moreira, I. S., Fernandes, P. A., & Ramos, M. J. (2010). Protein–protein docking dealing with the unknown. *Journal of Computational Chemistry*, *31*, 317–342.

Morris, G. M., Goodsell, D. S., Halliday, R. S., Huey, R., Hart, W. E., Belew, R. K., et al. (1998). Automated docking using a Lamarckian genetic algorithm and an empirical binding free energy function. *Journal of Computational Chemistry*, *19*, 1639–1662.

Morris, G. M., Huey, R., Lindstrom, W., Sanner, M. F., Belew, R. K., Goodsell, D. S., et al. (2009). Autodock4 and AutoDockTools4: Automated docking with selective receptor flexibility. *Journal of Computational Chemistry*, *16*, 2785–2791.

Muegge, I. (2006). PMF scoring revisited. *Journal of Medicinal Chemistry*, *49*, 5895–5902.

Murphy, J. A., Barrantes-Reynolds, R., Kocherlakota, R., Bond, J. P., & Greenblatt, M. S. (2004). The CDKN2A database: Integrating allelic variants with evolution, structure, function, and disease association. *Human Mutation*, *24*, 296–304.

Nabuurs, S. B., Wagener, M., & de Vlieg, J. (2007). A flexible approach to induced fit docking. *Journal of Medicinal Chemistry*, *50*(26), 6507–6518.

Nagasundaram, N., & George Priya Doss, C. (2013). Predicting the impact of single-nucleotide polymorphisms in CDK2-flavopiridol complex by molecular dynamics analysis. *Cell Biochemistry and Biophysics*, *66*(3), 681–695.

Ng, P. C., & Henikoff, S. (2001). Predicting deleterious amino acid substitutions. *Genome Research*, *11*(5), 863–874.

Ng, P. C., & Henikoff, S. (2006). Predicting the effects of amino-acid substitutions on protein function. *Annual Review of Genomics and Human Genetics*, *7*, 61–80.

Notredame, C., Higgins, D. G., & Heringa, J. (2000). T-Coffee: A novel method for fast and accurate multiple sequence alignment. *Journal of Molecular Biology*, *302*(1), 205–217.

Novikov, F. N., Stroylov, V. S., Zeifman, A. A., Stroganov, O. V., Kulkov, V., & Chilov, G. G. (2012). Lead Finder docking and virtual screening evaluation with Astex and DUD test sets. *Journal of Computer-Aided Molecular Design*, *26*(6), 725–735.

Nuytemans, K., Theuns, J., Cruts, M., & Van Broeckhoven, C. (2010). Genetic etiology of Parkinson disease associated with mutations in the SNCA, PARK2, PINK1, PARK7, and LRRK2 genes: A mutation update. *Human Mutation*, *31*, 763–780.

Ode, H., Matsuyama, S., Hata, M., Neya, S., Kakizawa, J., Sugiura, W., et al. (2007). Computational characterization of structural role of the non-active site mutation M36I of human immunodeficiency virus type 1 protease. *Journal of Molecular Biology*, *370*, 598–607.

Olatubosun, A., Valiaho, J., Harkonen, J., Thusberg, J., & Vihinen, M. (2012). PON-P: Integrated predictor for pathogenicity of missense variants. *Human Mutation*, *33*, 1166–1174.

Ortiz, M. A., Light, J., Maki, R. A., & Assa-Munt, N. (1999). Mutation analysis of the pip interaction domain reveals critical residues for protein-protein interactions. *Proceedings of the National Academy of Sciences of the United States of America*, *96*(6), 2740–2745.

O'Sullivan, O., Suhre, K., Abergel, C., Higgins, D. G., & Notredame, C. (2004). 3DCoffee: Combining protein sequences and structures within multiple sequence alignments. *Journal of Molecular Biology*, *340*, 385–395.

Ozbabacan, S. E. A., Gursoy, A., Keskin, O., & Nussinov, R. (2010). Conformational ensembles, signal transduction and residue hot spots: Application to drug discovery. *Current Opinion in Drug Discovery and Development*, *13*(5), 527–537.

Pace, C. N., Fu, H., Fryar, K. L., Landua, J., Trevino, S. R., Shirley, B. A., et al. (2011). Contributions of hydrophobic interactions to protein stability. *Journal of Molecular Biology*, *408*, 514–528.

Pak, Y., & Wang, S. (2000). Application of a molecular dynamics simulation method with a generalized effective potential to the flexible molecular docking problems. *The Journal of Physical Chemistry. B*, *104*(2), 354–359.

Park, H., Lee, J., & Lee, S. (2006). Critical assessment of the automated AutoDock as a new docking tool for virtual screening. *Proteins: Structure, Function, and Bioinformatics*, *65*(3), 549–554.

Parthiban, V., Gromiha, M. M., Hoppe, C., & Schomburg, D. (2007). Structural analysis and prediction of protein mutant stability using distance and torsion potentials: Role of secondary structure and solvent accessibility. *Proteins*, *66*, 41–52.

Parthiban, V., Gromiha, M. M., & Schomburg, D. (2006). CUPSAT: Prediction of protein stability upon point mutations. *Nucleic Acids Research*, *34*, W239–W242.

Pastinen, T., Ge, B., & Hudson, T. J. (2006). Influence of human genome polymorphism on gene expression. *Human Molecular Genetics*, *15*, 9–16.

Pearson, W. R., & Lipman, D. J. (1988). Improved tools for biological sequence comparison. *Proceedings of the National Academy of Sciences of the United States of America*, *85*, 2444–2448.

Pedretti, A., Villa, L., & Vistoli, G. (2004). VEGA—An open platform to develop chemo-bio-informatics applications, using plug-in architecture and script" programming. *Journal of Computer-Aided Molecular Design*, *18*, 167–173.

Peter, C., Rueping, M., Worner, H. J., Jaun, B., Seebach, D., & van Gunsteren, W. E. F. (2003). Molecular dynamics simulations of small peptides: Can one derive conformational preferences from ROESY spectra? *Chemistry*, *9*, 5838–5849.

Petersen, B., Petersen, T. N., Andersen, P., Nielsen, M., & Lundegaard, C. (2009). A generic method for assignment of reliability scores applied to solvent accessibility predictions. *BMC Structural Biology*, *9*, 51.

Peterson, T. A., Doughty, E., & Kann, M. G. (2013). Towards precision medicine: Advances in computational approaches for the analysis of human variants. *Journal of Molecular Biology*, *425*(21), 4047–4063.

Petrey, D., Xiang, Z., Tang, C. L., Xie, L., Gimpelev, M., Mitros, T., et al. (2003). Using multiple structure alignments, fast model building, and energetic analysis in fold recognition and homology modeling. *Proteins*, *53*(6), 430–435.

Pettersen, E. F., Goddard, T. D., Huang, C. C., Couch, G. S., Greenblatt, D. M., Meng, E. C., et al. (2004). UCSF Chimera—A visualization system for exploratory research and analysis. *Journal of Computational Chemistry*, *25*(13), 1605–1612.

Phillips, J. C., Braun, R., Wang, W., Gumbart, J., Tajkhorshid, E., Villa, E., et al. (2005). Scalable molecular dynamics with NAMD. *Journal of Computational Chemistry*, *26*(16), 1781–1802.

Pierri, C. L., Parisi, G., & Porcelli, V. (2010). Computational approaches for protein function prediction: A combined strategy from multiple sequence alignment to molecular docking-based virtual screening. *Biochimica et Biophysica Acta*, *1804*, 1695–1712.

Plimpton, S. (1995). Fast parallel algorithms for short-range molecular dynamics. *Journal of Computational Physics*, *117*(1), 1–19.

Pollastri, G., Baldi, P., Fariselli, P., & Casadio, R. (2002). Prediction of coordination number and relative solvent accessibility in proteins. *Proteins*, *47*, 142–153.

Pons, J. L., & Labesse, G. (2009). TOME-2: A new pipeline for comparative modeling of protein–ligand complexes. *Nucleic Acids Research*, *37*(2), W485–W491.

Prilusky, J., Felder, C. E., Zeev-Ben-Mordehai, T., Rydberg, E. H., Man, O., Beckmann, J. S., et al. (2005). FoldIndex: A simple tool to predict whether a given protein sequence is intrinsically unfolded. *Bioinformatics*, *21*(16), 3435–3438.

Proia, R. L., & Neufeld, E. F. (1982). Synthesis of beta-hexosaminidase in cell-free translation and in intact fibroblasts: An insoluble precursor alpha chain in a rare form of Tay-Sachs disease. *Proceedings of the National Academy of Sciences of the United States of America, 79,* 6360–6364.

Prokunina, L., & Alarcon-Riquelme, M. E. (2004). Regulatory SNPs in complex diseases: Their identification and functional validation. *Expert Reviews in Molecular Medicine, 6,* 1–15.

Pruitt, K. D., Tatusova, T., Brown, G. R., & Maglott, D. R. (2012). NCBI Reference Sequences (RefSeq): Current status, new features and genome annotation policy. *Nucleic Acids Research, 40,* D130–D135.

Pupko, T., Bell, R. E., Mayrose, I., Glaser, F., & Ben-Tal, N. (2002). Rate4Site: An algorithmic tool for the identification of functional regions in proteins by surface mapping of evolutionary determinants within their homologues. *Bioinformatics, 18,* S71–S77.

Qu, X., Swanson, R., Day, R., & Tsai, J. (2009). A guide to template based structure prediction. *Current Protein & Peptide Science, 10,* 270–285.

Radivojac, P., Baenziger, P. H., Kann, M. G., Mort, M. E., Hahn, M. W., & Mooney, S. D. (2008). Gain and loss of phosphorylation sites in human cancer. *Bioinformatics, 24*(16), i241–i247.

Ramensky, V., Bork, P., & Sunyaev, S. (2002). Human non-synonymous SNPs: Server and survey. *Nucleic Acids Research, 30,* 3894–3900.

Rao, V. S., & Srinivas, K. (2011). Modern drug discovery process: An in silico approach. *Journal of Bioinformatics and Sequence Analysis, 2*(5), 89–94.

Rarey, M., Kramer, B., Lengauer, T., & Klebe, G. (1996). A fast flexible docking method using an incremental construction algorithm. *Journal of Molecular Biology, 261,* 470–489.

Rask-Andersen, M., Almen, M. S., & Schioth, H. B. (2011). Trends in the exploitation of novel drug targets. *Nature Reviews. Drug Discovery, 10,* 579–590.

Rasmussen, B. F., Stock, A. M., Ringe, D., & Petsko, G. A. (1992). Crystalline ribonuclease A loses function below the dynamical transition at 220 K. *Nature, 357,* 423–424.

Raychaudhuri, S. (2011). Mapping rare and common causal alleles for complex human diseases. *Cell, 147,* 57–69.

Rees, D. C., Congreve, M., Murray, C. W., & Carr, R. (2004). Fragment-based lead discovery. *Nature Reviews. Drug Discovery, 3*(8), 660–672.

Ren, J., Xie, L., Li, W. W., & Bourne, P. E. (2010). SMAP-WS: A parallel web service for structural proteome-wide ligand-binding site comparison. *Nucleic Acids Research, 38,* W441–W444.

Rennell, D., Bouvier, S. E., Hardy, L. W., & Poteete, A. R. (1991). Systematic mutation of bacteriophage T4 lysozyme. *Journal of Molecular Biology, 222,* 67–88.

Reumers, J., Schymkowitz, J., & Rousseau, F. (2009). Using structural bioinformatics to investigate the impact of non-synonymous SNPs and disease mutations: Scope and limitations. *BMC Bioinformatics, 27,* 10.

Reva, B., Antipin, Y., & Sander, C. (2011). Predicting the functional impact of protein mutations: Application to cancer genomics. *Nucleic Acids Research, 39,* e118.

Ridder, L., Wang, H., de Vlieg, J., & Wagener, M. (2011). Revisiting the rule of five on the basis of pharmacokinetic data from rat. *ChemMedChem, 6*(11), 1967–1970.

Rignall, T. R., Baker, J. O., McCarter, S. L., Adney, W. S., Vinzant, T. B., Decker, S. R., et al. (2002). Effect of single active-site cleft mutation on product specificity in a thermostable bacterial cellulase. *Applied Biochemistry and Biotechnology, 98–100,* 383–394.

Risch, N. J. (2000). Searching for genetic determinants in the new millennium. *Nature, 405,* 847–856.

Rodriguez-Casado, A. (2012). In silico investigation of functional nsSNPs—An approach to rational drug design. *Research and Reports in Medicinal Chemistry, 2,* 31–42.

Rohl, C. A., Strauss, C. E., Misura, K. M., & Baker, D. (2004). Protein structure prediction using Rosetta. *Methods in Enzymology*, *383*, 66–93.

Rose, G. D., & Wolfenden, R. (1993). Hydrogen bonding, hydrophobicity, packing, and protein folding. *Annual Review of Biophysics and Biomolecular Structure*, *22*, 381–415.

Rost, B., Fariselli, P., & Casadio, R. (1996). Topology prediction for helical transmembrane proteins at 86% accuracy. *Protein Science*, *5*(8), 1704–1718.

Ruiz-Pesini, E., Lott, M. T., Procaccio, V., Poole, J. C., Brandon, M. C., Mishmar, D., et al. (2007). An enhanced MITOMAP with a global mtDNA mutational phylogeny. *Nucleic Acids Research*, *35*, D823–D828.

Ryu, G. M., Song, P., Kim, K. W., Oh, K. S., Park, K. J., & Kim, J. H. (2009). Genome-wide analysis to predict protein sequence variations that change phosphorylation sites or their corresponding kinases. *Nucleic Acids Research*, *37*(4), 1297–1307.

Sadee, W., & Dai, Z. (2005). Pharmacogenetics/genomics and personalized medicine. *Human Molecular Genetics*, *14*, R207–R214.

Salsbury, F. R., Jr. (2010). Molecular dynamics simulations of protein dynamics and their relevance to drug discovery. *Current Opinion in Pharmacology*, *10*, 738–744.

Samudrala, R., & Moult, J. (1998). Determinants of side chain conformational preferences in protein structures. *Protein Engineering*, *11*(11), 991–997.

Saunders, C. T., & Baker, D. (2002). Evaluation of structural and evolutionary contributions to deleterious mutation prediction. *Journal of Molecular Biology*, *322*(4), 891–901.

Sayle, R., & White, E. J. M. (1995). RasMol: Biomolecular graphics for all. *Trends in Biochemical Sciences*, *20*(9), 374.

Schaefer, C., Meier, A., Rost, B., & Bromberg, Y. (2012). SNPdbe: Constructing an nsSNP functional impacts database. *Bioinformatics*, *28*(4), 601–602.

Schlick, T., Collepardo-Guevara, R., Halvorsen, L. A., Jung, S., & Xiao, X. (2011). Biomolecular modeling and simulation: A field coming of age. *Quarterly Reviews of Biophysics*, *44*, 191–228.

Schneidman-Duhovny, D., Inbar, Y., Nussinov, R., & Wolfson, H. J. (2005a). Geometry based flexible and symmetric protein docking. *Proteins*, *60*, 224–231.

Schneidman-Duhovny, D., Inbar, Y., Nussinov, R., & Wolfson, H. J. (2005b). PatchDock and SymmDock: Servers for rigid and symmetric docking. *Nucleic Acids Research*, *33*, W363–W367.

Schwarz, J. M., Rodelsperger, C., Schuelke, M., & Seelow, D. (2010). MutationTaster evaluates disease-causing potential of sequence alterations. *Nature Methods*, *7*, 575–576.

Seiler, K. P., George, G. A., Happ, M. P., Bodycombe, N. E., Carrinski, H. A., Norton, S., et al. (2008). ChemBank: A small-molecule screening and cheminformatics resource database. *Nucleic Acids Research*, *36*, D351–D359.

Shan, Y., Kim, E. T., Eastwood, M. P., Dror, R. O., Seeliger, M. A., Shaw, D. E., et al. (2011). How does a drug molecule find its target binding site? *Journal of the American Chemical Society*, *133*, 9181–9183.

Shastry, B. S. (2006a). Pharmacogenetics and the concept of individualized medicine. *The Pharmacogenomics Journal*, *6*, 16–21.

Shastry, B. S. (2006b). Role of SNPs and haplotypes in human disease and drug development. In M. Ozkan, M. J. Heller, & M. Ferrari (Eds.), *Micro/nano technology in genomics and proteomics: 2*, (pp. 447–458). New York: Springer.

Sherry, S. T., Ward, M. H., Kholodov, M., Baker, J., Phan, L., Smigielski, E. M., et al. (2001). dbSNP: The NCBI database of genetic variation. *Nucleic Acids Research*, *29*, 308–311.

Shihab, H. A., Gough, J., Cooper, D. N., Stenson, P. D., Barker, G. L., Edwards, K. J., et al. (2013). Predicting the functional, molecular, and phenotypic consequences of amino acid substitutions using hidden Markov models. *Human Mutation*, *34*(1), 57–65.

Shirley, B. A., Stanssens, P., Hahn, U., & Pace, C. N. (1992). Contribution of hydrogen bonding to the conformational stability of ribonuclease T1. *Biochemistry*, *31*, 725–732.

Sickmeier, M., Hamilton, J. A., LeGall, T., Vacic, V., Cortese, M. S., Tantos, A., et al. (2007). DisProt: The database of disordered proteins. *Nucleic Acids Research*, *35*, D786–D793.

Siepel, A., Bejerano, G., Pedersen, J. S., Hinrichs, A. S., Hou, M., Rosenbloom, K., et al. (2005). Evolutionarily conserved elements in vertebrate, insect, worm, and yeast genomes. *Genome Research*, *15*, 1034–1050.

Sievers, F., Wilm, A., Dineen, D., Gibson, T. J., Karplus, K., Li, W., et al. (2011). Fast, scalable generation of high-quality protein multiple sequence alignments using Clustal Omega. *Molecular Systems Biology*, *7*, 539.

Singh, T., Biswas, D., & Jayaram, B. (2011). AADS—An automated active site identification, docking and scoring protocol for protein targets based on physico-chemical descriptors. *Journal of Chemical Information and Modeling*, *51*(10), 2515–2527.

Sippl, M. J. (1993). Recognition of errors in three-dimensional structures of proteins. *Proteins*, *17*, 355–362.

Smith, E. P., Boyd, J., Frank, G. R., Takahashi, H., Cohen, R. M., Specker, B., et al. (1994). Estrogen resistance caused by a mutation in the estrogen-receptor gene in a man. *The New England Journal of Medicine*, *331*, 1056–1061.

Smith, R. D., Engdahl, A. L., Dunbar, J. B., Jr., & Carlson, H. A. (2012). Biophysical limits of protein–ligand binding. *Journal of Chemical Information and Modeling*, *52*(8), 2098–2106.

Söding, J., Biegert, A., & Lupas, A. N. (2005). The HHpred interactive server for protein homology detection and structure prediction. *Nucleic Acids Research*, *33*, W244–W248.

Song, E. S., Daily, A., Fried, M. G., Juliano, M. A., Juliano, L., & Hersh, L. B. (2005). Mutation of active site residues of insulin degrading enzyme alters allosteric interactions. *The Journal of Biological Chemistry*, *280*(18), 17701–17706.

Song, X., Geng, Z., Zhu, J., Li, C., Hu, X., Bian, N., et al. (2009). Structure-function roles of four cysteine residues in the human arsenic (+3 oxidation state) methyltransferase (hAS3MT) by site-directed mutagenesis. *Chemico-Biological Interactions*, *179*, 321–328.

Song, C. M., Lim, S. J., & Tong, J. C. (2009). Recent advances in computer-aided drug design. *Briefings in Bioinformatics*, *10*(5), 579–591.

Sousa, S. F., Fernandes, P. A., & Ramos, M. J. (2006). Protein-ligand docking: Current status and future challenges. *Proteins: Structure, Function, and Bioinformatics*, *65*(1), 15–26.

Steen, M., Miteva, M., Villoutreix, B. O., Yamazaki, T., & Dahlback, B. (2003). Factor V new brunswick: Ala221Val associated with FV deficiency reproduced in vitro and functionally characterized. *Blood*, *102*(4), 1316–1322.

Stenson, P. D., Ball, E. V., Mort, M., Phillips, A. D., Shaw, K., & Cooper, D. N. (2012). The human gene mutation database (HGMD) and its exploitation in the fields of personalized genomics and molecular evolution. *Current Protocols in Bioinformatics*, chap: unit 6.

Stenson, P. D., Ball, E. V., Mort, M., Phillips, A. D., Shiel, J. A., Thomas, N. S., et al. (2003). Human gene mutation database (HGMD): 2003 update. *Human Mutation*, *21*, 577–581.

Stevanin, G., Hahn, V., Lohmann, E., Bouslam, N., Gouttard, M., Soumphonphakdy, C., et al. (2004). Mutation in the catalytic domain of protein kinase C γ and extension of the phenotype associated with spinocerebellar ataxia type 14. *Archives of Neurology*, *61*(8), 1242–1248.

Stitziel, N. O., Tseng, Y. Y., Pervouchine, D., Goddeau, D., Kasif, S., & Liang, J. (2003). Structural location of disease-associated single-nucleotide polymorphisms. *Journal of Molecular Biology*, *327*, 1021–1030.

Stryer, L. (1995). *Biochemistry* (4th ed.). New York: W.H. Freeman.

Sunyaev, S., Ramensky, V., Koch, I., Lathe, W., III, Kondrashov, A. S., & Bork, P. (2001). Prediction of deleterious human alleles. *Human Molecular Genetics*, *10*(6), 591–597.

Sushma, B., & Suresh, C. V. (2012). Docking—A review. *Journal of Applicable Chemistry, 1*(2), 167–173.

Taboureau, O., Baell, J. B., Fernández-Recio, J., & Villoutreix, B. O. (2012). Established and emerging trends in computational drug discovery in the structural genomics era. *Chemistry & Biology, 19*(1), 29–41.

Taillon-Miller, P., Gu, Z., Li, Q., Hillier, L., & Kwok, P. Y. (1998). Overlapping genomic sequences: A treasure trove of single-nucleotide polymorphisms. *Genome Research, 8*(7), 748–754.

Takamiya, O., Seta, M., Tanaka, K., & Ishida, F. (2002). Human factor VII deficiency caused by S339C mutation located adjacent to the specificity pocket of the catalytic domain. *Clinical and Laboratory Haematology, 24*(4), 233–238.

Tang, K. E. S., & Dill, K. A. (1998). Native protein fluctuations: The conformational-motion temperature and the inverse correlation of protein flexibility with protein stability,. *Journal of Biomolecular Structure and Dynamics, 16*(2), 397–411.

Tavtigian, S. V., Deffenbaugh, A. M., Yin, L., Judkins, T., Scholl, T., Samollow, P. B., et al. (2006). Comprehensive statistical study of 452 BRCA1 missense substitutions with classification of eight recurrent substitutions as neutral. *Journal of Medical Genetics, 43*(4), 295–305.

Taylor, R. D., Jewsbury, P. J., & Essex, J. W. (2002). A review of protein-small molecule docking methods. *Journal of Computer-Aided Molecular Design, 16*(3), 151–166.

Teng, S., Madej, T., Panchenko, A., & Alexov, E. (2009). Modeling effects of human single-nucleotide polymorphisms on protein-protein interactions. *Biophysical Journal, 96*(6), 2178–2188.

Tennessen, J. A., Bigham, A. W., O'Connor, T. D., Fu, W., Kenny, E. E., Gravel, S., et al. (2012). Evolution and functional impact of rare coding variation from deep sequencing of human exomes. *Science, 337*, 64–69.

The PyMOL Molecular Graphics System, Version 1.2r3pre, Schrödinger, LLC.

Thomas, M., Dadgar, N., Aphale, A., Harrell, J. M., Kunkel, R., Pratt, W. B., et al. (2004). Androgen receptor acetylation site mutations cause trafficking defects, misfolding, and aggregation similar to expanded glutamine tracts. *The Journal of Biological Chemistry, 279*, 8389–8395.

Thomas, R., McConnell, R., Whittacker, J., Kirkpatrick, P., Bradley, J., & Sandford, R. (1999). Identification of mutations in the repeated part of the autosomal dominant polycystic kidney disease type 1 gene PKD1, by long-range PCR. *American Journal of Human Genetics, 65*, 39–49.

Thomas, P. J., Qu, B. H., & Pedersen, P. L. (1995). Defective protein folding as a basis of human disease. *Trends in Biochemical Sciences, 20*, 456–459.

Thomsen, R., & Christensen, M. H. (2006). MolDock:? A new technique for high-accuracy molecular docking. *Journal of Medicinal Chemistry, 49*(11), 3315–3321.

Thusberg, J., Olatubosun, A., & Vihinen, M. (2011). Performance of mutation pathogenicity prediction methods on missense variants. *Human Mutation, 32*(4), 358–368.

Thusberg, J., & Vihinen, M. (2009). Pathogenic or not? And if so, then how? Studying the effects of missense mutations using bioinformatics methods. *Human Mutation, 30*, 703–714.

Tian, J., Wu, N., Guo, X., Guo, J., Zhang, J., & Fan, Y. (2007). Predicting the phenotypic effects of non-synonymous single nucleotide polymorphisms based on support vector machines. *BMC Bioinformatics, 8*, 450.

Tietze, S., & Apostolakis, T. (2007). Glamdock: Development and validation of a new docking tool on several thousand protein-ligand complexes. *Journal of Chemical Information and Modeling, 47*(4), 1657–1672.

Tiffin, N., Okpechi, I., Perez-Iratxeta, C., Andrade-Navarro, M. A., & Ramesar, R. (2008). Prioritization of candidate disease genes for metabolic syndrome by computational analysis of its defining phenotypes. *Physiological Genomics, 35*(1), 55–64.

Tolkacheva, T., Boddapati, M., Sanfiz, A., Tsuchida, K., Kimmelman, A. C., & Chan, A. M. (2001). Regulation of PTEN binding to MAGI-2 by two putative phosphorylation sites at threonine 382 and 383. *Cancer Research*, *61*(13), 4985–4989.

Tomalik-Scharte, D., Lazar, A., Fuhr, U., & Kirchheiner, J. (2008). The clinical role of genetic polymorphisms in drug-metabolizing enzymes. *Pharmacogenomics*, *J8*, 4–15.

Tovchigrechko, A., & Vakser, I. A. (2006). GRAMM-X public web server for protein-protein docking. *Nucleic Acids Research*, *34*, W310–W314.

Tress, M. L., Jones, D., & Valencia, A. (2003). Predicting reliable regions in protein alignments from sequence profiles. *Journal of Molecular Biology*, *330*(4), 705–718.

Trott, O., & Olson, A. J. (2010). AutoDock Vina: Improving the speed and accuracy of docking with a new scoring function, efficient optimization and multithreading. *Journal of Computational Chemistry*, *31*, 455–461.

Trovato, A., Seno, F., & Tosatto, S. C. (2007). The PASTA server for protein aggregation prediction. *Protein Engineering, Design & Selection*, *20*, 521–523.

Tsai, T. Y., Chang, K. W., & Chen, C. Y. (2011). iScreen: World's first cloud-computing web server for virtual screening and de novo drug design based on TCM database@Taiwan. *Journal of Computer-Aided Molecular Design*, *25*(6), 525–531.

Tsolis, A. C., Papandreou, N. C., Iconomidou, V. A., & Hamodrakas, S. J. (2013). A consensus method for the prediction of 'aggregation-prone' peptides in globular proteins. *PLoS ONE*, *8*(1), e54175.

Tuckerman, M. E., Yarne, D. A., Samuelson, S. O., Hughes, A. L., & Martyna, G. J. (2000). Exploiting multiple levels of parallelism in molecular dynamics based calculations via modern techniques and software paradigms on distributed memory computers. *Computer Physics Communications*, *128*(1), 333–376.

Ung, M. U., Lu, B., & McCammon, J. A. (2006). E230Q mutation of the catalytic subunit of cAMP-dependent protein kinase affects local structure and the binding of peptide inhibitor. *Biopolymers*, *81*, 428–439.

Utesch, T., Daminelli, G., & Mroginski, M. A. (2011). Molecular dynamics simulations of the adsorption of bone morphogenetic protein-2 on surfaces with medical relevance. *Langmuir*, *27*(21), 13144–13153.

Uzun, A., Leslin, C. M., Abyzov, A., & Ilyin, V. (2007). Structure SNP (StSNP): A web server for mapping and modeling nsSNPs on protein structures with linkage to metabolic pathways. *Nucleic Acids Research*, *35*, W384–W392.

Valerio, M., Colosimo, A., Conti, F., Giuliani, A., Grottesi, A., Manetti, C., et al. (2005). Early events in protein aggregation: Molecular flexibility and hydrophobicity/charge interaction in amyloid peptides as studied by molecular dynamics simulations. *Proteins*, *58*(1), 110–118.

Van Durme, J., Maurer-Stroh, S., Gallardo, R., Wilkinson, H., Rousseau, F., & Schymkowitz, J. (2009). Accurate prediction of DnaK-peptide binding via homology modelling and experimental data. *PLoS Computational Biology*, *5*(8), e1000475.

van Wijk, R., Rijksen, G., Huizinga, E. G., Nieuwenhuis, H. K., & van Solinge, W. W. (2003). HK Utrecht: Missense mutation in the active site of human hexokinase associated with Hexo-kinase deficiency and severe nonspherocytic hemolytic anemia. *Blood*, *101*(1), 345–347.

Vatsis, K. P., Martell, K. J., & Weber, W. W. (1991). Diverse point mutations in the human gene for polymorphic N-acetyltransferase. *Proceedings of the National Academy of Sciences of the United States of America*, *88*, 6333–6337.

Vazquez, F. (2000). Phosphorylation of the PTEN tail regulates protein stability and function. *Molecular and Cellular Biology*, *20*, 5010–5018.

Venkatesan, R. N., Treuting, P. M., Fuller, E. D., Goldsby, R. E., Norwood, T. H., Gooley, T. A., et al. (2007). Mutation at the polymerase active site of mouse DNA

polymerase delta increases genomic instability and accelerates tumorigenesis. *Molecular and Cellular Biology*, *27*, 7669–7682.
Venselaar, H., Te Beek, T. A., Kuipers, R. K., Hekkelman, M. L., & Vriend, G. (2010). Protein structure analysis of mutations causing inheritable diseases. An e-Science approach with life scientist friendly interfaces. *BMC Bioinformatics*, *11*, 548.
Verdonk, M. L., Cole, J. C., Hartshorn, M. J., Murray, C. W., & Taylor, R. D. (2003). Improved protein–ligand docking using GOLD. *Proteins: Structure, Function, and Bioinformatics*, *52*(4), 609–623.
Villoutreix, B. O., Renault, N., Lagorce, D., Sperandio, O., Montes, M., & Miteva, M. A. (2007). Free resources to assist structure-based virtual ligand screening experiments. *Current Protein and Peptide Science*, *8*(4), 381–411.
Vitkup, D., Sander, C., & Church, G. M. (2003). The amino acid mutational spectrum of human genetic disease. *Genome Biology*, *4*, R72–R80.
Vogt, G., Vogt, B., Chuzhanova, N., Julenius, K., Cooper, D. N., & Casanova, J. L. (2007). Gain-of-glycosylation mutations. *Current Opinion in Genetics & Development*, *17*, 245–251.
Vriend, G. (1990). WHAT IF: A molecular modeling and drug design program. *Journal of Molecular Graphics*, *8*, 52–56.
Vullo, A., Bortolami, O., Pollastri, G., & Tosatto, S. C. (2006). Spritz: A server for the prediction of intrinsically disordered regions in protein sequences using kernel machines. *Nucleic Acids Research*, *34*, W164–W168.
Wallner, B., & Elofsson, A. (2005). All are not equal: A benchmark of different homology modeling programs. *Protein Science*, *14*(5), 1315–1327.
Wallner, R. N., Lindahl, E., & Elofsson, A. (2008). Using multiple templates to improve quality of homology models in automated homology modeling. *Protein Science*, *17*, 990–1002.
Walsh, C. T. (2006). *Posttranslational modification of proteins: Expanding nature's inventory*. Englewood, CO: Roberts and Company Publishers.
Wang, J. C., Chu, P. Y., Chen, C. M., & Lin, J. H. (2012). idTarget: A web server for identifying protein targets of small chemical molecules with robust scoring functions and a divide-and-conquer docking approach. *Nucleic Acids Research*, *40*, W393–W399.
Wang, Q., Curran, M. E., Splawski, I., Burn, T. C., Millholland, J. M., VanRaay, T. J., et al. (1996). Positional cloning of a novel potassium channel gene: KVLQT1 mutations cause cardiac arrhythmias. *Nature Genetics*, *12*(1), 17–23.
Wang, Z., Eickholt, J., & Cheng, J. (2010). MULTICOM: A multi-level combination approach to protein structure prediction and its assessments in CASP8. *Bioinformatics*, *26*, 882–888.
Wang, R., Fang, X., Lu, Y., & Wang, S. (2004). The PDBbind database: Collection of binding affinities for protein–ligand complexes with known three-dimensional structures. *Journal of Medicinal Chemistry*, *47*, 2977–2980.
Wang, Z., & Moult, J. (2001). SNPs, protein structure, and disease. *Human Mutation*, *7*, 263–270.
Wang, Z., & Moult, J. (2003). Three-dimensional structural location and molecular functional effects of missense SNPs in the T cell receptor Vbeta domain. *Proteins*, *53*, 748–757.
Wang, J., Ronaghi, M., Chong, S. S., & Lee, C. G. (2011). pfSNP: An integrated potentially functional SNP resource that facilitates hypotheses generation through knowledge syntheses. *Human Mutation*, *32*, 19–24.
Wang, M., Sun, Z., Akutsu, T., & Song, J. (2013). Recent advances in predicting functional impact of single amino acid polymorphisms: A review of useful features, computational methods and available tools. *Current Bioinformatics*, *8*, 161–176.
Wang, J., Wolf, R. M., Caldwell, J. W., Kollman, P. A., & Case, D. A. (2004). Development and testing of general amber force field. *Journal of Computational Chemistry*, *25*, 1157–1174.

Wang, Y., Xiao, J., Suzek, T. O., Zhang, J., Wang, J., & Bryant, S. H. (2009). PubChem: A public information system for analyzing bioactivities of small molecules. *Nucleic Acids Research, 37*, W623–W633.

Wang, L. L., Yang, A. K., Li, Y., Liu, J. P., & Zhou, S. F. (2010). Phenotype prediction of deleterious nonsynonymous single nucleotide polymorphisms in human alcohol metabolism-related genes: A bioinformatics study. *Alcohol, 44*(5), 425–438.

Wass, M. N., Kelley, L. A., & Sternberg, M. J. (2010). 3DLigandSite: Predicting ligand-binding sites using similar structures. *Nucleic Acids Research, 38*, W469–W473.

Weaire, D., & Aste, T. (2010). The pursuit of perfect packing. *Contemporary Physics, 51*, 1.

Weigelt, J. (2010). Structural genomics—Impact on biomedicine and drug discovery. *Experimental Cell Research, 316*, 1332–1338.

Weisel, M., Proschak, E., & Schneider, G. (2007). PocketPicker: Analysis of ligand binding sites with shape descriptors. *Chemistry Central Journal, 13*(1), 7.

Werner, T., Morris, M. B., Dastmalchi, S., & Church, W. B. (2012). Structural modelling and dynamics of proteins for insights into drug interactions. *Advanced Drug Delivery Reviews, 64*(4), 323–343.

Wiederstein, M., & Sippl, M. J. (2007). ProSA-web: Interactive web service for the recognition of errors in three-dimensional structures of proteins. *Nucleic Acids Research, 35*, W407–W410.

Wilke, R. A., & Dolan, M. E. (2011). Genetics and variable drug response. *Journal of the American Medical Association, 306*, 306–307.

Wilkinson, G. R. (2005). Drug metabolism and variability among patients in drug response. *The New England Journal of Medicine, 352*, 2211–2221.

Wishart, D. S. (2007). Human Metabolome Database: Completing the 'human parts list'. *Pharmacogenomics, 8*(7), 683–686.

Witham, S., Takano, K., Schwartz, C., & Alexov, E. (2011). missense mutation in CLIC2 associated with intellectual disability is predicted by in silico modeling to affect protein stability and dynamics. *Proteins, 79*(8), 2444–2454.

Wjst, M. (2004). Target SNP selection in complex disease association studies. *BMC Bioinformatics, 5*, 92.

Wong, P., Fritz, A., & Frishman, D. (2005). Designability, aggregation propensity and duplication of disease-associated proteins. *Protein Engineering, Design & Selection, 18*, 503–508.

Wong, C. F., & McCammon, J. A. (2003). Protein flexibility and computer-aided drug design. *Annual Review of Pharmacology and Toxicology, 43*, 31–45.

Wright, J. D., & Lim, C. (2007). Mechanism of DNA-binding loss upon single-point mutation in p53. *Journal of Biosciences, 32*(5), 827–839.

Xiang, Z. (2006). Advances in homology protein structure modeling. *Current Protein & Peptide Science*, 7(3), 217–227.

Xiang, Z., Soto, C. S., & Honig, B. (2002). Evaluating conformational free energies: The colony energy and its application to the problem of loop prediction. *Proceedings of the National Academy of Sciences of the United States of America, 99*(11), 7432–7437.

Xu, J., & Berger, B. (2006). Fast and accurate algorithms for protein side-chain packing. *Journal of the ACM, 53*(4), 533–557.

Xu, H., Gregory, S. G., Hauser, E. R., Stenger, J. E., Pericak-Vance, M. A., Vance, J. M., et al. (2005). SNPselector: A web tool for selecting SNPs for genetic association studies. *Bioinformatics, 21*, 4181–4186.

Xu, J., Li, M., Kim, D., & Xu, Y. (2003). RAPTOR: Optimal protein threading by linear programming. *Journal of Bioinformatics and Computational Biology, 1*, 95–117.

Yamada, Y., Banno, Y., Yoshida, H., Kikuchi, R., Akao, Y., Murate, T., et al. (2006). Catalytic inactivation of human phospholipase D2 by a naturally occurring Gly901Asp mutation. *Archives of Medical Research, 37*, 696–699.

Yang, S. Y. (2010). Pharmacophore modeling and applications in drug discovery: Challenges and recent advances. *Drug Discovery Today, 15*(11), 444–450.

Yang, J. M., & Chen, C. C. (2004). GEMDOCK: A generic evolutionary method for molecular docking. *Proteins: Structure, Function, and Bioinformatics, 55*, 288–304.

Yang, Z. R., Thomson, R., McNeil, P., & Esnouf, R. M. (2005). RONN: The bio-basis function neural network technique applied to the detection of natively disordered regions in proteins. *Bioinformatics, 21*, 3369–3376.

Yoshida, A., Huang, I. Y., & Ikawa, M. (1984). Molecular abnormality of an inactive aldehyde dehydrogenase variant commonly found in Orientals. *Proceedings of the National Academy of Sciences of the United States of America, 81*, 258–261.

Young, M. A., Gonfloni, S., Superti-Furga, G., Roux, B., & Kuriyan, J. (2001). Dynamic coupling between the SH2 and SH3 domains of c-Src and Hck underlies their inactivation by C terminal tyrosine phosphorylation. *Cell, 105*(1), 115–126.

Yuan, H. Y., Chiou, J. J., Tseng, W. H., Liu, C. H., Liu, C. K., Lin, Y. J., et al. (2006). FASTSNP: An always up-to-date and extendable service for SNP function analysis and prioritization. *Nucleic Acids Research, 34*, 35–41.

Yue, P., & Moult, J. (2006). Identification and analysis of deleterious human SNPs. *Journal of Molecular Biology, 356*, 1263–1274.

Zemla, A., Venclovas, C., Reinhardt, A., Fidelis, K., & Hubbard, T. J. (1997). Numerical criteria for the evaluation of ab initio predictions of protein structure. *Proteins, 1*, 140–150.

Zhang, Y. (2008a). Progress and challenges in protein structure prediction. *Current Opinion in Structural Biology, 18*(3), 342–348.

Zhang, Y. (2008b). I-TASSER server for protein 3D structure prediction. *BMC Bioinformatics, 9*, 40.

Zhang, Y. (2009). Protein structure prediction: When is it useful? *Current Opinion in Structural Biology, 19*, 145–155.

Zhang, Z., Norris, J., Schwartz, C., & Alexov, E. (2011). In-silico and in vitro investigations of the mutability of disease-causing missense mutation sites in spermine synthase. *PLoS One, 6*(5), e20373.

Zhang, Z., Teng, S., Wang, L., Schwartz, C. E., & Alexov, E. (2010). Computational analysis of missense mutations causing Snyder-Robinson syndrome. *Human Mutation, 31*(9), 1043–1049.

Zhang, J., Wang, Q., Barz, B., He, Z., Kosztin, I., Shang, Y., et al. (2010). MUFOLD: A new solution for protein 3D structure prediction. *Proteins, 78*, 1137–1152.

Zhao, Y., & Sanner, M. F. (2007). FLIPDock: Docking flexible ligands into flexible receptors. *Proteins: Structure, Function, and Bioinformatics, 68*(3), 726–737.

Zhou, H., & Zhou, Y. (2002). Distance-scaled, finite ideal-gas reference state improves structure-derived potentials of mean force for structure selection and stability prediction. *Protein Science, 11*(Suppl. 11), 2714–2726.

Zhou, S. F., Chan, E., Zhou, Z. W., Xue, C. C., Lai, X., & Duan, W. (2009). Insights into the structure, function, and regulation of human cytochrome P450 1A2. *Current Drug Metabolism, 10*(7), 713–729.

Zsoldos, Z., Reid, D., Simon, A., Sadjad, B. S., & Johnson, A. P. (2006). eHiTS: An innovative approach to the docking and scoring function problems. *Current Protein and Peptide Science*, 7(5), 421–435.

Zsoldos, Z., Reid, D., Simon, A., Sadjad, S. B., & Johnson, A. P. (2007). eHiTS: A new fast, exhaustive flexible ligand docking system. *Journal of Molecular Graphics and Modelling, 26*(1), 198–212.

AUTHOR INDEX

Note: Page numbers followed by "*f*" indicate figures and "*t*" indicate tables.

A

Abadie, V., 366–368
Abagyan, R., 85–89, 94–95, 384–390, 385*t*
Abbasi, S. W., 383
Abdelhay, E., 29, 31*t*, 32
Abdi, F., 45, 48–49
Abel, B. L., 394–396
Abel, R., 299
Abergel, C., 379–380
Abeysinghe, S., 366–368
Abola, E. E., 379–380, 385*t*
Abramavicius, D., 316–317
Abyzov, A., 374–375
Adachi, T., 22
Adams, R. M., 368–370, 371–372
Addington, A. M., 53–54
Adkins, J. N., 8*t*, 10, 14
Adney, W. S., 366–370
Adzhubei, I. A., 181, 182, 183–184, 258–259, 372–374
Aebersold, R. H., 4, 8*t*, 11–13, 14, 44–45
Aflalo, C., 85–89, 86*t*
Agarwal, G., 124–126, 128–130
Agostini, M., 366–368
Agrawal, P., 379–380
Ahmad, S., 97
Ahmed, T., 178–179
Aicher, L., 366–368
Akao, Y., 368–370
Akhavan, S., 368–370
Akke, M., 393–394
Aksianov, E. A., 78–79
Akutsu, T., 374–375
Al Hasan, M., 86*t*
Al-Shahrour, F., 372–374
Alam-Faruque, Y., 374–375
Alamanova, D., 79, 98–99
Alarcon-Riquelme, M. E., 366–368
Alber, F., 84–85
Alder, B., 274–275
Aldini, G., 26
Aleksiev, T., 133*t*, 137
Alexander, N., 385*t*
Alexeevski, A. V., 78–79
Alexopoulos, P., 62–63
Alexov, E., 259–260, 368–370, 394–396
Allen, M. P., 293–294
Allinger, N. L., 306–307
Almen, M. S., 380–382
Almo, S. C., 377–378
Almond, H. R., 384–390
Alonso, H., 383, 384–390
Aloy, P., 78–79, 80, 81, 82, 83–84, 90–92, 94–95, 100–102, 105
AlQuraishi, M., 99–100
Altevogt, P., 30–31
Altman, R. B., 368–370, 372–375, 384–390
Altschul, S. F., 97, 100–102, 128–130, 147, 379–380
Alves, P., 12
Amberger, J., 374–375
Amenta, N., 86*t*
Amess, B., 61
Amino, N., 368–370
Amode, M. R., 183
Amos, B., 183
Amos-Binks, A., 93–94
Anand, P., 385*t*
Andersen, J. N., 122, 155–156
Andersen, P., 376–377
Anderson, J. B., 148–149
Anderson, L., 65
Anderson, N. G., 47–48
Anderson, N. L., 47–48
Ando, K., 63–64
Andrade-Navarro, M. A., 374–375
Andreasen, N., 42–43, 54
Andries, K., 270
Andrusier, N., 86*t*
Angarica, V. E., 79, 98–99
Annala, M., 100
Ansong, C., 14
Ansseau, M., 50
Antipin, Y., 372–374

Antonioli, P., 31*t*
Antoon, J., 92
Antunes, R., 374–375
Aphale, A., 368–370
Apol, E., 323
Apostolakis, J., 385*t*
Apostolakis, T., 385*t*
Appel, R. D., 46–47
Apweiler, R., 47–48, 374–375, 377–378
Arakaki, A. K., 93–94
Aranome, A. M., 29, 31*t*
Arbiza, L., 262–263, 368–370, 376–377
Arjun, P., 374–375
Arndt, C., 178–179
Arnold, K., 379–380
Arnold, R. J., 12
Aronow, B. J., 372–374
Arriaga, E. A., 21, 23
Arruda, M. A., 65–66
Arshinova, T. V., 179–180
Artemyev, N. O., 326
Arturi, F., 30–31, 31*t*
Ashkenazy, H., 93–94, 181, 182, 376–377
Aspinall-O'Dea, M., 45
Asplund, M. C., 316–317
Assa-Munt, N., 368–370
Aste, T., 394–396
Astner, H., 31*t*
Atkins, W. M., 26
Aviles, F. X., 375–376
Avril, T., 30–31, 31*t*
Avron, B., 375–376
Axenopoulos, A., 86*t*
Azam, S. S., 383

B

B-Rao, C., 384–390, 394–396
Baase, W. A., 375–376
Babu, M. M., 343
Bachi, A., 46
Bachschmid, M. M., 22–23
Baell, J. B., 380–382
Baenziger, P. H., 368–370, 375–376
Baglioni, M., 54, 62–63
Bahadur, R. P., 91*t*
Bahn, S., 40–41, 47–50, 51*t*, 55, 56*t*, 61, 62, 65–66
Bailey, S., 380–382
Bailey, T. L., 102–105
Bairoch, A., 377–378
Baiz, C. R., 316–317
Bajorath, J., 390, 391
Bakall, B., 178–179
Bakayan, A., 341–342
Baker, 259–260
Baker, C. I., 53–54
Baker, D., 92, 124–126, 125*t*, 368–370, 371–372, 379–380
Baker, J. O., 183, 366–370, 374–375
Bakowies, D., 394–396
Balaji, V., 394–396
Balasubramanian, S., 259–260
Baldessarini, R. J., 53
Baldi, P., 220–221, 372–374, 375–377
Baldwin, E. P., 375–376
Balius, T. E., 385*t*
Ball, E. V., 183, 366–368, 374–375
Ballatore, C., 49–50
Balog, C. I., 64–65
Balogh, L. M., 26
Banach, M., 321, 322–323, 340, 343
Bandeira, N., 10
Banerjee-Basu, S., 374–375
Banks, J. L., 384–390
Banno, Y., 368–370
Banzato, C. E., 65–66
Bao, L., 368–370, 372–374
Barbas, C. F. III., 348
Barber, J. D., 376–377
Barik, A., 91*t*
Barkan, D. T., 83–84
Barker, G. L., 372–374
Barker, W. C., 377–378
Barnes, J., 283–284
Barnes, S., 26
Barney, B., 285
Barnickel, G., 385*t*
Baron, R., 394–396
Barrantes-Reynolds, R., 368–370
Barrell, D., 183
Barrett, J. E., 50, 51*t*, 60–61
Barroso, I., 366–368
Barton, G. J., 376–377
Bartunik, H., 330–331
Barz, B., 368–370, 379–380
Bashford, D., 302, 303–304, 394–396

Bastiani, P., 54, 62–63
Basu, S. N., 374–375
Bates, P. A., 79, 85–89, 86*t*, 90–92, 379–380, 382
Batzoglou, S., 122, 155–156
Baù, D., 84–85
Baucom, A., 374–375
Bauer, T. M., 49–50
Baummann, M., 21
Bayly, C. I., 90, 299–301
Bazenet, C., 63–64
Beach, T., 42–43
Beal, K., 183
Beasley, C. L., 65, 66
Beaudet, J. G., 259–260
Beavis, R. C., 7, 8*t*, 9–10
Beckmann, J. S., 375–376
Bedell, V. M., 348
Beeman, D., 278
Behan, A., 66
Behler, J., 307
Bejerano, G., 376–377
Bélarbi, K., 63–64
Belew, R. K., 384–390, 385*t*
Bell, R. E., 376–377
Bellew, M., 13–14
Bellis, L. J., 93–94
Bellott, M., 302, 303–304, 394–396
Bembom, O., 109*f*
Ben-Tal, N., 93–94, 97, 182, 376–377
Bencurova, E., 300
Bender, C., 306–307
Bendtsen, J. D., 375–376
Benhar, M., 25
Benkovic, S. J., 393–394
Benos, P. V., 99–100, 105
Benros, C., 128–130
Benson, M. L., 385*t*
Bento, A. P., 93–94
Benyamini, H., 156–157, 158–161, 159*t*
Berendsen, H. J. C., 283–284, 323
Berezin, C., 376–377
Berger, B., 122–123, 130–132, 138–139, 143, 147–148, 159*t*, 165, 379–380
Berman, H. M., 78–79
Bernard, S., 307–308
Bernasconi, M., 307–308
Berne, B. J., 283–284
Bernlohr, D. A., 21, 23, 26, 27
Bernstein, F. C., 377–378
Bertho, G., 385*t*
Best, F., 178–179
Best, S. A., 306
Betz, S. F., 375–376
Bhakta, M. S., 349
Bhandarkar, M., 394–396
Bhat, T. N., 78–79
Bhide, M., 300
Bhushan, K., 379–380
Bian, N., 262–263, 376–377
Biancardi, A., 317–318
Biegert, A., 379–380
Bigham, A. W., 366–368
Bikadi, Z., 385*t*
Billeter, S. R., 305
Binkowski, A., 385*t*
Birg, I. N., 60–61
Birzele, F., 123–124, 132, 133*t*, 137–138, 139, 143–144
Bisca, A., 30–31, 31*t*
Biswas, D., 385*t*
Bizzarri, A. R., 282–283, 284–285
Blaber, M., 375–376
Black, G. C., 178–179
Blackledge, M., 394–396
Blackshaw, S., 98
Blackshields, G., 379–380
Blackwell, T., 28
Blagg, J., 380–382
Blagoev, B., 4, 45–46
Blaisdell, B. E., 213–220, 376–377
Blake, M. R., 305–306
Blennow, K., 42–43
Bliznyuk, A. A., 383, 384–390
Block, J. N., 379–380
Blom, N., 375–376
Blucher, C., 349, 357–358
Blundell, T. L., 165, 379–382
Bo, L., 168
Board, P. G., 368–370
Bocchetta, M., 54
Bocchini, C., 374–375
Bocchio-Chiavetto, L., 39-76
Boch, J., 348, 349, 357–358
Boddapati, M., 368–370
Bode, W., 330–331

Bodycombe, N. E., 385*t*
Boeckmann, B., 377–378
Boelens, R., 85–89, 92, 385*t*
Bogaerts, P., 372–374
Bogdanove, A. J., 348–349
Boisvieux, J., 124–126, 125*t*
Bollen, Y. J., 341–342
Bolton, E., 385*t*
Bonaldi, T., 46
Bond, D. J., 53
Bond, J. P., 261–262, 368–370
Bonet, J., 78–79, 83, 84–85, 93
Bonnardeaux, A., 366–368
Bonnet, J. P., 300
Bonvin, A. M. J. J., 85–89, 86*t*, 91*t*, 92, 385*t*
Boomsma, W., 169
Borchers, C. H., 27–28
Bordoli, L., 379–380
Borg, M., 169
Bork, P., 181, 182, 183–184, 258–260, 366–368, 371–375, 376–377
Bornot, A., 124–130
Bortolami, O., 375–376
Bosco, D. A., 393–394
Bose, S., 379–380
Bossers, K., 64–65
Bougueleret, L., 4–5, 14
Boulling, A., 368–370
Bourne, P. E., 133*t*, 137, 141, 142–143, 147, 148–149, 158, 159*t*, 385*t*
Bourreau, A., 45
Bouslam, N., 368–370
Bouvier, G., 385*t*
Bouvier, S. E., 368–370
Bowie, J. U., 379–380
Boyd, J., 366–368
Braberg, H., 83–84
Bradbury, M. E., 45–46
Bradley, J., 366–368
Bradley, P., 122, 130–132, 348–349, 379–380
Brandon, M. C., 374–375
Brannetti, B., 133*t*, 169
Branson, H. R., 124–126
Braun, R., 394–396
Brendel, V., 213–220, 376–377
Brenke, R., 85–89, 86*t*
Brenner, S. E., 148–149, 168, 211–213
Brent, S., 183
Bretteville, A., 63–64
Brice, M. D., 377–378
Briggs, S. P., 10
Brody, E., 366–368
Bromberg, Y., 181, 182, 183–184, 258–259, 372–374
Broniatowska, E., 339–340
Broniowska, K. A., 22–23
Brookes, A. J., 374–375
Brooks, B. R., 84, 90, 302
Brooks, C. L. III., 78–79, 98–99, 284–285
Brouillet, S., 133*t*, 139–140, 169
Brown, F. K., 384–390
Brown, G. R., 377–378
Brown, N. P., 379–380
Brown, S. J., 94
Brozell, S. R., 385*t*
Bruccoleri, R. E., 84, 90, 302
Brugge, J. M., 394–396
Brunak, S., 375–376
Brunner, R., 394–396
Bruschweiler, R., 394–396
Brutlag, D., 148–149
Bryant, S. H., 122–123, 132, 133*t*, 148–149, 155–156, 385*t*
Brylinski, M., 93–94, 122, 168, 319, 321, 339–340, 385*t*
Bucher, P., 213–220, 376–377
Buckland, P. R., 179–180
Budowski-Tal, I., 124–126, 125*t*
Bujnicki, J. M., 79, 92, 377–378, 379–380
Bulyk, M. L., 98, 99–100, 105
Bunck, M., 60–61
Buning, C., 384–390
Bürgi, R., 394–396
Burguet, J., 262–263, 368–370, 376–377
Burkhard, P. R., 48–49
Burn, T. C., 366–368
Burns, J. K., 53–54
Burrill, J. S., 21, 23
Bush, G., 53–54
Busse, D., 12
Bustamante, C. D., 366–368
Butterfield, D. A., 56*t*, 64
Buxbaum, J. D., 179–180
Bycroft, M., 375–376
Byers, H. L., 42–43

Byrnes, B., 259–260
Byron, M., 84
Bystritsky, A., 40–41
Bystroff, C., 86*t*, 124–126, 125*t*

C

C, G. P., 394–396
Cai, J., 21, 54, 62–64
Cai, Y., 97
Cai, Z., 371–372
Cairns, N. J., 42–43
Calabrese, R., 181, 182, 183–184, 258–259, 372–374
Caldwell, J. W., 296*t*, 299, 300–301, 385*t*, 394–396
Calladine, C., 106–107
Callén, E., 368–370
Camacho, C. J., 86*t*, 90, 385*t*
Cammi, R., 317–318
Campbell, I. D., 300
Campbell, J. M., 42–43, 348
Campbell, S. J., 390
Campos, F. C., 29
Camproux, A.-C., 124–126, 125*t*, 128–130
Cannistraro, S., 282–283, 284–285
Canterbury, J. D., 8*t*, 9
Cao, J., 97
Cappelli, C., 317–318
Capriotti, E., 84, 181, 182, 183–184, 258–259, 368–370, 372–374, 377–378, 384–390
Carare, R. O., 368–370, 375–376
Carbone, D. L., 26
Carini, M., 26
Carloni, P., 393–394
Carlson, H. A., 271, 385*t*, 391
Carmona, S. J., 385*t*
Carpentier, M., 133*t*, 139–140, 169
Carr, R., 380–382
Carrette, O., 48–49
Carrinski, H. A., 385*t*
Carrol, K. S., 27
Carson, M. B., 94
Cartegni, L., 179–180, 366–368
Casadio, R., 181, 182, 183–184, 258–259, 372–374, 376–378
Casado, J. A., 368–370
Casanova, J. L., 368–370, 375–376
Case, D. A., 296*t*, 299, 300–301, 385*t*, 394–396
Case TAD, D. A., 350–351
Castagna, A., 31*t*
Castaño, E. M., 42–43
Castella, M., 368–370
Castellana, S., 374–375
Castellanos, J. I., 169–170
Catherinot, V., 385*t*
Caudle, W. M., 48–49, 66–67
Caulfield, M. J., 376–377
Causevic, M., 42–43
Cawkwell, L., 31*t*, 32
Ceaser, E. K., 23–24
Cecchetti, R., 54, 62–63
Cecchini, R., 29, 31*t*, 32
Cecconi, D., 30–31, 31*t*
Celano, M., 30–31, 31*t*
Céol, A., 78–79, 82, 83–84, 100–102
Ceriani, C., 307–308
Cernadas, R. A., 348–349
Cervantes, C. F., 394–396
Chabot, J. G., 62
Chahed, K., 30–31, 31*t*
Chai, J., 348–349
Chait, B. T., 45–46
Chakraborty, C., 377–378, 394–396
Chakravarty, A., 46–47
Chakravarty, S., 55, 56*t*, 60
Chaleil, R. A. G., 86*t*
Chambers, J., 93–94
Chambers, M. C., 8*t*, 10
Chan, A. M., 368–370
Chan, E., 368–370, 372–374
Chan, P. A., 261–262
Chan, S. W., 53–54
Chandonia, J. M., 211–213
Chandra, N., 385*t*
Chandralekha, R., 64
Chandramouli, K., 44
Chandrasekhar, J., 284–285
Chanfreau, G., 341
Chang, D., 79, 98–99
Chang, H., 374–375
Chang, K. W., 385*t*
Chang, W. C., 375–376
Chao, M. V., 45–46
Chapdelaine, P., 348

Charru, A., 366–368
Chasman, D., 368–370, 371–372
Chaudhary, K., 385*t*
Cheatham, T. E. III., 295–299, 350–351, 394–396
Chen, A., 40–41
Chen, C. C., 385*t*
Chen, C. M., 385*t*
Chen, C. W., 372–374
Chen, C.-Y., 79, 98–99, 100, 385*t*
Chen, H.-L., 83, 376–377
Chen, J. M., 368–370, 372–374
Chen, L., 394–396
Chen, R., 85–89, 86*t*, 90–92, 391
Chen, S. C., 53–54
Chen, S. X., 30–31
Chen, V. B., 379–380
Chen, X., 30, 45–46, 385*t*
Chen, Y., 92, 183
Chen, Y. C., 79, 94–95, 98–99
Chen, Y. Z., 385*t*
Cheng, A., 306
Cheng, H., 148–149, 379–380
Cheng, J., 220–221, 372–377
Cheng, K., 55, 56*t*, 60
Cheng, T. M., 382
Cheng, X., 272
Chenna, R., 379–380
Cherukuri, P. F., 148–149
Chételat, G., 54
Chevrolat, J., 124–126, 125*t*
Chia, K. S., 371–372
Chiappe, D., 4–5, 14
Chiappetta, G., 28
Chiarotti, G., 307–308
Chien, T.-Y., 79, 98–99
Chilov, G. G., 385*t*
Chiou, J. J., 183, 372–374
Chiranjib, C., 374–375
Chitipiralla, S., 374–375
Chivian, D., 379–380
Cho, Y., 368–370, 374–375, 385*t*
Chodera, J. D., 375–376
Choe, L. H., 42–43
Choi, H., 8*t*, 9, 11–12, 14
Chong, L., 349–350
Chong, L. T., 273
Chong, S. S., 372–374
Chothia, C., 148–149, 168
Chouchane, L., 30–31, 31*t*
Choudhury, P., 374–375
Chowdhury, S., 299–300, 350–351
Christen, M., 394–396
Christensen, M. H., 385*t*
Christiansson, M., 42–43
Chu, B., 341–342
Chu, P. Y., 385*t*
Chu, Y. W., 372–374
Chuang, C. C., 53–54
Chun, S., 261–262
Chung, H. S., 343
Church, D., 374–375
Church, G. M., 262–263, 376–377
Church, W. B., 377–378
Chuzhanova, N. A., 179–180, 368–370, 375–376
Ciccotti, G., 283–284
Cieplak, P., 90, 299–301
Clauser, E., 366–368
Claussen, H., 384–390
Clavreul, N., 22
Clementi, E., 274–275
Clemmer, D. E., 12
Clerens, S., 62
Climenti, E., 274–275
Clough, T., 8*t*, 14
Coccia, R., 63–64
Cocciolo, A., 63–64
Coffey, R. J., 42–43
Coffman, A. C., 385*t*
Coggan, M., 368–370
Cohen, E., 385*t*
Cohen, R. A., 22
Cohen, R. M., 366–368
Cole, C., 376–377
Cole, J. C., 385*t*
Coleman, S. L., 179–180
Colin, Y., 128–130
Colinge, J., 4–5, 14
Colizzi, F., 385*t*
Collado-Vides, J., 79, 98–99
Collepardo-Guevara, R., 394–396
Collet, B., 30–31, 31*t*
Colosimo, A., 368–370, 375–376
Comeau, S. R., 85–89, 86*t*, 90, 385*t*
Conaway, R. C., 12

Conchillo-Sole, O., 375–376
Conde, L., 372–374
Cong, L., 348, 349, 357–358
Cong, Q., 379–380
Congreve, M., 380–382
Conrad, C. A., 61
Conrads, T. P., 40–41, 48–49
Conti, F., 368–370, 375–376
Contreras-Moreira, B., 79, 98–99
Cooper, D. N., 179–180, 368–370, 372–376
Cooper, G. M., 53–54
Coram, M., 13–14
Corey, R. B., 124–126
Cornell, W. D., 90, 299–301
Corona, R. I., 79, 91*t*, 92
Cortens, J. C., 8*t*, 9–10
Cortese, M. S., 316–317, 375–376
Costa, J., 30–31
Cote, A., 100
Cotter, D. R., 65–66
Cottrell, J. S., 7, 8*t*
Couch, G. S., 101*f*, 385*t*
Coulombe, Z., 348
Courty, M., 300
Coward, L., 26
Cowburn, D., 45–46
Cowen, L., 122–123, 132, 138–139, 143, 147–148, 159*t*, 165
Cowley, G. S., 366–368
Cox, J., 8*t*
Cozzini, S., 133*t*, 137
Craft, G. E., 40–41
Craig, R., 7, 8*t*, 9–10
Crawford, J., 21
Creasy, D. M., 7, 8*t*
Crofton, A., 64–65
Croghan, T. W., 55
Crooks, G. E., 211–213
Cross, F. R., 45–46
Crowley, V. E., 366–368
Crowther, G. J., 385*t*
Cruts, M., 374–375
Csaba, G., 123–124, 132, 133*t*, 137–138, 139, 143–144
CsizmLok, V., 375–376
Csorba, A., 60–61
Cueto, M., 385*t*
Cui, M., 390, 391
Cui, Y., 368–370, 372–374
Cunniff, M. M., 348, 349, 357–358
Curran, M. E., 366–368
Curtis, D., 371–372
Curtis, J. M., 21, 23
Cutrufello, N. J., 66
Czibere, L., 48–49, 60–61

D

Da Cruz, S., 30
Dadgar, N., 368–370
Daga, P. R., 377–378
Dahlback, B., 394–396
Dai, H.-K., 98
Dai, J., 30
Dai, Z., 380–382
Daiber, A., 22–23
Daily, A., 368–370
Dalke, A., 385*t*
Dalle-Donne, I., 64
Dalton, J. A., 379–380
Dambosky, J., 368–370
Daminelli, G., 393–394
Dammer, E. B., 64–65
Daneshjou, R., 368–370
Daniel, V., 49–50
Daniels, C. R., 300, 301*t*
Daniels, N., 143, 147–148
Daniluk, P., 132, 133*t*, 137
Daras, P., 86*t*
Darden, T., 279, 394–396
Darie, C. C., 45–46
Darley-Usmar, V., 26
Das, K., 270
Das, M. K., 98
Das, R., 379–380
Das, T., 55, 56*t*, 60
Dasari, S., 8*t*, 10
Dastmalchi, S., 377–378
d'Aubenton-Carafa, Y., 366–368
Daura, X., 375–376
Davidchack, R. L., 281–282
Davidsson, P., 42–43
Davies, E., 366–368
Davies, M., 93–94
Davis, B. P., 78–79
Davis, I. W., 379–380

Day, R., 377–378
Dayhoff, M. O., 145–146
Dayon, L., 27
De Baets, G., 366–368, 375–376
De Brevern, A. G., 122–123, 124–130, 125*t*, 132, 133*t*
De Cesco, S., 385*t*
de Groot, N. S., 375–376
de Haan, L., 53–54
de la Cruz, X., 98, 368–370, 372–374, 376–377
de la Mata, M., 28–29
De Laey, J. J., 178–179
de Leon, M. J., 54
de Vlieg, J., 380–382, 385*t*
De Vries, S. J., 86*t*
Dean, P. M., 391
Deane, C. M., 165, 379–380
Debajyoti, C., 394–396
Debulpaep, M., 375–376
Decker, S. R., 366–370
Decornez, H., 390, 391
Decrescenzo, G. A., 380–382
Deelder, A. M., 64–65
Deffenbaugh, A. M., 182, 368–370, 372–374, 376–377
Defoort-Dhellemmes, S., 178–179
Dehouck, Y., 372–374
Deinhardt, K., 45–46
Del Zompo, M., 50–53, 51*t*, 65
Delaby, C., 51*t*, 62–63
Delacroix, S., 300
Delarue, M., 379–380
Delia, D., 368–370
DeLisi, C., 272
Dellheden, B., 42–43
Demarco, M. L., 300
Deng, D., 348–349, 350–353
Deng, L., 83–84
Denicola, A., 23
Déon, C., 30–31, 31*t*
deRossi, T., 29
Desiderio, D. M., 48–49
Deslandes, S., 385*t*
Dessailly, B. H., 385*t*
Deutsch, E. W., 8*t*, 9–10
Devignes, M.-D., 86*t*, 385*t*
Devos, D., 84–85
Devraj, R. V., 380–382
DeWeese-Scott, C., 148–149
Dewhirst, M. W., 22–23
Dhaenens, C. M., 178–179
Di Domenico, F., 63–64
Di Giovanni, C., 380–382
Di Girolamo, G., 55
Di Stefano, F., 54
Dias-Neto, E., 47–49, 50, 51*t*, 56*t*
Dicker, P., 65–66
Dickinson, D. A., 23–24
Diekhans, M., 259–260, 368–370
Diers, A. R., 28–29
Dietz, K. J., 24, 27
Dihazi, H., 65–66
Dill, K. A., 316, 317–318, 340, 368–370, 375–376
DiMasi, J. A., 380–382
DiMatteo, M. R., 55
Dimmic, M. W., 366–368
Dineen, D., 379–380
Ding, J., 270
Dinur, U., 304
Dittmar, G., 12
Ditzen, C., 48–49
Dixit, A., 368–370
Dixon, S. E., 4
Dobbins, S. E., 89
Dobson, C. M., 368–370
Dobson, R. J., 376–377
Doerksen, R. J., 377–378
Doherty-Kirby, A., 5, 8*t*
Dokholyan, N. V., 27–28
Dokudovskaya, S., 84–85
Dolan, M. E., 380–382
Domenici, E., 48–49
Domingues, F. S., 133*t*, 137, 142–143, 147
Dominguez, C., 85–89, 385*t*
Donadelli, M., 30–31, 31*t*
Donovan, L. E., 64–65
Doolittle, R., 379–380
Doorn, J. A., 26
Dopazo, J., 366–368, 375–376
Doro, G., 30–31, 31*t*
Dosztányi, Z., 316–317, 372–374, 375–376
Doughty, E., 374–375
Doulias, P. T., 29
Dourlen, P., 63–64

Downing, A. K., 300
Drachkova, I. A., 179–180
Drake, P. G., 122, 155–156
Drew, P. J., 31*t*, 32
Dror, O., 156–157, 158–161, 159*t*
Dror, R. O., 275–276, 394–396
Dryja, T. P., 366–368
D'Souza, B. T., 26
Duan, W., 368–370, 372–374
Duan, Y., 299–300, 350–351
Dubois, B., 54
Duchi, S., 368–370, 376–377
Dudev, M., 124–126, 128–130
Dudley, E., 40–41
Duke, R. E., 350–351
Dunbar, J. B. Jr., 391
Dunbar, R. I. M., 53–54
Dunbrack, L. D. Jr., 379–380
Dunbrack, R. L., 261–262, 302, 303–304
Dunbrack, R. L. Jr., 78–79, 122, 155–156
Dundas, J., 385*t*
Dunham, I., 366–368
Dunkel, M., 385*t*
Dunker, A. K., 316
Dunn, M. J., 65–66
Duong, D. M., 24
Duraisamy, S., 261–262
Durrant, J., 382
Durrant, J. D., 273
D'Urso, D., 368–370

E

Eagle, G. L., 31*t*, 32
Eastwood, J. W., 283–284
Eastwood, M. P., 394–396
Ebenazer, A., 394–396
Eckweiler, D., 100
Eddes, J. S., 8*t*, 9–10
Edelman, M., 384–390, 385*t*
Edgar, R. C., 122, 155–156, 211–213, 379–380
Edwards, K. J., 372–374
Ehrenreich, H., 65–66
Ehrlich, L. P., 391
Eickholt, J., 374–375
Eidhammer, I., 122
Eisenberg, D., 379–380
Eisenmesser, E. Z., 393–394
Eisenstein, M., 85–89, 86*t*
El Filali, Z., 48–49
El Hassan, M., 106–107
Elber, R., 277–278
Eliuk, S. M., 26
Elkayam, T., 368–370
Ellden, J., 341–342
Elles, L. M. S., 368–370
Elliot, V., 45
Elliott, J. R., 393–394
Elms, P., 27–28
Elofsson, A., 379–380
Emekli, U., 169
Emmett, M. R., 61
Eng, J. K., 7, 8*t*, 9–10
Engdahl, A. L., 391
Engelsen, S. B., 300
English, J. A., 65–66
Engstrom, P. G., 179–180
Epelbaum, S., 54
Eramian, D., 83, 100–102, 379–380
Erdin, S., 366–368
Erez, E., 93–94, 182, 376–377
Eriksson, A. E., 375–376
Eriksson, M. A., 271
Erman, B., 94–95
Ernoult, E., 45
Esh, C. L., 42–43
Esnouf, R. M., 372–374
Essex, J. W., 384–390
Eswar, N., 83, 100–102, 259–260, 368–370, 379–380
Etchebest, C., 124–130, 125*t*
Evanseck, J., 302, 303–304
Evanseck, R. L. J. D., 394–396
Evrard-Todeschi, N., 385*t*
Ewig, C. S., 304
Ewing, T. J., 384–390
Exner, T. E., 385*t*
Eyck, L. T., 89

F

Fabbri, C., 55
Facino, R. M., 26
Fagan, A., 65
Faghihi, M. A., 179–180
Fajardo, J. E., 379–380
Falkai, P., 47–49, 50, 51*t*, 56*t*

Fan, H., 394–396
Fan, Y., 372–374
Fang, X., 385*t*, 394–396
Fariselli, P., 181, 182, 183–184, 258–259, 372–374, 376–378
Fathi, G., 385*t*
Fay, J. C., 261–262
Fedorov, A. A., 377–378
Fedorov, E., 377–378
Feig, M. III., 78–79, 98–99
Feigon, J., 341
Felder, C. E., 375–376
Feldmann, R. E. Jr., 42–43
Feliu, E., 78–79, 83, 84–85, 90–92, 94–95, 105
Feng, G., 348
Feng, Z., 78–79
Fennen, J., 305
Fenton, I., 366–368
Fenyo, D., 8*t*, 9–10
Feolo, M., 374–375
Ferec, C., 368–370
Ferguson, D. M., 90, 299–300
Fermin, D., 8*t*, 14
Fernald, G. H., 368–370
Fernandes, P. A., 385*t*, 393
Fernández- Rodriguez, R., 27–28
Fernandez-Escamilla, A. M., 375–376
Fernandez-Fuentes, N., 379–380
Fernández-Recio, J., 79, 85–89, 86*t*, 90–92, 91*t*, 94–95, 98, 380–382, 385*t*
Ferrada, E., 79
Ferrer-Costa, C., 368–370, 372–374, 376–377
Ferro, S., 377–378
Fersht, A. R., 375–376
Féry, I., 366–368
Fidelis, K., 377–378
Field, M. J., 302, 303–304, 394–396
Filiou, M. D., 46, 47–48, 60, 65, 66–67
Findlay, J. B., 178–179
Finehout, E. J., 42–43
Fiorini, A., 63–64
Fiorucci, S., 86*t*
Fischer, M., 83–84
Fischer, S., 394–396
Fiser, A., 377–378, 379–380
Fitzgibbon, M., 13–14
Flicek, P., 183
Flocco, M. M., 126–128
Flores, G., 62
Flurchick, K. M., 306–307
Föcking, M., 65–66
Folch, B., 372–374
Foley, B. L., 300, 301*t*
Folkesson, S., 42–43
Forman, M. S., 67
Fornes, O., 78–79, 93, 100–102
Fornili, A., 124–126, 125*t*
Forrester, M. T., 25
Förster, F., 84
Foss, E., 13–14
Fourrier, L., 128–130
Fraenkel, E., 101*f*
Franck, Z., 42–43
Frank, A. M., 5, 8*t*, 10
Frank, E., 46, 47–48, 66–67
Frank, G. R., 366–368
Fraternali, F., 284–285
Frauenfelder, H., 393–394
Frazer, K. A., 366–368
Fredman, D., 183, 374–375
Freed, K. F., 92
Freedman, R., 53–54
Freimuth, R. R., 372–374
Freinkman, E., 259–260
Freitas, L. F., 29
French, B., 55
Fried, M. G., 368–370
Friedhoff, P., 377–378
Friedland, R. P., 368–370, 375–376
Friedman, D. B., 42–43
Friesem, A. A., 85–89, 86*t*
Friesner, R. A., 384–390
Frishman, D., 368–370, 375–377
Frisoni, G. B., 54
Fritz, A., 368–370, 375–376
Frolkis, A., 93–94, 385*t*
Frommel, C., 385*t*
Frousios, K., 374–375
Fryar, K. L., 375–376
Fu, H., 375–376
Fu, W., 366–368
Fuhr, U., 368–370
Fujihara, J., 179–180
Fujita, T., 374–375

Fujiwara, H., 368–370
Fuller, E. D., 368–370
Fuller, R. P., 348
Furr, J. R., 390, 391
Fütterer, C. D., 42–43

G

Gabb, H. A., 85–89, 86*t*, 90–92
Gabdoulline, R., 100
Gabelle, A., 51*t*, 62–63
Gabreëls-Festen, A. A., 368–370
Gaj, T., 348
Gallardo, R., 375–376
Gallat, F. X., 341–342
Gallion, S., 306–307
Gallis, B., 13
Galzitskaya, O. V., 375–376
Gamelin, E., 45
Gammeltoft, S., 375–376
Gandhimathi, A., 385*t*
Gane, P. J., 391
Gao, H., 348–349
Gao, M., 79, 83, 93–94, 97
Gao, Z., 385*t*
Garbuzynskiy, S. O., 375–376
Garcia-Garcia, J., 78–79, 93
Garner, J., 374–375
Garrels, J., 78–79
Garzon, J. I., 85–89, 86*t*
Gass, J. D., 178–179
Gatchell, D. W., 86*t*, 90, 385*t*
Gaulton, A., 93–94
Gautam, A., 385*t*
Ge, B., 366–368
Ge, H., 78–79
Gearing, M., 64–65
Geenen, L., 62
Geer, L. Y., 148–149
Geerke, D. P., 394–396
Gehrcke, F., 28
Gelb, M. H., 4, 44–45
Gelly, J.-C., 124–126, 128–130
Gelpi, J. L., 372–374
Gemma, A., 30–31, 31*t*
Geng, Z., 262–263, 376–377
George, G. A., 385*t*
George Priya Doss, C., 374–375, 377–378, 394–396
Gerasimova, A., 181, 182, 183–184, 258–259, 372–374
Gerber, S. A., 4, 44–45
Gerlach, M., 40–41
Gersbach, C. A., 348
Gerstein, M., 122–123, 133*t*, 137, 138–139, 141–142, 259–260
Gewehr, J. E., 132, 133*t*, 137, 139, 143–144
Ghersi, D., 384–390
Ghoorah, A. W., 385*t*
Giaccone, G., 48–49
Gianazza, E., 21
Giardine, B., 374–375
Gibrat, J., 122–123, 132, 133*t*, 148–149
Gibrat, J. F., 385*t*
Gibson, B. W., 24
Gibson, T. J., 372–374, 375–376, 379–380
Gilis, D., 372–374
Gille, H., 368–370
Gilliland, G., 78–79
Gilson, M. K., 385*t*
Gimpelev, M., 379–380
Ginalski, K., 78–79
Giovinazzo, D., 29
Giovine, A., 56*t*, 60
Giron, P., 27
Gish, W., 379–380
Gitter, A., 100–102
Giuliani, A., 368–370, 375–376
Glaser, F., 376–377
Glaserm, F., 376–377
Glen, R. C., 384–390
Glover, J. N. M., 101*f*
Gnad, F., 374–375
Go, Y. M., 24
Goate, A. M., 42–43
Goddard, T. D., 101*f*, 385*t*
Goddard, W. A., 296*t*, 304
Goddeau, D., 368–370
Godin, K. S., 259–260
Godzik, A., 132, 133*t*, 138–139, 143, 156–157
Goede, A., 385*t*
Goedhart, S., 53–54
Gohlke, H., 393, 394–396
Golab, J., 49–50
Golaz, O., 46–47
Gold, A. K., 383

Gold, N. D., 390
Goldberg, M. P., 368–370
Goldsby, R. E., 368–370
Gomes, C., 30–31
Gomez, A., 368–370
Gonfloni, S., 368–370
Gong, S., 374–375
Gonik, M., 48–49
Gonzalez-Outeirino, J., 300, 301*t*
Goo, Y. A., 13
Goodlett, D. R., 10, 13, 44–45
Goodsell, D. S., 384–390, 385*t*
Gooley, A. A., 375–376
Gooley, T. A., 368–370
Gormanns, P., 47–49, 50, 51*t*, 56*t*, 60
Gottlieb, A. A., 285
Gotz, A. W., 350–351
Gou, Z., 94
Gough, J., 372–374
Gould, I. R., 90, 299–300
Gouttard, M., 368–370
Gowrisankar, S., 372–374
Grabowski, H. G., 380–382
Grant, J. A., 384–390
Grasbon-Frodl, E., 368–370, 375–376
Grassl, J., 47–48, 62–63, 66–67
Grau, J., 98
Gravel, S., 366–368
Gray, J. J., 85–89, 86*t*
Gray, V. E., 374–375
Gready, J. E., 383, 384–390
Green
Green, J. R., 93–94
Greenblatt, D. M., 101*f*, 385*t*
Greenblatt, M. S., 259–260, 368–370
Gregory, F., 368–370
Gregory, P. D., 105–106
Gregory, S. G., 372–374
Griffin, N. M., 12
Griffin, T. J., 23, 26, 27
Griffith, M., 385*t*
Griffith, O. L., 385*t*
Grigorovich, D. A., 385*t*
Grimsrud, P. A., 23, 26, 27
Grinter, S. Z., 384–390
Grishin, N. V., 148–149, 379–380
Gritsenko, M. A., 4
Groenhof, G., 323
Gromiha, M. M., 97, 368–370, 372–375
Grosdidier, A., 385*t*
Grosfils, A., 372–374
Gross, J., 42–43
Grosse, I., 98
Grosveld, J., 341–342
Grottesi, A., 368–370, 375–376
Grunze, H., 53
Gruza, J., 300
Gu, S., 86*t*
Gu, X., 376–377
Gu, Z., 366–368
Guda, C., 158, 159*t*
Gudiseva, H. V., 28
Guerois, R., 90
Guest, P. C., 47–50, 51*t*, 55, 56*t*, 61
Guette, C., 45
Gui, Y., 93–94
Guibas, L., 124–128, 125*t*
Guitton, N., 30–31, 31*t*
Gulati, S., 382
Gumbart, J., 394–396
Gump, J. R., 259–260
Guney, E., 78–79, 83, 93
Guo, J., 79, 91*t*, 92, 100–102, 106, 372–374
Guo, J.-T., 79, 98–99
Guo, L. H., 56*t*, 62–63, 64
Guo, L. W., 326
Guo, X., 372–374
Gupta, A., 385*t*
Gupta, N., 11–12
Gupta, R., 375–376
Gupta, S., 102–105, 385*t*
Gurnell, M., 366–368
Gursoy, A., 78–79, 83–85, 368–370, 394–396
Gustafson, J. L., 285–286
Gutteridge, J. M. C., 20, 21, 22, 29
Guvench, O., 294–295
Guy, C. A., 179–180
Gwadz, M., 148–149
Gygi, S. P., 4, 44–45
Gyllensten, U., 374–375

H

Habrian, C. H., 349
Hahn, L. B., 366–368
Hahn, M. W., 368–370, 375–376

Hahn, S., 348, 357–358
Hahn, U., 368–370
Hahn, V., 368–370
Hainaut, P., 372–374
Hairer, E., 278
Hakes, L., 391
Halgren, T. A., 296*t*, 301*t*, 304, 305, 306, 384–390
Haliloglu, T., 94–95, 169
Hall, S. C., 4
Halliday, R. S., 384–390
Halliwell, B., 20–21, 22, 29
Halvorsen, L. A., 394–396
Ham, A.-J. L., 8*t*, 10
Hambsch, B., 48–49
Hamdan, M., 30–31, 31*t*
Hamelryck, T., 169
Hamilton, J. A., 316–317, 375–376
Hamilton, T. G., 13–14
Hamlat, A., 30–31, 31*t*
Hammel, M., 84
Hammes-Schiffer, S., 393–394
Hamodrakas, S. J., 375–376
Hamosh, A., 374–375
Hamrita, B., 30–31, 31*t*
Han, A., 374–375
Han, Z., 348–349
Handy, D. E., 23, 28–29
Hanemann, C. O., 368–370
Hanh, W. S., 21, 23
Hanks, D. L., 50, 51*t*, 55
Hannon, L., 274–275
Hansen, J., 375–376
Hansen, R. W., 380–382
Hansson, O., 54
Hansson, S. F., 42–43
Hansson, T., 277
Happ, M. P., 385*t*
Harbottle, A., 30
Harder, T., 169
Hardt, M., 4, 368–370
Hardy, L. W., 368–370
Hariprasad, G., 64
Harkonen, J., 372–374
Harrell, J. M., 368–370
Harris, L. W., 47–50, 51*t*, 55, 56*t*, 61
Harrison, S. C., 101*f*
Harsha, H. C., 45–46
Hart, W. E., 384–390
Hartman, E., 341
Hartmann, B., 92
Hartshorn, M. J., 385*t*
Harvey, S. C., 259–260
Hasegawa, H., 122–124, 132, 137, 138–139, 147–148, 169
Hashimoto, A., 126–128
Hass, J., 86*t*
Hassett, C., 366–368
Hata, J., 179–180
Hata, M., 368–370
Hattan, S., 45
Hauser, E. R., 372–374
Hausladen, A., 25
Hazai, E., 385*t*
Hazout, S., 124–130, 125*t*
He, F., 66
He, Z., 368–370, 379–380
Heering, H. D., 53–54
Hefferon, T., 374–375
Heinig, M., 376–377
Heinz, D. W., 375–376
Hekkelman, M. L., 374–375
Held, J. M., 24
Henderson, M. C., 368–370
Hendlich, M., 385*t*
Hendrie, C., 5, 8*t*
Henikoff, J. G., 124, 126, 144–145
Henikoff, S., 124, 126, 144–145, 181, 183–185, 366–368, 371–374
Henninger, N., 42–43
Henrick, K., 89, 133*t*, 137
Henter, I. D., 50
Hererra, A. C. S. A., 29
Heringa, J., 376–377, 379–380
Hermann, J. C., 377–378
Hermans, J., 323
Hernansanz-Augustín, P., 27–28
Herrera, A. C. S. A., 29, 31*t*, 32
Herschbach, D. R., 394–396
Hersey, A., 93–94
Hersh, L. B., 368–370
Hess, B., 323
Hess, D. T., 25
Heuft, G., 50
Hickie, I. B., 50–53
Higginbotham, L., 64–65

Higgins, D. G., 379–380
Hildebrandt, A., 385*t*
Hilger, M., 8*t*
Hill, S., 42–43
Hillier, L., 366–368
Hinrichs, A. S., 376–377
Hinrichsen, I., 377–378
Hirst, S. J., 385*t*
Hirtz, C., 51*t*, 62–63
Hoatson, G. L., 341–342
Hochstrasser, D. F., 30–31, 31*t*, 48–49
Hochstrasser, R. M., 316–317
Hockney, R. W., 283–284
Hodgkinson, V. C., 31*t*, 32
Hoekman, K., 48–49
Hofmann, J. P., 65
Hogg, N., 22–23, 28–29
Holm, L., 122–124, 126, 130, 132, 133*t*, 137, 138–139, 140–141, 147–148, 158, 168, 169, 379–380
Holmes, M. C., 105–106
Holzhütter, H. G., 47–48
Hon, G., 211–213
Hong, B., 91*t*
Honig, B. H., 83–84, 379–380
Hooft, R. W., 379–380, 385*t*
Hoogendoorn, B., 179–180
Hoppe, C., 372–374
Hornak, V., 299
Horng, J. T., 375–376
Horovitz, A., 375–376
Hosseini, A., 42–43
Hota, P., 374–375
Hou, J., 30
Hou, M., 376–377
Houstis, E. N., 86*t*
Hovatta, I., 55
Howells, K., 183
Hsiung, P. C., 53–54
Hsu, F., 374–375
Hsu, J. B., 375–376
Hsu, P. C., 375–376
Hu, L., 385*t*
Hu, S., 98
Hu, X., 262–263, 376–377
Hu, Y., 42–43
Huang, A., 31–32, 31*t*
Huang, B., 94, 385*t*
Huang, C. C., 31–32, 31*t*, 94, 101*f*, 385*t*
Huang, H. D., 375–376
Huang, I. Y., 366–368
Huang, J.-W., 14, 124–130, 125*t*, 133*t*, 139–140, 169, 233–259
Huang, K., 45–46
Huang, L. J., 30–31
Huang, N., 385*t*
Huang, S.-Y., 91*t*, 384–390
Huang, Y. N., 45
Hubbard, T. J., 148–149, 168, 377–378
Huber, R., 330–331
Hudson, T. J., 179–180, 366–368
Huey, R., 384–390, 385*t*
Hug, S., 271–272
Hughes, A. L., 394–396
Hughes, J. D., 380–382
Hughes, S., 48–49
Huizinga, E. G., 368–370
Humphrey, W., 385*t*
Hünenberger, P. H., 305, 394–396
Hünnerkopf, R., 47–48, 62–63, 66–67
Huo, S., 349–350
Hut, P., 283–284
Hwang, H., 63–64, 90–92, 91*t*, 94–95
Hwang, J., 379–380
Hwang, M. J., 304
Hwang, S., 94
Hwu, H. G., 53–54
Hye, A., 42–43
Hyrc, K. L., 368–370

I

Iconomidou, V. A., 375–376
Igbaria, A., 28
Igra, M., 13–14
Iida, R., 179–180
Ikawa, M., 366–368
Ilbert, M., 28
Ilinkin, I., 123, 156–157, 158, 159*t*, 161, 165
Iliopoulos, C. S., 374–375
Ilkowski, B., 377–378
Ilyin, V., 374–375
Imberty, A., 300
Immermann, F. W., 64
Impey, R. W., 284–285
Inbar, Y., 86*t*, 385*t*

Ince, P. G., 368–370, 375–376
Ingelman-Sundberg, M., 368–370
Ingman, M., 374–375
Ionescu, D. F., 50
Irving, J. A., 122, 155–156
Irwin, J. J., 385*t*, 394–396
Ishida, F., 368–370
Ishida, T., 179–180, 375–376
Ishihama, Y., 12
Ishii, T., 26
Ishikawa, T., 179–180
Ishioka, C., 372–374
Isom, A. L., 26
Issaq, H. J., 40–41, 48–49
Ivanisenko, V. A., 385*t*
Ivanov, I., 272
Iversen, L. F., 122, 155–156
Ivetac, A., 394–396
Izquierdo-Alvarez, A., 27–28

J

Jacak, R., 349–350
Jackson, R. M., 85–89, 86*t*, 379–380, 390
Jacupciak, J. P., 30
Jadczyk, T., 322–323
Jain, A. N., 385*t*
Jaitly, N., 8*t*, 10
Jakob, U., 25
Janáky, T., 60–61
Janardan, R., 123, 156–157, 158, 159*t*, 161, 165
Jang, W., 374–375
Janin, J., 85–89, 90–92, 91*t*, 94–95
Jankovic, J., 45, 48–49
Janssen, P. A., 270
Jaruzelska, J., 366–368
Jatzek, A., 25
Jaun, B., 394–396
Jayaram, B., 379–380, 385*t*
Jegga, A. G., 372–374
Jellema, T., 53–54
Jelveh, S., 13–14
Jenkins, R. E., 45
Jensen, L. J., 372–374
Jensen, O. N., 25
Jeong, J. S., 29
Jernigan, R. L., 80, 169
Jeunemaitre, X., 366–368
Jewison, T., 93–94, 385*t*
Jewsbury, P. J., 384–390
Ji, Z. L., 385*t*
Jiang, H. H., 30–31
Jiang, L., 98
Jiang, R., 258–259
Jiang, X. S., 30
Jiang, Y., 341–342
Jimenez, C. R., 48–49
Jimenez-Barbero, J., 300
Jiménez-García, B., 86*t*, 91*t*
John, A. M., 394–396
Johnson, A. P., 384–390, 385*t*
Johnston-Wilson, N. L., 65
Jonassen, I., 122
Jones, D., 379–380
Jones, D. P., 24
Jones, D. T., 130–132, 168, 379–380
Jones, G., 384–390
Jones, K. C., 316–317
Jones, R., 368–370
Jones, S., 93–94, 130–132, 168
Jonsson, M., 54
Joo, H., 30
Jordan, I. K., 259–260
Jorgensen, W. L., 284–285, 296*t*, 380–382, 394–396
Jorissen, R. N., 385*t*
Joseph, A. P., 124–130
Jothi, A., 377–378
Judkins, T., 182, 368–370, 372–374, 376–377
Julenius, K., 368–370, 375–376
Julian, A., 54
Juliano, L., 368–370
Juliano, M. A., 368–370
Jung, E., 375–376
Jung, J., 148–149
Jung, S., 394–396
Jungblut, P. R., 47–48
Jurkowski, W., 321, 323, 339–340, 343

K

Kabbage, M., 30–31, 31*t*
Kabsch, W., 141–142, 147, 163, 376–377
Kagerbauer, S. M., 56*t*, 64
Kailasanathan, A., 178–179
Kakizawa, J., 368–370

Kalé, L., 394–396
Kalko, S. G., 295–299
Kall, L., 8*t*, 9
Källberg, M., 168
Kamati, K. K., 372–374
Kamuf, J., 22–23
Kaname, T., 179–180
Kang, H. C., 29
Kang, H. J., 374–375
Kang, L., 385*t*
Kann, M. G., 368–370, 374–376
Kapral, G. J., 379–380
Kaptein, R., 92
Karanicolas, J. III., 78–79, 98–99
Karchin, R., 259–260, 368–370, 372–374
Karczewski, K. J., 368–370
Karlin, S., 213–220, 376–377
Karpievitch, Y., 14
Karplus, K., 379–380
Karplus, M., 84, 90, 275–276, 277, 284–285, 302, 368–370, 375–376, 393–394
Karyagina, A. S., 78–79
Kasif, S., 368–370
Kasprzak, J. M., 379–380
Kastritis, P. L., 92
Katchalski-Katzir, E., 85–89, 86*t*
Kato, S., 372–374
Kauwe, J. S., 42–43
Kauzmann, W., 317–318
Kawabata, T., 148–149, 374–375
Kay, L. E., 393–394
Kay, S., 348, 357–358
Kaye, J. A., 44–45, 54
Keage, H. A., 368–370, 375–376
Keilwagen, J., 98
Kel, A., 79, 98–99, 100
Keller, A., 8*t*, 11–12
Keller, J. W., 42–43
Keller, M. B., 50, 51*t*, 55
Keller, S., 30–31
Kelley, L. A., 130–132, 259–260, 368–370, 372–374, 379–380, 385*t*
Kemp, G. J. L., 85–89, 90
Kendrew, J., 368–370
Kenny, E. E., 366–368
Kern, D., 393–394
Keskin, O., 83–84, 368–370
Kessler, M. S., 46, 47–48, 60–61, 66–67
Keszler, A., 28–29
Khadjinova, A., 341–342
Khan, I. A., 179–180
Khan, S., 374–375, 376–377
Khawaja, X., 50, 51*t*, 60–61
Kholodov, M., 183, 374–375
Kiaei, M., 23–24
Kiebler, Z., 26
Kiemer, L., 375–376
Kieselbach, T., 24
Kihara, D., 86*t*
Kikuchi, R., 368–370
Kim, B. H., 148–149
Kim, C., 148–149
Kim, C. J., 31*t*
Kim, C. M., 368–370, 385*t*
Kim, D., 379–380
Kim, D. E., 379–380
Kim, D. S., 368–370, 385*t*
Kim, E. T., 30, 368–370, 394–396
Kim, H., 26
Kim, H. G., 40–41, 60
Kim, J. H., 368–370
Kim, J. K., 368–370, 385*t*
Kim, K. L., 40–41, 60
Kim, K. W., 368–370
Kim, M.-S., 349
Kim, P. S., 122, 130–132
Kim, R., 91*t*, 100–102, 106
Kim, T., 30
Kim, Y. J., 374–375
Kimmelman, A. C., 368–370
Kimura, M., 258–259, 366–370
Kimura-Kataoka, K., 179–180
Kinch, L., 379–380
King, N., 8*t*, 9–10
Kinoshita, A., 179–180
Kinoshita, K., 375–376
Kipper, J., 84–85
Kirchheiner, J., 368–370
Kirchhoff, P. D., 271
Kirk, M., 26
Kirkpatrick, P., 366–368
Kirsanov, D. D., 78–79
Kirsch, W., 64–65
Kirschner, K. N., 300, 301*t*
Kitchen, D. B., 390, 391

Kjong-Van, L., 258–259
Klebe, G., 384–390, 393, 394–396
Klein, J. B., 54, 62–63
Klein, M. L., 284–285
Klein, T. E., 259–260
Klein-Seetharaman, J., 78–79
Kleina, L. G., 368–370
Kleinjung, J., 124–126, 125*t*
Klepeis, J. L., 275–276
Klicic, J. J., 384–390
Klutstein, M., 100–102
Knapp, E.-W., 132, 133*t*
Knegtel, R. M. A., 92
Knol, J. C., 48–49
Knox, C., 93–94, 385*t*
Koca, J., 368–370
Koch, I., 259–260, 366–368, 371–372, 376–377
Kochańczyk, M., 339–340
Kocher, J.-P. A., 126–128
Kocherlakota, R., 368–370
Koehl, P., 86*t*, 122–123, 124–132, 125*t*, 170, 379–380
Koetzle, T. F., 377–378
Koh, J., 259–260
Kohane, I. S., 371–372
Kohlbacher, O., 385*t*
Koisti, M. J., 178–179
Kokjohn, T. A., 42–43
Kolb, H. C., 380–382
Kolbeck, B., 132, 133*t*
Kolchanov, N. A., 385*t*
Kolinski, A., 377–378
Kolker, E., 8*t*, 11–12
Kollman, P. A., 271, 295–299, 296*t*, 300–301, 349–350, 384–390, 385*t*, 394–396
Kollu, R., 374–375
Kolodny, R., 124–128, 125*t*
Komives, E. A., 394–396
Kompagne, H., 60–61
Konagurthu, A. S., 123, 158, 159*t*, 165
Kondo, T., 30–31, 31*t*
Kondrashov, A. S., 259–260, 366–368, 371–372, 376–377
Konieczny, L., 319, 321, 322–323, 339–340, 343
Kono, H., 368–370, 374–375
Kopp, J., 379–380
Koppensteiner, W. A., 133*t*, 137, 142–143, 147
Korb, O., 385*t*
Korkin, D., 84
Korner, S., 23–24
Kortemme, T., 92
Kosinski, J., 377–378, 379–380
Kossida, S., 274–275
Kosztin, I., 368–370, 379–380
Koukouritaki, S. B., 368–370
Koumandou, V. L., 274–275
Kovacs, J., 85–89
Koymans, L., 270
Kozakov, D., 78–79, 85–89, 86*t*
Krainer, A. R., 179–180, 366–368
Kramer, B., 384–390
Kratchmarova, I., 4, 45–46
Krawczak, M., 366–368
Krawetz, N., 394–396
Kretzschmar, H. A., 368–370, 375–376
Krishnan, V. G., 371–374
Krissinel, E., 133*t*, 137
Kristensen, D. B., 4, 45–46
Krivov, G., 379–380
Kriwacki, R. W., 343
Kroes, R. A., 61
Krömer, S. A., 60–61
Krueger, S. K., 368–370
Krug, R. G. II., 348
Kruger, W. D., 261–262
Kruyt, F. A., 48–49
Kryshtafovych, A., 377–378
Kubo, M., 179–180
Kudoh, S., 30–31, 31*t*
Kuemmerer, N., 375–376
Kuhlman, B., 85–89, 317–318
Kuhn, B., 349–350
Kuhn, M., 64
Kuipers, R. K., 374–375
Kukurba, K. R., 374–375
Kulkov, V., 385*t*
Kullback, S., 319–320, 323
Kumar, C., 28
Kumar, M., 97
Kumar, P., 181, 183–184, 372–374
Kumar, R., 385*t*
Kumar, S., 259–260, 374–375, 385*t*

Kumazawa, S., 26
Kunkel, R., 368–370
Kuntz, I. D., 384–390
Kuo, Y. C., 349, 357–358
Kurien, R., 26
Kuriyan, J., 368–370, 375–376
Kurono, H., 26
Kurz, A., 56*t*, 62–63, 64
Kuznetsov, D., 384–390
Kuznetsov, I. B., 94
Kwok, P. Y., 366–368

L

Labesse, G., 385*t*
Lackner, P., 133*t*, 137, 142–143, 147
LaConte, L. E. W., 394–396
Laezza, F., 61
Lafferty, J., 162–163
Lagache, S., 4–5, 14
Lagorce, D., 380–382
Lahti, J. L., 384–390
Lai, X., 368–370, 372–374
Laine, R. A., 368–370
Laio, A., 307–308
Lajoie, B. R., 84
Lam, H., 8*t*, 9–10
Lan, T. H., 53–54
Landar, A., 23–24
Landgraf, A., 348, 349, 357–358
Landua, J., 375–376
Lane, C. S., 45
Lange, O. F., 349–350
Langlois, R., 94
Larkin, M. A., 379–380
Lasagna, L., 380–382
Lasker, K., 84–85
Laskowski, R. A., 78, 322, 379–380, 385*t*
Lasters, I., 148–149, 165
Lathe, W., 259–260
Lathe, W. III., 366–368, 371–372, 376–377
Lau, A., 20–21
Launay, G., 385*t*
Laurila, K., 368–370
Laval, F., 22–23
Laval, J., 22–23
Lavecchia, A., 380–382
Law, V., 93–94, 385*t*
Lawrence, J. B., 84
Lazar, A., 368–370
Le, Q., 128–130
Le Grand, S., 350–351
le Marechal, C., 368–370
Le Van Kim, C., 128–130
Leach, A. R., 293–294, 295–299, 306–307, 384–390
Leaver-Fay, A., 349–350, 379–380
Lee, B., 148–149
Lee, C., 379–380
Lee, C. G., 372–374
Lee, C. Y., 53–54
Lee, H. S., 83–84, 385*t*
Lee, J., 391–392
Lee, K. H., 31*t*, 42–43
Lee, M. C., 299–300, 350–351
Lee, P. H., 183, 372–374
Lee, S. B., 31*t*, 374–375, 391–392
Lee, T. Y., 375–376
Lee, V. M., 49–50, 67
Lee, Y. I., 29
Lenhof, H. P., 385*t*
LeGall, T., 316–317, 375–376
Lehmann, S., 51*t*, 62–63
Lehväslaiho, H., 374–375
Leibler, R. A., 319–320, 323
Leichert, L. I., 24, 28
Lengauer, T., 384–390
Lenhard, B., 179–180, 183
Lensink, M. F., 89, 385*t*
Leo-Macias, A., 156–157, 159*t*, 165
Leonard, S. E., 27
Lepper, H. S., 55
Lerner, M. G., 385*t*
Leroy, B. P., 178–179
Lesk, A. M., 122, 123, 155–156, 158, 159*t*, 165, 377–378
Lesk, V. I., 86*t*, 89
Leslie, C., 97
Leslin, C. M., 374–375
Lesueur, F., 259–260
Lesyng, B., 132, 133*t*, 137
Lethier, M., 341–342
Levan, D., 385*t*
Leverenz, J. B., 45, 48–49
Levin, Y., 56*t*

Levitt, M., 122–123, 124–128, 125*t*, 133*t*, 137, 138–139, 141–142, 317–319, 377–378, 379–380
Levonen, A. L., 23–24
Levy, S., 233–259
Levy, S. E., 42–43
Lewis, S. M., 349–350
Li, B., 372–374
Li, C., 262–263, 376–377
Li, H., 385*t*
Li, J., 30
Li, K., 233–259
Li, L., 279
Li, M., 5, 8*t*, 379–380
Li, N., 12
Li, Q., 11–12, 366–368
Li, S., 168
Li, T., 79, 98–99
Li, W. W., 178–179, 379–380, 385*t*
Li, X. J., 8*t*, 12–13
Li, Y. H., 12, 97, 379–380
Li, Z., 83–84
Liang, C., 5, 8*t*
Liang, H., 79, 92, 97, 98–99, 108
Liang, J., 368–370, 385*t*
Liang, J. J., 50, 51*t*, 60–61
Liang, S., 379–380
Liao, L. J., 45–46
Liao, W., 379–380
Liao, Y., 379–380
Licht, R. W., 53
Lichtarge, O., 366–368
Lidow, M. S., 66
Lie, G., 274–275
Light, J., 368–370
Lii, J. H., 306–307
Lill, H., 341–342
Lim, C., 79, 94–95, 98–99, 124–126, 128–130, 368–370
Lim, S. J., 376–377
Lin, B., 94
Lin, C.-K., 79, 98–99, 100
Lin, C.-W., 79, 98–99
Lin, D., 23
Lin, J. H., 98, 372–374, 385*t*
Lin, K., 31–32, 31*t*, 376–377
Lin, Y. J., 183, 372–374, 385*t*
Lin, Z., 323
Linard, B., 374–375
Lind, M. J., 31*t*, 32
Lindahl, E., 323, 379–380
Lindahl, M., 24
Lindbjer, M., 42–43
Lindemann, C., 28
Linding, R., 372–374, 375–376
Lindorff-Larsen, K., 275–276
Lindstrom, W., 385*t*
Linke, K., 25
Lipinski, C. A., 380–382
Lipman, D. J., 379–380
Liu, C. H., 183, 372–374
Liu, C. K., 183, 372–374
Liu, H., 4–5, 14
Liu, J. P., 31–32, 31*t*, 379–380
Liu, P., 93–94, 385*t*
Liu, S., 98–99
Liu, T., 4, 377–378, 384–390, 385*t*
Liu, X., 126, 130, 385*t*
Liu, Z., 26, 79, 98–99
Ljung, E., 56*t*, 62
Lobanov, M. Y., 375–376
Lohmann, E., 368–370
London, N., 385*t*
Long, E. K., 21, 23
Long, F., 12
Lopez, M. F., 30
Lopéz-Blanco, J. R., 85–89
Lopez de la Paz, M., 375–376
López-Sánchez, L. M., 28–29
Lorenz, H., 368–370, 375–376
Loscalzo, J., 23, 28–29
Lott, M. T., 374–375
Love, S., 368–370, 375–376
Lovell, S. C., 391
Lovestone, S., 63–64
Lu, B., 366–368
Lu, H., 79, 83, 94, 168
Lu, L., 79, 83
Lu, S., 158, 159*t*
Lu, X.-J., 78–79, 98–99, 100–102
Lu, Y., 385*t*, 394–396
Lubich, C., 278
Lui, L., 25
Lundberg, Y. W., 368–370
Lundegaard, C., 376–377
Luo, R., 350–351, 394–396

Luo, W. J., 30–31
Luo, X., 385*t*
Lupas, A. N., 379–380
Lupyan, D., 156–157, 159*t*, 165
Luque, F. J., 295–299
Luscombe, N. M., 94, 99–100
Lüthy, R., 379–380
Luu, P., 53–54
Luu, T. D., 374–375
Ly, S., 93–94, 385*t*
Lynham, S., 42–43
Lyskov, S., 86*t*
Lysova, M. V., 179–180

M

Ma, B., 5, 8*t*, 368–370
Ma, D., 61
Ma, J., 123–124, 130–132, 133*t*, 137–138, 144, 168
Ma, W., 62
MacArthur, M. W., 379–380
MacCallum, J. L., 316, 340
MacCallum, R. M., 379–380
Maccarrone, G., 46, 47–49, 50, 51*t*, 56*t*, 60, 66–67
MacCoss, M. J., 8*t*, 9, 11–12
Machius, M., 317–318
Macindoe, G., 86*t*
MacKerell, A. D. Jr., 294–295, 296*t*, 302, 303–304, 394–396
Mackereth, C. D., 84
Madden, T. L., 97, 128–130, 147, 379–380
Madej, T., 122–123, 132, 133*t*, 148–149, 259–260, 368–370
Madera, M., 49–50
Madhusudhan, M. S., 83, 100–102, 379–380
Madian, A. G., 21, 27
Madl, T., 84
Madrid-Aliste, C. J., 379–380
Madura, J. D., 284–285
Magariños, M. P., 385*t*
Magesh, R., 394–396
Maginn, E. J., 393–394
Maglott, D. R., 374–375, 377–378
Magrane, M., 374–375
Magyar, C., 316–317, 372–374
Mah, J. T. L., 371–372
Mahajan, S., 124–126, 128–130
Mahfouz, M., 348–349, 350–353
Mainz, D. T., 384–390
Mairesse, J., 56*t*, 60
Mak, A. N., 348–349
Maki, R. A., 368–370
Makino, S., 384–390
Malinowska, K., 30–31
Man, O., 375–376
Manasa, P., 91*t*
Manchia, M., 50–53, 65
Manetti, C., 368–370, 375–376
Mann, M., 2, 8*t*
Mann, U., 368–370, 375–376
Mannick, J. B., 25
Manning, G., 374–375
Mannucci, P. M., 368–370
Manoukian, S., 368–370
Manral, P., 64
Manson, F. D., 178–179
Mantsyzov, A. B., 385*t*
Manuel A., 78–79, 83, 84–85
Mao, Y., 13–14
Maple, J., 304
Marchese, J. N., 45
Marchewka, D., 321, 323
Marchler-Bauer, A., 122, 148–149, 155–156
Marialke, J., 385*t*
Marie, J., 366–368
Marin Lopez, 78–79, 83, 84–85
Marinescu, V. D., 371–372
Mark, A. E., 305, 323
Markley, J. L., 326
Marknell, T., 178–179
Markwick, P. R. L., 394–396
Marrocco, J., 56*t*, 60
Martell, K. J., 366–368
Martelli, P. L., 181, 182, 183–184, 258–259, 372–374
Marti-Arbona, R., 377–378
Martí-Renom, M. A., 79, 83, 100–102, 377–378, 379–380
Martin, J., 56*t*, 64
Martin, L., 385*t*
Martin, M. J., 374–375
Martin, P., 56*t*, 62
Martinou, J. C., 30
Martins, I. C., 375–376

Martins-de-Souza, D., 47–50, 51*t*, 55, 56*t*, 61, 65, 66–67
Martonák, R., 307–308
Martyna, G. J., 283–284, 394–396
Martz, E., 93–94, 182, 376–377
Mashiach, E., 86*t*, 385*t*
Masso, M., 366–368, 372–374
Massova, I., 349–350
Mászáros, B., 316–317
Mata-Cabana, A., 24
Mathe, E., 372–374
Matic, I., 8*t*
Matsuda, H., 126–128
Matsuda, K., 179–180
Matsumoto, K., 26
Matsuyama, S., 368–370
Matthews, L. R., 78–79
Mattson, W., 278–279
Mattsson, N., 54
Maurer, M. H., 42–43, 44
Maurer-Stroh, S., 366–368, 375–376
Mavridis, L., 86*t*
Maxwell, D. S., 394–396
May, P., 132, 133*t*
Mayo, S. L., 296*t*, 304
Mayrose, I., 376–377
Mazeau, K., 300
Mazza, T., 374–375
McAdams, H. H., 99–100
McBride, C., 261–262
McCallum, A., 162–163
McCammon, J. A., 126–128, 271, 273, 275–276, 277, 366–368, 382, 393–396
McCarter, S. L., 366–370
McCarthy, S. E., 53–54
McCarty, N. A., 259–260
McClellan, J. M., 53–54
McConkey, B. J., 384–390, 385*t*
McConnell, R., 366–368
McCormack, A. L., 7, 8*t*
McDonnell, L. A., 64–65
McGann, M. R., 384–390
McGee, T. L., 366–368
McGettigan, P. A., 379–380
McHaourab, H. S., 385*t*
McIntosh, M., 45, 48–49
McLeod, H. L., 372–374
McMichael, J. F., 385*t*
McNeil, P., 372–374
McWilliam, H., 379–380
Mechref, Y., 49–50
Meckler, J. F., 349
Meghah, V., 55, 56*t*, 60
Meier, A., 372–374
Meiler, J., 385*t*
Meissner, G., 27–28
Melo, F., 79, 169–170, 377–378
Melov, S., 30
Meng, E. C., 101*f*, 385*t*
Meng, X. Y., 390, 391
Menke, M., 122–123, 132, 138–139, 143, 147–148, 159*t*, 165
Mennucci, B., 317–318
Menon, S., 394–396
Mercer, A. C., 348
Merchant, N. B., 42–43
Meri, S., 21
Merz, K. M. Jr., 90, 299–300, 306, 394–396
Mester, N., 385*t*
Mészáros, B., 316–317
Meunier, I., 178–179
Meyer, E. F. Jr., 377–378
Mezei, M., 390, 391
Mi, H., 181, 182, 183–184, 372–374
Miao, Q. X., 25
Micheletti, C., 133*t*, 137, 156–157, 159*t*
Michie, A., 130–132, 168
Michielin, O., 385*t*
Michon, A., 54
Mihala, N., 27
Mihasan, M., 393
Miley, M. J., 317–318
Milfay, D., 60–61
Miller, I., 21
Miller, J. H., 368–370
Miller, J. L., 295–299
Miller, M. P., 259–260
Miller, P. J., 261–262
Miller, R. T., 379–380
Miller, W., 97, 128–130, 147, 379–380
Millholland, J. M., 366–368
Milli, A., 30–31, 31*t*
Mills, J. L., 317–318
Minin, V. N., 10
Minkler, P. E., 26
Mintseris, J., 85–89, 86*t*, 90–92, 94–95

Miriyala, S., 33
Mishmar, D., 374–375
Misura, K. M., 379–380
Mitchell, J. B., 22–23
Mitchell, P. B., 50–53, 51*t*
Miteva, M. A., 368–370, 380–382, 394–396
Mitros, T., 379–380
Miyazawa, S., 80
Mizuguchi, K., 165
Mizuno, Y., 179–180
Mlynarcik, P., 300
Moal, I. H., 79, 85–89, 86*t*, 90–92
Mocarelli, P., 30–31, 31*t*
Moereels, H., 270
Mohabatkar, H., 368–370
Molina, H., 45–46
Moll, A., 385*t*
Moniatte, M., 4–5, 14
Monroe, M. E., 8*t*, 10
Montaner, D., 262–263, 368–370, 376–377
Montes, M., 380–382
Monticelli, L., 271–272
Mooney, S. D., 259–260, 366–370, 374–376
Moont, G., 90–92
Moosawi, F., 368–370
Moreira, I. S., 393
Mori, Y., 30–31, 31*t*
Morris, G. M., 384–390, 385*t*
Morris, M. B., 377–378
Mort, M., 179–180, 183, 366–368, 374–375
Mort, M. E., 368–370, 372–374, 375–376
Mortensen, O. H., 122, 155–156
Mosca, R., 78–79, 83–84, 100–102, 133*t*, 169
Moscou, M. J., 348
Moskal, J. R., 61
Moss, D. S., 379–380
Mottagui-Tabar, S., 179–180
Moughon, S., 85–89
Moulin, S., 368–370
Moulinier, L., 374–375
Moult, J., 89, 366–370, 372–374, 379–380
Mowbray, S. L., 126–128
Mroginski, M. A., 393–394
Muegge, I., 385*t*
Mukamel, S., 316–317
Mukherjee, S., 385*t*
Mukhyala, K., 374–375
Müller, G., 60–61
Müller, H. W., 368–370
Muller, J., 22–23, 374–375
Mumby, M., 8*t*, 10
Munnich, A., 366–368
Munns, G., 183
Munro, S., 273
Munroe, P. B., 376–377
Muntané, J., 28–29
Muradov, H., 326
Muramatsu, T., 300
Murate, T., 368–370
Murphy, G. S., 317–318
Murphy, J. A., 368–370
Murphy, R. B., 384–390
Murray, C. W., 380–382, 385*t*
Murray, S. S., 45, 366–368
Muruganujan, A., 182, 183–184, 372–374
Murzin, A., 148–149, 168
Myers, E. W., 379–380
Mysinger, M. M., 385*t*

N

Nabuurs, S. B., 385*t*
Nadimpalli, S., 143, 147–148
Nagaraj, N., 8*t*
Nagasu, T., 12
Nagasundaram, N., 374–375, 377–378, 394–396
Nah
Nair, A., 394–396
Nairn, A. C., 40–41
Naismith, S. L., 50–53
Nakagome, K., 53–54
Nakajima, T., 179–180
Nakano, T., 179–180
Nakayama, T., 26
Narang, P., 379–380
Nasr, H. B., 30–31, 31*t*
Nathans, J., 178–179
Naujokat, C., 49–50
Ndiaye, S., 28
Neduva, V., 372–374, 375–376
Nelson, W. D., 394–396
Nesvizhskii, A. I., 8*t*, 9, 10–12, 14
Neubert, T. A., 45–46

Neufeld, E. F., 366–368
Nevett, C. L., 178–179
Newell, J. A., 261–262
Neya, S., 368–370
Ng, P. C., 181, 183–185, 211–213, 233–259, 366–368, 371–374
Nicholls, A., 384–390
Niciu, M. J., 50
Nicolaes, G. A. F., 394–396
Nicoletti, F., 56*t*, 60
Nicoll, J. A., 368–370, 375–376
Nielsen, J. E., 90
Nielsen, M., 376–377
Nieuwenhuis, H. K., 368–370
Nikki, E., 23
Nilges, M., 84
Nilsson, C. L., 42–43, 56*t*, 62
Nimrod, G., 97
Ninokata, A., 179–180
Ninomiya, T., 179–180
Nishikawa, K., 374–375
Nithin, C., 91*t*
Nithipongvanitch, R., 33
Nitsch, R. M., 368–370, 375–376
Noble, W. S., 8*t*, 9, 10, 102–105
Noel, T., 33
Nolen, W. A., 53
Norel, R., 83, 100
Norris, J., 368–370
North, B. V., 371–372
Norton, S., 385*t*
Norwood, T. H., 368–370
Nose, O., 368–370
Notredame, C., 379–380
Nourbakhsh, I., 213–220, 376–377
Nov, Y., 124–126, 125*t*
Novikov, F. N., 385*t*
Novikova, S. I., 66
Novotny, M. V., 12, 49–50
Nussinov, R., 83–84, 86*t*, 123, 133*t*, 137, 156–157, 158–161, 159*t*, 165, 169, 368–370, 385*t*
Nuytemans, K., 374–375

O

Oberley, T., 33
O'Connor, T. D., 366–368
Oda, Y., 12, 45–46
Ode, H., 368–370
O'Donovan, C., 374–375
O'Donovan, M. C., 179–180
Offmann, B., 122–123, 124–126, 128–130, 132, 133*t*
Oh, J. Y., 27
Oh, K. S., 368–370
Oh, P., 12
Okpechi, I., 374–375
Okur, A., 299
Olafson, B. D., 84, 90, 296*t*, 302, 304
Olatubosun, A., 261–262, 372–375
Oliva, B., 78–79, 80, 81, 83, 84–85, 90–92, 93, 94–95, 100–102, 105
Olivella, R., 78–79, 83–84
Oliver, S. G., 391
Olivier, M., 372–374
Olmea, O., 137, 142–143
Olsen, J. V., 8*t*
Olsen, O. H., 122, 155–156
Olson, A. J., 384–390, 385*t*
Olson, W. K., 78–79, 98–99, 100–102
Olsson, J. E., 366–368
Omiecinski, C. J., 366–368
Ong, S. E., 4, 45–46
Onishi, A., 98
Oobatake, M., 368–370, 374–375
Oostenbrink, C., 277, 305
Opdenakker, G., 300
Orengo, C. A., 130–132, 133*t*, 137, 143–144, 168, 385*t*
Orland, H., 156–157, 159*t*
Orozco, M., 85–89, 98, 295–299, 368–370, 372–374, 376–377
Orr, M., 24
Ortiz, A. R., 137, 142–143, 156–157, 159*t*, 165
Ortiz, M. A., 368–370
Ostrovsky, D., 341–342
O'Sullivan, O., 379–380
Ota, M., 374–375
Otten, R., 341–342
Ou, K., 46–47
Ousterout, D. G., 348
Ouyang, Z., 385*t*
Ovadia, R., 349
Overington, J. P., 165

Ozbabacan, S. E. A., 368–370
Ozbek, P., 94–95
Ozkan, S. B., 375–376

P

Pabo, C. O., 101*f*
Pacchiarotti, I., 53
Pace, C. N., 368–370, 375–376
Pain, R. H., 126–128
Pajot-Augy, E., 385*t*
Pak, Y., 382
Palmieri, M., 30–31, 31*t*
Pampin, S., 179–180
Pan, X., 348–349, 350–353
Panchenko, A., 122, 155–156, 259–260, 368–370
Pandey, A., 4, 45–46
Pandini, A., 124–126, 125*t*
Pandit, S. B., 93–94
Panis, C., 29, 31*t*, 32
Panjkovich, A., 79, 85–89
Pantoja-Uceda, D., 262–263, 368–370, 376–377
Paoletti, A. C., 12
Papadopoulos, G. E., 86*t*
Papandreou, N. C., 375–376
Pappin, D. J. C., 7, 8*t*
Pappu, R. V., 343
Pariante, C. M., 65
Parisi, G., 377–378
Parisien, M., 92
Park, B. H., 126–128
Park, H., 391–392
Park, J. S., 31*t*
Park, K. J., 368–370
Park, S. J., 30
Park, S. K., 45–46
Parker, K., 45
Parmely, T. J., 12
Parnetti, L., 54
Parone, P. A., 30
Parr, R. L., 30
Parraga, I., 372–374
Parrinello, M., 307–308, 393–394
Parthiban, V., 372–374
Paschalidis, I. C., 89
Pasek, D., 27–28
Pasquali, C., 46–47
Pastinen, T., 179–180, 366–368
Patel, R. Y., 377–378
Patterson, S. D., 4
Patulea, C., 93–94
Pauling, L., 124–126
Paulson, L., 56*t*, 62
Paz, I., 376–377
Pearson, W. R., 379–380
Pedersen, J. S., 376–377
Pedersen, L., 279
Pedersen, P. L., 368–370
Pedretti, A., 385*t*
Peng, C. S., 316–317
Peng, J., 123–124, 130–132, 133*t*, 137–138, 144, 155–157, 159*t*, 161, 168
Pennington, K., 65, 66
Perego, P., 31*t*
Pereira, F., 162–163
Perera, L., 279
Pérez, A. G., 79, 98–99
Perez, M. S., 86*t*
Pérez, S., 300
Pérez-Cano, L., 79, 91*t*, 92, 94–95
Perez-Iratxeta, C., 374–375
Perez-Pinera, P., 348
Pericak-Vance, M. A., 372–374
Perkins, D. N., 7, 8*t*
Perneczky, R., 56*t*, 62–63, 64
Perrett, D. I., 53–54
Persson, A., 56*t*, 62
Pervouchine, D., 368–370
Peshkin, L., 181, 182, 183–184, 258–259, 372–374
Peskind, E., 44–45, 48–49
Peter, C., 394–396
Peters, G. H., 122, 155–156
Petersen, B., 376–377
Petersen, D. R., 26
Petersen, T. N., 376–377
Peterson, T. A., 374–375
Petrey, D., 83–84, 379–380
Petri, S., 23–24
Petritis, A. D., 4
Petrotchenko, E. V., 27–28
Petrukhin, K., 178–179
Petsko, G. A., 393–394
Pettersen, E. F., 101*f*, 385*t*
Petyuk, V. A., 4

Pevzner, P. A., 5, 8*t*, 11–12
Peyvandi, F., 368–370
Phan, L., 183, 374–375
Phillips, A. D., 183, 366–368, 374–375
Phillips, J. C., 394–396
Phillips, J. L., 84
Pich, E. M., 48–49
Pickering, E., 64
Pieper, U., 83–84, 259–260, 368–370
Pierce, B., 85–89, 86*t*, 90–92, 94–95
Pierce, K., 368–370
Pierce, S. B., 53–54
Pierce, W. M., 63–64
Pierri, C. L., 377–378
Pineau, C., 30–31, 31*t*
Pineda-Lucena, A., 262–263, 368–370, 376–377
Pique, M. E., 86*t*
Pitchot, W., 50
Pitera, J., 271
Pitre, S., 93–94
Piwowar, M., 323
Pizzatti, L., 29, 31*t*, 32
Plabplueng, C. D., 33
Planas, J., 78–79, 93
Planas-Iglesias, J., 78–79, 83, 84–85
Plimpton, S., 279–281
Plotz, G., 377–378
Poch, M. T., 368–370
Poidevin, L., 374–375
Pollastri, G., 128–130, 375–377
Ponomarenko, M. P., 179–180
Ponomarenko, P. M., 179–180
Pons, C., 79, 85–89, 86*t*, 90–92
Pons, J. L., 385*t*
Pontiggia, F., 133*t*, 137
Poole, D., 350–351
Poole, J. C., 374–375
Porcelli, V., 377–378
Posch, S., 98
Poshusta, T. L., 348
Posner, M. I., 53–54
Postel, D., 300
Postma, J. P. M., 323
Postmes, L., 53–54
Poteete, A. R., 368–370
Potestio, R., 133*t*, 137
Pothier, J., 133*t*, 139–140, 169
Potter, M. J., 271
Potter, W. Z., 64
Poulain, P., 92
Pratt, W. B., 368–370
Preissner, R., 385*t*
Prévost, C., 86*t*, 92
Price, D. A., 380–382
Priebe, W., 61
Prieto, D., 40–41, 48–49
Prilusky, J., 375–376
Priyakumar, U. D., 317–318
Procaccio, V., 374–375
Proia, R. L., 366–368
Prokop, M., 368–370
Prokunina, L., 366–368
Proschak, E., 385*t*
Pruitt, K. D., 377–378
Prymula, K., 321, 322–323, 339–340, 343
Pu, S., 53–54
Puchades, M., 42–43
Puech, B., 178–179
Pujadas, G., 372–374
Pujol, R., 368–370
Pupko, T., 93–94, 182, 376–377

Q

Qian, B., 379–380
Qian, J., 98
Qian, P.-Y., 44
Qian, W. J., 4
Qiang, L., 83–84
Qu, B. H., 368–370
Qu, X., 377–378
Quinn, J. F., 44–45, 48–49

R

Rabinovici, G. D., 54
Radi, R., 23
Radivojac, P., 368–370, 375–376
Radulovic, D., 13–14
Ragg, S., 8*t*, 14
Raghava, G. P. S., 97
Raguenes, O., 368–370
Rahmoune, H., 47–48, 49–50, 55, 61
Rai, B. K., 379–380
Raiteri, P., 307–308
Rajaratnam, S., 394–396
Rajasekaran, R., 374–375, 394–396

Rajith, B., 377–378, 394–396
Ralph, S. A., 385*t*
Ramachandra, A., 23–24
Raman, S., 379–380
Ramensky, V. E., 181, 182, 183–184, 258–260, 366–368, 371–374, 376–377
Ramesar, R., 374–375
Ramoni, M. F., 371–372
Ramos, E., 27–28
Ramos, M. J., 385*t*, 393
Rana, S., 30–31
Randall, A. Z., 220–221, 372–374, 375–377
Randolph, T., 13–14
Ranish, J. A., 8*t*, 12–13
Rao, S., 50
Rao, V. S., 382–383
Rappsilber, J., 12
Rarey, M., 384–390
Rask-Andersen, M., 380–382
Rasmussen, B. F., 296*t*, 393–394
Raveh, B., 385*t*
Raychaudhuri, S., 366–368
Rebar, E. J., 105–106
Reboul, J., 78–79
Reckow, S., 46, 47–48, 60, 66–67
Reddy, B. R., 55, 56*t*, 60
Rees, D. C., 380–382
Rees, H. D., 64–65
Regnier, F. E., 21, 27
Reichel, E., 366–368
Reid, D., 384–390, 385*t*
Reinhardt, A., 377–378
Relkin, N., 42–43
Ren, J., 385*t*
Renault, N., 380–382
Renfrow, M. B., 26
Rennell, D., 368–370
Reppert, M. E., 316–317
Reumers, J., 366–368, 374–376
Reva, B., 372–374
Reyes, C., 349–350
Reynolds, C. H., 306
Ricart, K., 27
Rice, B. M., 278–279
Ridder, L., 380–382
Riemer, C., 374–375
Righetti, S. C., 31*t*
Rignall, T. R., 366–370
Rijksen, G., 368–370
Riley, T. R., 100
Ringe, D., 393–394
Rios, D., 183
Ripp, R., 374–375
Rippmann, F., 385*t*
Risch, N. J., 366–368
Rishishwar, L., 259–260
Rist, B., 4, 44–45
Ritchie, D. W., 85–89, 86*t*, 90, 385*t*
Riva, A., 371–372
Rizzo, R. C., 385*t*
Roberts, V. A., 86*t*
Robertson, D. L., 391
Robertson, T. A., 79, 92, 98–99
Robillard, R., 50–53
Robledo, M., 372–374
Robson, B., 126–128
Rödelsperger, C., 258–259, 261–262, 372–374
Rodgers, J. R., 26, 377–378
Rodriguez-Antona, C., 368–370
Rodríguez-Ariza, A., 28–29
Rodriguez-Casado, A., 368–370
Rodriguez-Rey, J. C., 179–180
Roe, D. R., 385*t*
Roede, J. R., 24
Røgen, P., 169
Roher, A. E., 42–43
Rohl, C. A., 85–89, 379–380
Roitberg, A., 299
Rolf, A., 183
Ronaghi, M., 372–374
Rooman, M., 372–374
Rooman, M. J., 126–128
Roos, D. S., 385*t*
Rose, G.D., 368–370
Rose, K., 27
Rose, N., 46–47
Rosén, E., 54
Rosenberg, J., 376–377
Rosenbloom, K., 376–377
Rosenström, P., 140–141
Rosing, J., 394–396
Ross, P. L., 45
Rossi, I., 181, 182, 183–184, 372–374
Rost, B., 83, 181, 182, 183–184, 258–259, 372–374, 377–378

Roterman, I., 319, 321, 322–323, 339–340, 343
Rothlisberger, U., 393–394
Rotkiewicz, P., 377–378
Rousseau, F., 374–376
Rousseau, J., 348
Roux, B., 368–370
Roy, A., 130–132, 168
Roy, K., 341
Ruas, M., 368–370
Rudd, P. M., 300
Rueda, M., 85–89, 295–299
Rueping, M., 394–396
Ruiz-Llorente, S., 372–374
Ruiz-Pesini, E., 374–375
Rullmann, C., 92
Ruoho, A. E., 326
Rush, A. J., 50
Russel, D., 84–85
Russell, R. B., 79, 83–84, 372–374, 375–376
Russo, D., 30–31, 31*t*
Rusu, A. M., 374–375
Ryckaert, J.-P., 283–284
Rydberg, E. H., 375–376
Ryu, G. M., 368–370
Ryu, J., 368–370, 385*t*
Ryu, S., 10, 13–14

S

Sadee, W., 380–382
Sadek, M., 273
Sadjad, B. S., 385*t*
Sadjad, S. B., 384–390
Sadowski, M. I., 168, 169
Sadus, R. J., 283–284
Sadygov, R. G., 4–5, 14
Sael, L., 86*t*
Saikali, S., 30–31, 31*t*
Sakamoto, M., 30–31, 31*t*
Saladin, A., 86*t*, 92
Sali, A., 79, 84, 259–260, 368–370, 372–374, 377–378, 394–396
Salles, F. T., 368–370
Salsbury, F. R. Jr., 394–396
Sambo, A. V., 63–64
Samollow, P. B., 182, 368–370, 372–374, 376–377
Samudrala, R., 379–380
Samuelson, S. O., 394–396
Sanchez, J. C., 27, 30–31, 31*t*, 46–47, 48–49
Sanchez, R., 384–390
Sánchez, R., 377–378
Sander, C., 122–123, 126, 130, 132, 133*t*, 137, 139, 140–141, 147–148, 158, 168, 169, 262–263, 372–374, 376–377, 379–380, 385*t*
Sandford, R., 366–368
Sanfiz, A., 368–370
Sanjana, N. E., 348
Sanner, M. F., 385*t*
Santos, A., 54
Santoyo, J., 372–374
Sanyal, A., 84
Sanz, G., 385*t*
Saqi, M. A. S., 376–377
Sarai, A., 97, 368–370, 374–375
Sarto, C., 30–31, 31*t*
Sato, S., 12
Sato, T., 12
Sattler, M., 84
Saunders, C. T., 259–260, 368–370, 371–372
Saville, G., 283–284
Savinkova, L. K., 179–180
Saxena, S., 55, 56*t*, 60
Sayle, R., 385*t*
Sayre, L. M., 23, 26
Scandolo, S., 307–308
Scantamburlo, G., 50
Scarpa, A., 30–31, 31*t*
Schaefer, C., 372–374
Schäffer, A. A., 97, 128–130, 147, 379–380
Scheeff, E. D., 158, 159*t*
Schildknecht, S., 22–23
Schioth, H. B., 380–382
Schleif, R., 276–277
Schleker, S., 78–79
Schlessinger, A., 83–84
Schlick, T., 394–396
Schlitt, T., 374–375
Schlüter, H., 47–48
Schmidt, R., 385*t*
Schmidt, S., 181, 182, 183–184, 258–259, 372–374
Schmidt-Goenner, T., 132, 133*t*

Schmitt, A., 47–49, 50, 51*t*, 56*t*
Schneider, G., 50, 385*t*
Schneider, S., 86*t*
Schneider, T., 133*t*, 169
Schneidman-Duhovny, D., 83–85, 86*t*, 169, 385*t*
Schnellmann, R. G., 22
Schoenrock, A., 93–94
Scholl, T., 182, 368–370, 372–374, 376–377
Scholze, H., 348, 357–358
Schomburg, D., 372–374
Schork, N. J., 366–370
Schornack, S., 348, 357–358
Schrag, M., 64–65
Schrempp, C., 42–43
Schrödinger, L., 100–102
Schroeder, M., 94, 385*t*
Schubert, K. O., 65–66
Schuchhardt, J., 12
Schueler-Furman, O., 85–89, 385*t*
Schuelke, M., 258–259, 261–262, 372–374
Schulten, K., 385*t*
Schushan, M., 97
Schwabe, J. W., 366–368
Schwanhausser, B., 12
Schwartz, C., 368–370, 394–396
Schwartz, C. E., 394–396
Schwartz, R. M., 145–146
Schwarz, J. M., 258–259, 261–262, 372–374
Schwede, T., 379–380
Schymkowitz, J., 374–376
Scipioni, A., 30–31, 31*t*
Scott, W. R., 305
Seebach, D., 394–396
Seeliger, M. A., 394–396
Seelow, D., 258–259, 261–262, 372–374
Seike, M., 30–31, 31*t*
Seiler, K. P., 385*t*
Selbach, M., 8*t*
Sénéchal, A., 178–179
Seno, F., 375–376
Serrano, L., 90, 375–376
Serretti, A., 55
Seshadri, M. S., 394–396
Seta, M., 368–370
Sethumadhavan, R., 374–375
Sethuraman, M., 22
Seyfried, N. T., 24
Shaffer, S. A., 13
Shaikh, S. A., 275–276
Sham, P. C., 371–372
Shamir, R., 78–79
Shan, Y., 394–396
Shang, Y., 368–370, 379–380
Shanmugam, D., 385*t*
Shanmugham, A., 341–342
Shao, M., 168
Shapiro, J., 148–149
Shapovalov, M., 379–380
Sharan, R., 78–79
Shariv, I., 85–89, 86*t*
Sharma, P., 64, 385*t*
Sharma, S. D., 384–390, 394–396
Sharpless, K. B., 380–382
Shastry, B. S., 258–259, 366–368
Shatkay, H., 183, 372–374
Shatsky, M., 123, 133*t*, 137, 156–157, 159*t*, 165
Shaw, D. E., 275–276, 394–396
Shaw, J. L., 8*t*, 10
Shaw, K., 374–375
Shen, M. Y., 79, 83, 84, 100–102, 379–380
Shen, Y., 89
Shen, Z., 10
Sheng, Q. H., 30
Shenoi, S., 26
Shenoy, S. R., 379–380
Shentu, Z., 86*t*
Shera, D. M., 64
Sherry, S. T., 183, 374–375
Shi, M., 48–49, 63–64, 66–67
Shi, T., 97
Shi, Y., 83–84
Shiel, J. A., 366–368
Shien, D. M., 375–376
Shihab, H. A., 372–374
Shindyalov, I. N., 133*t*, 137, 141, 142–143, 147, 148–149, 158, 159*t*
Shirley, B. A., 368–370, 375–376
Shoichet, B. K., 377–378, 385*t*, 394–396
Shonsey, E. M., 26
Shore, A. D., 65
Shore, S., 12
Showalter, S. A., 394–396
Shu, H., 8*t*, 10

Si, J., 94
Sibille, E., 55
Sicheritz-Ponten, T., 375–376
Sickmeier, M., 316–317, 375–376
Siddens, L. K., 368–370
Sidhu, J. S., 366–368
Siegfried, M., 183, 374–375
Siegfried, Z., 100–102
Siepel, A., 376–377
Sies, H., 20
Sievers, F., 379–380
Sikich, L., 53–54
Silkov, A., 83–84
Sim, S. C., 368–370
Simmerling, C. L., 299, 350–351
Simon, A., 384–390, 385*t*
Simon, B., 84
Simon, I., 100–102, 316–317, 372–374, 375–376
Simossis, V. A., 376–377
Simpson, M. A., 374–375
Sims, C. D., 65
Singh, G., 30
Singh, H., 385*t*
Singh, I., 46–47
Singh, T. P., 330–331
Sinha, P., 374–375
Sippl, M. J., 79, 80, 133*t*, 137, 142–143, 147, 169–170, 379–380
Sjöholm, F., 183
Skeel, R., 394–396
Skillman, A. G., 384–390
Skolnick, J., 79, 83, 84, 93–94, 97, 122–123, 133*t*, 137, 138–139, 141–142, 147–148, 168, 377–378, 385*t*
Skorodumov, C., 94–95
Skovronsky, D. M., 67
Slater, A. W., 169–170
Slebos, R. J., 8*t*, 10
Sligar, S. G., 393–394
Smaïl-Tabbone, M., 385*t*
Smigielski, E. M., 183, 374–375
Smith, E. P., 366–368
Smith, L. M., 45–46
Smith, M. A., 22
Smith, M. R., 27
Smith, R. D., 8*t*, 10, 385*t*, 391
Smith, S. K., 179–180
Sno, H. N., 53–54
Soares, H. D., 64
Sobolev, V., 384–390, 385*t*
Söding, J., 379–380
Sokolowska, E., 55
Sokolowska, I., 40–41
Solernou, A., 79
Soner, S., 94–95
Song, C. M., 376–377
Song, E. S., 368–370
Song, G., 169
Song, J., 326, 374–375
Song, P., 368–370
Song, X. Q., 262–263, 376–377
Soos, M. A., 366–368
Sørensen, M. R., 307
Sosnick, T. R., 92
Soto, C. S., 379–380
Soumphonphakdy, C., 368–370
Sousa, S. F., 385*t*
Souza, J. M., 23
Speck, J., 368–370
Specker, B., 366–368
Spellman, D. S., 45–46
Sperandio, O., 380–382
Spicket, C. M., 26
Spies, N. C., 385*t*
Spirin, S. A., 78–79
Splawski, I., 366–368
Spouge, J., 122–123, 132, 133*t*, 148–149
Squassina, A., 50–53, 65
Srajan, J., 374–375, 394–396
Srinivas, K., 382–383
Srinivasan, A., 64
Srinivasan, N., 122–123, 130, 132, 133*t*
Stamatoyannopoulos, J. A., 102–105
Stamler, J. S., 25
Standley, D. M., 379–380
Stanley, J., 14
Stanssens, P., 368–370
Starker, C. G., 348
States, D. J., 84, 90
Stawowczyk, E., 339–340
Steen, H., 4, 45–46
Steen, M., 394–396
Stegmaier, P., 79, 98–99, 100
Steigemann, W., 330–331
Stein, A., 78–79, 82, 83–84, 85–89

Stein, S. E., 8*t*, 9–10
Steinke, T., 132, 133*t*
Stelzhammer, V., 61
Stenger, J. E., 372–374
Stenson, P. D., 183, 258–259, 366–368, 372–375
Stephens, M., 11–12
Sternberg, M. J. E., 85–89, 86*t*, 90–92, 130–132, 379–380, 385*t*
Stevanin, G., 368–370
Stitziel, N. O., 368–370
Stock, A. M., 393–394
Stockfisch, T. P., 304
Stockwell, T. B., 233–259
Stoddard, B. L., 348–349
Stormo, G. D., 99–100, 105, 372–374
Strauss, C. E., 137, 142–143, 379–380
Streett, W., 283–284
Streubel, J., 349, 357–358
Strey, A., 285
Strockbine, B., 299
Stroganov, O. V., 385*t*
Stroher, E., 24, 27
Stroylov, V. S., 385*t*
Stryer, L., 270, 276, 368–370
Stuart, A. C., 377–378
Stuckey, P. J., 123, 158, 159*t*, 165
Stühmer, W., 65–66
Stützle, T., 385*t*
Subbiah, S., 379–380
Subramanian, J., 384–390, 394–396
Sudandiradoss, C., 374–375
Sudhakar, S. R., 55, 56*t*, 60
Sugiura, W., 368–370
Suhre, K., 379–380
Sultana, R., 54, 56*t*, 62–63
Sun, C. W., 53–54
Sun, H., 178–179
Sun, Z., 20–21, 374–375
Sunyaev, S., 258–260, 366–368, 371–374, 376–377
Superti-Furga, G., 368–370
Suresh, C. V., 384–390
Sushma, B., 384–390
Sussulini, A., 65–66
Suzek, T. O., 385*t*
Swaab, D. F., 64–65
Swaminathan, S., 84, 90, 302
Swanson, R., 377–378
Swapna, L. S., 124–126, 128–130
Sweredoski, M. J., 220–221, 375–377
Swindells, M., 130–132, 168
Szabó, Z., 60–61
Szego, E. M., 56*t*, 60–61
Szilágyi, A., 97
Szyperski, T., 317–318

T

Tabata, T., 12
Tabb, D. L., 8*t*, 10
Taboureau, O., 380–382
Tai, C. H., 149
Taillon-Miller, P., 366–368
Tajkhorshid, E., 275–276, 394–396
Takahashi, H., 366–368
Takamiya, O., 368–370
Takano, K., 394–396
Takeda, T., 79, 92
Takeshita, H., 179–180
Talamo, F., 30–31, 31*t*
Talavera, D., 90–92
Tan, C., 341–342
Tan, Y., 341–342
Tanaka, K., 368–370
Tanaka, S., 368–370
Tang, C. L., 379–380
Tang, G. W., 55, 56*t*, 60, 377–378, 384–390
Tang, H., 12
Tang, K. E. S., 368–370
Tang, X., 23
Taniguchi, F., 126–128
Tanner, S., 8*t*, 10
Tantos, A., 316–317, 375–376
Targosz, B. S., 48–49
Tatsuda, E., 26
Tatsumi, K. I., 368–370
Tatusova, T., 377–378
Taurines, R., 40–41
Taverner, T., 14
Tavtigian, S. V., 182, 259–260, 368–370, 372–374, 376–377
Taylor, R. D., 384–390, 385*t*
Taylor, W. R., 122, 133*t*, 137, 143–144, 168, 169, 376–377
Te Beek, T. A., 374–375
Téletchéa, S., 385*t*

Tello, D., 27–28
Ten Eyck, L. F., 86*t*
Teng, S., 259–260, 368–370, 394–396
Tennessen, J. A., 366–368
Tessier, M. B., 300, 301*t*
Tezel, G., 21
Thambisetty, M., 42–43
Thaminy, S., 8*t*, 14
Therrien, E., 385*t*
Theuns, J., 374–375
Thiessen, P. A., 385*t*
Thomas, D., 374–375
Thomas, D. D., 394–396
Thomas, D. J., 259–260, 368–370
Thomas, L. R., 4
Thomas, M., 368–370
Thomas, N. S., 183, 366–368
Thomas, P. D., 182, 183–184, 372–374
Thomas, P. J., 368–370
Thomas, R., 366–368
Thome, J., 40–41, 47–48, 62–63, 66–67
Thompson, D. R., 53–54
Thompson, E. E., 86*t*
Thompson, J., 349–350, 379–380
Thomsen, R., 385*t*
Thomson, R., 372–374
Thongboonkerd, V., 42–43, 44
Thornton, J. M., 78, 93–94, 99–100, 130–132, 168, 379–380
Thu, C. A., 83–84
Thusberg, J., 259–260, 261–262, 368–370, 372–375
Tian, J., 372–374
Tieleman, D. P., 271–272
Tietze, S., 385*t*
Tiffin, N., 374–375
Tildesley, D. J., 283–284, 293–294
Ting, C., 258–259
Tirado-Rives, J., 394–396
Tironi, I. G., 305
Tjioe, E., 84–85
Tjong, H., 94–95
Tkaczuk, K. L., 379–380
Tokmakoff, A., 316–317, 343
Toledano, M. B., 28
Tolić, N., 8*t*, 10
Tolkacheva, T., 368–370
Tomalik-Scharte, D., 368–370
Tomasi, J., 317–318
Tomomori-Sato, C., 12
Tompa, P., 375–376
Tonack, S., 45
Tong, J. C., 376–377
Topf, M., 84
Topol, E. J., 366–368
Torchala, M., 79, 86*t*, 89, 90–92
Torkamani, A., 368–370
Torrey, E. F., 65
Tosatti, E., 307–308
Tosatto, S. C., 375–376
Totrov, M., 94–95, 384–390, 385*t*
Touchon, J., 51*t*, 62–63
Tournamille, C., 128–130
Tovchigrechko, A., 86*t*, 385*t*
Tremblay, J. P., 348
Tress, M. L., 379–380
Treuting, P. M., 368–370
Trevino, S. R., 375–376
Tripathi, M., 64
Trivedi, R., 64–65
Trojanowski, J. Q., 49–50, 67
Trombley, L., 259–260
Trott, O., 385*t*
Trouve, P., 368–370
Trovato, A., 375–376
Trujillo, J. P., 368–370
Tsai, J., 377–378
Tsai, T. Y., 385*t*
Tschampel, S. M., 300, 301*t*
Tseng, J., 385*t*
Tseng, S. W., 375–376
Tseng, W. H., 183, 372–374
Tseng, Y. Y., 368–370
Tsolis, A. C., 375–376
Tsuchida, K., 368–370
Tsunenari, T., 178–179
Tsung, E. F., 371–372
Tuckerman, M. E., 283–284, 394–396
Tuffery, P., 124–126, 125*t*
Tuncbag, N., 83–84
Tung, C.-H., 124–130, 125*t*, 133*t*, 139–140, 169
Turck, C. W., 47–49, 50, 51*t*, 56*t*, 65
Turecek, F., 4, 44–45
Turner, D., 100–102, 106
Turpaz, Y., 385*t*

Tusnády, G. E., 372–374
Tuszynska, I., 79, 92
Tyagi, E., 259–260
Tyagi, M., 122–123, 130, 132, 133*t*
Tyers, M., 2
Tyka, M., 349–350

U

Uchida, K., 26
Uedaira, H., 368–370, 374–375
Ueki, M., 179–180
Uhlenbeck, O. C., 368–370
Ulitsky, I., 78–79
Ulrich, V., 22–23
Umlauf, E., 47–48, 56*t*, 62–63, 66–67
Ung, M. U., 366–368
Ungard, R., 30
Urbic, T., 317–318
Urday, S., 61
Urnov, F. D., 105–106
Utesch, T., 393–394
Uversky, V. N., 316
Uzun, A., 374–375

V

Vacic, V., 316–317, 375–376
Vaglio, P., 78–79
Vaisman, I. I., 366–368, 372–374
Vajda, S., 78–79, 85–89, 86*t*, 90, 385*t*
Vakili, P., 89
Vakser, I. A., 85–89, 86*t*
Valadié, H., 128–130
Valdar, W. S. J., 93–94
Valencia, A., 379–380
Valerio, M., 368–370, 375–376
Valiaho, J., 372–374
Van Broeckhoven, C., 374–375
van Buuren, A. R., 323
van der Spoel, D., 323
van der Stel, J., 53–54
van Dijk, M., 86*t*, 91*t*, 92
van Drunen, R., 323
Van Durme, J., 366–368, 375–376
van Eijck, L., 341–342
van Gunsteren, W. E. F., 394–396
van Gunsteren, W. F., 270, 277, 284–285, 296*t*, 305, 323
van Heel, A. J., 341–342
Van Nhien, A. N., 300
van Solinge, W. W., 368–370
Van Walle, I., 148–149, 165
van Wijk, R., 368–370
Vanattou-Saifoudine, N., 47–48, 49–50, 55, 61
Vance, J. M., 372–374
Vander Velden, K., 376–377
Vanderklish, P., 45–46
VanDyke, J. E., 368–370
Vanhee, P., 366–368, 375–376
VanRaay, T. J., 366–368
Vaquerizas, J. M., 372–374
Varadarajulu, J., 48–49
Varani, G., 79, 92, 98–99
Vargas, G., 48–49
Varghese, N., 259–260
Vasconcelos, A. T., 79, 98–99
Vatsis, K. P., 366–368
Vayalil, P. K., 27
Vazquez, F., 368–370
Veenhoff, L. M., 84–85
Veenstra, T. D., 40–41, 48–49
Veitinger, M., 62–63, 66–67
Velázquez-Muriel, J., 84–85
Velez, J. M., 33
Venclovas, C., 377–378
Vendrell, J., 375–376
Venisse, L., 368–370
Venkatesan, R. N., 368–370
Venkatraman, V., 86*t*
Venner, E., 366–368
Venselaar, H., 374–375
Ventura, S., 375–376
Vercauteren, F. G., 62
Vercoutter-Edouart, A. S., 56*t*, 60
Verdonk, M. L., 385*t*
Verkhivker, G., 368–370
Verlet, L., 274–275, 278–279
Vernon, J., 380–382
Vernon, R., 379–380
Verschoor, M. L., 30
Victorino, V. J., 29, 31*t*
Vihinen, M., 259–260, 261–262, 368–370, 372–375, 376–377
Villa, A., 305
Villa, E., 394–396
Villa, L., 385*t*

W

Weston, J., 8*t*, 9
Whisstock, J. C., 122, 123, 155–156, 158, 159*t*, 165
White, E. J. M., 385*t*
Whiteman, M., 20–21
Whittacker, J., 366–368
Wicker, B., 53–54
Wiederstein, M., 79, 169–170, 379–380
Wiehe, K., 85–89, 86*t*, 90–92
Wijte, D., 64–65
Wikström, H., 306–307
Wilke, R. A., 380–382
Wilkins, M. R., 46–47
Wilkinson, A., 270
Wilkinson, G. R., 380–382
Wilkinson, H., 375–376
Wilkinson, T., 270
Willett, P., 384–390
Williams, G. J., 377–378
Williams, K. L., 375–376
Williamson, B., 45
Williamson, M. J., 350–351
Wilm, A., 379–380
Wilson, L., 26
Windl, O., 368–370, 375–376
Wink, D. A., 22–23
Winter, J., 25
Wishart, D. S., 383
Witham, S., 394–396
Witkowska, H. E., 4
Wjst, M., 372–374
Wodak, S. J., 85–89, 126–128, 385*t*
Wolf, J., 12
Wolf, R. M., 296*t*, 299, 300–301, 385*t*, 394–396
Wolfenden, R., 368–370
Wolfson, H. J., 84, 86*t*, 123, 133*t*, 137, 156–157, 158–161, 159*t*, 165, 169, 368–370, 385*t*
Wolynes, P. G., 393–394
Won, C. I., 368–370, 385*t*
Wong, C. F., 393–394
Wong, H., 128–130
Wong, P. K., 20–21, 368–370, 375–376
Wood, K., 341–342
Woods, A. G., 40–41
Woods, R. J., 300
Wormald, M. R., 300
Worner, H. J., 394–396
Wright, J. D., 79, 94–95, 98–99, 368–370
Wrigley, J. D., 178–179
Wu, C. H., 299–300, 350–351, 377–378
Wu, J. R., 30, 258–259
Wu, N., 372–374
Wu, X., 348–349
Wyns, L., 148–149, 165

X

Xia, Q. C., 30, 64–65
Xia, Y., 259–260
Xiang, Z., 379–380
Xiao, J., 385*t*
Xiao, X., 394–396
Xiao, Z., 40–41, 48–49
Xie, G., 178–179
Xie, H., 23, 26, 27
Xie, L., 379–380, 385*t*
Xie, Z., 30, 98
Xin, F., 372–374
Xiong, G., 299–300, 350–351
Xu, B., 79, 92, 97, 98–99, 108
Xu, D., 350–351
Xu, H., 372–374
Xu, J., 50, 51*t*, 60–61, 123–124, 130–132, 133*t*, 137–138, 142, 144, 155–157, 159*t*, 161, 164, 168, 379–380
Xu, T., 45–46
Xu, Y., 79, 98–99, 379–380
Xue, C. C., 368–370, 372–374
Xun, Z., 12

Y

Yamada, Y., 368–370
Yamazaki, T., 394–396
Yan, C., 348–349, 350–353
Yang, 64–65
Yang, A. K., 379–380
Yang, D., 55, 56*t*, 60
Yang, F., 30
Yang, J., 130–132, 168
Yang, J.-M., 124–130, 125*t*, 133*t*, 139–140, 169, 385*t*
Yang, J. Y., 94
Yang, K., 385*t*
Yang, L., 169
Yang, M. Q., 94

Yang, S. Y., 393–394
Yang, X., 21
Yang, Y. D., 55, 56*t*, 60, 79, 86*t*, 92, 97, 98–99, 108, 133*t*, 137, 141–142, 349–350
Yang, Z. R., 372–374
Yarne, D. A., 394–396
Yates, J. R. III., 4–5, 7, 8*t*, 14, 45–46
Yau, K. W., 178–179
Ye, J., 123, 156–157, 158, 159*t*, 161, 165
Ye, Y., 31–32, 31*t*, 132, 133*t*, 138–139, 143, 156–157, 159*t*
Yeturu, K., 385*t*
Yi, H., 30–31
Yim, E. K., 31*t*
Yin, L., 182, 368–370, 372–374, 376–377
Ying, J., 22
Yonemoto, K., 179–180
Yong Shi, S., 379–380
Yongye, A. B., 300, 301*t*
Yoshida, A., 366–368
Yoshida, H., 368–370
Yoshiura, K., 179–180
Young, M. A., 368–370
Yu, J., 12
Yu, K., 385*t*
Yu, X., 97, 98
Yuan, H. Y., 183, 372–374
Yuan, Q., 23
Yuan, X., 48–49, 168
Yuan, Y. P., 374–375
Yue, P., 366–368, 372–374

Z

Zacharias, M., 86*t*
Zaki, M. J., 86*t*
Zamakola, L., 372–374
Zandi, E., 8*t*, 10
Zanegina, O. N., 78–79
Zanni, M. T., 316–317
Zanoni, G., 23–24
Zarate, C. A., 50
Zeev-Ben-Mordehai, T., 375–376
Zeifman, A. A., 385*t*
Zellner, M., 62–63, 66–67
Zemla, A., 133*t*, 138–139, 142–143, 377–378
Zeng, M., 25
Zhai, J., 379–380
Zhan, J., 133*t*, 137, 141–142
Zhang, 45
Zhang, C., 98–99, 379–380
Zhang, D. D., 20–21
Zhang, F., 348, 349, 357–358
Zhang, H., 8*t*, 12–13, 385*t*
Zhang, H. S., 105–106
Zhang, H. X., 390, 391
Zhang, J., 44–45, 48–49, 63–64, 66–67, 97, 128–130, 147, 368–370, 372–374, 379–380, 385*t*
Zhang, K., 5, 8*t*, 64–65
Zhang, L., 30
Zhang, P. F., 30–31
Zhang, Q., 341
Zhang, Q. C., 83–84
Zhang, W., 84–85, 299–300, 350–351
Zhang, X. J., 341–342, 375–376
Zhang, Y., 46, 47–48, 60, 66–67, 83, 93–94, 97, 122–123, 130–132, 133*t*, 137, 138–139, 141–142, 147–148, 164, 168, 377–378, 379–380, 385*t*
Zhang, Z., 94, 97, 128–130, 147, 368–370, 374–375, 379–380, 394–396
Zhao, F., 130–132, 168
Zhao, H., 79, 97, 133*t*, 137, 141–142, 349–350
Zhao, R., 341–342
Zhao, Y., 100, 385*t*
Zheng, D., 379–380
Zheng, S., 79, 92
Zheng, W. M., 123, 124–128, 125*t*, 130, 133*t*, 137, 142–143, 144, 146–147, 155–157, 159*t*, 161, 165, 167–168
Zhou, C., 55, 56*t*, 60
Zhou, F., 27
Zhou, H.-X., 79, 93–95, 98–99, 376–377
Zhou, J., 55, 56*t*, 60
Zhou, M., 372–374
Zhou, R., 349, 357–358
Zhou, S. F., 368–370, 372–374, 379–380
Zhou, Y., 44–45, 79, 92, 97, 98–99, 108, 133*t*, 137, 141–142, 348, 349–350
Zhou, Z. W., 368–370, 372–374
Zhu, D., 12
Zhu, H., 98, 394–396
Zhu, J. K., 148–149, 262–263, 348–349, 350–353, 376–377

Zhu, Q., 98–99
Zhu, X., 23
Zielenski, J., 374–375
Zimmer, R., 123–124, 132, 133*t*, 137–138, 139, 143–144
Zimmerman, L. J., 8*t*, 10
Zisook, S., 50
Zoete, V., 385*t*
Zoller, M., 30–31
Zou, X., 91*t*, 384–390
Zsoldos, Z., 384–390, 385*t*
Zuena, A. R., 56*t*, 60
Zwier, M. C., 273
Zykovich, A., 349

SUBJECT INDEX

Note: Page numbers followed by "*f*" indicate figures and "*t*" indicate tables.

A

AFPs. *See* Aligned fragment pairs (AFPs)
Algorithms, MD simulations
- calculations, 279–281
- Ewald summation, 279
- global communication, 279–281
- inter-node communication, 279–281
- Newton's equation, 278
- van der Waals forces and electrostatic interactions, 278–279
- Verlet integrator, 278–279

Aligned fragment pairs (AFPs)
- alignment path, 141
- chaining methodology, 141

Align-GVGD, 190*t*, 209, 210*t*
Alzheimer's disease
- blood-based biomarker approach, 62–63
- characteristic features and treatment, 54, 56*t*
- epigenetics, 64–65
- imaging biomarkers, 54
- iTRAQ method, 63–64
- pathophysiology, 62–63
- Pin1, 63–64
- protein separation techniques, 62–63
- proteins/possible biomarkers, 56*t*
- redox proteomics, 64

AMBER. *See* Assisted Model Building with Energy Refinement (AMBER)
Anxiety and depression
- characteristic features and treatment, 50, 51*t*
- 2D DIGE and MALDI-TOF, 60
- glyoxalase-I, 60–61
- mitochondrial function, 60
- PFC of CUMS rat model, 60
- proteins modulation, 60–61
- proteins/possible biomarkers, 56*t*
- unipolar depression, 55
- unpredictable stress factors, 55

Application-specific integrated circuits (ASICs), 288
ASICs. *See* Application-specific integrated circuits (ASICs)
Assisted Model Building with Energy Refinement (AMBER)
- description, 295–299
- equation, 299
- GAFF, 300–301
- GLYCAM force, 300
- hydrogen-bond energy, 295–299
- MMFF94 and GLYCAM06 force field, 300, 301*t*
- parameterization, 300–301
- versions, 299–300

AutoDock, 391–392, 392*f*

B

Binary complexes, protein interactions
- Interactome3D, 83–84
- MODELLER, 83
- M-TASSER, 83
- PrePPI, 83–84
- PRISM, 83–84

Binding free energy evaluation, RVDs
- AMBERTOOLS, 351–352
- binary code, DDNA3, 351–352
- description, 362–363
- DNA-bound dHAX3 X-ray structure, 351–352
- hydrogen bonding, 348–349
- NG-T interaction, 359–360
- parameterization, 361
- PBSA, VDW, Rosetta and DDNA3, 357–359, 358*t*
- protein–DNA interactions, 362
- ranking performance, VDW, 359
- side-chain orientations, 362
- template selection, 361–362

Biogenesis of lysosome-related organelles complex 1 (BLOC-1), 62

Biomarkers
definition, 46–47
genomics tools, 47–48
peptide, 48–49
proteomic approach, 47–48
PTMs, 49–50
Bipolar disorder
antidepressants, 53
bipolar disorder I (BDI), 53
bipolar disorder II (BDII), 53
characteristic features, 50–53, 51*t*
lithium, 65
molecular fingerprint, 65–66
proteins/possible biomarkers, 56*t*
SELDI-TOF-MS ProteinChip profiling, 66
BLOC-1. *See* Biogenesis of lysosome-related organelles complex 1 (BLOC-1)

C

CAPRI. *See* Critical Assessment of Predicted Interactions (CAPRI)
Carbonylation
4-HNE, 23
inhibitory biological effects, 23
Keap1-Nrf 2-ARE axis, activation, 23–24
CDD. *See* Conserved domain database (CDD)
CHARMM. *See* Chemistry at HARvard Macromolecular Mechanics (CHARMM)
Chemistry at HARvard Macromolecular Mechanics (CHARMM), 302–304
CID. *See* Collision-induced dissociation (CID)
CLE. *See* Conformational letter (CLE)
CLE-based pairwise alignment of protein structure (CLEPAPS)
DeepAlign, 144–147
initial and refinement stage, 142–143
and ProSup, 142–143
SFPs, 144–147
CLEPAPS. *See* CLE-based pairwise alignment of protein structure (CLEPAPS)
Collision-induced dissociation (CID), 2
Combinatorial extension (CE)
AFPs, 141
dRMS, 141
Combinatorial extension Monte Carlo (CE-MC), 158
Computer molecular simulations
complexity, 281–285
description, 272, 273
"The Molecular Dynamics of Proteins", 275–276
molecules arrangement, 274–275
Newton's laws, 275–276
Verlet time integrator, 274–275
X-ray crystallography, 275–276
Conformational letter (CLE)
contiguous C-alpha atoms, 128
mixture model, 128, 129*t*
Conserved domain database (CDD)
human-curated alignments, 148–149
and MALIDUP, 152
performance, 149
Critical Assessment of Predicted Interactions (CAPRI), 89
Cysteine/disulfide oxidation
caspase-3, 25
cysteine–thiol chemistry, 24–25
dysfunctional redox signaling, 24
Hsp33, 25
nucleophilic properties, 24
oxidative stress, 24
oxoforms, 24

D

DALI. *See* Distance matrix alignment (DALI)
DeepAlign algorithm
CLEPAPS, 144
description, 146, 146*f*
gap elimination, 147
iterative refinement, 147
scoring functions, 144–146
SFPs, 147
similar fragment pairs, 146–147
Depression. *See* Anxiety and depression
2DGE. *See* Two-dimensional gel electrophoresis (2DGE)

Distance matrix alignment (DALI)
scoring function, 140–141
and TMalign alignments, 152
and Vorolign, 132–137
Distance root mean squared (dRMS) deviation
CE, 141
DALI, 140–141
DNA/RNA-binding proteins, prediction
DBD-Hunter, 97
DBD-Threader, 97
Docking
advantages, 392–393
algorithms, 384–390, 385*t*
AutoDock, 391–392, 392*f*
binding sites, 383, 384–390
CAPRI, 89
drug banks, 383
drug designing, 382–383
drug-discovery process, 380–382
HIV protease inhibitor, 382
in silico tools, 380–382
interface (I-RMSD), 89
lead optimization, drugs, 383
ligand-RMSD, 89
limitations, 393
methods, 85–89, 86*t*
protein interactions, 85–92
protein–ligand, 391
protein–nucleic acid, 92
protein–protein, 90–92, 391
rigid-body, 384–390
scoring functions (*see* Scoring functions, docking)
virtual screening, 380–382
DREAM5 targets
3DTF, 100
knowledge-based potentials, 100–105
modeling transcription factors, 100–102, 101*f*
PiDNA, 100
protein–DNA interactions, 106–108
PWM predictions, 102–105, 103*t*
TF–DNA complexes, 100–102
dRMS. *See* Distance root mean squared (dRMS) deviation

E
Elasticity, ligand-free TALE
conformational sampling, 356–357, 356*f*
features, 356
MD simulation trajectory, 355–356, 355*f*
RMSD (*see* Root mean squared deviation (RMSD))
Electrospray ionization (ESI), 26, 44
ESI. *See* Electrospray ionization (ESI)
Evolutionary-based prediction methods, SAPs
deleterious SAPs, 188–208, 190*t*
PANTHER, 184–185
PhD-SNP, 184–185
SIFT, 184–185

F
FASTSNP. *See* Function analysis and selection tool for single nucleotide polymorphisms (FASTSNP)
Flexible alignment by AFP-chaining allow twists (FATCAT), 143
Force fields, MD simulations
AMBER, 295–301
CHARMM, 302–304
computer simulations, 293–294
consistent, 304
description, 295
DREIDING, 304
GROMOS, 305
MMFF, 305–306
MM2/MM3, 306–307
molecular properties, 294–295
potential energy surface, 293–294
properties, 295, 296*t*
Functional SNP (F-SNP)
BEST1 and *PRPH2* genes, 222–233, 234*t*
coding and noncoding regions, 222–233
FASTSNP, 187, 222–233, 224*t*
score system, 187
Function analysis and selection tool for single nucleotide polymorphisms (FASTSNP), 187, 222–233, 224*t*
Fuzzy oil drop model
closeness, measuring, 320
definition, 317–318
3D Gaussian, 319

Fuzzy oil drop model (*Continued*)
external factors, 321
hydrophobic core structure, 319
hydrophobic density distribution, 318–319
internal force field, 323
Kullback–Leibler's entropy criterion, 319–320
Levitt's formula, 318–319
O/T and O/R values, 322–323
real-world proteins, 319–321
relative distance (RD), 320, 321*f*
value of D_{KL}, 320

G

GAFF. *See* General AMBER force field (GAFF)
General AMBER force field (GAFF), 300–301
Genome-wide association (GWA), 181, 233–258
Graphics processing units (GPUs)
algorithms, 289
description, 288–289
dynamic parallelism, 289–290
multithreaded streaming multiprocessors, 289–290
SIMT, 289–290
GROMOS. *See* GROningen Molecular Simulation (GROMOS)
GROningen Molecular Simulation (GROMOS), 305
GWA. *See* Genome-wide association (GWA)

H

Heat shock protein 33 (Hsp33), 25
High-performance computing (HPC)
Amdahl's law, 285–286
ASICs, 288
CPUs, 287–288
field-programmable gate arrays, 288
instruction-level parallelism, 287
MDGRAPE-3, 288
memory architectures, 286–287
Moore's law, 285
parallel computing, 285
parallel slowdown, 285–286
4-HNE. *See* 4-Hydroxy-trans-2,3-nonenal (4-HNE)
HPC. *See* High-performance computing (HPC)
Hsp33. *See* Heat shock protein 33 (Hsp33)
4-Hydroxy-trans-2,3-nonenal (4-HNE), 23

I

ICATs. *See* Isotope-coded affinity tags (ICATs)
IMAC. *See* Immobilized metal ion affinity chromatography (IMAC)
Immobilized metal ion affinity chromatography (IMAC), 49–50
IMP. *See* Integrative Modeling Platform (IMP)
In silico methods. *See* Single amino acid polymorphisms (SAPs)
Integrative Modeling Platform (IMP), 84–85
Internal force field
Gromacs program, 323
procedure, 323–324
protein molecule interaction, 323
Intrinsically disordered proteins
antifreeze (class II) proteins, 340
bioinformatics, 316
characteristics, 339–340
DisProt fragments, 316–317, 324*t*, 325
Drosophila melanogaster, 325
fast-folding (downhill) proteins, 340
flexible linkers/spacers, 325
fuzzy oil drop model (*see* Fuzzy oil drop model)
internal interactions, optimization, 342
2JU4 (*see* 2JU4 protein)
PDB data, 317
poly-peptide chain, 342
properties, disordered fragments, 316
protein folding, 316
1QO9 protein, 325, 326*f*, 327*f*
2TPI (*see* 2TPI protein)
unfolding (denaturation), 342
water environment, 317
1YDV (*see* 1YDV homodimer)
Isobaric tag for relative and absolute quantitation (iTRAQ), 40–41, 45
Isotope-coded affinity tags (ICATs), 4, 40–41

ITRAQ. *See* Isobaric tag for relative and absolute quantitation (iTRAQ)

J

2JU4 protein
3D presentation, 328–329, 329*f*
hydrophobicity distribution, 328, 328*f*, 329
O/T and O/R analysis, 328
PDEgamma, 327
RD values, 327–328
structure, 327

K

Knowledge-based potentials. *See also* Split-statistical potentials
definition, 80
PMF, 80, 81
protein interactions, analysis, 79, 80

L

Label-free liquid chromatography-mass spectrometry (LC-MS(E)), 49–50
LC-MS(E). *See* Label-free liquid chromatography-mass spectrometry (LC-MS(E))

M

Major depressive disorder (MDD), 40–41
MALDI. *See* Matrix-assisted laser desorption/ionization (MALDI)
MAPSCI. *See* Multialign of protein structure and consensus identification (MAPSCI)
MASS. *See* Multiple alignment by secondary structure (MASS)
Mass spectrometry (MS) data
bioinformatics algorithms, 5
CID, 2
ICATs, 4
isotope dilution theory, 4
label-free approach, 4
mass-to-charge ratios *(m/z)*, 2–3
peptide sequence, 3–4
proteomics workflow, 2, 3*f*
spectral counting, 4–5
tandem MS, 3–4
Matrix-assisted laser desorption/ionization (MALDI), 26, 44
MATT. *See* Multiple alignment with translations and twists (MATT)
MD. *See* Molecular dynamics (MD)
MDD. *See* Major depressive disorder (MDD)
Merck molecular force field (MMFF), 305–306
MMFF. *See* Merck molecular force field (MMFF)
Molecular dynamics (MD)
advantages, 394–396
algorithms (*see* Algorithms, MD simulations)
artificial forms, 272
computer-aided drug design, 393–394
computer molecular (*see* Computer molecular simulations)
drug-discovery process, 393–394
equations of motion, atoms, 270
force fields (*see* Force fields, MD)
GPUs, 288–290
Hook's law, 270
HPC, 285–288
limitations, 307–308
modeled protein structures, 394–396
Newtonian dynamics, 271, 394–396
physics, 276–278
protein–ligand interactions, 271
protein structures, conventional methods, 393–394
quantum effects, 271–272
software, 290–292, 394–396
MSA. *See* Multiple structure alignment (MSA)
Multialign of protein structure and consensus identification (MAPSCI), 161
Multimeric complexes, protein interactions
IMP, 84–85
physical-chemical information, 84
spatial restraints, 84, 85
Multiple alignment by secondary structure (MASS)
advantage, 158–161
and SSEs, 158–161

Multiple alignment with translations and twists (MATT)
and FATCAT, 132–137, 143
and TMalign, 149–152
Multiple structure alignment (MSA)
BLOMAPS, 161
CE-MC, 158
components, 156–157
CORE-LEN, 165–167
3DCOMB, 161–162
description, 155–156
HOMSTRAD, 165
HSFBs, 162–163
MAPSCI, 161
MASS, 158–161
MUSTANG, 158
pairwise alignment adjustment, 164
refinement, 164–165
RMSD, 164
SABmark, 167
scoring function, 163–164
SFBs, 157
structures, 156
TMscore, 165
tools, 158, 159*t*

N

Neuropsychiatric disorders
Alzheimer's disease, 54, 62–65
anxiety and depression, 50, 55–61
bipolar disorder, 50–53, 65–66
schizophrenia, 53–54, 61–62
Nitrosoproteome approach
protein microarrays, 29
S-nitrosoproteins, 28–29
NOxICAT method, 28

O

ODA. *See* Optimal Docking Area (ODA)
OPRA. *See* Optimal Protein-RNA Area (OPRA)
Optimal Docking Area (ODA), 94–95
Optimal Protein-RNA Area (OPRA), 94–95
O/T and O/R values, hydrophobicity density distribution
chain, 2TPI protein, 334, 335*t*
DisProt, 322
domain, 2TPI protein, 334–336, 337*t*
fragments, 2TPI protein, 332, 333*t*
hydrophobicity density distribution, 322
PDBSum criteria, 322
protein structures, analysis, 322–323
status of units, 322
structural role of fragment, 322
value of D_{KL}, 322
OxiCAT strategy, 28
Oxidative stress and redox homeostasis
antioxidant enzyme synthesis, 20–21
nitrosative stress, 20–21
reactive species (RS), 20

P

PDB. *See* Protein Data Bank (PDB)
Peptide identification
accurate mass and time (AMT) tag approach, 10
clustering approach, 10
database-searching software, 7, 8*t*
decoy database, 9
de novo approach, 5
hybrid approach, 10
monoisotopic residue masses, 5, 6*t*
m/z values of peaks, 6–7
PepNovo and PEAKS, 5
peptide identification paradigm, 6
PeptideProphet, 9
spectral library searching, 9–10
structure, 6–7
tandem mass spectra, 7–9
top-ranked peptides, 7–9
Peptidomics, 48–49
Peptidyl-prolyl *cis/trans* isomerase (Pin1), 63–64
PhD-SNP. *See* Predictor of human deleterious SNP (PhD-SNP)
Phosphodiesterase (PDE6) inhibitory gamma subunit (PDEgamma), 327
Pin1. *See* Peptidyl-prolyl *cis/trans* isomerase (Pin1)
PMF. *See* Potential of mean force (PMF)
Polymorphism phenotyping (PolyPhen-2), 185–186
PolyPhen-2. *See* Polymorphism phenotyping (PolyPhen-2)

Position-specific-independent counts (PSIC), 185–186
Positive predictive value (PPV), 96–97, 96*f*
Posttranslational modifications (PTMs), 42–43, 49–50
Potential of mean force (PMF), 80, 81
Predictor of human deleterious SNP (PhD-SNP), 184–185, 189–208, 209–211, 260
Protein carbonylation/4-HNE adducts, MS
  2D electrophoresis, 26
  ESI, 26
  4-HNE, 26
  hydrazine/hydrazide, 26
  in vitro reports, 26
  MALDI, 26
  serum proteomics, 27
  TOF-MS, 26
Protein Data Bank (PDB), 78–79
Protein–DNA interactions
  DREAM5 targets, 106–108
  "E_{local}" terms, 105–106
  residue–environment combinations, 105
  split-statistical potentials, 105–109, 107*f*
Protein identification
  "best peptide", 11–12
  confidence scores, 10–11
  multiple proteins, 10–11
  ProteinProphet, 11
Protein interfaces, identification
  BindN, 94
  "BS-E_{local}" statistical potential, 96–97
  Consurf, 93–94
  DISPLAR, 94–95
  DP-Bind, 94
  "E_{local}" statistical potential, 95–96, 105–106
  experimental approaches, 93
  FINDSITE, 93–94
  metaDBSite, 94
  NAPS, 94
  ODA, 94–95
  OPRA, 94–95
  PIPE-Sites, 93–94
  PPV, 96–97, 96*f*
  PSIFR, 93–94
  sequence, methods, 93–94
  split-statistical potentials, 95–97
  structure, methods, 94–95
Protein–nucleic acid docking
  protein–DNA docking, 92
  protein–RNA docking, 92
Protein pairwise structure alignment
  C-alpha, 132
  CDD, 149, 150*t*
  CE, 141
  CLEPAPS and ProSup, 142–143
  C-map, 132
  components, 137
  DALI, 140–141
  DeepAlign, 137–138, 144–147, 153–154, 154*t*
  description, 130–132
  evaluation tests, 147–148
  implicit and explicit flexible, 132–137
  MALIDUP, 149–152, 150*t*
  MALISAM, 150*t*, 152
  MATT and FATCAT, 143
  one dimensional objects, 139–140
  rigid perspective, 132–137
  SABmark, 148–149, 152–153
  scoring functions, 138
  SSE, 132
  symmetry proteins alignment, 153–154, 155*f*
  three-dimensional objects, 138–139
  TMalign, 141–142
  tools, 132, 133*t*
  two dimensional objects, 139
  vorolign, 143–144
Protein–protein docking
  AMBER, 90
  benchmarking, 90, 91*t*
  CHARMM, 90
  FOLD-X, 90
  I-RMSD, 90
  ligand-RMSD, 90
  "MixRank", 90–92
  SIPPER, 90–92
  split-statistical potentials, decoys
  ZRANK, 90–92
  ZRANK2, 90–92
Protein quantification
  biological samples, 12
  ICAT data, 12–13

Protein quantification (*Continued*)
label-free data, 13–14
missing values, peptide-level, 14
model-based approach, 14
MS1-based label-free data, 13
peptide features, 13–14
peptide intensities, 12
single-ion chromatograms, 13
spectral count, 14
Proteins, oxidative modification
carbonylation, 23–24
carbonyl proteins, 22
cysteine/disulfide, 24–25
posttranslational modifications, 21
protein oxidative injury, 21, 22
proteolytic mechanisms, 22
redox modifications, 21
S-nitroso modifications, 22–23
Protein structure alignment
components, 122–123
MSA, 122, 123
pairwise alignment programs, 123–124, 130–154
structural alphabets (*see* Structural alphabets)
Protein-tyrosine kinases (PTPs), 32–33
Proteomics, biomarker discovery
2DGE and liquid chromatography, 42–43, 43*f*
ESI, MALDI and SELDI, 44
isotope-coded affinity tag, 44–45
iTRAQ, 45
neuropsychiatric disorders, 50–54, 55–66
^{15}N metabolic labeling, 46, 47*f*
peptides, 48–49
PTMs, 49–50
SILAC, 45–46
PSIC. *See* Position-specific-independent counts (PSIC)
PTMs. *See* Posttranslational modifications (PTMs)
PTPs. *See* Protein-tyrosine kinases (PTPs)

R

Reactive species (RS), 20
Redox homeostasis network
cancer studies, components, 30, 31*t*
ICAT approach, 30
MALDI-TOF/TOF technology, 31–32
MS, 31
oxidative stress, 29–30
proteomic mapping, 29, 30
PTPs, 32–33
redox-related proteins, 31*t*, 32
redox signaling, 30
SOD, 30–31
Redox proteomics, oxidation-induced protein damage
cellular function, 25
definition, 25
nitrosoproteome approach, 28–29
protein carbonylation/4-HNE adducts, MS, 26–27
screening technologies, 26–27
thiol–cysteine state, 27–28
Repeat-variable di-residues (RVDs)
binding free energy evaluation (*see* Binding free energy evaluation, RVDs)
C-terminal capping, helix, 348–349
semiquantitative experiments, 349
RMSD. *See* Root mean squared deviation (RMSD)
Root mean squared deviation (RMSD)
constant oscillation, 354
DNA removal, 355
50-ns MD simulations, 352–353, 352*f*, 353*f*
200-ns MD simulations, 353, 354*f*
radius of gyration (Rg), 354–355
sampling data, 356–357, 356*f*
RVDs. *See* Repeat-variable di-residues (RVDs)

S

SAPs. *See* Single amino acid polymorphisms (SAPs)
Schizophrenia
BLOC-1, 62
characteristic features and treatment, 56*t*
MALDI-TOF peptide mass fingerprinting analysis, 61
neurotransmitter systems, 62
protein phosphorylation patterns, 61
proteins/possible biomarkers, 56*t*
self-disorder, 53–54

Scoring functions, docking
consensus, 391
empirical free-energy, 390
force-field, 390
knowledge, 390
ligand-protein conformations, 384–390
rigid-body, 384–390
Secondary structure elements (SSEs)
MASS, 158–161
physical meaning, 124–126
SELDI. *See* Surface-enhanced laser desorption/ionization (SELDI)
SFBs. *See* Similar fragment blocks (SFBs)
SIFT. *See* Sorting intolerant from tolerant (SIFT)
SILAC. *See* Stable isotope labeling with amino acids in cell culture (SILAC)
Similar fragment blocks (SFBs), 157
SIMT. *See* Single-instruction, multiple-thread (SIMT)
Single amino acid polymorphisms (SAPs)
Align-GVGD, 190*t*, 209, 210*t*
BEST1 and PRPH2 protein sequence, 211–213, 214*t*, 215*t*, 216*f*, 218*f*, 219*f*
biophysical validation, 186–187
computational methods, 258–259
concordance, 209–211, 211*f*, 212*t*
conservation profiling, 187
csSNPs, 258–259
Cys residues, protein sequence, 262–263
data information, 183
dbSNPs, distributions, 188, 189*f*
disease-causing/deleterious mutations, 259–260
disulfide bonds, 220–221
evolutionary-based prediction methods, 184–185
F-SNP, 187, 222–233, 224*t*, 234*t*
GWA studies, 233–258
in vitro studies and animal models, 181
I-Mutant 3.0, 208–209
native and mutant state of BEST1 and PRPH2, 213–220, 221*f*
PolyPhen-2 and PhD-SNP, 260, 261–262
population genetic studies, 262–263
protein stability analysis, 186
protein structure and function, 259–260
ranking, 233, 240*t*, 262
sequence-based methods, 260
and SNPs, 179–180, 181–183, 263–264
statistical analysis, 188, 222, 223*t*
structure-based prediction methods, 185–186, 208
tools, 261–262
VMD, 178–179
Single amino acid substitutions. *See also* Single amino acid polymorphisms (SAPs)
bioinformatics, 370
classification, 372–374
clinical practices, 398–399
computational methods, 371–374, 377–380
database resources, 374–375
description, 371–372, 373*f*
disease-causing, 368–370
DNA variation, SNPs, 366–368
docking (*see* Docking)
homology modeling, 377–378, 379–380
Human Genome Project (HGP), 371–372, 396–397, 398–399
kinetic properties, 368–370
MD (*see* Molecular dynamics (MD))
molecular phenotypic effect, 375–376
personalized medicine, 397–398
predictions, 371–372
protein 3D structure, 370
protein products, 368–370
protein structures, 377–378
sequence information, 376–377
sickle-cell anemia, 368–370
traditional methods, 371–372
Single-instruction, multiple-thread (SIMT), 289–290
Single-nucleotide polymorphisms (SNPs)
alleles, 366–368
BEST1 and PRPH2 genes, 188, 189*f*
dbSNP, 183
and disease susceptibility, 180
distribution, 374–375

Single-nucleotide polymorphisms (SNPs) (*Continued*)
experimental methods, 396–397
FASTSNP, 224*t*
F-SNP (*see* Functional SNP (F-SNP))
in silico methods, 181
monogenic and complex disorders, 366–368
noncoding regions, 179–180
PhD-SNP, 184–185
World Wide Web, 179–180
SNPs. *See* Single-nucleotide polymorphisms (SNPs)
Sorting intolerant from tolerant (SIFT), 184–185, 188–189, 190*t*, 240*t*, 261–262
Split-statistical potentials
amino acid and DNA, 107–108
definition, 81–82
dinucleotides, 105
DNA groove, 106–107
DNA parameters, 106, 107*f*
DREAM5 targets, 106–109
protein–DNA interactions, 105–109, 107*f*
protein interfaces, identification, 93–97
protein–protein docking, 90–92
protein–protein interactions, 82
PWM prediction, 108–109, 109*f*
quality of decoys, 81
Z-scores, 82
SSEs. *See* Secondary structure elements (SSEs)
Stable isotope labeling with amino acids in cell culture (SILAC), 40–41, 45–46
Structural alphabets
applications, 128–130
CLE, 126–128, 127*f*
principle, 126–128, 127*f*
protein local structures, 124–126
substitution matrix, 126, 130, 131*t*
Structure-based prediction methods, SAPs
deleterious SAPs, 190*t*, 208
PolyPhen-2, 185–186, 208
PSIC, 185–186
SNAP, 185–186, 208
Surface-enhanced laser desorption/ionization (SELDI), 44

T

TALEs. *See* Transcription activator-like effectors (TALEs)
Thiol–cysteine state, proteomics
MALDI detection, 27–28
NOxICAT method, 28
OxiCAT strategy, 28
oxidative cysteine modifications, 27
protein oxidation, cell physiology, 28
thiol modification mapping, 27
Time-of-flight mass spectrometry (TOF-MS), 26
TMalign score (TMscore), 141–142, 149, 165
TOF-MS. *See* Time-of-flight mass spectrometry (TOF-MS)
2TPI protein
Chain I, 330–331, 332, 334
Chain Z, 330–331, 332, 334
characteristics, 334–336
complex, 332–333
DisProt1 fragment, 336–338
domains, 334–336
electrostatic and van der Waals density, 339
external factors, 332
high RD values, 339
internal force field, 338–339
O/T and O/R values (*see* O/T and O/R values, hydrophobicity density distribution)
stages in formation, 336
structural units, 331–332
Transcription activator-like effectors (TALEs)
binding free energy evaluation (*see* Binding free energy evaluation, RVDs)
ligand-free, elasticity, 352–357
low free energy barrier, 360–361
MD simulations, 349–351
NH-G binding, 349

recognition code, 348
X-ray structures, 348–349
Two-dimensional gel electrophoresis (2DGE), 26, 42–43, 43*f*

V

Vitelliform macular dystrophy (VMD)
BEST1 and *PRPH2* gene mutations, 181–183
description, 178–179
MEDLINE and PubMed, 183
and SNPs, 262–263
VMD. *See* Vitelliform macular dystrophy (VMD)
Vorolign, 132–137, 143–144

Y

1YDV homodimer
DisProt fragments, 330
protein–protein interface, 330, 331*f*
structural units, RD values, 329–330, 330*t*
tri-osephosphate isomerase, 329